CAMBRIDGE STUDIES IN
ADVANCED MATHEMATICS 57

NATURAL DUALITIES FOR THE WORKING ALGEBRAIST

The theory of natural dualities, as presented in this text, is broad enough to encompass many known dualities through a rich assortment of substantive theorems yet concrete enough to be used to generate an array of previously undiscovered dualities. Stone's duality between Boolean algebras and Boolean spaces and Priestley's duality between distributive lattices and Priestley spaces are prototypical natural dualities. This text will serve as a user manual for algebraists, for category theorists and for those who use algebra in their work, particularly mathematicians and computer scientists interested in non-classical logics. It will also give the specialist a complete account of the foundations, leading to the research frontier of this rapidly developing field. As the first text devoted to the theory of natural dualities, it provides an efficient path through a large body of results, examples and applications in this subject which is otherwise available only in scattered research papers. To enable the book to be used in courses, each chapter ends with an extensive exercise set. Several fundamental unsolved problems are included.

David M. Clark is Full Professor of Mathematics, SUNY, College at New Paltz, New York.

Brian A. Davey is Reader and Associate Professor of Mathematics, La Trobe University, Bundoora.

Already published

1 W.M.L. Holcombe *Algebraic automata theory*
2 K. Petersen *Ergodic theory*
3 P.T. Johnstone *Stone spaces*
4 W.H. Schikhof *Ultrametric calculus*
5 J.-P. Kahane *Some random series of functions, 2nd edition*
6 H. Cohn *Introduction to the construction of class fields*
7 J. Lambek & P.J. Scott *Introduction to higher-order categorical logic*
8 H. Matsumura *Commutative ring theory*
9 C.B. Thomas *Characteristic classes and the cohomology of finite groups*
10 M. Aschbacher *Finite group theory*
11 J.L. Alperin *Local representation theory*
12 P. Koosis *The logarithmic integral I*
13 A. Pietsch *Eigenvalues and s-numbers*
14 S.J. Patterson *An introduction to the theory of the Riemann zeta-function*
15 H.J. Baues *Algebraic homotopy*
16 V.S. Varadarajan *Introduction to harmonic analysis on semisimple Lie groups*
17 W. Dicks & M. Dunwoody *Groups acting on graphs*
18 L.J. Corwin & F.P. Greenleaf *Representations of nilpotent Lie groups and their applications*
19 R. Fritsch & R. Piccinini *Cellular structures in topology*
20 H Klingen *Introductory lectures on Siegel modular forms*
22 M.J. Collins *Representations and characters of finite groups*
24 H. Kunita *Stochastic flows and stochastic differential equations*
25 P. Wojtaszczyk *Banach spaces for analysts*
26 J.E. Gilbert & M.A.M. Murray *Clifford algebras and Dirac operators in harmonic analysis*
27 A. Frohlich & M.J. Taylor *Algebraic number theory*
28 K. Goebel & W.A. Kirk *Topics in metric fixed point theory*
29 J.F. Humphreys *Reflection groups and Coxeter groups*
30 D.J. Benson *Representations and cohomology I*
31 D.J. Benson *Representations and cohomology II*
32 C. Allday & V. Puppe *Cohomological methods in transformation groups*
33 C. Soulé et al *Lectures on Arakelov geometry*
34 A. Ambrosetti & G. Prodi *A primer of nonlinear analysis*
35 J. Palis & F. Takens *Hyperbolicity, stability and chaos at homoclinic bifurcations*
37 Y. Meyer *Wavelets and operators I*
38 C. Weibel *An introduction to homological algebra*
39 W. Bruns & J. Herzog *Cohen-Macaulay rings*
40 V. Snaith *Explicit Brauer induction*
41 G. Laumon *Cohomology of Drinfeld modular varieties I*
42 E.B. Davies *Spectral theory and differential operators*
43 J. Diestel, H. Jarchow & A. Tonge *Absolutely summing operators*
44 P. Mattila *Geometry of sets and measures in Euclidean spaces*
45 R. Pinsky *Positive harmonic functions and diffusion*
46 G. Tenenbaum *Introduction to analytic and probabilistic number theory*
47 C. Peskine *An algebraic introduction to complex projective geometry I*
48 Y. Meyer & R. Coifman *Wavelets and operators II*
49 R. Stanley *Enumerative combinatories*
50 I. Porteous *Clifford algebras and the classical groups*
51 M. Audin *Spinning tops*
52 V. Jurdjevic *Geometric control theory*
53 H. Voelklein *Groups as Galois groups*
54 J. Le Potier *Lectures on vector bundles*
55 D. Bump *Automorphic forms*
56 G. Laumon *Cohomology of Drinfeld modular varieties II*
60 M. Brodmann & R. Sharp *Local cohomology*

Natural Dualities for the Working Algebraist

David M. Clark and Brian A. Davey

CAMBRIDGE
UNIVERSITY PRESS

PUBLISHED BY THE PRESS SYNDICATE OF THE UNIVERSITY OF CAMBRIDGE
The Pitt Building, Trumpington Street, Cambridge CB2 1RP, United Kingdom

CAMBRIDGE UNIVERSITY PRESS
The Edinburgh Building, Cambridge CB2 2RU, UK http://www.cup.cam.ac.uk
40 West 20th Street, New York, NY 10011-4211, USA http://www.cup.org
10 Stamford Road, Oakleigh, Melbourne 3166, Australia

First published 1998

Printed in the United Kingdom at the University Press, Cambridge

Typeset in Computer Modern by the author, in TEX

A catalogue record of this book is available from the British Library

Library of Congress Cataloguing in publication data

ISBN 0 521 45415 8 hardback

This book is dedicated to the memory of

Garrett Birkhoff

Lev Semenovič Pontryagin

Marshall Harvey Stone

who laid the foundations upon which it is built.

Acknowledgements

The production of this book represents the collective contributions of a number of people. The first author would like to thank Brian Davey and La Trobe University as well as Jürg Schmid and the University of Bern, while the second author would like to thank Hilary Priestley and Oxford University, for their financial support in the initial stages. We are both indebted to David Tranah and the staff of Cambridge University Press for their help and patience with our repeated underestimates of the time required to complete the project. A special thanks goes to Miroslav Haviar and George McNulty for their thorough reading and many helpful suggestions, and to graduate students Chris Daniels, Michael Jerie, Jane Pitkethly and Claudia Wegener who also proofread various parts of the manuscript. Finally, we owe our deepest gratitude to our wives,

Joanne and Helen,

for their support and indulgence during this lengthy endeavour, and to our children,

Elisa, Evan, Bonnie, Robin, Jessica, Owen, Erin and Caitlin,

for their forbearance as they grew up doing their artwork, scoring their games and printing their own essays on the backs of the many discarded versions of this work.

D.M.C. and B.A.D.
January 1998

Contents

Preface

In 1936 Marshall Stone published a truly novel theorem. Stone wanted to understand Boolean algebras. Like many other classes of algebras, Boolean algebras could be described in two different ways. They are normally defined abstractly as the class of all models of a familiar set of axioms. Alternatively, a class of algebras is often described by giving some concrete way to represent its members. For example, Stone was aware of the work of Boole and Tarski who had shown that certain Boolean algebras are representable as a subalgebra of the Boolean algebra of subsets of some set. Interesting as they are, these descriptions do not always provide a direct and usable insight into the structure of the individual algebras themselves. What Stone discovered was a representation for all Boolean algebras which gave algebraists a usable understanding of their structure. Using topological spaces to construct the representations, he proved that every Boolean algebra is isomorphic to the algebra of all clopen subsets of some totally disconnected compact space.

In fact, Stone [Sto36] proved much more than a representation theorem. He showed that the homomorphisms between Boolean algebras correspond precisely to the continuous maps between the corresponding spaces. In modern language, he proved that the category of Boolean algebras is dually equivalent to the category of Boolean spaces. Two similar results appeared almost simultaneously. Birkhoff [Bir37] showed that every finite distributive lattice is isomorphic to the lattice of all decreasing subsets of a finite ordered set, and established a close correspondence between finite distributive lattices and finite ordered sets. Looking at the category of abelian groups, L. Pontryagin [Pon34a, Pon34b] discovered that it was dually equivalent to the category of compact abelian groups. The beauty of these theorems is that they translate algebraic problems, normally stated in abstract symbolic language, into dual, topological problems where our geometric intuitions can be brought to bear.

These three seminal results planted the seeds for a considerable assortment of new dualities. Hu [Hu71] realised that Stone's theorem extended to the variety generated by any primal algebra. Still the most widely used of these new dualities is Hilary Priestley's duality [Pri70, Pri72] which combined the theorems of Stone and Birkhoff into a dual equivalence between the variety of *all* bounded distributive lattices and the category of all totally order-disconnected compact spaces. This discovery initiated a plethora of parallel results: Hofmann, Mislove and Stralka [HMS74] (semilattices with 1), Davey [Dav76] (relative Stone algebras), Cornish and Fowler [CF77] (de Morgan algebras), Isbell [Isb80] and Werner [Wer81] (median algebras), Davey [Dav82] (Stone and double

Stone algebras) and Davey and Werner [DW83] (Kleene algebras, abelian groups, quasi-primal algebras). During this period, Davey [Dav78] began to use the methods of universal algebra to formulate a general theory of duality. The early eighties saw this method crystallised in the papers of Davey and Werner [DW83] and subsequently, Clark and Krauss [CK84].

Out of these foundations has evolved a powerful theory of what have come to be known as **natural dualities**. A class of algebras is a candidate for a natural duality provided that it is the quasi-variety generated by some finite algebra. A natural duality for a quasi-variety gives us a uniform method to represent each algebra in the quasi-variety as the algebra of all continuous homomorphisms over some structured Boolean space. When we are lucky, the association between algebras and corresponding structured spaces forms a dual equivalence between the quasi-variety of algebras and the dual category of structured Boolean spaces. If this happens in a special canonical way we say that we have a **strong duality**. This book presents a detailed exposition of natural dualities up to the point that the theory had developed by early 1998. It is primarily intended for the working algebraist who, like Marshall Stone, wants to understand some special finitely generated quasi-variety.

Two appendices summarise the prerequisite background in universal algebra and topology. We rely heavily on foundational results of Birkhoff, Jónsson and Pixley. In Chapter 1 we give a bare-hands proof of Priestley's duality for distributive lattices, and then draw on it in the rest of the chapter to formulate the general concepts out of which natural dualities arise. In Chapter 2 we develop the method of Davey and Werner [DW83] for identifying a category of structured Boolean spaces dual to a given quasi-variety. Chapter 3 investigates strong dualities and culminates in two major theorems which tell exactly how, if the generating algebra satisfies certain conditions, we can directly construct a strong duality. The NU Strong Duality Theorem 3.3.8 utilises Baker and Pixley [BP75] to generalise Priestley [Pri72], while the Two-for-One Strong Duality Theorem 3.3.2 builds on the theorem of Pontryagin [Pon34a, Pon34b]. Each of the above listed strong dualities is a direct consequence of one of these two theorems, and Stone's theorem is a consequence of both. This fact is established in Chapter 4 where we give a vast assortment of examples of strong dualities for many familiar quasi-varieties. Chapters 2 and 3 might be viewed as the theoretical core of the subject. They have their roots in Davey and Werner [DW83] and Clark and Krauss [CK84] and their joint extension, Clark and Davey [CD95]. We expect that many readers will choose to scan much of the initial material and jump to the theorems in the last section of Chapter 3. The examples in Chapter 4 illustrate how these theorems can be directly applied with minimal reference to the lengthy development needed to establish them.

It is our underlying contention that natural dualities, and particularly strong dualities, can provide a very useful way to understand a quasi-variety. We substantiate this claim in Chapter 5, where we give a diverse selection of applications of our dualities. In each case, we show how one of the dualities obtained in the previous chapter can be applied to solve a particular algebraic problem that arises, quite independently of duality theory, within the given quasi-variety. We refine the method of Adams and Clark [AC90] to determine endomorphism monoid uniqueness, the method of Clark [Cla89] and Clark and Schmid [CS96] to characterise algebraic and existential closure, the method of Davey [Dav76] to characterise injectives and its variations, and the method of Davey and Werner [DW83] to characterise category equivalence.

Our success in Chapter 5 depends on using dualities in which the dual of an algebra is a structured space which is in some tangible sense simpler or more tractable than the original algebra. For example, finite distributive lattices are dual under Birkhoff–Priestley duality to finite ordered sets while rings in the quasi-variety generated by $\mathbf{Z}_6$ are equivalent to Boolean rectangular bands. On the other hand, dualities frequently arise in which the structured spaces are much more complex than the associated algebras. In general, such a duality is unlikely to be any use to the algebraist. In Chapter 6, an updated version of [CD96], we identify a number of different criteria which can make a strong duality useful, and in each case we give a characterisation of those quasi-varieties which admit a strong duality satisfying that criterion. Chapters 8 and 9 address the problem of complexity of the dual structures from a different point of view. Given a natural duality, how can we simplify the type of the dual category to obtain a more useful natural duality? In Chapter 8, based on Davey and Priestley [DP96a], we present an algorithmic method for reducing the type of a dualising category to a type that is, in an appropriate sense, minimal among all dualising categories. Chapter 9 presents a different method of optimising dualities which, though not algorithmic, is generally more convenient to use when the example is small. The results of this chapter first appeared in Davey, Haviar and Priestley [DHP95a].

The notions of natural duality and strong duality apply to quasi-varieties generated by a single finite algebra. In Chapter 7 we show how Davey and Priestley [DP87] broadened the scope of these notions to include quasi-varieties generated by finitely many finite algebras. Given a finite set of finite algebras which have distributive lattice reducts, we describe the method of constructing a broadened duality which rides piggyback on the underlying Priestley duality as it evolved from Davey and Werner [DW85, DW86] to Davey [Dav93] and Priestley [Pri95a]. We exhibit applications of the piggyback technique taken from Davey and Priestley [DP93a], Davey [Dav96] and Davey and Pitkethly [DP00].

Which finite algebras generate a dualisable quasi-variety? This question is at the forefront of research in the theory of natural dualities. In Chapter 10 we review what is known. As the previous chapters have dealt with finite algebras that generate dualisable quasi-varieties, we focus here on methods of showing that a particular quasi-variety is *not* dualisable. A central result is the Big NU Obstacle Theorem 10.2.2 of Davey, Heindorf and McKenzie [DHM95] which asserts that near-unanimity is necessary for dualisability in a congruence join-semidistributive quasi-variety. Several methods of demonstrating non-dualisability are presented: one based on Heindorf [Hei93], one first used in Davey and Werner [DW83] and one based on Clark and Krauss [CK84]. We conclude by determining exactly which clones on a two-element set generate a dualisable quasi-variety.

The tome now before you might easily convey the misimpression that it contains a complete account of the theory of natural dualities. As an account of the part of that theory that is known to date, it is indeed quite complete. But in absolute terms our present knowledge is very limited. Outside of congruence distributive varieties, we know little of which finite algebras generate a dualisable quasi-variety. We do not even yet know which finite groups, finite rings or finite semigroups generate a dualisable quasi-variety. Even within congruence distributive varieties we have no general method to determine whether or not a particular finite algebra generates a dualisable quasi-variety. We can extend many dualities to strong dualities, but we do not know if this is always possible. There still remain many fundamental questions that we cannot yet answer.

It is our hope that working algebraists will be able to use the strong duality theorems of Chapter 3 to produce new dualities for the quasi-varieties they study, to use the reduction methods of Chapters 2, 8 and 9 to simplify those dualities, and then to invoke the resulting dualities to help solve real problems in their chosen quasi-varieties. We further hope that some particularly hard working algebraists will be undaunted when the theorems of Chapter 2, 3 or 7 do not directly apply, and will continue to find new methods of producing dualities and strong dualities. Given success in these endeavours, we can then hope that some of the very hardest working algebraists will join our effort to resolve some of the still outstanding problems in the theory of natural dualities itself.

1

Dual Equivalences and Where to Find Them

Distributive lattices, Boolean algebras, median algebras, Stone algebras, semilattices, Łukasiewicz algebras of order n and vector spaces over a finite field: these are but a few familiar and important classes of algebras which can be described as the class of all isomorphic copies of subalgebras of direct powers of some single finite algebra $\underline{\mathbf{M}}$. The class $\mathcal{A} := \mathbb{ISP}\underline{\mathbf{M}}$ is called the **quasi-variety generated by the algebra** $\underline{\mathbf{M}}$. Interest in such classes arises from several points of view. To the algebraist, $\mathcal{A}$ is the smallest class of algebras containing $\underline{\mathbf{M}}$ and closed under $\mathbb{I}$, $\mathbb{S}$ and $\mathbb{P}$. To the logician, $\mathcal{A}$ is the class of all algebras which satisfy all equations and implications that hold in $\underline{\mathbf{M}}$. To the category theorist, $\mathcal{A}$ is the category of all algebras whose elements can be separated by homomorphisms into the 'cogenerator' $\underline{\mathbf{M}}$.

This book will develop a tool for studying $\mathcal{A}$ that exploits its properties as a category. Our goal will be to give a uniform method to construct a dual category $\mathcal{X}$ of structured Boolean spaces which effectively mirrors the structure of the category $\mathcal{A}$ in much the same way that a semigroup anti-homomorphism mirrors the structure of one semigroup within another while reversing the order of operation. This construction we will call a **dual adjunction**. When we are lucky, our method will give us an exact (but reversed) copy of $\mathcal{A}$ within $\mathcal{X}$ just as a semigroup anti-embedding gives an exact (but reversed) copy of one semigroup in another. A dual adjunction with this further property will be called a **dual representation**. Better yet will be to have a full (anti-)equivalence between the categories $\mathcal{A}$ and $\mathcal{X}$ analogous to an anti-isomorphism between two semigroups. In the language of category theory this is called a **dual equivalence**. Ultimately we will be interested in dual equivalences in which the dual category $\mathcal{X}$ is in some concrete sense simpler, more tractable or more easily understood than the category $\mathcal{A}$.

In the five sections of this chapter we will gather the background required to develop the theory of natural dualities. We begin by looking at categories, presenting the notions, terminology and basic facts needed to formalise the ideas we want to express. Armed with this small amount of background, we will be ready to examine in detail a particular, and historically significant, example of a natural duality: the Priestley duality for distributive lattices. This example illustrates just what it is that we would like our theory ultimately to provide for us. Drawing on it we will be able to formulate a general description of the dual categories $\mathcal{X}$ of structured Boolean spaces that we will need and to sort out some of the simple properties that are common to all such categories. Putting these observations together, we will lay out the foundational ideas upon which we can build a theory of natural dualities.

Most of the proofs of the assertions presented in this chapter are fairly standard and straightforward, although in many cases specialised and tailored to our needs. For this reason we have not included them in detail, but instead have listed them as exercises at the end of the chapter. The reader who does these exercises will finish the chapter with a thorough understanding of the background needed to read the rest of the book.

1.1 Categories

Natural duality is fundamentally a category-theoretic concept. As a result we turn to category theory to find a context where the necessary notions, terminology and facts are available to build a general theory of natural dualities. At the same time we hasten to point out that the amount of category theory *per se* that is actually required to develop our theory is very small. In this section we present that small piece of category theory.

A category $\mathcal{C}$ has three components:

- a class whose members are called the **objects** of $\mathcal{C}$;
- a set $\mathcal{C}(\mathbf{A}, \mathbf{B})$, for each pair of objects $\mathbf{A}$ and $\mathbf{B}$ of $\mathcal{C}$, called the **morphisms** from $\mathbf{A}$ to $\mathbf{B}$;
- a partial binary operation $\circ$, called **composition**, on the class of all morphisms of $\mathcal{C}$ such that if $u \in \mathcal{C}(\mathbf{A}, \mathbf{B})$ and $v \in \mathcal{C}(\mathbf{B}, \mathbf{C})$, then $v \circ u \in \mathcal{C}(\mathbf{A}, \mathbf{C})$.

We write $u\colon \mathbf{A} \to \mathbf{B}$ to mean that $u \in \mathcal{C}(\mathbf{A}, \mathbf{B})$. We say that $\mathcal{C}$ is a **category** if it satisfies the following axioms:

(i) for each object $\mathbf{A}$ of $\mathcal{C}$ there is a morphism $\mathrm{id}_{\mathbf{A}} : \mathbf{A} \to \mathbf{A}$, called the **identity**on $\mathbf{A}$, such that $\mathrm{id}_{\mathbf{A}} \circ u = u$ and $v \circ \mathrm{id}_{\mathbf{A}} = v$ whenever $\mathbf{B}$ and $\mathbf{C}$ are objects of $\mathcal{C}$ and $u\colon \mathbf{B} \to \mathbf{A}$ and $v\colon \mathbf{A} \to \mathbf{C}$ are morphisms;

(ii) for objects $\mathbf{A}$, $\mathbf{B}$, $\mathbf{C}$ and $\mathbf{D}$ of $\mathcal{C}$ and morphisms $u\colon \mathbf{A} \to \mathbf{B}$, $v\colon \mathbf{B} \to \mathbf{C}$ and $w\colon \mathbf{C} \to \mathbf{D}$ we have $w \circ (v \circ u) = (w \circ v) \circ u$.

While the notion of a category is quite general, all of the categories considered in this book will have as their objects algebraic structures (perhaps with a compatible topology), and will have all (continuous) homomorphisms between them as their morphisms.

A morphism $u\colon \mathbf{A} \to \mathbf{B}$ is called an **isomorphism** if there is a morphism $v\colon \mathbf{B} \to \mathbf{A}$ such that $v \circ u = \mathrm{id}_{\mathbf{A}}$ and $u \circ v = \mathrm{id}_{\mathbf{B}}$. We then say that $\mathbf{A}$ and $\mathbf{B}$ are **isomorphic** and write $\mathbf{A} \cong \mathbf{B}$. Since these equations uniquely determine v, we call v the **inverse** of u and denote it by u^{-1}.

Given a category $\mathcal{A}$ and objects $\mathbf{A}$ and $\mathbf{B}$ of $\mathcal{A}$, we say that an object $\mathbf{C}$ of $\mathcal{A}$ is a **product** of $\mathbf{A}$ and $\mathbf{B}$ (in $\mathcal{A}$) if there are associated (**projection**) **morphisms** $p_1 \colon \mathbf{C} \to \mathbf{A}$ and $p_2 \colon \mathbf{C} \to \mathbf{B}$ such that for any object $\mathbf{D}$ of $\mathcal{A}$ and any morphisms $q_1 \colon \mathbf{D} \to \mathbf{A}$ and $q_2 \colon \mathbf{D} \to \mathbf{B}$, there is a unique morphism $u \colon \mathbf{D} \to \mathbf{C}$ such that $p_1 \circ u = q_1$ and $p_2 \circ u = q_2$. We shall denote the morphism u by $q_1 \sqcap q_2 \colon \mathbf{D} \to \mathbf{A} \times \mathbf{B}$. It is straightforward to check that products of two objects, if they exist at all, are determined up to isomorphism (Exercise 1.1). As a result we denote *the* product of objects $\mathbf{A}$ and $\mathbf{B}$ by $\mathbf{A} \times \mathbf{B}$ and we write

$$\mathbf{A} \underset{p_1}{\leftarrow} \mathbf{A} \times \mathbf{B} \underset{p_2}{\rightarrow} \mathbf{B}$$

where p_1 and p_2 are the associated projection morphisms. It is important to bear in mind that a category may not have a product of any two of its objects. Even when it does, these products may not be **concrete**, that is, they may not be the set-theoretic Cartesian products with their usual underlying projection maps. We will often extend direct products to finite direct powers. If $p_i \colon \mathbf{A}^n \to \mathbf{A}$ is the ith projection and $q_i \colon \mathbf{D} \to \mathbf{A}$, for $1 \le i \le n$, then $q_1 \sqcap q_2 \sqcap \cdots \sqcap q_n \colon \mathbf{D} \to \mathbf{A}^n$ denotes the unique morphism u such that $p_i \circ u = q_i$ for each i.

Dual to the notion of product is the notion obtained by reversing the directions of the morphisms in its definition. A **coproduct** of objects $\mathbf{X}$ and $\mathbf{Y}$ in a category $\mathfrak{X}$ is an object $\mathbf{Z}$ in $\mathfrak{X}$ with associated morphisms $\sigma_1 \colon \mathbf{X} \to \mathbf{Z}$ and $\sigma_2 \colon \mathbf{Y} \to \mathbf{Z}$ such that, for any object $\mathbf{W}$ of $\mathfrak{X}$ and any morphisms $\rho_1 \colon \mathbf{X} \to \mathbf{W}$ and $\rho_2 \colon \mathbf{Y} \to \mathbf{W}$, there is a unique morphism $\tau = \rho_1 \sqcup \rho_2 \colon \mathbf{Z} \to \mathbf{W}$ such that $\tau \circ \sigma_1 = \rho_1$ and $\tau \circ \sigma_2 = \rho_2$. Coproducts, when they exist, are unique up to isomorphism as well. We write

$$\mathbf{X} \underset{\sigma_1}{\rightarrow} \mathbf{X} * \mathbf{Y} \underset{\sigma_2}{\leftarrow} \mathbf{Y}$$

where $\mathbf{X} * \mathbf{Y}$ denotes the coproduct of $\mathbf{X}$ and $\mathbf{Y}$ and σ_1 and σ_2 denote the associated morphisms. The coproduct of two sets (in the category of sets and all functions between them) is simply their disjoint union—this is a convenient but uncommon occurrence (see Chapter 6).

Let $\mathcal{A}$ and $\mathfrak{X}$ be categories. Then $D \colon \mathcal{A} \to \mathfrak{X}$ is a **contravariant functor** provided that the following hold:

(i) for any objects $\mathbf{A}$ and $\mathbf{B}$ in $\mathcal{A}$ and any morphism $u \colon \mathbf{A} \to \mathbf{B}$ in $\mathcal{A}(\mathbf{A}, \mathbf{B})$, the objects $D(\mathbf{A})$ and $D(\mathbf{B})$ are in $\mathfrak{X}$ and the corresponding morphism $D(u) \colon D(\mathbf{B}) \to D(\mathbf{A})$ is in $\mathfrak{X}(D(\mathbf{B}), D(\mathbf{A}))$,

(ii) $D(\mathrm{id}_{\mathbf{A}}) = \mathrm{id}_{D(\mathbf{A})}$ for all $\mathbf{A}$ in $\mathcal{A}$,

(iii) if $\mathbf{A}$, $\mathbf{B}$ and $\mathbf{C}$ are objects in $\mathcal{A}$ and $u \colon \mathbf{A} \to \mathbf{B}$ and $v \colon \mathbf{B} \to \mathbf{C}$ are morphisms, then $D(v \circ u) = D(u) \circ D(v) \colon D(\mathbf{C}) \to D(\mathbf{A})$ is a morphism. (See Figure 1.1.)

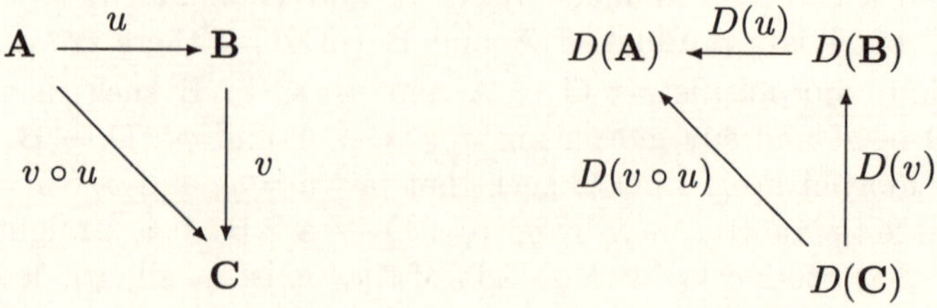

Figure 1.1 a contravariant functor

1.1 Lemma *Let $D\colon \mathcal{A} \to \mathcal{X}$ be a contravariant functor,* $\mathbf{A}$ *and* $\mathbf{B}$ *be objects of* $\mathcal{A}$ *and* $u\colon \mathbf{A} \to \mathbf{B}$ *be an isomorphism. Then* $D(u)\colon D(\mathbf{B}) \to D(\mathbf{A})$ *is an isomorphism and* $D(u)^{-1} = D(u^{-1})$*. In particular,* $\mathbf{A} \cong \mathbf{B}$ *implies that* $D(\mathbf{A}) \cong D(\mathbf{B})$.

We say that $D\colon \mathcal{A} \to \mathcal{X}$ is a **covariant functor** if instead it takes $\mathcal{A}(\mathbf{A},\mathbf{B})$ to $\mathcal{X}(D(\mathbf{A}), D(\mathbf{B}))$ and, in part (iii), $D(v \circ u) = D(v) \circ D(u)$. In either case the functor D is **faithful** if it is one-to-one on each homset $\mathcal{A}(\mathbf{A},\mathbf{B})$ and it is **full** if it maps each homset $\mathcal{A}(\mathbf{A},\mathbf{B})$ onto the appropriate homset between the images $D(\mathbf{A})$ and $D(\mathbf{B})$. Note that the appropriate analogue of Lemma 1.1.1 holds for covariant functors as well. The functors that arise in duality theory are almost exclusively contravariant.

We will be interested in pairs of contravariant functors $D\colon \mathcal{A} \to \mathcal{X}$ and $E\colon \mathcal{X} \to \mathcal{A}$ which are in a general sense inverse to one another. To explain exactly what this means, suppose that for each $\mathbf{A} \in \mathcal{A}$ and each $\mathbf{X} \in \mathcal{X}$ there are morphisms

$$e_{\mathbf{A}}\colon \mathbf{A} \to ED(\mathbf{A}) \quad \text{and} \quad \varepsilon_{\mathbf{X}}\colon \mathbf{X} \to DE(\mathbf{X}).$$

We say that $\langle D, E, e, \varepsilon\rangle$ is a **dual adjunction** between $\mathcal{A}$ and $\mathcal{X}$, that D and E are **dually adjoint**, that E is a **dual adjoint to** D, and that D is a **dual adjoint to** E if the following conditions hold:

(i) for $u\colon \mathbf{A} \to \mathbf{B}$ in $\mathcal{A}$ and $\varphi\colon \mathbf{X} \to \mathbf{Y}$ in $\mathcal{X}$, the two squares in Figure 1.2 **commute**, that is, $e_{\mathbf{B}} \circ u = ED(u) \circ e_{\mathbf{A}}$ and $\varepsilon_{\mathbf{Y}} \circ \varphi = DE(\varphi) \circ \varepsilon_{\mathbf{X}}$,

(ii) for $\mathbf{A} \in \mathcal{A}$ and $\mathbf{X} \in \mathcal{X}$ there is a bijection between $\mathcal{A}(\mathbf{A}, E(\mathbf{X}))$ and $\mathcal{X}(\mathbf{X}, D(\mathbf{A}))$ associating u and φ as given in the commuting triangles of Figure 1.2, that is, $u = E(D(u) \circ \varepsilon_{\mathbf{X}}) \circ e_{\mathbf{A}}$ and $\varphi = D(E(\varphi) \circ e_{\mathbf{A}}) \circ \varepsilon_{\mathbf{X}}$.

Those already experienced with categories will notice that $\mathrm{id}_{\mathcal{A}}\colon \mathcal{A} \to \mathcal{A}$ and $ED\colon \mathcal{A} \to \mathcal{A}$ are covariant functors and that the left square of Figure 1.2 says precisely that $e\colon \mathrm{id}_{\mathcal{A}} \to ED$ is a **natural transformation**. A similar observation can be made about the right square.

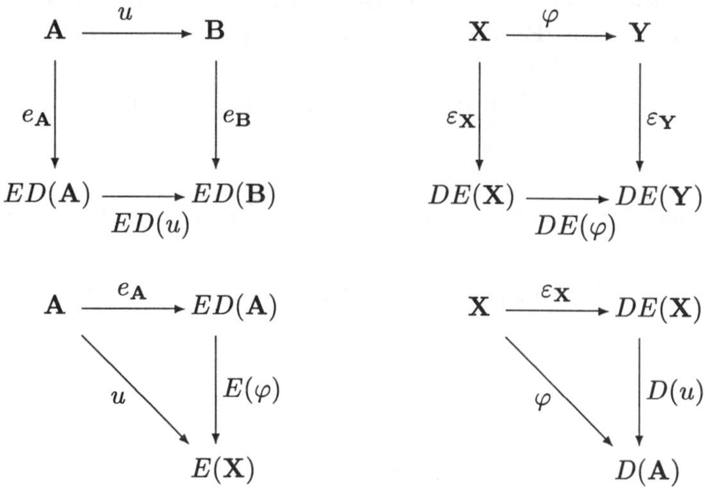

Figure 1.2 a dual adjunction

The property of having a dual adjunction is a symmetric relation between the categories $\mathcal{A}$ and $\mathcal{X}$. In practice we usually view the two categories of a dual adjunction from different perspectives. For example, $\mathcal{A}$ may be a category in which we have a prior interest. We then discover a dual adjunction which tells us that $\mathcal{A}$ is tied in a special way to a new category $\mathcal{X}$. If we are looking for information about the objects in $\mathcal{A}$, our first requirement must be that each object $\mathbf{A}$ of $\mathcal{A}$ is represented as $E(\mathbf{X})$ for some object $\mathbf{X}$ of $\mathcal{X}$, that is, $\mathbf{A} \cong E(\mathbf{X})$. But how are we to find the object $\mathbf{X}$ and the required isomorphism? A dual adjunction provides us with a pair of canonical candidates: we take $\mathbf{X} = D(\mathbf{A})$ and ask that $e_{\mathbf{A}}$ be an isomorphism. This observation leads us to the following definition. We say that the dual adjunction $\langle D, E, e, \varepsilon\rangle$ between categories $\mathcal{A}$ and $\mathcal{X}$ is a **dual representation** if, for every object $\mathbf{A}$ of $\mathcal{A}$, the morphism

$$e_{\mathbf{A}} \colon \mathbf{A} \to ED(\mathbf{A})$$

is an isomorphism.

1.2 Lemma *If $\langle D, E, e, \varepsilon\rangle$ is a dual representation, then D is both full and faithful.*

Proof To see that D is faithful, let $D(u) = D(v)$ where $u, v \colon \mathbf{A} \to \mathbf{B}$. Applying the first square diagram of Figure 1.2 and the fact that $e_{\mathbf{B}}$ is an isomorphism, we obtain

$$u = e_{\mathbf{B}}^{-1} \circ ED(u) \circ e_{\mathbf{A}} = e_{\mathbf{B}}^{-1} \circ ED(v) \circ e_{\mathbf{A}} = v.$$

To see that D is full, let $\psi\colon D(\mathbf{B}) \to D(\mathbf{A})$. We claim that $\psi = D(v)$ where $v\colon \mathbf{A} \to \mathbf{B}$ is defined as $v = e_{\mathbf{B}}^{-1} \circ E(\psi) \circ e_{\mathbf{A}}$. We first apply the triangles in Figure 1.2 with $\mathbf{X} = D(\mathbf{B})$, with $\varphi = \mathrm{id}_{D(\mathbf{B})}$ and with $\mathbf{A}$ replaced by $\mathbf{B}$. This gives us $D(e_{\mathbf{B}}) \circ \varepsilon_{D(\mathbf{B})} = \mathrm{id}_{D(\mathbf{B})}$ (since $E(\mathrm{id}_{D(\mathbf{B})}) = \mathrm{id}_{ED(\mathbf{B})}$), and hence

$$\varepsilon_{D(\mathbf{B})} = D(e_{\mathbf{B}})^{-1}.$$

Next we apply the triangles again, this time with $\mathbf{X} = D(\mathbf{B})$ and $\varphi = \psi$, to get $\psi = D(E(\psi) \circ e_{\mathbf{A}}) \circ \varepsilon_{D(\mathbf{B})}$. Using Lemma 1.1.1 we obtain

$$\begin{aligned}\psi &= D(E(\psi) \circ e_{\mathbf{A}}) \circ \varepsilon_{D(\mathbf{B})} = D(E(\psi) \circ e_{\mathbf{A}}) \circ D(e_{\mathbf{B}})^{-1}\\ &= D(E(\psi) \circ e_{\mathbf{A}}) \circ D(e_{\mathbf{B}}^{-1}) = D(e_{\mathbf{B}}^{-1} \circ E(\psi) \circ e_{\mathbf{A}}) = D(v),\end{aligned}$$

as required. ■

A dual representation gives us a way to construct the objects of $\mathcal{A}$ from the objects of $\mathfrak{X}$ and the contravariant functor E. What it does not give us is a close connection between $\mathfrak{X}$ and $\mathcal{A}$ as categories. This we do obtain if we also require that each object $\mathbf{X}$ of $\mathfrak{X}$ is represented as $D(\mathbf{A})$ for some object $\mathbf{A}$ of $\mathcal{A}$. Again a dual adjunction offers a uniform way in which this wish can be fulfilled. We say that a dual representation $\langle D, E, e, \varepsilon\rangle$ between categories $\mathcal{A}$ and $\mathfrak{X}$ is a **dual equivalence** if, for every object $\mathbf{X}$ of $\mathfrak{X}$,

$$\varepsilon_{\mathbf{X}}\colon \mathbf{X} \to DE(\mathbf{X})$$

is an isomorphism. It is our contention that dual equivalence is a strong and appropriate notion of 'equivalence' of categories. We substantiate this contention by showing that, under a dual equivalence, D and E preserve all of the purely category-theoretic properties of $\mathcal{A}$ and $\mathfrak{X}$. We suggest that the reader construct appropriate commuting diagrams to illustrate the verifications of the items in the lemma below.

1.3 Lemma *Assume that $\langle D, E, e, \varepsilon\rangle$ is a dual equivalence between categories $\mathcal{A}$ and $\mathfrak{X}$. Let $\mathbf{A}$ and $\mathbf{B}$ be objects of $\mathcal{A}$, let $\mathbf{X}$ and $\mathbf{Y}$ be objects of $\mathfrak{X}$ and let $u\colon \mathbf{A} \to \mathbf{B}$ and $\varphi\colon \mathbf{X} \to \mathbf{Y}$ be morphisms.*

(i) *There is an object $\mathbf{X}$ of $\mathfrak{X}$, namely $\mathbf{X} = D(\mathbf{A})$, such that $\mathbf{A} \cong E(\mathbf{X})$.*
(ii) *There is an object $\mathbf{A}$ of $\mathcal{A}$, namely $\mathbf{A} = E(\mathbf{X})$, such that $\mathbf{X} \cong D(\mathbf{A})$.*
(iii) *Both D and E are full and faithful.*
(iv) *u is an isomorphism if and only if $D(u)$ is an isomorphism, and consequently $\mathbf{A} \cong \mathbf{B}$ if and only if $D(\mathbf{A}) \cong D(\mathbf{B})$.*
(v) *φ is an isomorphism if and only if $E(\varphi)$ is an isomorphism, and consequently $\mathbf{X} \cong \mathbf{Y}$ if and only if $E(\mathbf{X}) \cong E(\mathbf{Y})$.*

Proof Although parts (i) and (ii) are restatements of the definitions, we have included them for emphasis. Part (iii) follows from Lemma 1.1.2

since both $\langle D, E, e, \varepsilon\rangle$ and $\langle E, D, \varepsilon, e\rangle$ are dual representations. Parts (iv) and (v) now follow from Lemma 1.1.1 and the fact D and E are full. ■

Building on these facts we find that, under a dual equivalence, products and coproducts are reversed. In particular one category has coproducts if and only if the other has products. We ask the reader to use the definitions and facts that we have established to prove the following result. (See Exercise 1.2 and Exercise 1.3.)

1.4 Lemma *Under a dual adjunction coproducts go to products, and under a dual equivalence products go to coproducts. More explicitly, assume that $\langle D, E, e, \varepsilon\rangle$ is a dual adjunction between categories $\mathcal{A}$ and $\mathcal{X}$. Let $\mathbf{A}$ and $\mathbf{B}$ be objects of $\mathcal{A}$ and let $\mathbf{X}$ and $\mathbf{Y}$ be objects of $\mathcal{X}$.*

(i) *If $\mathbf{X} \underset{\sigma_1}{\rightarrow} \mathbf{X} * \mathbf{Y} \underset{\sigma_2}{\leftarrow} \mathbf{Y}$ is a coproduct of $\mathbf{X}$ and $\mathbf{Y}$ in $\mathcal{X}$, then*

$$E(\mathbf{X}) \underset{E(\sigma_1)}{\leftarrow} E(\mathbf{X} * \mathbf{Y}) \underset{E(\sigma_2)}{\rightarrow} E(\mathbf{Y})$$

is a product of $E(\mathbf{X})$ and $E(\mathbf{Y})$ in $\mathcal{A}$.

(ii) *If $\langle D, E, e, \varepsilon\rangle$ is a dual equivalence and $\mathbf{A} \underset{p_1}{\leftarrow} \mathbf{A} \times \mathbf{B} \underset{p_2}{\rightarrow} \mathbf{B}$ is a product of $\mathbf{A}$ and $\mathbf{B}$ in $\mathcal{A}$, then*

$$D(\mathbf{A}) \underset{D(p_1)}{\rightarrow} D(\mathbf{A} \times \mathbf{B}) \underset{D(p_2)}{\leftarrow} D(\mathbf{B})$$

is a coproduct of $D(\mathbf{A})$ and $D(\mathbf{B})$ in $\mathcal{X}$.

In Chapters 5 and 6 we will see examples of dual equivalences in which coproducts in $\mathcal{X}$ are very simple to construct, and as a result Lemma 1.1.4(i) will give us a convenient representation of products in $\mathcal{A}$. In fact dual adjunctions satisfy something much more general: colimits go to limits—see Mac Lane [Mac:CW].

1.2 Distributive lattices

The early development of different natural dualities grew primarily out of the prototype published by Hilary Priestley in 1970 for the category $\mathcal{D}$ of distributive lattices. Priestley found a dual equivalence between $\mathcal{D}$ and a special category $\mathcal{X}_{01}$ of ordered Boolean spaces. We will present her dual equivalence here, and then use it later as a prototype to formulate our general theory as well.

A **distributive lattice** is an algebra $\mathbf{L} = \langle L; \vee, \wedge\rangle$ which satisfies all equations, such as

$$x \wedge (y \vee z) \approx (x \wedge y) \vee (x \wedge z),$$

that hold whenever the variables are replaced by arbitrary sets, $\vee$ (called **join**) is interpreted as union, and $\wedge$ (called **meet**) is interpreted as intersection. For example, any collection of sets closed under union and intersection forms a distributive lattice.

While this rather glib definition of 'distributive lattice' conveys the significance of the notion, it grants us precious little assistance in deciding whether or not an arbitrary algebra with two binary operations is indeed a distributive lattice. Our first goal will be to show that a small finite subset of these equations implies them all. The subset we choose is a useful one, though still highly redundant. We refer to this subset as (DL) and list it below.

$$\left.\begin{array}{l}(x \vee y) \vee z \approx x \vee (y \vee z)\\ (x \wedge y) \wedge z \approx x \wedge (y \wedge z)\end{array}\right\} \quad \text{(associative laws)}$$

$$\left.\begin{array}{l}x \vee y \approx y \vee x\\ x \wedge y \approx y \wedge x\end{array}\right\} \quad \text{(commutative laws)}$$

$$\left.\begin{array}{l}x \vee x \approx x\\ x \wedge x \approx x\end{array}\right\} \quad \text{(idempotency laws)}$$

$$\left.\begin{array}{l}x \vee (x \wedge y) \approx x\\ x \wedge (x \vee y) \approx x\end{array}\right\} \quad \text{(absorption laws)}$$

$$\left.\begin{array}{l}x \wedge (y \vee z) \approx (x \wedge y) \vee (x \wedge z)\\ x \vee (y \wedge z) \approx (x \vee y) \wedge (x \vee z)\end{array}\right\} \quad \text{(distributive laws)}$$

As a simple exercise we ask the reader to check that any model $\mathbf{A}$ of (DL) has a reflexive ordering $\leqslant$ defined by

$$a \leqslant b \quad \text{if and only if} \quad a \wedge b = a.$$

Relative to this ordering, $a \wedge b$ is the greatest lower bound of a and b and $a \vee b$ is the least upper bound of a and b (Exercise 1.5). We adopt the notation

$$\downarrow a = \{\, c \mid c \in A \text{ and } c \leqslant a \,\} \quad \text{and} \quad \uparrow b = \{\, d \mid d \in A \text{ and } b \leqslant d \,\}$$

for all $a, b \in A$.

We would like to show that any model $\mathbf{A}$ of (DL) is a distributive lattice in the sense that we have defined. To do so we will show that there are a set X and an embedding of $\mathbf{A}$ into the lattice $\mathbf{P}(X)$ of all subsets of X, telling us that $\mathbf{A}$ is isomorphic to some distributive lattice of sets. The set X must be chosen in such a way that there is a natural pairing of distinct elements of $\mathbf{A}$ with distinct subsets of X. To do this we need several definitions. A subset x of an ordered set is **decreasing** if $b \in x$ whenever $b \leqslant a$ and $a \in x$. Similarly, a subset y of an ordered set is **increasing** if $a \in y$ whenever $b \leqslant a$ and $b \in y$. Now let $\mathbf{A}$ be a model of (DL). A decreasing subset of $\mathbf{A}$ is called an **ideal** if it is closed under $\vee$, and an increasing subset of $\mathbf{A}$ is called a **filter** if it is closed under $\wedge$. For example, $\downarrow a$ is an ideal and $\uparrow b$ is a filter. An ideal (filter) x of $\mathbf{A}$ is called **prime** if $a \in x$ or $b \in x$ whenever $a \wedge b \in x$ ($a \vee b \in x$), that is, if the complement of x is a filter (an ideal). Note that in this particular context we regard the empty set and the entire set A as being prime. We first prove the fundamental existence theorem for prime ideals.

2.1 Prime Ideal Theorem *Let $\mathbf{A}$ be a model of* (DL), *let x be an ideal in $\mathbf{A}$ and let y be a filter in $\mathbf{A}$ that is disjoint from x. Then there is a prime ideal containing x and not intersecting y.*

Proof By Zorn's Lemma, there is an ideal z which is maximal with respect to containing x and not intersecting y. We will show that z is a prime ideal. To do so, suppose that $a, b \notin z$ but that $a \wedge b \in z$.

The ideal generated by $z \cup \{a\}$ equals $\bigcup\{\downarrow (c \vee a) \mid c \in z\}$ and, by the maximality of z, must intersect y. Thus there is a $c \in z$ such that $c \vee a \in y$ and similarly there is a $d \in z$ such that $d \vee b \in y$. Now

$$(c \vee a) \wedge (d \vee b) \leq (c \vee d \vee a) \wedge (c \vee d \vee b) = (c \vee d) \vee (a \wedge b).$$

But the left-hand side is in y as $c \vee a, d \vee b \in y$ and the right-hand side is in z as $c, d, a \wedge b \in z$. This implies that $y \cap z \neq \varnothing$, a contradiction. ■

We will show that each element $a \in A$ is uniquely determined by the set of all prime ideals not containing it. More specifically, let X denote the set of all prime ideals of $\mathbf{A}$ and, for $a \in A$, define

$$e_{\mathbf{A}}(a) = \{x \in X \mid a \notin x\} \subseteq X.$$

2.2 Lemma *If $\mathbf{A}$ satisfies* (DL), *then the map $e_{\mathbf{A}} \colon \mathbf{A} \to \mathbf{P}(X)$ is an embedding of $\mathbf{A}$ into the distributive lattice $\mathbf{P}(X)$ of all subsets of X. In particular, any model $\mathbf{A}$ of* (DL) *is itself a distributive lattice.*

Proof From the definition of prime ideal it follows immediately that $e_{\mathbf{A}}$ is a homomorphism, that is, for any $a, b \in A$ we have

$$e_{\mathbf{A}}(a \vee b) = e_{\mathbf{A}}(a) \cup e_{\mathbf{A}}(b) \quad \text{and} \quad e_{\mathbf{A}}(a \wedge b) = e_{\mathbf{A}}(a) \cap e_{\mathbf{A}}(b).$$

To prove that $e_{\mathbf{A}}$ is one-to-one consider two elements a and b of $\mathbf{A}$ where $a \nleq b$. By the Prime Ideal Theorem there is a prime ideal z containing the ideal $\downarrow b$ and not intersecting the filter $\uparrow a$. Thus z is in $e_{\mathbf{A}}(a)$ but not in $e_{\mathbf{A}}(b)$.

Since $\mathbf{A}$ is isomorphic to a sublattice of the distributive lattice $\mathbf{P}(X)$, it satisfies every equation satisfied by every distributive lattice of sets. ■

As a result of this fact we could, as is usually done, take (DL) as a definition of 'distributive lattice'.

The success that we have achieved in representing the distributive lattice **A** as a lattice of sets suggests that we press further and see if we can discover even more. Can we obtain a tighter representation of **A** by saying more exactly just which subsets of X form the image of **A** under $e_{\mathbf{A}}$? For example, X is ordered by inclusion and each image $e_{\mathbf{A}}(a)$, where $a \in A$, is easily seen to be a decreasing subset of the ordered set $\langle X; \subseteq\rangle$. Thus we have shown that every distributive lattice is isomorphic to a sublattice of the lattice of *decreasing* subsets of some ordered set.

It turns out that we can give a precise and useful description of the sublattice of $\mathbf{P}(X)$ that is isomorphic to **A**. The key to doing so is the following curious property of distributive lattices.

2.3 Compactness Lemma *Let s and t be subsets of the distributive lattice* **A** *with the property that every prime ideal containing s also intersects t. Then there are finite subsets $s_0 \subseteq s$ and $t_0 \subseteq t$ such that every prime ideal containing s_0 also intersects t_0.*

Proof Suppose that no such finite subsets $s_0 \subseteq s$ and $t_0 \subseteq t$ exist. We first observe that there cannot be finite subsets $s_0 \subseteq s$ and $t_0 \subseteq t$ such that the join of s_0 is greater than or equal to the meet of t_0. If there were, then every prime ideal containing s_0 would contain the meet of t_0 and therefore would have to contain a member of t_0. It follows that the ideal x consisting of all elements dominated by the join of some finite subset of s is disjoint from the filter y consisting of all elements which dominate the meet of some finite subset of t. By the Prime Ideal Theorem there is a prime ideal z containing x and not intersecting y. This means that z contains s but does not intersect t, a contradiction. ■

The name of this lemma arises from the observation that it appears to be saying that a certain topological space is compact. To see that it is indeed saying this, we need to introduce a topology on the set X. We refer the reader to Appendix B for a review of the background from topology that we will need. Suppose that $\mathbf{Y} = \langle Y; 0, 1, \leqslant, \mathcal{T}\rangle$ is a bounded, ordered compact space, that is, $\langle Y; \mathcal{T}\rangle$ is a compact space and $\leqslant$ is an order relation on the set Y with minimum element 0 and maximum element 1. We say that $\mathbf{Y}$ is a **bounded Priestley space** if $0 \neq 1$ and the space is **totally order-disconnected**, that is, for any $x, y \in Y$ where $x \nleqslant y$, there is a clopen decreasing set containing y but not containing x. We can now say exactly which sublattice of $\mathbf{P}(X)$ is isomorphic to the distributive lattice **A**.

2.4 Representation Theorem *Let* **A** *be a distributive lattice and let $D(\mathbf{A}) = \langle X; 0, 1, \subseteq, \mathcal{T}\rangle$, where X is the set of prime ideals of* **A** *ordered by inclusion $\subseteq$, with minimum element $0 = \varnothing$ and maximum element $1 = A$,* *and $\mathcal{T}$ is the topology on X having as subbasis all sets $e_{\mathbf{A}}(a)$ and their complements for a ranging over A. Then*

(i) *$D(\mathbf{A})$ is a bounded Priestley space, and*

(ii) *$\{e_{\mathbf{A}}(a) \mid a \in A\}$ is exactly the set of proper, non-empty, clopen decreasing subsets of $D(\mathbf{A})$.*

In particular, every distributive lattice is isomorphic to the lattice of all proper, non-empty, clopen decreasing subsets of some bounded Priestley space.

Proof (i) Clearly $\langle X; 0, 1, \subseteq\rangle$ is a bounded ordered set. To see that $D(\mathbf{A})$ is totally order-disconnected, choose $x, y \in X$ where $x \nsubseteq y$. Let a be in x but not y. Then $e_{\mathbf{A}}(a)$ contains y but not x.

In order to show that $D(\mathbf{A})$ is compact we use Alexander's Subbasis Lemma B.2. Let $s, t \subseteq A$ be such that

$$\{\, e_{\mathbf{A}}(a) \mid a \in s \,\} \cup \{\, X \backslash e_{\mathbf{A}}(b) \mid b \in t \,\}$$

is a subbasic cover of $D(\mathbf{A})$. A prime ideal x containing s is not in $e_{\mathbf{A}}(a)$ for any $a \in s$. Consequently $x \notin e_{\mathbf{A}}(b)$ for some $b \in t$, that is, x intersects t. By the Compactness Lemma 1.2.3 there are finite subsets $s_0 \subseteq s$ and $t_0 \subseteq t$ such that a prime ideal x containing s_0 must intersect t_0, that is,

$$\{\, e_{\mathbf{A}}(a) \mid a \in s_0 \,\} \cup \{\, X \backslash e_{\mathbf{A}}(b) \mid b \in t_0 \,\}$$

covers $D(\mathbf{A})$. Thus $D(\mathbf{A})$ is compact.

(ii) Assume that c is a clopen, decreasing subset of $D(\mathbf{A})$ containing 0 but not 1. We first argue that, for $y \notin c$, there is an $a \in A$ such that $y \notin e_{\mathbf{A}}(a) \supseteq c$. Choose any $x \in c$. Since c is decreasing, $y \nsubseteq x$, so there is a b in y but not in x. Then $y \notin e_{\mathbf{A}}(b)$ but $x \in e_{\mathbf{A}}(b)$. Thus $\{e_{\mathbf{A}}(b) \mid y \notin e_{\mathbf{A}}(b)\}$ covers c. Since c is a closed subset of X, it is compact. Consequently there is a finite set $\{b_1, b_2, \dots, b_n\} \subseteq A$ such that

$$y \notin e_{\mathbf{A}}(b_1) \cup e_{\mathbf{A}}(b_2) \cup \cdots \cup e_{\mathbf{A}}(b_n) = e_{\mathbf{A}}(b_1 \vee b_2 \vee \cdots \vee b_n) \supseteq c.$$

Since this can be done for every $y \notin c$, we have $c = \bigcap\{e_{\mathbf{A}}(a) \mid c \subseteq e_{\mathbf{A}}(a)\}$. Since c is open, $X \backslash c$ is compact. As $X \backslash c = \bigcup\{\, X \backslash e_{\mathbf{A}}(a) \mid c \subseteq e_{\mathbf{A}}(a)\}$, there is a finite set $\{a_1, a_2, \dots, a_m\} \subseteq A$ such that $X \backslash c = X \backslash e_{\mathbf{A}}(a_1) \cup \cdots \cup X \backslash e_{\mathbf{A}}(a_m)$ and therefore

$$c = e_{\mathbf{A}}(a_1) \cap e_{\mathbf{A}}(a_2) \cap \cdots \cap e_{\mathbf{A}}(a_m) = e_{\mathbf{A}}(a_1 \wedge a_2 \wedge \cdots \wedge a_m).$$

Thus c is of the form $e_{\mathbf{A}}(a)$ for some $a \in A$, as claimed. ∎

Implicit in the Representation Theorem 1.2.4 is a statement about two categories. The class of distributive lattices forms a familiar category $\mathfrak{D}$ with homomorphisms as morphisms. On the other side we have the class of bounded Priestley spaces. A map φ from a bounded Priestley space $\mathbf{X}$ to a bounded Priestley space $\mathbf{Y}$ is a **continuous homomorphism** if it is continuous, $\varphi(0) = 0$ and $\varphi(1) = 1$, and $\varphi(x) \leqslant \varphi(y)$ whenever $x, y \in X$ and $x \leqslant y$. The class of bounded Priestley spaces forms a category $\mathfrak{X}_{01}$ with continuous homomorphisms as morphisms (Exercise 1.6). Connecting these two categories are the correspondences

$$D\colon \mathfrak{D} \to \mathfrak{X}_{01} \quad \text{and} \quad E\colon \mathfrak{X}_{01} \to \mathfrak{D}$$

where, for $\mathbf{A} \in \mathfrak{D}$ and $\mathbf{X} \in \mathfrak{X}_{01}$, we have $D(\mathbf{A})$ as the prime ideal space of $\mathbf{A}$ and $E(\mathbf{X})$ as the lattice of proper, non-empty, clopen decreasing subsets of $\mathbf{X}$. Moreover, D and E extend in a natural and simple way to morphisms, provided we are willing to reverse the directions of the morphisms. For a distributive-lattice homomorphism $u\colon \mathbf{A} \to \mathbf{B}$ in $\mathfrak{D}$ we define

$$D(u)\colon D(\mathbf{B}) \to D(\mathbf{A}) \quad \text{by} \quad D(u)(x) = u^{-1}(x) \quad \text{for} \quad x \in D(\mathbf{B}).$$

Similarly, for a continuous homomorphism $\varphi\colon \mathbf{X} \to \mathbf{Y}$ in $\mathfrak{X}_{01}$ we define

$$E(\varphi)\colon E(\mathbf{Y}) \to E(\mathbf{X}) \quad \text{by} \quad E(\varphi)(\alpha) = \varphi^{-1}(\alpha) \quad \text{for} \quad \alpha \in E(\mathbf{Y}).$$

(See Exercise 1.7(i), (ii).) In this terminology the Representation Theorem 1.2.4 simply says that, for any distributive lattice $\mathbf{A} \in \mathfrak{D}$, its **dual** $D(\mathbf{A})$ is in $\mathfrak{X}_{01}$ and $e_{\mathbf{A}} : \mathbf{A} \to ED(\mathbf{A})$ is an isomorphism.

A similar construction can be done in the opposite direction. Let $\mathbf{X} \in \mathfrak{X}_{01}$ and let $x \in X$. Then the set $\varepsilon_{\mathbf{X}}(x)$ of all proper, non-empty, clopen decreasing subsets of $\mathbf{X}$ that do not contain x is clearly a prime ideal in $E(\mathbf{X})$. This gives us the corresponding map $\varepsilon_{\mathbf{X}} : \mathbf{X} \to DE(\mathbf{X})$.

2.5 Priestley Dual Equivalence Theorem *$\langle D, E, e, \varepsilon \rangle$ is a dual equivalence between the category $\mathfrak{D}$ of distributive lattices and the category $\mathfrak{X}_{01}$ of bounded Priestley spaces.*

Proof The proof of this theorem is now a matter of putting together what we have established and verifying the necessary details. It is outlined in Exercise 1.7. ∎

This theorem gives us a tight equivalence between the category $\mathfrak{D}$, which we originally set out to investigate, and the new category $\mathfrak{X}_{01}$, which might at first appear rather strange and unfamiliar to the algebraist. It is our contention that the category $\mathfrak{X}_{01}$ is in many respects an easier and more tractable environment in which to work than the category $\mathfrak{D}$. In Chapter 5 we will present several specific examples of algebraic problems which arise in $\mathfrak{D}$ but seem to be most readily solved by transferring them to the category $\mathfrak{X}_{01}$, solving them there, and then transferring back. In order to facilitate our work in the category $\mathfrak{X}_{01}$, we need to take a closer look at the structure of its members. Our knowledge of the structure of distributive lattices will serve as a useful guide.

Consider the two-element distributive lattice

$$\underline{\mathbf{D}} = \langle \{0,1\}; \vee, \wedge \rangle$$

where $\vee$ and $\wedge$ are least upper bound and greatest lower bound, respectively. If X is any set, then there is a natural bijection χ from the subsets of X to the maps from X into $\{0,1\}$ which takes each subset of X to its **characteristic function**: for each subset c of X and each element x of X, we define $\chi_c(x)$ to be 1 if x is in c and 0 otherwise. The map χ is a bijection from the set of all subsets of X to the set of all functions from X into $\{0,1\}$. We can now check that χ takes unions and intersections to pointwise joins and meets, respectively, in the direct power lattice $\underline{\mathbf{D}}^X$ (Exercise 1.8). As a result we conclude that the lattice $\mathbf{P}(X)$ of all subsets of X is isomorphic to the lattice $\underline{\mathbf{D}}^X$. Combining this fact with Lemma 1.2.2 we have the following alternative description of distributive lattices.

2.6 Corollary *Let* $\mathbf{L} = \langle L; \vee, \wedge \rangle$. *Then* $\mathbf{L}$ *is a distributive lattice if and only if it is isomorphic to a sublattice of a direct power of* $\underline{\mathbf{D}} = \langle \{0,1\}; \vee, \wedge \rangle$.

It will be useful for us to state this fact more compactly in symbols. Given an arbitrary class $\mathcal{B}$ of *algebras*, we denote by

$$\mathbb{I}\mathcal{B}, \quad \mathbb{S}\mathcal{B}, \quad \mathbb{P}\mathcal{B},$$

respectively, the classes of isomorphic copies, subalgebras and direct products (over possibly empty index sets) of members of $\mathcal{B}$. In these symbols the above corollary asserts that the category of distributive lattices is given as

$$\mathcal{D} = \mathbb{ISP}\underline{\mathbf{D}} \quad \text{where} \quad \underline{\mathbf{D}} = \langle \{0,1\}; \vee, \wedge \rangle.$$

Given the dual equivalence between the categories $\mathcal{D}$ and $\mathfrak{X}_{01}$, it is natural to look for a comparable description of $\mathfrak{X}_{01}$. Can $\mathfrak{X}_{01}$ be generated in a similar manner within the larger category $\mathcal{W}$ of all topological structures $\mathbf{W} = \langle W; 0, 1, \leqslant, \mathcal{T} \rangle$ of the same type? The category $\mathfrak{X}_{01}$ also has a two-element structure

$$\underset{\sim}{\mathbf{D}} = \langle \{0,1\}; 0, 1, \leqslant, \mathcal{T} \rangle$$

where $\leqslant$ is the usual order and $\mathcal{T}$ is the discrete topology. We can form the direct power $\underset{\sim}{\mathbf{D}}^I$ of $\underset{\sim}{\mathbf{D}}$ over a *non-empty* index set I in the usual manner.

The order on $\underset{\sim}{\mathbf{D}}^I$ is the pointwise order inherited from $\underset{\sim}{\mathbf{D}}$; the constants of $\underset{\sim}{\mathbf{D}}^I$ are the two constant maps, and the topology is the product topology (see Appendix B). With this structure $\underset{\sim}{\mathbf{D}}^I$ forms a bounded Priestley space (Exercise 1.9). We can also form substructures of bounded Priestley spaces by relativising the order and the topology to any subset containing 0 and 1. In order to ensure that our substructures are again compact, and therefore are again bounded Priestley spaces, we will consider only *closed substructures* of bounded Priestley spaces (Exercise 1.10). Accordingly we define, for a subclass $\mathcal{Y}$ of $\mathcal{W}$,

$$\mathbb{I}\mathcal{Y}, \quad \mathbb{S}_c\mathcal{Y}, \quad \mathbb{P}^+\mathcal{Y},$$

respectively, the classes of isomorphic copies (within $\mathcal{W}$), closed substructures and direct products over *non-empty* index sets of members of $\mathcal{Y}$. We would like to show that every bounded Priestley space can be embedded into a power of $\underset{\sim}{\mathbf{D}}$. The next lemma makes this notion precise.

2.7 Lemma *Suppose a bounded Priestley space morphism* $\tau\colon \mathbf{X} \to \mathbf{Y}$ *has the property that* $x \leqslant x'$ *whenever* $x, x' \in X$ *and* $\tau(x) \leqslant \tau(x')$*. Then the image* $\tau(\mathbf{X}) \subseteq \mathbf{Y}$ *is also a bounded Priestley space and is isomorphic to* $\mathbf{X}$.

Proof By Lemma B.1(i), the image $\tau(\mathbf{X})$ is closed, and is therefore easily seen to be a bounded Priestley space. If $\tau(x) = \tau(x')$, then $\tau(x) \leqslant \tau(x')$ and $\tau(x') \leqslant \tau(x)$, whence $x \leqslant x'$ and $x' \leqslant x$ and hence $x = x'$. Thus τ is one-to-one. By Lemma B.1(ii), the inverse τ^{-1} is a continuous map from $\tau(\mathbf{X})$ onto $\mathbf{X}$. By our hypothesis τ^{-1} preserves $\leqslant$ as well as 0 and 1, and is therefore a bounded Priestley space morphism. It follows that τ is an isomorphism when viewed as a map from $\mathbf{X}$ onto $\tau(\mathbf{X})$. ■

The morphism τ described in Lemma 1.2.7 is called a **bounded Priestley space embedding**.

2.8 Lemma *Let* $\mathbf{X} = \langle X; 0, 1, \leqslant, \mathcal{T}\rangle$*. Then* $\mathbf{X}$ *is a bounded Priestley space if and only if it is isomorphic to a closed substructure of a direct power of* $\underset{\sim}{\mathbf{D}}$*; in symbols,* $\mathcal{X}_{01} = \mathbb{I}\mathbb{S}_c\mathbb{P}^+\underset{\sim}{\mathbf{D}}$ *where* $\underset{\sim}{\mathbf{D}} = \langle \{0,1\}; 0, 1, \leqslant, \mathcal{T}\rangle$.

Proof In Exercises 1.9 and 1.10 we ask the reader to check that bounded Priestley spaces are closed under the formation of closed substructures and direct powers. Consequently every member of $\mathbb{I}\mathbb{S}_c\mathbb{P}^+\underset{\sim}{\mathbf{D}}$ is a bounded Priestley space.

Conversely, let $\mathbf{X}$ be a bounded Priestley space. We have seen that $\varepsilon_{\mathbf{X}}\colon \mathbf{X} \to DE(\mathbf{X})$ is an isomorphism pairing each member $x \in \mathbf{X}$ with a subset $\varepsilon_{\mathbf{X}}(x) \subseteq E(\mathbf{X})$. Our strategy is to pair each $x \in \mathbf{X}$ instead with the characteristic function of $\varepsilon_{\mathbf{X}}(x)$. We will show that this is an embedding of $\mathbf{X}$ into the direct power $\underset{\sim}{\mathbf{D}}^{E(\mathbf{X})}$. Define $\tau\colon \mathbf{X} \to \underset{\sim}{\mathbf{D}}^{E(\mathbf{X})}$ by $\tau = \chi \circ \varepsilon_{\mathbf{X}}$ where χ maps each subset of $E(\mathbf{X})$ to its characteristic function. Written out explicitly we have, for $x \in X$ and $\alpha \in E(\mathbf{X})$,

$$\tau(x)(\alpha) = \begin{cases} 1 & \text{if } \; x \notin \alpha, \\ 0 & \text{if } \; x \in \alpha. \end{cases}$$

We will verify the conditions of Lemma 1.2.7. To see that τ is continuous, we apply Lemma B.3. According to Lemma B.6(i), a subbasic clopen subset of $\underset{\sim}{\mathbf{D}}^{E(\mathbf{X})}$ has the form $U_{\alpha,a} = \{\varphi \in D^{E(\mathbf{X})} \mid \varphi(\alpha) = a\}$ where a is 0 or 1 and $\alpha \in E(\mathbf{X})$. First assume that $a = 0$. Then, for $x \in X$, we have

$$x \in \tau^{-1}(U_{\alpha,0}) \iff \tau(x) \in U_{\alpha,0} \iff \tau(x)(\alpha) = 0 \iff x \in \alpha.$$

Thus $\tau^{-1}(U_{\alpha,0}) = \alpha$, a clopen set. Similarly, $\tau^{-1}(U_{\alpha,1})$ is the complement of α, also a clopen set. Thus τ is continuous.

From the definitions, $\tau(0)$ is the map with constant value 0 and $\tau(1)$ is the map with constant value 1. Finally, for $x, y \in X$, we have

$$x \leqslant y \iff \varepsilon_{\mathbf{X}}(x) \subseteq \varepsilon_{\mathbf{X}}(y) \iff \chi_{\varepsilon_{\mathbf{X}}(x)} \leqslant \chi_{\varepsilon_{\mathbf{X}}(y)} \iff \tau(x) \leqslant \tau(y).$$

From Lemma 1.2.7 we conclude that $\tau(\mathbf{X})$ is a bounded Priestley space which is isomorphic to $\mathbf{X}$. ∎

1.3 Quasi-varieties

In this section we will describe the categories of algebras that will play the role of the category $\mathcal{A}$ in the dual adjunctions, dual representations and dual equivalences that we want to study. Much of the material that we present here relies heavily on the universal algebra background that is outlined in Appendix A.

The example of distributive lattices suggests a way to choose the appropriate categories. Let F be a set of finitary operation symbols, each carrying a natural number called its arity and defined by a fixed arity function on F. The class of all algebras of type F, together with their homomorphisms, forms the umbrella category in which we will work. Consider an arbitrary finite algebra

$$\underline{\mathbf{M}} := \langle M; F \rangle$$

of type F. Natural dualities provide a tool for understanding the category

$$\mathcal{A} := \mathbb{ISP}\underline{\mathbf{M}},$$

consisting of all isomorphic copies of subalgebras of direct powers of $\underline{\mathbf{M}}$. Since a power of an algebra $\underline{\mathbf{M}}$ can be indexed by the empty set, $\mathcal{A}$ always includes the one-element algebras. Such categories are of interest from several points of view, as we now describe. Our first description of $\mathcal{A}$ is an algebraic one, as it shows that the members of $\mathcal{A}$ are characterised by their homomorphisms into $\underline{\mathbf{M}}$.

3.1 Algebraic Separation Theorem *An algebra* $\mathbf{A}$ *is in* $\mathcal{A} = \mathbb{ISP}\underline{\mathbf{M}}$ *if and only if, for each* $a, b \in A$ *where* $a \neq b$*, there is a homomorphism* $u_{ab} \colon \mathbf{A} \to \underline{\mathbf{M}}$ *such that* $u_{ab}(a) \neq u_{ab}(b)$.

Proof First assume $\mathbf{A} \in \mathcal{A}$ where $\mathbf{A} \leqslant \underline{\mathbf{M}}^I$, and that $a, b \in A$ where $a \neq b$. Then $I \neq \varnothing$ and consequently $a(i) \neq b(i)$ for some $i \in I$. This means that the projection homomorphism taking each element of $\mathbf{A}$ to its ith coordinate will separate a and b. Conversely, suppose that a separating homomorphism $u_{ab} \colon \mathbf{A} \to \underline{\mathbf{M}}$ can be found for each pair $a, b \in A$ where $a \neq b$. Let $I = \{(a, b) \in A \times A \mid a \neq b\}$ and define

$$t \colon \mathbf{A} \to \underline{\mathbf{M}}^I \quad \text{where} \quad t(c)(a, b) = u_{ab}(c).$$

As an exercise we ask the reader to verify that t is an embedding. ■

Using the Algebraic Separation Theorem 1.3.1 we will obtain a different description of $\mathcal{A}$ by looking at its logical properties. Let $\{v_s \mid s \in S\}$ be a fixed set of pairwise distinct **variables** indexed by the set S. An **equation** is an expression of the form $t_1 \approx t_2$ where t_1 and t_2 are terms, and an **implication** is an expression of the form

$$\beta_1 \wedge \ldots \wedge \beta_m \implies \alpha$$

where $\beta_1, \ldots, \beta_m$ and α are equations. A **quasi-equation** is an expression which is either an equation or an implication. An algebra $\mathbf{A}$ **satisfies** a quasi-equation if the quasi-equation is true for every substitution of elements of $\mathbf{A}$ for its variables. An algebra $\mathbf{A}$ is a **model** of a set of quasi-equations provided that it satisfies every quasi-equation in the set. The **quasi-equational theory** of $\mathbf{A}$ is the set of all quasi-equations that are satisfied by $\mathbf{A}$. The **quasi-variety generated by** $\mathbf{A}$ is the class of all algebras of the same type as $\mathbf{A}$ that are models of the quasi-equational theory of $\mathbf{A}$.

Our next lemma gives a property of $\mathcal{A}$ which we will frequently use. An algebra $\mathbf{A}$ is **locally finite** if every non-empty finite subset of A is contained in (and therefore generates) a finite subalgebra of $\mathbf{A}$. A class $\mathcal{B}$ of algebras is **locally finite** if each member of $\mathcal{B}$ is locally finite. We refer the reader to Appendix A for a discussion of free algebras.

3.2 Lemma *Every model of the equational theory of a finite algebra $\underline{\mathbf{M}}$ is locally finite. In particular, $\mathcal{A} = \mathbb{ISP}\underline{\mathbf{M}}$ is locally finite.*

Proof Let $\mathbf{B}$ be a model of the equational theory of $\underline{\mathbf{M}}$ and let $T = \{a_1, a_2, \ldots, a_n\} \subseteq B$ be finite. By Proposition A.3 the subalgebra of $\mathbf{B}$ generated by T is a homomorphic image of the free algebra of n-ary term functions over $\underline{\mathbf{M}}$ and is therefore finite. Thus $\mathbf{B}$ is locally finite. From the $\mathbb{HSP}$-Theorem A.4 we see that each member of $\mathcal{A}$ is a model of the equational theory of $\underline{\mathbf{M}}$ and is therefore locally finite. ■

Next we introduce a powerful technique that draws on Tychonoff's Product Theorem B.4. Let $\langle I; \leqslant \rangle$ be a directed ordered set, that is, $I \neq \varnothing$ and each pair of elements in I has an upper bound in I. Assume that for each $i \in I$ we have a *non-empty* set A_i and for $j \geqslant i$ we have a map $\varphi_{ji} \colon A_j \to A_i$ such that for all $i, j, k \in I$,

$$\text{(a)}\varphi_{ii} = \mathrm{id}_{A_i} \quad \text{and} \quad \text{(b) } k \geqslant j \geqslant i \implies \varphi_{ji} \circ \varphi_{kj} = \varphi_{ki}.$$

Such a system of sets and maps is called an **inverse system**. The **inverse limit** of this inverse system is

$$\varprojlim_{i \in I} A_i := \{\, a \in \prod_{i \in I} A_i \mid (\forall i, j \in I)\,(j \geqslant i \implies \varphi_{ji}(a_j) = a_i)\}.$$

In general the inverse limit of an inverse system may well be empty. For example, this is the case if we let $I = \mathbb{N}$ with the usual order, let $A_i = \mathbb{N}$ for all $i \in \mathbb{N}$ and for $j \geqslant i$ define $\varphi_{ji}(n) := n + j - i$. A simple condition guarantees that the inverse limit is not empty.

3.3 Inverse Limit Theorem *The inverse limit of an inverse system of finite sets is non-empty.*

Proof Assume that the sets A_i and the maps φ_{ji} form an inverse system with A_i finite for all $i \in I$. For each $k \in I$, define

$$X_k := \{\, a \in \prod_{i \in I} A_i \mid (\forall i \in I)\,(k \geqslant i \implies \varphi_{ki}(a_k) = a_i)\}.$$

If we fix any element $a_k \in A_k$ and define $a_i = \varphi_{ki}(a_k)$ for all $i \in I$ where $k \geqslant i$, then every extension of a to a member of $\prod\{A_i \mid i \in I\}$ is in X_k. Thus each $X_k \neq \varnothing$, and the intersection of the sets X_k for $k \in I$ is exactly the inverse limit.

To see that the inverse limit is not empty, we endow each set A_i with the discrete topology. By Lemma B.6(iii), each X_k is closed in the product topology on $\prod_{i \in I} A_i$. Now consider any finite subset $\{k_1, k_2, \ldots, k_\ell\}$ of I. If $j \in I$ is an upper bound for this set, then $X_j \subseteq X_{k_1} \cap \cdots \cap X_{k_\ell} \neq \varnothing$. By

Tychonoff's Product Theorem B.4, the product space $\prod_{i\in I} A_i$ is compact, and we conclude that $\bigcap_{k\in I} X_k \neq \varnothing$. ■

We can now prove the analogue of the $\mathbb{HSP}$-Theorem A.4 for finitely generated quasi-varieties.

3.4 $\mathbb{ISP}$-Theorem *Let $\underline{\mathbf{M}}$ be a finite algebra. For any algebra $\mathbf{B}$ of the same type as $\underline{\mathbf{M}}$, the following are equivalent:*

(i) $\mathbf{B} \in \mathcal{A} = \mathbb{ISP}\underline{\mathbf{M}}$;

(ii) $\mathbf{B}$ *is a model of the quasi-equational theory of* $\underline{\mathbf{M}}$;

(iii) $\mathbf{B}$ *is obtainable from* $\underline{\mathbf{M}}$ *by repeated applications of* $\mathbb{I}$, $\mathbb{S}$ *and* $\mathbb{P}$.

In particular, $\mathcal{A} = \mathbb{ISP}\underline{\mathbf{M}}$ *is exactly the quasi-variety generated by* $\underline{\mathbf{M}}$.

Proof Trivially (i) implies (iii). It is easy to verify that (iii) implies (ii), for if a quasi-equation holds in an algebra, then it will hold in an isomorphic copy or a subalgebra of that algebra, and if it holds in a collection of algebras, it will also hold in their product. Now assume (ii). We first prove (i) if $\mathbf{B}$ is finite and then apply Lemma 1.3.2 and the Inverse Limit Theorem 1.3.3 to extend the result to infinite $\mathbf{B}$.

Let $b_1, b_2 \in B$ where $b_1 \neq b_2$, and $B = \{b_1, b_2, \dots, b_n\}$ is finite. Suppose that, contrary to (i) and the Algebraic Separation Theorem 1.3.1, there is no homomorphism $u\colon \mathbf{B} \to \underline{\mathbf{M}}$ separating b_1 and b_2. Let $x_1, x_2, \dots, x_n$ be a sequence of variables corresponding to $b_1, b_2, \dots, b_n$. For each map $h\colon B \to M$ which is not a homomorphism, let $f_h \in F$ be an operation not preserved by h. Then the set G of all f_h, where $h\colon B \to M$ is not a homomorphism, is a finite subset of F. Moreover, a map $g\colon B \to M$ is a homomorphism if and only if g preserves each operation in G. Let β be the conjunction of all equations $f(x_{i_1}, x_{i_2}, \dots, x_{i_k}) \approx x_j$ such that $f \in G$ and, in $\mathbf{B}$, the assertion $f(b_{i_1}, b_{i_2}, \dots, b_{i_k}) = b_j$ is valid. Consider the quasi-equation

$$\psi := [\beta \implies (x_1 \approx x_2)].$$

We claim that $\underline{\mathbf{M}}$ must satisfy ψ. To see that it does, let $p_1, p_2, \dots, p_n \in M$ satisfy β, that is, $f(p_{i_1}, p_{i_2}, \dots, p_{i_k}) = p_j$ whenever $f \in G$ and $f(b_{i_1}, b_{i_2}, \dots, b_{i_k}) = b_j$. Then $b_i \mapsto p_i$ is a homomorphism from $\mathbf{B}$ to $\underline{\mathbf{M}}$ which must not separate b_1 and b_2. Thus $p_1 = p_2$, and ψ is satisfied. Now substitute b_i for x_i in ψ. Clearly β is satisfied. By (ii) the algebra $\mathbf{B}$ also satisfies ψ and therefore $b_1 = b_2$, a contradiction. Thus every finite model of the quasi-equational theory of $\underline{\mathbf{M}}$ is in $\mathbb{ISP}\underline{\mathbf{M}}$.

Now consider an infinite model $\mathbf{B}$ of the quasi-equational theory of $\underline{\mathbf{M}}$. To show that $\mathbf{B} \in \mathbb{ISP}\underline{\mathbf{M}}$ we choose $c, d \in B$ where $c \neq d$. Let I be the set of all finite subalgebras of $\mathbf{B}$ containing c and d and ordered by inclusion. By Lemma 1.3.2 any two members of I are contained in a third, and therefore $\langle I, \subseteq \rangle$ is a directed set. For each $\mathbf{F} \in I$ let $A_\mathbf{F}$ be the (finite) set of all homomorphisms from $\mathbf{F}$ into $\underline{\mathbf{M}}$ which separate c and d. As we have just seen, each $A_\mathbf{F}$ is non-empty. If $\mathbf{F}, \mathbf{G} \in I$ and $\mathbf{F} \subseteq \mathbf{G}$, let $\varphi_{\mathbf{GF}}\colon A_\mathbf{G} \to A_\mathbf{F}$ restrict each homomorphism (from $\mathbf{G}$ into $\underline{\mathbf{M}}$) to $\mathbf{F}$. This gives us an inverse limit system of sets and maps. By the Inverse Limit Theorem 1.3.3 there is an element $u \in \varprojlim_{\mathbf{F} \in I} A_\mathbf{F}$. It is now easy to check that v is a homomorphism from $\mathbf{B}$ into $\underline{\mathbf{M}}$ separating c and d where, for each $a \in B$, we define $v(a) = u_\mathbf{F}(a)$ for any $\mathbf{F} \in I$ containing a. By the Algebraic Separation Theorem 1.3.1 we see that $\mathbf{B} \in \mathbb{ISP}\underline{\mathbf{M}}$. ■

In the context of studying an algebra $\mathbf{A} \in \mathcal{A}$, we are often interested only in those congruences $\theta \in \mathrm{Con}(\mathbf{A})$ for which $\mathbf{A}/\theta$ is again in $\mathcal{A}$. Such a congruence is called an **$\mathcal{A}$-congruence**, and we write

$$\mathrm{Con}_\mathcal{A}(\mathbf{A}) := \{\, \theta \in \mathrm{Con}(\mathbf{A}) \mid \mathbf{A}/\theta \in \mathcal{A} \,\}$$

for the set of all $\mathcal{A}$-congruences on $\mathbf{A}$. While $\mathrm{Con}_\mathcal{A}(\mathbf{A})$ is not in general a sublattice of $\mathrm{Con}(\mathbf{A})$, the following result will prove useful.

3.5 Theorem *If $\mathbf{A} \in \mathcal{A}$, then $\mathrm{Con}_\mathcal{A}(\mathbf{A})$ forms a complete lattice where, for a non-empty subset $\Sigma \subseteq \mathrm{Con}_\mathcal{A}(\mathbf{A})$,*

$$\bigwedge \Sigma = \bigcap \Sigma \quad \text{and} \quad \bigvee \Sigma = \bigcap \{\, \theta \in \mathrm{Con}_\mathcal{A}(\mathbf{A}) \mid \bigcup \Sigma \subseteq \theta \,\}.$$

Proof We prove that $\mathrm{Con}_\mathcal{A}(\mathbf{A})$ is closed under arbitrary intersection, from which the rest immediately follows. Let $\Sigma \subseteq \mathrm{Con}_\mathcal{A}(\mathbf{A})$ and let $\theta := \bigcap \Sigma$. Suppose that $a, b \in A$ where $(a, b) \notin \theta$. Then there is a $\psi \in \Sigma$ such that $(a, b) \notin \psi$. By the Algebraic Separation Theorem 1.3.1 there is a homomorphism $u\colon \mathbf{A} \to \underline{\mathbf{M}}$ such that $(a, b) \notin \ker(u) \supseteq \psi$. Since $\psi \supseteq \theta$, we obtain a homomorphism from $\mathbf{A}/\theta$ into $\underline{\mathbf{M}}$ separating a and b. Thus $\theta = \bigcap \Sigma \in \mathrm{Con}_\mathcal{A}(\mathbf{A})$. ■

In general the quasi-variety $\mathbb{ISP}\underline{\mathbf{M}}$ forms a proper subclass of the variety $\mathbb{HSP}\underline{\mathbf{M}}$, but it often happens that the two coincide for those choices of $\underline{\mathbf{M}}$ that occur in duality theory. We state here the most common conditions under which this is the case.

3.6 Theorem *Let $\mathcal{V} = \mathbb{HSP}\underline{\mathbf{M}}$ and $\mathcal{A} = \mathbb{ISP}\underline{\mathbf{M}}$ be the variety and quasi-variety generated by the finite algebra $\underline{\mathbf{M}}$, and let $\mathcal{V}_{\mathrm{si}}$ denote the class of subdirectly irreducible members of $\mathcal{V}$.*

(i) $\mathcal{V} = \mathbb{ISP}\mathcal{V}_{\mathrm{si}}$.

(ii) *$\mathcal{V} = \mathcal{A}$ if $\mathcal{V}_{\mathrm{si}} \subseteq \mathbb{IS}\underline{\mathbf{M}}$.*

(iii) *$\mathcal{V} = \mathcal{A}$ if $\mathcal{V}$ is congruence distributive and $\mathbb{HS}\underline{\mathbf{M}} \subseteq \mathbb{IS}\underline{\mathbf{M}}$.*

Proof (i) By Theorem A.10 every member $\mathbf{A}$ of $\mathcal{V}$ is a subdirect product of subdirectly irreducible algebras which, being homomorphic images of $\mathbf{A}$, are in $\mathcal{V}_{\text{si}}$. To prove (ii) we use (i) and the fact, from the $\mathbb{ISP}$-Theorem, that $\mathcal{A}$ is closed under $\mathbb{I}$, $\mathbb{S}$ and $\mathbb{P}$. Thus $\mathcal{V} \subseteq \mathbb{ISP}\mathcal{V}_{\text{si}} \subseteq \mathbb{ISP}(\mathbb{IS}\underline{\mathbf{M}}) \subseteq \mathcal{A}$. Similarly, (iii) follows from (ii) and Jónsson's Second Theorem A.16. ∎

1.4 Structured topological spaces

A natural duality is a special kind of dual representation that can exist between a finitely generated quasi-variety $\mathcal{A}$ and a category $\mathfrak{X}$ of structured Boolean spaces. In this section we describe the kinds of structured Boolean spaces that arise in this way. The quasi-variety of distributive lattices, for example, is dual to a category of Boolean spaces which carry two nullary total operations (constants) and one binary relation. Other dualities will require spaces with different kinds of structure. To lay the groundwork for these dualities we consider the most general kind of structure that we might want to impose.

We begin with three sets of symbols:

(i) a set G of finitary total operation symbols,
(ii) a set H of finitary partial operation symbols,
(iii) a set R of finitary relation symbols.

Each symbol carries an arity, which is a natural number defined by a fixed arity function on these sets of symbols. Operations may be nullary, but partial operations and relations must have positive arities. By a **structured topological space** of type $\langle G, H, R\rangle$ we mean a structure

$$\mathbf{X} = \langle X; G^{\mathbf{X}}, H^{\mathbf{X}}, R^{\mathbf{X}}, \mathcal{T}^{\mathbf{X}}\rangle$$

where

(i) $G^{\mathbf{X}}$ consists of an n-ary total operation $g^{\mathbf{X}} \colon X^n \to X$ for each n-ary total operation symbol $g \in G$,
(ii) $H^{\mathbf{X}}$ consists of an n-ary partial operation $h^{\mathbf{X}} \colon \operatorname{dom}(h^{\mathbf{X}}) \to X$ for each n-ary partial operation symbol $h \in H$, where $\operatorname{dom}(h^{\mathbf{X}}) \subseteq X^n$,
(iii) $R^{\mathbf{X}}$ consists of an n-ary relation $r^{\mathbf{X}} \subseteq X^n$ on X for each n-ary relation symbol $r \in R$,
(iv) $\langle X, \mathcal{T}^X\rangle$ is a topological space.

If G includes no nullary operations, then we will allow X to be empty. The **empty structure**, $\underset{\sim}{\varnothing}$, consists of the empty set with each operation, partial operation and relation being empty. In the actual cases that we will consider, the sets G, H and R will usually be either small or empty and the arities n will usually be small. Remember that the ultimate utility of natural dualities will hinge on our ability to construct a dual category

$\mathfrak{X}$ which is in a practical sense simpler than the corresponding original category $\boldsymbol{\mathcal{A}}$, and this will normally mean that its type must be small.

Many of the familiar algebraic notions and results discussed in the previous section have more than one natural extension to structures with relations and partial operations. It is therefore important that we formulate explicitly the extensions of these notions that will be used in this book. Let

$$\mathbf{X} = \langle X; G^{\mathbf{X}}, H^{\mathbf{X}}, R^{\mathbf{X}}, \mathcal{T}^{\mathbf{X}} \rangle$$

be a structured topological space. If $h^{\mathbf{X}} \in G^{\mathbf{X}} \cup H^{\mathbf{X}}$ is n-ary, then the **graph** of $h^{\mathbf{X}}$ is the $(n+1)$-ary relation

$$\operatorname{graph}(h^{\mathbf{X}}) = \{(x_1, x_2, \dots, x_n, y) \in X^{n+1} \mid h^{\mathbf{X}}(x_1, x_2, \dots, x_n) = y\}$$

on X. If $g \in G^{\mathbf{X}}$ is nullary, then $\operatorname{graph}(g) = \{g\}$.

If g is a k-ary partial operation and $h_1, \dots, h_k$ are n-ary partial operations on X, then the **composition** $g(h_1, \dots, h_k)$ is the n-ary partial operation whose domain consists of those $x \in X^n$ for which the expression $g(h_1(x), \dots, h_k(x))$ is defined. Of course, $g(h_1(x), \dots, h_k(x))$ could be the **empty map**, $\eta\colon \varnothing \to M$, even when g and $h_1, \dots, h_n$ have non-empty domains. We shall allow both $k = 0$ and $n = 0$ in this definition, but some care is required. When $n = 0$, even though g might be partial, this construct produces either η or a (total) nullary operation: if $c_1, \dots, c_k$ are nullary with $(c_1, \dots, c_k) \in \operatorname{dom}(g)$, then we obtain the nullary operation $g(c_1, \dots, c_k)$. When $k = 0$, the operation g is nullary and we declare that g composed with the empty set of n-ary total maps produces the constant total n-ary map with value g.

A set of partial operations on X is called an **enriched partial clone** on X if it includes the coordinate projections $\pi_i\colon X^n \to X$ for each $n \geqslant 1$ and is closed under composition. The **enriched partial clone generated** by a set P of partial operations on X is the smallest enriched partial clone on X containing P and is denoted by $[P]$. Note that, to guarantee closure under composition, $[P]$ may contain the empty map η. The **enriched partial clone of** $\mathbf{X} = \langle X; G^{\mathbf{X}}, H^{\mathbf{X}}, R^{\mathbf{X}}, \mathcal{T}^{\mathbf{X}} \rangle$ is defined to be $[G^{\mathbf{X}} \cup H^{\mathbf{X}}]$. Unlike clones on algebras, these clones are *enriched* to allow the possibility that they might include nullary operations as well.

Now consider another structured topological space

$$\mathbf{Y} = \langle Y; G^{\mathbf{Y}}, H^{\mathbf{Y}}, R^{\mathbf{Y}}, \mathcal{T}^{\mathbf{Y}} \rangle.$$

A continuous map $\varphi\colon X \to Y$ is a **morphism**, written $\varphi\colon \mathbf{X} \to \mathbf{Y}$, if it preserves each member of R and the graph of each member of $G \cup H$. Written out in detail, φ is a morphism if

(i) for each n-ary $g \in G$ and each $(x_1, x_2, \dots, x_n) \in \mathbf{X}$, we have

$$\varphi(g^{\mathbf{X}}(x_1, x_2, \dots, x_n)) = g^{\mathbf{Y}}(\varphi(x_1), \varphi(x_2), \dots, \varphi(x_n)),$$

(ii) for each n-ary $h \in H$ and each $(x_1, x_2, \ldots, x_n) \in \mathrm{dom}(h^{\mathbf{X}})$ we have $(\varphi(x_1), \varphi(x_2), \ldots, \varphi(x_n)) \in \mathrm{dom}(h^{\mathbf{Y}})$ and
$$\varphi(h^{\mathbf{X}}(x_1, x_2, \ldots, x_n)) = h^{\mathbf{Y}}(\varphi(x_1), \varphi(x_2), \ldots, \varphi(x_n)),$$
(iii) for each n-ary $r \in R$ and each $(x_1, x_2, \ldots, x_n) \in r^{\mathbf{X}}$ we have $(\varphi(x_1), \varphi(x_2), \ldots, \varphi(x_n)) \in r^{\mathbf{Y}}$, and
(iv) φ is continuous.

The class of all structured topological spaces of a given type $\langle G, H, R\rangle$ with these morphisms forms the broad category in which we will work. In the case that G contains no nullary operations and the empty structure is allowed, for each structure $\mathbf{X}$ of the type, there is a unique (empty) morphism from $\varnothing$ to $\mathbf{X}$ and there are no morphisms in the other direction when $\mathbf{X}$ is non-empty.

The structure $\mathbf{Y} = \langle Y; G^{\mathbf{Y}}, H^{\mathbf{Y}}, R^{\mathbf{Y}}, \mathcal{T}^{\mathbf{Y}}\rangle$ is called a **substructure** of the structured topological space $\mathbf{X} = \langle X; G^{\mathbf{X}}, H^{\mathbf{X}}, R^{\mathbf{X}}, \mathcal{T}^{\mathbf{X}}\rangle$, written $\mathbf{Y} \leqslant \mathbf{X}$, provided that $Y \subseteq X$ and

(i) for each n-ary $g \in G$ the operations $g^{\mathbf{Y}}$ and $g^{\mathbf{X}}$ agree on Y^n,
(ii) for each n-ary $h \in H$, we have $\mathrm{dom}(h^{\mathbf{Y}}) = \mathrm{dom}(h^{\mathbf{X}}) \cap Y^n$, and $h^{\mathbf{Y}}$ agrees with $h^{\mathbf{X}}$ on this set,
(iii) for each n-ary $r \in R$, we have $r^{\mathbf{Y}} = r^{\mathbf{X}} \cap Y^n$, and
(iv) $\mathcal{T}^{\mathbf{Y}}$ is the relative topology obtained from $\mathcal{T}^{\mathbf{X}}$.

Notice that the empty structure, $\varnothing$, is a substructure of $\mathbf{X}$ exactly when G includes no nullary operation symbols. Consequently the substructures of $\mathbf{X}$ are always closed under intersection and therefore form a complete lattice.

4.1 Warning! In the presence of partial operations, morphisms and substructures do not interact as they do in total algebras. A moment's reflection will show that the image $\varphi(\mathbf{Z})$ of a substructure $\mathbf{Z} \leqslant \mathbf{X}$ under a morphism $\varphi\colon \mathbf{X} \to \mathbf{Y}$ need not in general be a substructure of $\mathbf{Y}$ if partial operations are present. Many startling results can be obtained by overlooking this fact!

Products of structured topological spaces of the same type are defined over *non-empty index sets* in the usual manner, using pointwise total operations and relations and the product topology, and partial operations are defined pointwise whenever they are defined at each point. Concrete products defined in this way are products in the category of structured topological spaces in the category-theoretic sense (Exercise 1.12).

A morphism $\varphi\colon \mathbf{X} \to \mathbf{Y}$ is an **embedding** of $\mathbf{X}$ into $\mathbf{Y}$ if the image $\varphi(\mathbf{X})$ forms a substructure of $\mathbf{Y}$ and φ is an isomorphism from $\mathbf{X}$ onto $\varphi(\mathbf{X})$. While this definition requires verification of many details, a number of them follow automatically in the context where we will use it.

4.2 Lemma *Let* $\mathbf{X}$ *and* $\mathbf{Y}$ *be structured topological spaces and assume that* X *is a compact space and* Y *is a Hausdorff space. A morphism* $\varphi\colon \mathbf{X} \to \mathbf{Y}$ *is an embedding if and only if the following hold:*

(i) φ *is one-to-one,*

(ii) *for each* n*-ary* $h \in H$ *and for each* $(x_1, x_2, \dots, x_n) \in X^n$*, we can conclude that the* n*-tuple* $(x_1, x_2, \dots, x_n)$ *is in* $\mathrm{dom}(h^{\mathbf{X}})$ *whenever* $(\varphi(x_1), \varphi(x_2), \dots, \varphi(x_n))$ *is in* $\mathrm{dom}(h^{\mathbf{Y}})$*,*

(iii) *for each* n*-ary* $r \in R$ *and* $(x_1, x_2, \dots, x_n) \in X^n$*, we can conclude* $(x_1, x_2, \dots, x_n) \in r^{\mathbf{X}}$ *whenever* $(\varphi(x_1), \varphi(x_2), \dots, \varphi(x_n)) \in r^{\mathbf{Y}}$.

In particular, a morphism $\varphi\colon \mathbf{X} \to \mathbf{Y}$ *is an isomorphism if and only if it is bijective and satisfies (ii) and (iii).*

Proof Assume (i), (ii) and (iii). According to Lemma B.1, the map φ^{-1} is a continuous map from $\varphi(\mathbf{X})$ onto $\mathbf{X}$. Since φ preserves total operations, it follows easily that φ^{-1} does also. Conditions (ii) and (iii) now ensure that φ^{-1} is a morphism. Conversely, assume that φ is an embedding, and therefore an isomorphism from $\mathbf{X}$ onto $\varphi(\mathbf{X})$. Let $\psi\colon \varphi(\mathbf{X}) \to \mathbf{X}$ be a morphism such that $\psi \circ \varphi = \mathrm{id}_{\mathbf{X}}$. Then φ must be one-to-one and ψ must be φ^{-1}, a morphism. Now (ii) and (iii) hold since φ^{-1} preserves partial operations and relations. ■

From this point onward we will relax our use of the superscripts on $G^{\mathbf{X}}$, $H^{\mathbf{X}}$ and $R^{\mathbf{X}}$ and their members, using them only when there may be a danger of ambiguity.

Referring back to the example of Priestley's duality presented in the previous section, we can now see how to generate categories of structured topological spaces in a manner similar to the way we generate a quasi-variety $\mathcal{D} = \mathbb{ISP}\underline{\mathbf{D}}$ from a finite algebra $\underline{\mathbf{D}}$. Let

$$\underset{\sim}{\mathbf{M}} := \langle M; G, H, R, \mathcal{T} \rangle$$

be a structured topological space where M is finite and the topology $\mathcal{T}$ is discrete. For a class $\mathcal{Y}$ of structured topological spaces of the same type, we define

$$\mathbb{I}\mathcal{Y}, \quad \mathbb{S}_c\mathcal{Y}, \quad \mathbb{P}^+\mathcal{Y},$$

respectively, to be the class of isomorphic copies, topologically closed substructures and direct products (over non-empty index sets) of members of $\mathcal{Y}$. Beginning with $\underset{\sim}{\mathbf{M}}$ we generate the class

$$\mathcal{X} := \mathbb{IS}_c\mathbb{P}^+ \underset{\sim}{\mathbf{M}}$$

consisting of all isomorphic copies of closed substructures of direct powers of copies of $\underset{\sim}{\mathbf{M}}$. Because $\underset{\sim}{\mathbf{M}}$ is finite and discrete its topology is Boolean. As a result the topology on each other member of $\mathcal{X}$ is also Boolean (Lemma B.5). Thus $\mathcal{X}$ forms a category of **structured Boolean spaces**.

The dual of a finitely generated quasi-variety $\mathcal{A} = \mathbb{ISP}\underline{\mathbf{M}}$ under a natural duality will always be a category of the form $\mathfrak{X} = \mathbb{IS}_c\mathbb{P}^+ \underset{\sim}{\mathbf{M}}$ where $\underset{\sim}{\mathbf{M}}$ is a finite discrete structure. The next section and the next two chapters will focus on the problem of determining $\underset{\sim}{\mathbf{M}}$ when we are given $\underline{\mathbf{M}}$. Once that is achieved, we will need a useful description of the members of the dual category $\mathfrak{X}$. For example, if we choose $\underset{\sim}{\mathbf{D}} = \langle \{0,1\}; 0, 1, \leqslant, \mathcal{T}\rangle$, then we saw in the last section that $\mathfrak{X}_{01}$ would be the category of all compact, bounded, totally order-disconnected spaces. We now prove two theorems that will assist in establishing the correctness of such a description. These are the appropriate analogues of the $\mathbb{ISP}$-Theorem 1.3.4 and the Algebraic Separation Theorem 1.3.1, respectively. The Preservation Theorem will be used to show that an arbitrary structure in $\mathbb{IS}_c\mathbb{P}^+ \underset{\sim}{\mathbf{M}}$ meets a particular description while the Separation Theorem will be used to show that any structure meeting that description is in $\mathbb{IS}_c\mathbb{P}^+ \underset{\sim}{\mathbf{M}}$. The latter is a full analogue of its algebraic counterpart while the former, as we will see, extends in only one direction.

We will want to obtain a description of $\mathfrak{X}$ by looking for properties of $\underset{\sim}{\mathbf{M}}$ which are preserved under $\mathbb{I}$, $\mathbb{S}_c$ and $\mathbb{P}^+$. For example, being a Boolean space is such a property (Lemma B.5). Among the important preserved properties are those expressible by means of an appropriate analogue of quasi-equations. To say exactly what we mean by this in the presence of partial operations, we need a few definitions. As before, we let V be a fixed set of **variables**. We define a **term** of type $\langle G, H, R\rangle$ to be a term of type F, as defined in Appendix A, where $F = G \cup H$. Neither the relations R nor the fact that the symbols of H will be allowed to represent partial operations affects the notion of term. An **atomic formula** is an expression of type $\langle G, H, R\rangle$ of either of the forms

$$t_1 \approx t_2 \quad \text{and} \quad r(t_1, t_2, \ldots, t_n)$$

where $t_1, t_2, \ldots, t_n$ are terms and $r \in R$ is n-ary. A **quasi-atomic formula** of type $\langle G, H, R\rangle$ is an expression which is either an atomic formula, a **neg-atomic formula**, $\neg\alpha$, or an **implication**

$$\beta_1 \wedge \ldots \wedge \beta_m \implies \alpha$$

where $m \geqslant 1$ and $\beta_1, \ldots, \beta_m, \alpha$ are atomic formulæ.

We now provide these syntactic expressions with a semantic interpretation. Because of the increased complexity of the context and the possibility of more than one reasonable interpretation, we must proceed with more care and formality than was required in the previous section. Let $\mathbf{X}$ be a structured topological space of type $\langle G, H, R\rangle$. We denote by X^V the set of all maps from V into X. A variable assignment $x \in X^V$ extends to assign a value $x[t]$ in X to each term t for which the necessary partial operations are defined according to the following rules:

(i) $x[v] = x(v)$ if $v \in V$;

(ii) $x[c] = c^{\mathbf{X}}$ if c is a nullary operation symbol;

(iii) if h is an n-ary operation symbol in $G \cup H$, where $n > 0$, then $x[h(t_1, t_2, \ldots, t_n)] = h^{\mathbf{X}}(x[t_1], x[t_2], \ldots, x[t_n])$ provided that the elements $x[t_1], x[t_2], \ldots, x[t_n]$ are all defined and that their values are in the domain of $h^{\mathbf{X}}$, otherwise $x[h(t_1, t_2, \ldots, t_n)]$ is undefined.

Each quasi-atomic formula is assigned a **truth value** by x as follows:

(i) $t_1 \approx t_2$ is assigned **T** if $x[t_1]$ and $x[t_2]$ are both defined and are equal, otherwise it is assigned **F**;

(ii) $r(t_1, t_2, \ldots, t_n)$ is assigned **T** if $x[t_1], x[t_2], \ldots, x[t_n]$ are all defined and $(x[t_1], x[t_2], \ldots, x[t_n])$ is in $r^{\mathbf{X}}$, otherwise it is assigned **F**;

(iii) $\neg\alpha$ is assigned **T** if α is assigned **F**, otherwise it is assigned **F**;

(iv) $\beta_1 \wedge \ldots \wedge \beta_m \implies \alpha$ is assigned **T** unless $\beta_1, \ldots, \beta_m$ are all assigned **T** but α is assigned **F**, in which case it is assigned **F**.

Finally, we say that **X satisfies** a quasi-atomic formula if every x in X^V assigns it **T**. It is worth noting that the presence of partial operations does add a peculiar twist to these otherwise common notions. For example, if h is a partial binary operation symbol and $\mathbf{X}$ is a structure in which $\operatorname{dom}(h^{\mathbf{X}})$ is not all of X^2, then $\mathbf{X}$ does not satisfy the atomic formula $h(x, y) \approx h(x, y)$! The proof of the Preservation Theorem is entirely straightforward but somewhat tedious, and much more instructive to write than to read. We have therefore included it in the exercises. (See Exercises 1.13, 1.14 and 1.15.)

4.3 Preservation Theorem *Let $\underset{\sim}{\mathbf{M}}$ be a finite, discrete structured topological space and let $\mathbf{X} \in \mathbb{IS}_c\mathbb{P}^+ \underset{\sim}{\mathbf{M}}$.*

(i) *$\mathbf{X}$ is a structured Boolean space which satisfies every quasi-atomic formula that is satisfied by $\underset{\sim}{\mathbf{M}}$.*

(ii) *If $h \in G \cup H$ is n-ary, then $\operatorname{dom}(h^{\mathbf{X}})$ is a closed subset of X^n and $h^{\mathbf{X}} \colon \operatorname{dom}(h^{\mathbf{X}}) \to X$ is continuous.*

(iii) *If $r \in R$ is n-ary, then $r^{\mathbf{X}}$ is a closed subset of X^n.*

4.4 Separation Theorem *Let $\mathbf{X}$ be a compact structured topological space of the same type as the finite discretely topologised structure $\underset{\sim}{\mathbf{M}}$. Then $\mathbf{X} \in \mathbb{IS}_c\mathbb{P}^+ \underset{\sim}{\mathbf{M}}$ if and only if there is at least one morphism from $\mathbf{X}$ to $\underset{\sim}{\mathbf{M}}$ and the following conditions hold:*

(i) *for each $x, y \in X$ where $x \neq y$, there is an $\alpha \colon \mathbf{X} \to \underset{\sim}{\mathbf{M}}$ such that $\alpha(x) \neq \alpha(y)$,*

(ii) *for each n-ary $h \in H$ and $(x_1, x_2, \ldots, x_n) \in X^n \backslash \operatorname{dom}(h^{\mathbf{X}})$, there is an $\alpha \colon \mathbf{X} \to \underset{\sim}{\mathbf{M}}$ such that $(\alpha(x_1), \alpha(x_2), \ldots, \alpha(x_n)) \notin \operatorname{dom}(h^{\underset{\sim}{\mathbf{M}}})$,*

(iii) *for each n-ary $r \in R$ and $(x_1, x_2, \dots, x_n) \in X^n \backslash r^{\mathbf{X}}$, there is an $\alpha \colon \mathbf{X} \to \underset{\sim}{\mathbf{M}}$ such that $(\alpha(x_1), \alpha(x_2), \dots, \alpha(x_n)) \notin r^{\underset{\sim}{\mathbf{M}}}$.*

Proof Assume that $\varphi \colon \mathbf{X} \to \underset{\sim}{\mathbf{M}}^I$ is an embedding. Then Lemma 1.4.2 shows that the morphisms $\{\pi_i \circ \varphi \mid i \in I\}$ obtained from the projection maps $\{\pi_i \mid i \in I\}$ will satisfy conditions (i), (ii) and (iii).

Conversely, assume (i), (ii) and (iii). Let S be the set of all morphisms from $\mathbf{X}$ into $\underset{\sim}{\mathbf{M}}$ and note that $S \neq \varnothing$ by assumption. Define $\varphi \colon \mathbf{X} \to \underset{\sim}{\mathbf{M}}^S$ by $\varphi(x)(\alpha) = \alpha(x)$. We will apply the other direction of Lemma 1.4.2 to show that φ is an embedding. To see that φ is continuous, we use Lemma B.3 and consider the subbasic clopen subset $U_{\alpha,a}$ of $\underset{\sim}{\mathbf{M}}^S$ where $a \in M$ and $\alpha \in S$ (Lemma B.6). Then $\varphi^{-1}(U_{\alpha,a}) = \alpha^{-1}(a)$, which is open since α is continuous. Finally, φ preserves G, H and R since they are preserved by each $\alpha \in S$. Thus φ is a morphism. The conditions (i), (ii) and (iii) of Lemma 1.4.2 follow directly from the present conditions (i), (ii) and (iii), respectively. ∎

Through the next few chapters we will see how the laboriously acquired results of Section 2 follow easily from a general theory. As a first example we review the characterisation of $\mathfrak{X}_{01}$ as the category of bounded Priestley spaces in view of the Preservation and Separation Theorems.

4.5 Example: Bounded Priestley Spaces *If $\underset{\sim}{\mathbf{D}} = \langle \{0,1\}; 0, 1, \leqslant, \mathcal{T} \rangle$, then $\mathfrak{X} = \mathbb{IS}_{\mathrm{c}}\mathbb{P}^+ \underset{\sim}{\mathbf{D}}$ is exactly the class of bounded Priestley spaces.*

Proof Let $\mathbf{X} = \langle X; 0, 1, \leqslant, \mathcal{T} \rangle \in \mathbb{IS}_{\mathrm{c}}\mathbb{P}^+ \underset{\sim}{\mathbf{D}}$. We observe that the following quasi-atomic formulæ hold in $\underset{\sim}{\mathbf{D}}$:

(i) $x \lesssim x$,

(ii) $(x \lesssim y) \wedge (y \lesssim x) \implies (x \approx y)$,

(iii) $(x \lesssim y) \wedge (y \lesssim z) \implies (x \lesssim z)$,

(iv) $0 \lesssim x \lesssim 1$,

(v) $0 \not\approx 1$.

By the Preservation Theorem 1.4.3, the relation $\leqslant$ is an ordering with lower bound 0 and upper bound 1. By Tychonoff's Product Theorem B.4, X is compact. If $x, y \in X$ and $x \nleqslant y$, then, by the Separation Theorem 1.4.4, there is a morphism $\alpha \colon \mathbf{X} \to \underset{\sim}{\mathbf{D}}$ taking x to 1 and y to 0. Thus $\alpha^{-1}(0)$ is a clopen decreasing set containing y but not x.

Conversely, let $\mathbf{X}$ be a bounded Priestley space. We apply the Separation Theorem 1.4.4 to show that $\mathbf{X} \in \mathbb{IS}_{\mathrm{c}}\mathbb{P}^+ \underset{\sim}{\mathbf{D}}$. Suppose $x \nleqslant y$ in $\mathbf{X}$. Let U be a clopen decreasing set containing y but not x. Then $\alpha \colon \mathbf{X} \to \underset{\sim}{\mathbf{D}}$ is a morphism where α takes U to 0 and $X \backslash U$ to 1. If $x \neq y$, then $x \nleqslant y$ or $y \nleqslant x$ and we already have the required morphism. ∎

It is natural to wonder if a full analogue of the $\mathbb{ISP}$-Theorem 1.3.4 can be obtained by adding a converse to the Preservation Theorem. Can $\mathbb{IS}_{\mathrm{c}}\mathbb{P}^+ \underset{\sim}{\mathbf{M}}$

be described as the category of all structured Boolean spaces which are models of the quasi-atomic theory of $\langle M; G, H, R\rangle$ with continuous (partial) operations and closed relations? Unfortunately not, as shown by the following example of Stralka [Str80].

4.6 Counterexample *Let* $\underset{\sim}{\mathbf{D}} = \langle\{0,1\}; 0, 1, \leqslant, \mathcal{T}\rangle$. *Then there is a structured Boolean space* $\mathbf{X} = \langle X; 0, 1, \leqq, \mathcal{T}\rangle$ *such that* $\mathbf{X}$ *is a model of the quasi-atomic theory of* $\underset{\sim}{\mathbf{D}}$, *the order relation* $\leqq$ *is closed in* $X \times X$, *but* $\mathbf{X}$ *is not a bounded Priestley space.*

Proof Let $\mathbf{X}$ be the Cantor space, viewed as a subspace of the interval $[0,1]$. For $x, y \in X$, define $x \leqq y$ if either $0 = x$, $x = y$, $y = 1$, or y covers x under the usual order. Let $\mathbf{X}^*$ and $\underset{\sim}{\mathbf{D}}^*$ be obtained from $\mathbf{X}$ and $\underset{\sim}{\mathbf{D}}$ by deleting their topologies. Since $\mathbf{X}^*$ is a bounded ordered set, it follows from Exercise 1.16 that $\mathbf{X}^*$, and therefore also $\mathbf{X}$, is a model of the quasi-atomic theory of $\underset{\sim}{\mathbf{D}}^*$, and therefore also of $\underset{\sim}{\mathbf{D}}$.

Now suppose $x \nleqq y$. Then it is easy to find clopen sets U and V with $x \in U$ and $y \in V$ such that for every $u \in U$ and $v \in V$ we have $u \nleqq v$, showing that $\leqq$ is closed. However, suppose $y > x$ in the usual order but y does not cover x. Then $x \nleqq y$, but every clopen decreasing subset of $\mathbf{X}$ containing y also contains x, so $\mathbf{X}$ is not a bounded Priestley space. ■

The above example shows that the quasi-equational logic together with topological closure is not powerful enough to describe $\mathbb{IS}_c\mathbb{P}^+ \underset{\sim}{\mathbf{M}}$. A full description of these categories, and therefore a full analogue of the $\mathbb{ISP}$-Theorem 1.3.4, can be obtained by extending the language to include infinitary 'topological quasi-atomic formulæ'. Because the models of the 'topological quasi-atomic theory' of $\underset{\sim}{\mathbf{M}}$ are exactly the members of $\mathbb{IS}_c\mathbb{P}^+ \underset{\sim}{\mathbf{M}}$, this category is called a **topological quasi-variety**. We will not pursue this topic further, but refer the reader to Clark and Krauss [CK84].

The next example is similar to Example 1.4.5, and we leave the details of the proof as exercises. This example requires that we introduce a category-theoretic notion which we will need later as well. Given objects $\mathbf{X}$ and $\mathbf{Y}$ in a category $\mathcal{C}$, we say that Y is a **retract** of X if there exist morphisms $d\colon X \to Y$ and $b\colon Y \to X$ such that $d \circ b = \mathrm{id}_Y$. The morphism d is then called a **retraction** and b is called a **coretraction**. In most cases, we have $Y \subseteq X$ and b is the inclusion map.

4.7 Example: Priestley Spaces with a Lower Retract *Suppose that* $\underset{\sim}{\mathbf{S}} = \langle\{0, a, 1\}; d, \leqslant, \mathcal{T}\rangle$ *where* $\leqslant = \{(0,0), (a,a), (1,1), (1,a)\}$ *and* d *fixes* 0 *and takes* a *and* 1 *to* 1. *Then* $\mathcal{X} = \mathbb{IS}_c\mathbb{P}^+ \underset{\sim}{\mathbf{S}}$ *is exactly the class of compact, totally order-disconnected spaces with a continuous retraction* d *mapping each element (down) to the unique minimal element below it.*

Proof (Outline) According to the Preservation Theorem 1.4.3, every $\mathbf{X} \in \mathbb{IS}_c\mathbb{P}^+ \underset{\sim}{\mathbf{S}}$ satisfies the quasi-atomic formulæ

(i) $x \lesssim x$,

(ii) $(x \lesssim y) \wedge (y \lesssim x) \implies (x \approx y)$,

(iii) $(x \lesssim y) \wedge (y \lesssim z) \implies (x \lesssim z)$,

(iv) $d^2(x) \approx d(x) \lesssim x$,

(v) $(x \lesssim y) \implies (d(x) \approx d(y))$,

and is therefore a Priestley space with a lower retract. Conversely, let $\mathbf{X}$ be a Priestley space with a lower retract. To apply the Separation Theorem 1.4.4, choose $x, y \in X$ with $x \not\lesssim y$. Let U be a clopen decreasing set containing y but not x. Define a map φ by $\varphi(x) = 1$ if $x \in U$, $\varphi(x) = a$ if $x \in d^{-1}(U)\backslash U$ and $\varphi(x) = 0$ otherwise, and check that φ is a morphism from $\mathbf{X}$ to $\underset{\sim}{\mathbf{S}}$. ■

A third example calls upon quite different techniques.

4.8 Example: Abelian Groups of Exponent m *Suppose that* $\underset{\sim}{\mathbf{A}}_m = \langle \{0, 1, \ldots, m-1\}; +, 0, \mathcal{T}\rangle$ *is the discrete cyclic group of order* m. *Then* $\mathfrak{X} = \mathbb{IS}_c\mathbb{P}^+ \underset{\sim}{\mathbf{A}}_m$ *is exactly the class of Boolean spaces with a continuous abelian group operation satisfying* $mx \approx 0$.

Proof By the Preservation Theorem 1.4.3, every member of $\mathfrak{X}$ satisfies this description. Conversely, let $\mathbf{X} = \langle X; +, 0, \mathcal{T}\rangle$ be a Boolean topological group satisfying $mx \approx 0$ and $xy \approx yx$. To verify condition (i) of the Separation Theorem 1.4.4, choose $x \neq y$ in X.

We will first show that there is a clopen subgroup N of $\mathbf{X}$ such that x and y are separated in the quotient group $\mathbf{X}/N$. Let U be a clopen set containing 0 but not $x - y$, and let

$$N := \{\, t \in X \mid (\forall z \in X)[(t + z \in U) \iff (z \in U)]\,\}.$$

Then N is a subgroup of X since $0 \in N$ and, for all $u, v \in N$, we have $z \in U \iff v+z \in U \iff (u+v)+z \in U$, for all $z \in X$. Also, $y - x \notin N$ since $0 = (y-x) + (x-y) \in U$ but $x - y \notin U$. Thus $x + N \neq y + N$.

To see that N is closed, choose $r \notin N$. Then there is a $z \in X$ such that $(r+z, z)$ is not in the clopen subset $(U \times U) \cup [(X\backslash U) \times (X\backslash U)]$. Since $+$ is continuous, there are clopen sets R containing r and Z containing z such that $(R + Z) \times Z$ is disjoint from $(U \times U) \cup [(X\backslash U) \times (X\backslash U)]$. Thus R is a clopen set containing r but not intersecting N.

To see that N is open, choose $s \in N$. Then for each $z \in X$ we have $(s + z, z) \in (U \times U) \cup [(X\backslash U) \times (X\backslash U)]$. Again using the continuity of $+$, we find clopen sets S_z containing s and Z_z containing z such that

$$(S_z + Z_z) \times Z_z \subseteq (U \times U) \cup [(X\backslash U) \times (X\backslash U)]$$

for each $z \in X$. Since X is compact, we can find a finite set of choices of z so that the corresponding sets Z_z cover X. Consequently, the intersection of the corresponding sets S_z is a clopen subset of N containing s.

The quotient group $\mathbf{X}/N$ is a discrete group satisfying $mx = 0$, and therefore can be embedded into a direct power of $\underset{\sim}{\mathbf{A}}_m$. Following this embedding with an appropriate projection into $\underset{\sim}{\mathbf{A}}_m$ will provide the required morphism separating x and y. ∎

We will encounter many similar examples in Chapter 4 in which we can give a nice description of the category $\mathcal{X} = \mathbb{IS}_c\mathbb{P}^+ \underset{\sim}{\mathbf{M}}$.

1.5 Predualities

We are now ready to begin assembling the pieces. In this section we will present a uniform method whereby, starting with an arbitrary finite algebra $\underline{\mathbf{M}}$, we can always construct many different finite structures $\underset{\sim}{\mathbf{M}}$ so that $\mathcal{A} = \mathbb{ISP}\underline{\mathbf{M}}$ is dually adjoint to $\mathcal{X} = \mathbb{IS}_c\mathbb{P}^+ \underset{\sim}{\mathbf{M}}$. From this vast array of dual adjunctions we can hope to find many nice dual equivalences. As we proceed we will rely on the model of distributive lattices and bounded Priestley spaces to guide us in defining the structure $\underset{\sim}{\mathbf{M}}$, the contravariant functors D and E, and the special morphisms $e_{\mathbf{A}}$ and $\varepsilon_{\mathbf{X}}$ in a general setting. At the end of this section two questions will remain: for which choices of $\underset{\sim}{\mathbf{M}}$ is $\langle D, E, e, \varepsilon \rangle$ a dual representation and for which choices is it a dual equivalence? Answering these questions will be the goals of Chapters 2 and 3, respectively.

We begin with a fixed finite algebra $\underline{\mathbf{M}}$ generating a quasi-variety $\mathcal{A} = \mathbb{ISP}\underline{\mathbf{M}}$. The example of distributive lattices and bounded Priestley spaces suggests that $\underset{\sim}{\mathbf{M}}$ be a structure sharing the same underlying set M as $\underline{\mathbf{M}}$. Loosely speaking, we think of M as a **schizophrenic object** which has an algebraic personality $\underline{\mathbf{M}}$ and a topological personality $\underset{\sim}{\mathbf{M}}$. In its two personalities it generates two superficially different categories $\mathcal{A}$ and $\mathcal{X}$, but under the right conditions these two categories prove to be mirror images of one another.

How do we define the contravariant functors D and E in this setting? In the case that $\mathbf{A}$ was a distributive lattice, $D(\mathbf{A})$ was the space of prime ideals of $\mathbf{A}$. While 'prime ideal' is a notion very specialised to lattices, we note that the prime ideals of $\mathbf{A}$ are exactly the inverse images of 0 under homomorphisms from $\mathbf{A}$ into the two-element lattice $\underline{\mathbf{D}}$. Thus the members of $D(\mathbf{A})$ correspond exactly to the homomorphisms from $\mathbf{A}$ into $\underline{\mathbf{D}}$. This suggests that we try to 'define'

$$D(\mathbf{A}) := \mathcal{A}(\mathbf{A}, \underline{\mathbf{M}}) \subseteq M^A$$

in general. Viewing M as a discrete space, we notice that $\mathcal{A}(\mathbf{A}, \underline{\mathbf{M}})$ is always a closed subspace of the direct power M^A and is therefore a Boolean space (Exercise 1.19). It then remains to impose a structure on $\underset{\sim}{\mathbf{M}}$ that assures us that the homset $\mathcal{A}(\mathbf{A}, \underline{\mathbf{M}})$ is always a *substructure* of $\underset{\sim}{\mathbf{M}}^A$.

Before doing this, we consider the definition of E. In the case that $\mathbf{X}$ was a bounded Priestley space, recall that $E(\mathbf{X})$ was the lattice of proper, non-empty, clopen decreasing subsets of $\mathbf{X}$. Again this is a notion very specialised to a particular context, but we note that the proper, non-empty, clopen decreasing subsets of $\mathbf{X}$ are exactly the inverse images of 0 under continuous homomorphisms from $\mathbf{X}$ into the two-element bounded Priestley space $\underset{\sim}{\mathbf{D}}$. Thus the members of $E(\mathbf{X})$ correspond exactly to the morphisms from $\mathbf{X}$ into $\underset{\sim}{\mathbf{D}}$. As before, this suggests that we 'define'

$$E(\mathbf{X}) := \mathfrak{X}(\mathbf{X}, \underset{\sim}{\mathbf{M}}) \subseteq M^X$$

in general. Once again, we need to impose conditions on $\underset{\sim}{\mathbf{M}}$ which ensure that the homset $\mathfrak{X}(\mathbf{X}, \underset{\sim}{\mathbf{M}})$ is a subalgebra of the direct power $\underline{\mathbf{M}}^X$.

Exercise 1.18(i),(ii) show that, in the case of distributive lattices and bounded Priestley spaces, these homset definitions of $D(\mathbf{A})$ and $E(\mathbf{X})$ are fully justified, though with the slight twist that the order is reversed in each case! To avoid the need to stand on their heads when swapping from the *family of subsets* definitions to the *family of maps* definitions, many authors use prime filters of $\mathbf{A}$ instead of prime ideals and increasing subsets of $\mathbf{X}$ rather than decreasing subsets—see Exercise 1.18(iii),(iv).

In order to guarantee that these homsets determine substructures of the direct powers of the respective personalities of M, we must choose $\underset{\sim}{\mathbf{M}}$ so that its operations, partial operations and relations interact with the operations of $\underline{\mathbf{M}}$ in a very special way. The fundamental notion is that of an operation on M 'preserving' a relation on M.

5.1 Lemma *Let M be a non-empty set, $f\colon M^n \to M$ an n-ary operation and $r \subseteq M^m$ an m-ary relation where $m, n > 0$. Then the following are equivalent:*

(i) *the map $f\colon \langle M; r\rangle^n \to \langle M; r\rangle$ preserves r;*

(ii) *the set r is a subalgebra of $\langle M; f\rangle^m$.*

Proof Both of these assertions, written out in full, reduce to the following condition on every choice of $\{\, a_{ij} \in M \mid 1 \leqslant i \leqslant n;\ 1 \leqslant j \leqslant m \}$. If $(a_{i1}, a_{i2}, \dots, a_{im}) \in r$, for $i = 1, 2, \dots, n$, then $(b_1, b_2, \dots, b_m) \in r$, for $j = 1, 2, \dots, m$, where $b_j = f(a_{1j}, a_{2j}, \dots, a_{nj})$. ■

We say that the operation f **preserves** the relation r if the conditions of Lemma 1.5.1 hold. We illustrate this fundamental concept in Figure 1.3.

Given a type of algebras there is, up to isomorphism, a unique one-element algebra of that type. If we consider a type of structures with partial operations or relations, we notice that there are many non-isomorphic

$$\begin{array}{cccc c c} a_{11} & a_{12} & \cdots & a_{1n} & & b_1 \\ a_{21} & a_{22} & \cdots & a_{2n} & & b_2 \\ \vdots & \vdots & & \vdots & \xrightarrow{\ f\ } & \vdots \\ a_{m1} & a_{m2} & \cdots & a_{mn} & & b_m \\ \pitchfork & \pitchfork & & \pitchfork & & \pitchfork \\ r & r & & r & & r \end{array}$$

Figure 1.3 an operation f preserving a relation r

one-element structures, each determined by taking different partial operations and relations to be non-empty. We will refer to a one-element structure in which *all* partial operations and relations are non-empty as the **complete one-element structure** of that type.

The first and second conditions of the following theorem are the two conditions that we are seeking. These are together equivalent to the third condition, which is stated from the point of view of the structure on $\underline{\mathbf{M}}$, and the fourth condition, which is stated from the point of view of the structure on $\underset{\sim}{\mathbf{M}}$. Notice that in the fourth condition, if $n > 0$, then the graph of an n-ary (partial) operation h is a subalgebra of $\underline{\mathbf{M}}^{n+1}$ if and only if h is a homomorphism from a subalgebra of $\underline{\mathbf{M}}^n$ into $\underline{\mathbf{M}}$ (Exercise 1.20). Condition (iv) gives us a simple method of constructing *many* choices of $\underset{\sim}{\mathbf{M}}$ for any given algebra $\underline{\mathbf{M}}$.

5.2 Preduality Theorem *Let $\underline{\mathbf{M}}$ be a finite algebra and let $\underset{\sim}{\mathbf{M}}$ be a discrete topological structure such that each relation and the domain of each partial operation of $\underset{\sim}{\mathbf{M}}$ are non-empty and $\underset{\sim}{\mathbf{M}}$ has the same underlying set M as $\underline{\mathbf{M}}$. Then conditions* (ii), (iii) *and* (iv) *below are equivalent and imply condition* (i)*:*

(i) *$\mathcal{A}(\mathbf{A}, \underline{\mathbf{M}})$ is a closed substructure of $\underset{\sim}{\mathbf{M}}^A$, for each $\mathbf{A} \in \mathcal{A}$;*

(ii) *$\mathcal{X}(\mathbf{X}, \underset{\sim}{\mathbf{M}})$ is a subalgebra of $\underline{\mathbf{M}}^X$, for each $\mathbf{X} \in \mathcal{X}$;*

(iii) *each non-nullary operation of $\underline{\mathbf{M}}$ is a morphism from a power of $\underset{\sim}{\mathbf{M}}$ into $\underset{\sim}{\mathbf{M}}$, and each distinguished element of $\underline{\mathbf{M}}$ forms a complete substructure of $\underset{\sim}{\mathbf{M}}$;*

(iv) *each relation, the graph of each non-nullary operation and the graph of each partial operation of $\underset{\sim}{\mathbf{M}}$ are subalgebras of powers of $\underline{\mathbf{M}}$, and each distinguished element of $\underset{\sim}{\mathbf{M}}$ is a one-element subalgebra of $\underline{\mathbf{M}}$.*

Proof We first prove (iii) $\Rightarrow$ (ii). Let f be an n-ary operation of $\underline{\mathbf{M}}$ where $n > 0$. To see that $\mathcal{X}(\mathbf{X}, \underset{\sim}{\mathbf{M}})$ is closed under f, choose any $\alpha_1, \alpha_2, \ldots, \alpha_n \in \mathcal{X}(\mathbf{X}, \underset{\sim}{\mathbf{M}})$. Let $\alpha_1 \sqcap \alpha_2 \sqcap \cdots \sqcap \alpha_n \colon \mathbf{X} \to \underset{\sim}{\mathbf{M}}^n$ be the natural map (coming from the categorical definition of product). Then $f(\alpha_1, \alpha_2, \ldots, \alpha_n) = f \circ (\alpha_1 \sqcap \cdots \sqcap \alpha_n) \in \mathcal{X}(\mathbf{X}, \underset{\sim}{\mathbf{M}})$. If a is a distinguished element of $\underline{\mathbf{M}}$, then $\{a\}$ is a complete substructure of $\underset{\sim}{\mathbf{M}}$ and therefore the constant map $a^X : X \to \{a\}$ is in $\mathcal{X}(\mathbf{X}, \underset{\sim}{\mathbf{M}})$.

To prove that (ii) $\Rightarrow$ (iii), again let f be an n-ary operation of $\underline{\mathbf{M}}$, where $n > 0$. For $i = 1, 2, \ldots, n$, let $\pi_i \colon \underline{\mathbf{M}}^n \to \underline{\mathbf{M}}$ be the projection. Since $\mathcal{X}(\underset{\sim}{\mathbf{M}}^n, \underset{\sim}{\mathbf{M}})$ is a subalgebra of $\underline{\mathbf{M}}^{M^n}$, we have $f = f(\pi_1, \pi_2, \ldots, \pi_n) \in \mathcal{X}(\underset{\sim}{\mathbf{M}}^n, \underset{\sim}{\mathbf{M}})$. If a is a distinguished element of $\underline{\mathbf{M}}$, then, since $\mathcal{X}(\underset{\sim}{\mathbf{M}}, \underset{\sim}{\mathbf{M}})$ is a subalgebra of $\underline{\mathbf{M}}^M$, the constant map $\underset{\sim}{\mathbf{M}} \to \{a\}$ is a morphism. Since the relations and domains of partial operations of $\underset{\sim}{\mathbf{M}}$ are all non-empty, $\{a\}$ is a complete substructure of $\underset{\sim}{\mathbf{M}}$.

Next we prove (iii) $\Rightarrow$ (iv). By Lemma 1.5.1, each relation, the graph of each non-nullary operation and the graph of each partial operation of $\underset{\sim}{\mathbf{M}}$ are closed under each non-nullary operation of $\underline{\mathbf{M}}$. If a is a distinguished element of $\underline{\mathbf{M}}$, then $\{a\}$ is a complete substructure of $\underset{\sim}{\mathbf{M}}$ so that $(a, a, \ldots, a) \in r$, for each relation $r \in R$, and $h(a, a, \ldots, a) = a$, for each $h \in G \cup H$. Thus each relation, non-nullary operation and the graph of each partial operation are subalgebras of powers of $\underline{\mathbf{M}}$. If b is a distinguished element of $\underset{\sim}{\mathbf{M}}$, then $\{b\}$ must be closed under the non-nullary operations of $\underline{\mathbf{M}}$ since they are morphisms. If a is a distinguished element of $\underline{\mathbf{M}}$, then $\{a\}$ is a complete substructure of $\underset{\sim}{\mathbf{M}}$, by (iii), so $b \in \{a\}$. Thus $a = b$ and $\{b\}$ is a subalgebra of $\underline{\mathbf{M}}$.

Now assume (iv) to prove (iii). By Lemma 1.5.1, each non-nullary operation of $\underline{\mathbf{M}}$ preserves each relation, each non-nullary operation and each partial operation of $\underset{\sim}{\mathbf{M}}$, and it preserves each distinguished element b of $\underset{\sim}{\mathbf{M}}$ since $\{b\}$ is a subalgebra of $\underline{\mathbf{M}}$. Consequently it is a morphism. Let a be a distinguished element of $\underline{\mathbf{M}}$. If b is a distinguished element of $\underset{\sim}{\mathbf{M}}$, then $\{b\}$ is a subalgebra of $\underline{\mathbf{M}}$ so $a \in \{b\}$ and therefore $a = b$. Thus $b \in \{a\}$, and it follows from the first part of (iv) that $\{a\}$ is a complete substructure of $\underset{\sim}{\mathbf{M}}$.

The proof of (i) from (iii) and (iv) is illustrated in Figure 1.4. Consider an n-ary operation $h \in G \cup H$ where $n > 0$. Let $u_1, u_2, \ldots, u_n \in \mathcal{A}(\mathbf{A}, \underline{\mathbf{M}})$ with $(u_1, u_2, \ldots, u_n) \in \operatorname{dom}(h^{\underset{\sim}{\mathbf{M}}^A})$. To see that $h(u_1, u_2, \ldots, u_n)$ belongs to $\mathcal{A}(\mathbf{A}, \underline{\mathbf{M}})$, let $a_1, a_2, \ldots, a_m \in A$ and let f be an m-ary operation of $\underline{\mathbf{M}}$. First assume $m > 0$. Then

$$
\begin{aligned}
&h^{\underset{\sim}{\mathbf{M}}^A}(u_1, u_2, \ldots, u_n)(f^{\mathbf{A}}(a_1, a_2, \ldots, a_m)) \\
&\quad = h^{\underset{\sim}{\mathbf{M}}}(u_1(f^{\mathbf{A}}(a_1, \ldots)), u_2(f^{\mathbf{A}}(a_1, \ldots)), \ldots, u_n(f^{\mathbf{A}}(a_1, \ldots))) \\
&\quad = h^{\underset{\sim}{\mathbf{M}}}(f^{\underline{\mathbf{M}}}(u_1(a_1), \ldots, u_1(a_m)), \ldots, f^{\underline{\mathbf{M}}}(u_n(a_1), \ldots, u_n(a_m))) \\
&\quad = f^{\underline{\mathbf{M}}}(h^{\underset{\sim}{\mathbf{M}}}(u_1(a_1), \ldots, u_n(a_1)), \ldots, h^{\underset{\sim}{\mathbf{M}}}(u_1(a_m), \ldots, u_n(a_m))) \quad ((\text{iii})) \\
&\quad = f^{\underline{\mathbf{M}}}(h^{\underset{\sim}{\mathbf{M}}^A}(u_1, \ldots, u_n)(a_1), \ldots, h^{\underset{\sim}{\mathbf{M}}^A}(u_1, \ldots, u_n)(a_m)).
\end{aligned}
$$

Now suppose $m = 0$, that is, f is a nullary operation distinguishing an element $c \in M$. Then, again by (iii),

$$h^{\underset{\sim}{\mathbf{M}}^A}(u_1, \dots, u_n)(c^{\mathbf{A}}) = h^{\underset{\sim}{\mathbf{M}}}(u_1(c^{\mathbf{A}}), \dots, u_n(c^{\mathbf{A}})) = h^{\underset{\sim}{\mathbf{M}}}(c, c, \dots, c) = c.$$

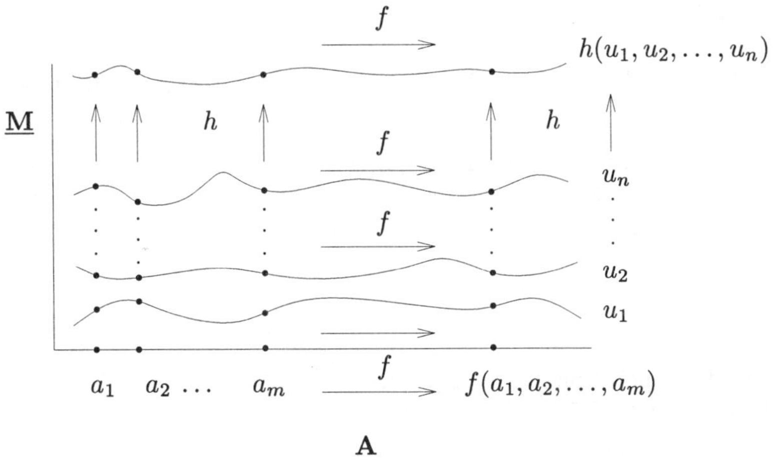

Figure 1.4 Preduality Theorem, proof of (i)

Next consider the case that $n = 0$, that is, $h^{\underset{\sim}{\mathbf{M}}}$ is a nullary operation of $\underset{\sim}{\mathbf{M}}$ distinguishing an element $b \in M$. Then $h^{\underset{\sim}{\mathbf{M}}^A}$ distinguishes the constant map $A \to \{b\}$ which is a homomorphism since, by (iv), we know that $\{b\}$ is a subalgebra of $\underline{\mathbf{M}}$. ■

An n-ary relation on M is called **algebraic over $\underline{\mathbf{M}}$** if it forms a subalgebra of $\underline{\mathbf{M}}^n$. An n-ary operation g on M is called **algebraic over $\underline{\mathbf{M}}$** if it is a homomorphism from $\underline{\mathbf{M}}^n$ to $\underline{\mathbf{M}}$. In particular, a nullary operation is algebraic over $\underline{\mathbf{M}}$ if the element it distinguishes forms a one-element subalgebra of $\underline{\mathbf{M}}$. An n-ary partial operation h on M is called **algebraic over $\underline{\mathbf{M}}$** if it is a homomorphism from a subalgebra of $\underline{\mathbf{M}}^n$ to $\underline{\mathbf{M}}$. By Exercise 1.20, this is equivalent to saying that the graphs of g and h form subalgebras of $\underline{\mathbf{M}}^{n+1}$. In this language, condition (iv) of the Preduality Theorem 1.5.2 says precisely that each of the operations, partial operations and relations of $\underset{\sim}{\mathbf{M}}$ is algebraic over $\underline{\mathbf{M}}$. When this occurs, we say that $\underset{\sim}{\mathbf{M}}$ is **algebraic over $\underline{\mathbf{M}}$** and that $\underset{\sim}{\mathbf{M}}$ **yields a preduality on $\mathcal{A}$**.

Under a preduality the schizophrenia exhibited by the pair $\underline{\mathbf{M}}$–$\underset{\sim}{\mathbf{M}}$ proves to be contagious. We have sets which bear the identity of objects in one category but behave as relations in the other, or as morphisms in one category but as operations in the other. We will normally adopt notation appropriate to the identity under which the entity in question arises, while retaining the flexibility to shift to the other point of view whenever necessary. Many startling punches in duality theory result from such well-timed shifts.

Once $\underset{\sim}{\mathbf{M}}$ is algebraic over $\underline{\mathbf{M}}$, conditions (i) and (ii) of the Preduality Theorem 1.5.2 allow us to define $D\colon \mathcal{A} \to \mathcal{X}$ and $E\colon \mathcal{X} \to \mathcal{A}$ by

$$D(\mathbf{A}) = \mathcal{A}(\mathbf{A}, \underline{\mathbf{M}}) \leqslant \underset{\sim}{\mathbf{M}}^A \quad \text{and} \quad E(\mathbf{X}) = \mathcal{X}(\mathbf{X}, \underset{\sim}{\mathbf{M}}) \leqslant \underline{\mathbf{M}}^X.$$

Moreover, it turns out that there are simple and natural candidates for the remaining components of a dual adjunction in this setting. To complete the definitions of the contravariant functors D and E, consider $\mathbf{A}, \mathbf{B} \in \mathcal{A}$ and a homomorphism $u\colon \mathbf{A} \to \mathbf{B}$. We define

$$D(u)\colon D(\mathbf{B}) \to D(\mathbf{A}) \quad \text{by} \quad D(u)(x) = x \circ u.$$

Similarly, for $\mathbf{X}, \mathbf{Y} \in \mathcal{X}$ and $\varphi\colon \mathbf{X} \to \mathbf{Y}$, we define

$$E(\varphi)\colon E(\mathbf{Y}) \to E(\mathbf{X}) \quad \text{by} \quad E(\varphi)(\alpha) = \alpha \circ \varphi.$$

For each $\mathbf{A} \in \mathcal{A}$ and each $\mathbf{X} \in \mathcal{X}$ we define the **evaluation maps**

$$e_{\mathbf{A}}\colon \mathbf{A} \to ED(\mathbf{A}) \quad \text{by} \quad e_{\mathbf{A}}(a)(x) = x(a)$$

and

$$\varepsilon_{\mathbf{X}}\colon \mathbf{X} \to DE(\mathbf{X}) \quad \text{by} \quad \varepsilon_{\mathbf{X}}(x)(\alpha) = \alpha(x).$$

(See Figure 1.5.)

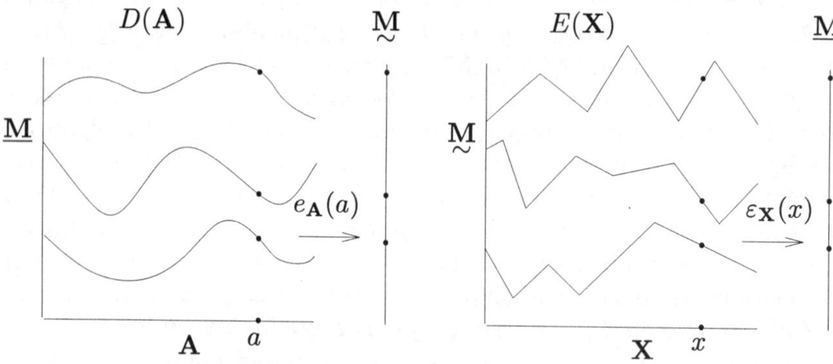

Figure 1.5 the evaluation maps $e_{\mathbf{A}}$ and $\varepsilon_{\mathbf{X}}$

The next theorem shows that in this context a large part of the construction we seek is present as soon as $\underset{\sim}{\mathbf{M}}$ is algebraic over $\underline{\mathbf{M}}$.

5.3 Dual Adjunction Theorem *Assume that* $\underset{\sim}{\mathbf{M}}$ *is algebraic over* $\underline{\mathbf{M}}$. *Then* $\langle D, E, e, \varepsilon\rangle$ *is a dual adjunction between* $\mathcal{A}$ *and* $\mathcal{X}$ *such that, for any* $u\colon \mathbf{A} \to \mathbf{B}$ *in* $\mathcal{A}$ *and any* $\varphi\colon \mathbf{X} \to \mathbf{Y}$ *in* $\mathcal{X}$, *the following additional properties hold:*

(i) *the evaluation maps* $e_{\mathbf{A}}$ *and* $\varepsilon_{\mathbf{X}}$ *are embeddings,*
(ii) D *and* E *are faithful,*
(iii) *if* u *is a surjection, then* $D(u)$ *is an embedding,*
(iv) *if* $D(u)$ *is a surjection, then* u *is an embedding,*
(v) *if* φ *is a surjection, then* $E(\varphi)$ *is an embedding,*
(vi) *if* $E(\varphi)$ *is a surjection, then* φ *is an embedding.*

Proof The parts of this proof are all direct consequences of the definitions and are listed in Exercise 1.22 through Exercise 1.28. ∎

The remainder of this book is devoted exclusively to the study of the setting described in the Dual Adjunction Theorem. In particular, there will never be occasion to refer to a choice of a structure $\underset{\sim}{\mathbf{M}}$ which is not algebraic over $\underline{\mathbf{M}}$. We therefore *assume without further mention* that all operations, partial operations and relations on $\underset{\sim}{\mathbf{M}}$ are algebraic over $\underline{\mathbf{M}}$.

Exercises 1

Exercises from the text. Prove Lemma 1.1.1. Complete the proof of Theorem 1.3.1 by verifying that t is an embedding. Complete the proof of the $\mathbb{ISP}$-Theorem 1.3.4 by showing that an isomorphic copy or a subalgebra of an algebra which satisfies a quasi-equation also satisfies that quasi-equation, and a product of algebras which satisfy a quasi-equation also satisfies that quasi-equation. Complete the proof of Example 1.4.7.

1.1 Show that any two products of objects $\mathbf{A}$ and $\mathbf{B}$ are isomorphic.

1.2 Assume that $\langle D, E, e, \varepsilon\rangle$ is a dual adjunction between $\mathcal{A}$ and $\mathcal{X}$. Use the triangles in Figure 1.2 to prove that, for all $\mathbf{A} \in \mathcal{A}$ and $\mathbf{X} \in \mathcal{X}$,

(i) $\mathrm{id}_{D(\mathbf{A})} = D(e_{\mathbf{A}}) \circ \varepsilon_{D(\mathbf{A})}$ (take $\mathbf{X} = D(\mathbf{A})$), and
(ii) $\mathrm{id}_{E(\mathbf{X})} = E(\varepsilon_{\mathbf{X}}) \circ e_{E(\mathbf{X})}$ (take $\mathbf{A} = E(\mathbf{X})$).

1.3 Use Exercise 1.2 to prove Lemma 1.1.4.

1.4 Assume that $\langle D, E, e, \varepsilon\rangle$ is a dual representation between categories $\mathcal{A}$ and $\mathcal{X}$. Use Exercise 1.2 to show that $E(\varepsilon_{\mathbf{X}})\colon EDE(\mathbf{X}) \to E(\mathbf{X})$ is an isomorphism in $\mathcal{A}$ for all $\mathbf{X} \in \mathcal{X}$.

1.5 Let $\mathbf{A}$ be a model of (DL). For $a, b \in A$ define $a \leqslant b$ if $a \wedge b = a$. Show that

(i) $a \leqslant b$ if and only if $a \vee b = b$,
(ii) $\leqslant$ is reflexive, antisymmetric and transitive,

(iii) $a \wedge b$ is the greatest lower bound of a and b, and

(iv) $a \vee b$ is the least upper bound of a and b.

1.6 Verify that both the class $\mathfrak{D}$ of distributive lattices with homomorphisms and the class $\mathfrak{X}_{01}$ of bounded Priestley spaces with continuous homomorphisms are categories.

1.7 Establish each of the following facts and then use them to prove the Priestley Dual Equivalence Theorem 1.2.5.

(i) For each $\mathbf{A}, \mathbf{B} \in \mathfrak{D}$ and each homomorphism $u: \mathbf{A} \to \mathbf{B}$, the map $D(u)$ is a continuous homomorphism from $D(\mathbf{B})$ to $D(\mathbf{A})$.

(ii) For each $\mathbf{X}, \mathbf{Y} \in \mathfrak{X}_{01}$ and each morphism $\varphi: \mathbf{X} \to \mathbf{Y}$, the map $E(\varphi)$ is a homomorphism from $E(\mathbf{Y})$ to $E(\mathbf{X})$.

(iii) $D: \mathfrak{D} \to \mathfrak{X}_{01}$ is a faithful contravariant functor.

(iv) $E: \mathfrak{X}_{01} \to \mathfrak{D}$ is a faithful contravariant functor.

(v) $\langle D, E, e, \varepsilon \rangle$ is a dual adjunction between $\mathfrak{D}$ and $\mathfrak{X}_{01}$.

(vi) For each $\mathbf{X} \in \mathfrak{X}_{01}$ the map $\varepsilon_{\mathbf{X}}: \mathbf{X} \to DE(\mathbf{X})$ is an isomorphism.

1.8 Let X be a set and let χ map each subset of X to its characteristic function. Show that χ is an isomorphism from the distributive lattice $\mathbf{P}(X)$ to the direct power $\langle \{0,1\}; \vee, \wedge \rangle^X$.

1.9 If $I \neq \varnothing$, show that $\underset{\sim}{\mathbf{D}}^I$ is a bounded Priestley space.

1.10 Verify that a closed subspace of a bounded Priestley space which contains 0 and 1 is again a bounded Priestley space.

1.11 (This is for those familiar with mathematical logic.) Use the Compactness Theorem of First-Order Logic to give an alternative proof that every model $\mathbf{A}$ of the quasi-equational theory of $\underline{\mathbf{M}}$ is in $\mathbb{ISP}\underline{\mathbf{M}}$ provided that the same is true of every finite model. Specifically, choose $c, d \in A$ where $c \neq d$. Now write down a set Σ of first order sentences which assert that $\{\, a^* \mid a \in A \,\}$ forms an algebra $\mathbf{A}^*$ isomorphic to $\mathbf{A}$, and that θ is a congruence which is the kernel of a homomorphism from $\mathbf{A}^*$ into $\underline{\mathbf{M}}$ which separates c and d. Show that every finite subset of Σ has a model, and use the Compactness Theorem to conclude that Σ has a model.

1.12 Let $\mathfrak{C}$ be the category of structured topological spaces of a fixed type $\langle G, H, R \rangle$ and let $\mathbf{X}$ and $\mathbf{Y}$ be objects of $\mathfrak{C}$. Show that the concrete product $\mathbf{X} \times \mathbf{Y}$ together with the projections $\pi_1: \mathbf{X} \times \mathbf{Y} \to \mathbf{X}$ and $\pi_2: \mathbf{X} \times \mathbf{Y} \to \mathbf{Y}$ is a product in the category $\mathfrak{C}$.

1.13 Prove part (i) of the Preservation Theorem by establishing each of the following facts and using Lemma B.5.

(i) An isomorphic copy of a structure which satisfies a quasi-atomic formula also satisfies that quasi-atomic formula.

(ii) A substructure of a structure which satisfies a quasi-atomic formula also satisfies that quasi-atomic formula.

(iii) A direct product of structures which satisfy a quasi-atomic formula also satisfies that quasi-atomic formula.

1.14 Use Lemma B.6(iii) to prove part (iii) of the Preservation Theorem. Again, show first that it is true when $\mathbf{X} \in \mathbb{S}_c\mathbb{P}^+\underset{\sim}{\mathbf{M}}$, and then show that it is preserved by isomorphism.

1.15 Use Lemma B.3 and Lemma B.6(iii) to prove part (ii) of the Preservation Theorem. First show that it is true when $\mathbf{X} \in \mathbb{S}_c\mathbb{P}^+\underset{\sim}{\mathbf{M}}$, and then show that it is preserved by isomorphism.

1.16 The Separation and $\mathbb{ISP}$-Theorems also hold when partial operations and relations are added. For example, let $\mathfrak{C}$ be the category of structures $\mathbf{P} = \langle P; 0, 1, \leqslant\rangle$ with two constants and a single binary relation. Notice that the category $\mathfrak{C}$ includes the two-element bounded ordered set $\underset{\sim}{\mathbf{D}}^* = \langle\{0,1\}; 0, 1, \leqslant\rangle$. Prove that the following are equivalent for $\mathbf{P} \in \mathfrak{C}$ by demonstrating that (i) $\Rightarrow$ (ii) $\Rightarrow$ (iii) $\Rightarrow$ (iv) $\Rightarrow$ (i):

(i) $\mathbf{P} \in \mathbb{ISP}\underset{\sim}{\mathbf{D}}^*$;

(ii) $\mathbf{P}$ is a model of the quasi-atomic theory of $\underset{\sim}{\mathbf{D}}^*$;

(iii) $\mathbf{P}$ is a bounded ordered set;

(iv) if $a \nleqslant b$ in $\mathbf{P}$, then there is a morphism $u\colon \mathbf{P} \to \underset{\sim}{\mathbf{D}}^*$ such that $u(a) = 1$ and $u(b) = 0$, and if $a \neq b$ in $\mathbf{P}$, then there is a morphism $u\colon \mathbf{P} \to \underset{\sim}{\mathbf{D}}^*$ such that $u(a) \neq u(b)$.

1.17 Show that, for any finite $\underset{\sim}{\mathbf{M}}$ with at least two elements and with G, H and R empty, $\mathbb{IS}_c\mathbb{P}^+\underset{\sim}{\mathbf{M}}$ is the category of Boolean spaces.

1.18 For a distributive lattice $\mathbf{A}$ and a bounded Priestley space $\mathbf{X}$, prove

(i) the bounded Priestley space $D(\mathbf{A})$ of prime ideals is isomorphic to the order-theoretic dual of $\mathfrak{D}(\mathbf{A}, \mathbf{D})$,

(ii) the distributive lattice $E(\mathbf{X})$ of non-empty, proper clopen decreasing subsets of X is isomorphic to the order-theoretic dual of the distributive lattice $\mathfrak{X}_{01}(\mathbf{X}, \underset{\sim}{\mathbf{D}})$,

(iii) the bounded Priestley space of prime filters of $\mathbf{A}$ is order-isomorphic to $\mathfrak{D}(\mathbf{A}, \mathbf{D})$,

(iv) the lattice of non-empty, proper clopen increasing subsets of $\mathbf{X}$ is order-isomorphic to $\mathfrak{X}_{01}(\mathbf{X}, \underset{\sim}{\mathbf{D}})$.

1.19 Let $\mathbf{A}$ and $\underline{\mathbf{M}}$ be algebras where $\underline{\mathbf{M}}$ is finite, $\mathcal{A} = \mathbb{ISP}\underline{\mathbf{M}}$. Use Lemma B.6(iii) to show that the homset $\mathcal{A}(\mathbf{A}, \underline{\mathbf{M}})$ is a closed subspace of the product space M^A if M has the discrete topology. More generally, show that if $\underline{\mathbf{M}} = \langle M; F\rangle$ is an algebra and $\mathcal{T}$ is a compact Hausdorff topology on M with respect to which the operations F

are continuous, then $\mathcal{A}(\mathbf{A}, \underline{\mathbf{M}})$ is a closed subspace of the product space M^A.

1.20 Let h be an n-ary partial operation on M. Show that $\mathrm{dom}(h)$ is a subalgebra of $\underline{\mathbf{M}}^n$ and $h\colon \mathrm{dom}(h) \to \underline{\mathbf{M}}$ is a homomorphism if and only if $\mathrm{graph}(h)$ is a subalgebra of $\underline{\mathbf{M}}^{n+1}$.

1.21 A morphism $\varphi\colon \mathbf{X} \to \mathbf{Y}$ is called **strong** if whenever r belongs to R or is the graph of a (partial) operation in $G \cup H$ and $x_1, \dots, x_n \in X$ with $(\varphi(x_1), \dots, \varphi(x_n)) \in r^{\mathbf{Y}}$, there exist $x'_1, \dots, x'_n \in X$ such that $\varphi(x'_i) = \varphi(x_i)$ for all i and $(x'_1, \dots, x'_n) \in r^{\mathbf{X}}$.

(i) Show that the condition for a morphism φ to be strong always holds with respect to the total operations in G.

(ii) Show that for every non-empty subset T of S, the restriction map $\pi_T\colon \underset{\sim}{\mathbf{M}}^S \to \underset{\sim}{\mathbf{M}}^T$ is strong.

(iii) Let $\mathbf{X}, \mathbf{Y}, \mathbf{Z} \in \mathcal{X}$, let $\varphi\colon \mathbf{X} \to \mathbf{Y}$ be a strong, surjective morphism and let $\psi\colon \mathbf{Y} \to \mathbf{Z}$ be a map such that $\psi \circ \varphi$ is a morphism. Show that ψ is a morphism.

In the remaining exercises, assume that $\underset{\sim}{\mathbf{M}}$ is algebraic over $\underline{\mathbf{M}}$. Prove the Dual Adjunction Theorem by verifying the following.

1.22 Show that $D\colon \mathcal{A} \to \mathcal{X}$ and $E\colon \mathcal{X} \to \mathcal{A}$ are contravariant functors.

1.23 Show that both of the squares of Figure 1.2 commute, that is,

(i) $e_{\mathbf{B}} \circ u = ED(u) \circ e_{\mathbf{A}}$ for all $u\colon \mathbf{A} \to \mathbf{B}$ in $\mathcal{A}$, and

(ii) $\varepsilon_{\mathbf{Y}} \circ \varphi = DE(\varphi) \circ \varepsilon_{\mathbf{X}}$ for all $\varphi\colon \mathbf{X} \to \mathbf{Y}$ in $\mathcal{X}$.

1.24 Establish the bijection given in the triangles of Figure 1.2 by verifying

(i) $u = E(D(u) \circ \varepsilon_{\mathbf{X}}) \circ e_{\mathbf{A}}$ for all $u\colon \mathbf{A} \to E(\mathbf{X})$ in $\mathcal{A}$, and

(ii) $\varphi = D(E(\varphi) \circ e_{\mathbf{A}}) \circ \varepsilon_{\mathbf{X}}$ for all $\varphi\colon \mathbf{X} \to D(\mathbf{A})$ in $\mathcal{X}$.

1.25 Show that $e_{\mathbf{A}}$ and $\varepsilon_{\mathbf{X}}$ are embeddings for all $\mathbf{A} \in \mathcal{A}$ and $\mathbf{X} \in \mathcal{X}$.

1.26 Show that D and E are both faithful.

1.27 Show that, for every homomorphism u of $\mathcal{A}$, if either u or $D(u)$ is a surjection, then the other is an embedding.

1.28 Show that, for every morphism φ of $\mathcal{X}$, if either φ or $E(\varphi)$ is a surjection, then the other is an embedding.

2

Natural Dualities

Having set the scene in the previous chapter, we can now begin to address the most immediate issues. As algebraists, our first aim is to obtain a representation of each algebra in $\mathcal{A} = \mathbb{ISP}\underline{\mathbf{M}}$ as an algebra of continuous, structure-preserving maps.

Given a discrete topological structure $\underset{\sim}{\mathbf{M}} = \langle M; G, H, R, \mathcal{T} \rangle$ which is algebraic over $\underline{\mathbf{M}}$, we have a dual adjunction $\langle D, E, e, \varepsilon \rangle$ between $\mathcal{A}$ and $\mathcal{X} := \mathbb{IS}_c\mathbb{P}^+ \underset{\sim}{\mathbf{M}}$ as described in the last section of the previous chapter. In particular, by the Dual Adjunction Theorem 1.5.3, the homomorphism $e_{\mathbf{A}} : \mathbf{A} \to ED(\mathbf{A})$ is an embedding for all $\mathbf{A} \in \mathcal{A}$. We shall say that $\underset{\sim}{\mathbf{M}}$ **yields a (natural) duality** on an algebra $\mathbf{A}$ in $\mathcal{A}$ if $e_{\mathbf{A}}$ is an isomorphism, that is, if the only continuous structure-preserving maps from $D(\mathbf{A})$ to $\underset{\sim}{\mathbf{M}}$ are the evaluations. If $\mathcal{C} \subseteq \mathcal{A}$ and $\underset{\sim}{\mathbf{M}}$ yields a duality on every algebra $\mathbf{A}$ in $\mathcal{C}$, then we say that $\underset{\sim}{\mathbf{M}}$ **yields a (natural) duality on** $\mathcal{C}$. Thus $\underset{\sim}{\mathbf{M}}$ yields a natural duality on $\mathcal{A}$ precisely when the preduality determined by $\underset{\sim}{\mathbf{M}}$ is a dual representation. Instead of saying that $\underset{\sim}{\mathbf{M}}$ yields a duality, we shall sometimes say that $G \cup H \cup R$ **yields a duality**. Putting the emphasis back on the algebra $\underline{\mathbf{M}}$, we say that $\underline{\mathbf{M}}$ **admits a (natural) duality**, or that $\underline{\mathbf{M}}$ is **dualisable**, if there exists some structure $\underset{\sim}{\mathbf{M}}$ which yields a duality on $\mathcal{A}$, in which case we often say simply that $\underset{\sim}{\mathbf{M}}$ (or $G \cup H \cup R$) **dualises** $\underline{\mathbf{M}}$.

Three issues, to be addressed in this chapter, present themselves.

- For a particular choice of $\underset{\sim}{\mathbf{M}}$, find necessary and sufficient conditions on $\underset{\sim}{\mathbf{M}}$ for it to yield a duality on $\mathcal{A}$.
- Find ways in which we may modify the structure on $\underset{\sim}{\mathbf{M}}$ without destroying an existing duality.
- Find useful sufficient conditions for $\underline{\mathbf{M}}$ to admit a duality.

The present chapter is based on Davey and Werner [DW83], and its precursor Davey [Dav78], the fundamental papers which gave birth to the theory presented in this monograph.

2.1 Brute force

We shall see in the next chapter that the operations in G and the partial operations in H play a vital role when we wish to upgrade a natural duality (= dual representation) to a full duality (= dual equivalence). Nevertheless, our first lemma implies that if we are trying to prove only that $\underset{\sim}{\mathbf{M}}$ yields a duality on $\mathcal{A}$, we may delete an operation from G or a partial operation from H provided we add its graph to the set R of relations. Let $\mathrm{dom}(h) \subseteq M^n$ and $h\colon \mathrm{dom}(h) \to M$. By Exercise 1.20, $\mathrm{graph}(h)$ is a subalgebra of $\underline{\mathbf{M}}^{n+1}$ if and only if $\mathrm{dom}(h)$ is a subalgebra of $\underline{\mathbf{M}}^n$ and $h\colon \mathrm{dom}(h) \to \underline{\mathbf{M}}$ is a homomorphism, that is, h is algebraic over $\underline{\mathbf{M}}$ if and only if $\mathrm{graph}(h)$ is algebraic over $\underline{\mathbf{M}}$. Thus it makes sense to delete h from G or H and to add $\mathrm{graph}(h)$ to R.

1.1 Lemma *Let* $\mathrm{dom}(h) \subseteq M^n$ *for some* $n \in \mathbb{N}$, *let* $h\colon \mathrm{dom}(h) \to M$, *let* S *and* T *be non-empty sets and extend both* h *and* $\mathrm{graph}(h)$ *pointwise to* M^S *and* M^T. *Then for all non-empty subsets* X *of* M^S *and* Y *of* M^T *which are closed under* h, *a map* $\alpha\colon X \to Y$ *preserves the (partial) operation* h *if and only if* α *preserves the relation* $\mathrm{graph}(h)$.

Proof Let $\mathrm{dom}(h)$, h, S, X, T and Y be as described above and assume that $\alpha\colon X \to Y$ preserves h. Denote the relation $\mathrm{graph}(h)$ by r and its pointwise extensions to M^S and M^T by r^S and r^T respectively. Let $x_1, \dots, x_n, x_{n+1} \in X$. Denote the n-tuples $(x_1, \dots, x_n)$, $(x_1(s), \dots, x_n(s))$ and $(\alpha(x_1), \dots, \alpha(x_n))$ by $\boldsymbol{x}$, $\boldsymbol{x}(s)$ and $\alpha(\boldsymbol{x})$ and denote the corresponding $(n+1)$-tuples by $(\boldsymbol{x}, x_{n+1})$, $(\boldsymbol{x}(s), x_{n+1}(s))$ and $(\alpha(\boldsymbol{x}), \alpha(x_{n+1}))$, respectively. Then

$$\begin{aligned}
&(\boldsymbol{x}, x_{n+1}) \in r^S \\
\Longrightarrow\ &(\forall s \in S)\,(\boldsymbol{x}(s), x_{n+1}(s)) \in r = \mathrm{graph}(h) \\
\Longrightarrow\ &(\forall s \in S)\,\boldsymbol{x}(s) \in \mathrm{dom}(h) \text{ and } h(\boldsymbol{x}(s)) = x_{n+1}(s) \\
\Longrightarrow\ &\boldsymbol{x} \in \mathrm{dom}(h^X) \text{ and } h(\boldsymbol{x}) = x_{n+1} \\
\Longrightarrow\ &\alpha(\boldsymbol{x}) \in \mathrm{dom}(h^Y) \text{ and } h(\alpha(\boldsymbol{x})) = \alpha(x_{n+1}) \\
\Longrightarrow\ &(\alpha(\boldsymbol{x}), \alpha(x_{n+1})) \in r^T.
\end{aligned}$$

Hence α preserves the relation $r = \mathrm{graph}(h)$. The converse we leave as an easy exercise. ■

It follows easily from this lemma that we can replace a (partial) operation by its graph without destroying a duality.

1.2 Lemma *Let* $\underset{\sim}{\mathrm{M}} = \langle M; G, H, R, \mathcal{T}\rangle$ *and let* $\underset{\sim}{\mathrm{M}}' = \langle M; R', \mathcal{T}\rangle$ *where* $R' := R \cup \{\mathrm{graph}(h) \mid h \in G \cup H\}$. *Then* $\underset{\sim}{\mathrm{M}}$ *yields a duality on* $\mathbf{A} \in \mathcal{A}$ *if and only if* $\underset{\sim}{\mathrm{M}}'$ *yields a duality on* $\mathbf{A}$.

Proof By definition, $\underset{\sim}{\mathrm{M}}$ yields a duality on $\mathbf{A} \in \mathcal{A}$ if the only continuous structure-preserving maps from $D(\mathbf{A}) \subseteq \underset{\sim}{\mathrm{M}}^A$ to $\underset{\sim}{\mathrm{M}}$ are the evaluation maps $e_{\mathbf{A}}(a)$ for $a \in A$. Since, by 1.5.2(i), $D(\mathbf{A})$ is closed under the (partial) operations in $G \cup H$, Lemma 2.1.1 implies that $\underset{\sim}{\mathrm{M}}$ yields a duality on $\mathbf{A}$ if and only if $\underset{\sim}{\mathrm{M}}'$ yields a duality on $\mathbf{A}$. ■

If $\underset{\sim}{\mathrm{M}} = \langle M; R, \mathcal{T}\rangle$ yields a duality on $\mathbf{A}$, for some set R of algebraic relations, then, since the evaluation maps $e_{\mathbf{A}}(a)\colon D(\mathbf{A}) \to M$ preserve every algebraic relation on $\underline{\mathbf{M}}$, for any set R' of algebraic relations which contains R, the structure $\underset{\sim}{\mathbf{M}}' = \langle M; R', \mathcal{T}\rangle$ will also yield a duality on $\mathbf{A}$. In particular, $\underset{\sim}{\mathbf{M}} = \langle M; \mathcal{B}, \mathcal{T}\rangle$ will yield a duality on $\mathbf{A}$ where

$$\mathcal{B} = \bigcup\{\, \mathbb{S}(\underline{\mathbf{M}}^n) \mid n \geqslant 1\,\}$$

is the set of *all* finitary algebraic relations on $\underline{\mathbf{M}}$. We refer to this as the **brute force** construction. Thus the issue of the existence of a duality may, on one level, be summed up as in the lemma below.

1.3 Lemma *Let $\mathbf{A} \in \mathcal{A}$. The following are equivalent:*

(i) *there is some structure $\underset{\sim}{\mathbf{M}} = \langle M; G, H, R, \mathcal{T}\rangle$ which yields a duality on $\mathbf{A}$;*

(ii) *there is some purely relational structure $\underset{\sim}{\mathbf{M}} = \langle M; R, \mathcal{T}\rangle$ which yields a duality on $\mathbf{A}$;*

(iii) *brute force yields a duality on $\mathbf{A}$ (that is, $\underset{\sim}{\mathbf{M}} = \langle M; \mathcal{B}, \mathcal{T}\rangle$ yields a duality on $\mathbf{A}$);*

(iv) *the evaluation maps $e_{\mathbf{A}}(a)\colon \mathcal{A}(\mathbf{A}, \underline{\mathbf{M}}) \to M$, where $a \in A$, are the only continuous maps from $\mathcal{A}(A, M)$ to M which preserve every finitary algebraic relation on $\underline{\mathbf{M}}$.*

The brute force construction is an important theoretical tool, but in practice we try to make the structure on $\underset{\sim}{\mathbf{M}}$ as simple as possible. Part of the beauty of Priestley's duality for distributive lattices is that it is given by a single, particularly simple relation and two constants. It is highly unlikely that anyone would use the brute force duality for the class $\mathcal{D}$ of distributive lattices. Nevertheless, brute force does yield a duality on $\mathcal{D}$.

2.2 Duality theorems

We return now to the situation where $\underset{\sim}{\mathbf{M}} = \langle M; G, H, R, \mathcal{T}\rangle$ and we seek readily verifiable and preferably finitary conditions which will guarantee that $\underset{\sim}{\mathbf{M}}$ yields a duality on $\mathcal{A}$. The idea behind the First Duality Theorem is quite simple: since every algebra in $\mathcal{A}$ is a homomorphic image of a free algebra in $\mathcal{A}$ (Theorem A.2), in order to prove that $\underset{\sim}{\mathbf{M}}$ yields a duality on $\mathcal{A}$ it suffices to

- show that $\underset{\sim}{\mathbf{M}}$ yields a duality on all the free algebras in $\mathcal{A}$,

and then

- show that the class of all algebras on which $\underset{\sim}{\mathbf{M}}$ yields a duality is closed under homomorphic images.

Obtaining necessary and sufficient finitary conditions for $\underset{\sim}{\mathbf{M}}$ to yield a duality on all free algebras in $\mathcal{A}$ is quite straightforward. We refer the reader to Appendix A for a review of free algebras.

2.1 Lemma *Let S be a non-empty set, let $\mathbf{F}_{\underline{\mathbf{M}}}(S)$ be the free algebra of S-ary term functions over $\underline{\mathbf{M}}$ and let*

$$\rho_S \colon D(\mathbf{F}_{\underline{\mathbf{M}}}(S)) = \mathcal{A}(\mathbf{F}_{\underline{\mathbf{M}}}(S), \underline{\mathbf{M}}) \to \underset{\sim}{\mathbf{M}}^S$$

be the map which restricts each homomorphism $x \colon \mathbf{F}_{\underline{\mathbf{M}}}(S) \to \underline{\mathbf{M}}$ to the generators, that is, $\rho_S(x)(s) = x(\pi_s)$. Then ρ_S is an isomorphism in $\mathfrak{X}$.

Proof Since $\mathbf{F}_{\underline{\mathbf{M}}}(S)$ is freely generated by $\Pi_S = \{\pi_s \mid s \in S\}$, the map ρ_S is a bijection and we must prove that both ρ_S and ρ_S^{-1} are $\mathfrak{X}$-morphisms. To see that they are continuous it suffices, by Lemma B.1, to show that ρ_S is continuous. For $s \in S$ and $a \in M$, let $U_{a,s} = \{y \in M^S \mid y(s) = a\}$. By Lemma B.3 and Lemma B.6(i) we need only observe that

$$\begin{aligned}\rho_S^{-1}(U_{a,s}) &= \{\, x \in D(\mathbf{F}_{\underline{\mathbf{M}}}(S)) \mid \rho_S(x)(s) = a \,\} \\ &= \{\, x \in D(\mathbf{F}_{\underline{\mathbf{M}}}(S)) \mid x(\pi_s) = a \,\}\end{aligned}$$

is a subbasic clopen set in $D(\mathbf{F}_{\underline{\mathbf{M}}}(S))$.

Since both $D(\mathbf{F}_{\underline{\mathbf{M}}}(S))$ and $\underset{\sim}{\mathbf{M}}^S$ are closed under every algebraic (partial) operation on $\underline{\mathbf{M}}$, by Lemma 2.1.1 it remains only to show that for every (n-ary) relation $r \in \mathcal{B}$ and for all $x_1, \dots, x_n \in D(\mathbf{F}_{\underline{\mathbf{M}}}(S))$,

$$(x_1, \dots, x_n) \in r \text{ on } D(\mathbf{F}_{\underline{\mathbf{M}}}(S)) \iff (\rho_S(x_1), \dots, \rho_S(x_n)) \in r \text{ on } \underset{\sim}{\mathbf{M}}^S.$$

To simplify the notation, denote $D(\mathbf{F}_{\underline{\mathbf{M}}}(S))$ by X and M^S by Y. So that it will be clear where a particular calculation is taking place, we shall denote the interpretations of a term t on $\mathbf{F}_{\underline{\mathbf{M}}}(S)$, on $\underline{\mathbf{M}}$ and on the *algebra* $\mathbf{r}$ by $t^{\mathbf{F}}$, $t^{\underline{\mathbf{M}}}$ and $t^{\mathbf{r}}$, respectively.

For all $a \in F_{\underline{\mathbf{M}}}(S)$, there exist $m \in \mathbb{N}$, an m-ary term t and $s_1, \dots, s_m \in S$ such that $t^{\mathbf{F}}(\pi_{s_1}, \dots, \pi_{s_m}) = a$. Let $r \in \mathcal{B}$ be any n-ary algebraic relation on $\underline{\mathbf{M}}$ and let $x_1, \dots, x_n \in X = \mathcal{A}(\mathbf{F}_{\underline{\mathbf{M}}}(S), \underline{\mathbf{M}})$.

$$\begin{aligned}&(x_1, \dots, x_n) \in r^X \\ \iff &(\forall a \in \mathbf{F}_{\underline{\mathbf{M}}}(S))\ (x_1(a), \dots, x_n(a)) \in r\end{aligned}$$

(quantifying over all m-ary terms t and all $s_1, \dots, s_m \in S$)

$$\begin{aligned}\iff &(x_1(t^{\mathbf{F}}(\pi_{s_1}, \dots, \pi_{s_m})), \dots, x_n(t^{\mathbf{F}}(\pi_{s_1}, \dots, \pi_{s_m}))) \in r \\ \iff &(t^{\underline{\mathbf{M}}}(x_1(s_1), \dots, x_1(s_m)), \dots, t^{\underline{\mathbf{M}}}(x_n(s_1), \dots, x_n(s_m))) \in r \\ \iff &t^{\mathbf{r}}((x_1(s_1), \dots, x_n(s_1)), \dots, (x_1(s_m), \dots, x_n(s_m))) \in r \\ \iff &(x_1(s_1), \dots, x_n(s_1)) \in r, \dots, (x_1(s_m), \dots, x_n(s_m)) \in r \\ \iff &(\forall s \in S)\ (x_1(s), \dots, x_n(s)) \in r \\ \iff &(\rho_S(x_1), \dots, \rho_S(x_n)) \in r^Y.\end{aligned}$$

(Note that the third-last equivalence—the only non-trivial one—holds because r is algebraic over $\underline{\mathbf{M}}$.) Hence ρ_S is an isomorphism in $\mathcal{X}$. ∎

The structure $\underset{\sim}{\mathbf{M}}$ is **injective** in the category $\mathcal{X}$ if, for every morphism $\alpha\colon \mathbf{X} \to \underset{\sim}{\mathbf{M}}$ and embedding $\varphi\colon \mathbf{X} \to \mathbf{Y}$ in $\mathcal{X}$, there is a morphism $\beta\colon \mathbf{Y} \to \underset{\sim}{\mathbf{M}}$ such that $\beta \circ \varphi = \alpha$. The injectivity of $\underset{\sim}{\mathbf{M}}$ will play a central role both here and in Chapter 3. Injectivity of $\underline{\mathbf{M}}$ in $\mathcal{A}$ will also be important in Chapter 3 but will not be considered here.

2.2 First Duality Theorem *The following are equivalent:*

(i) $\underset{\sim}{\mathbf{M}}$ *yields a duality on* $\mathcal{A}$;

(ii) *for all* $A \in \mathcal{A}$, *every morphism* $\alpha\colon D(\mathbf{A}) \to \underset{\sim}{\mathbf{M}}$ *extends to an* A*-ary term function* $t\colon M^A \to M$;

(iii) *the following two conditions hold—*

(INJ) $\underset{\sim}{\mathbf{M}}$ *is injective with respect to those embeddings in* $\mathcal{X}$ *which are of the form* $D(u)\colon D(\mathbf{A}) \to D(\mathbf{B})$ *where* $u\colon \mathbf{B} \to \mathbf{A}$ *is a surjective homomorphism, that is, for each morphism* $\alpha\colon D(\mathbf{A}) \to \underset{\sim}{\mathbf{M}}$ *there exists a morphism* $\beta\colon D(\mathbf{B}) \to \underset{\sim}{\mathbf{M}}$ *such that* $\beta \circ D(u) = \alpha$,

(CLO) *for each* $n \in \mathbb{N}$, *every morphism* $t\colon \underset{\sim}{\mathbf{M}}^n \to \underset{\sim}{\mathbf{M}}$ *is an* n*-ary term function on* $\underline{\mathbf{M}}$.

Proof We begin by proving that (CLO) does indeed capture the fact that $\underset{\sim}{\mathbf{M}}$ yields a duality on the free algebras in $\mathcal{A}$.

2.3 Proposition

(i) *For a fixed set* S, *the structure* $\underset{\sim}{\mathbf{M}}$ *yields a duality on the free algebra* $\mathbf{F} := \mathbf{F}_{\underline{\mathbf{M}}}(S)$ *if and only if the following condition holds:*

(CLO)$_S$ *every morphism* $t\colon \underset{\sim}{\mathbf{M}}^S \to \underset{\sim}{\mathbf{M}}$ *is an* S*-ary term function on* $\underline{\mathbf{M}}$.

(ii) *The following are equivalent:*

(a) (CLO) *holds;*

(b) (CLO)$_S$ *holds for every non-empty set* S;

(c) $\underset{\sim}{\mathbf{M}}$ *yields a duality on the finitely generated free algebras in* $\mathcal{A}$;

(d) $\underset{\sim}{\mathbf{M}}$ *yields a duality on all free algebras in* $\mathcal{A}$.

Proof To prove (i) we first give an alternative formulation of (CLO)$_S$. Since $\rho_S\colon D(\mathbf{F}_{\underline{\mathbf{M}}}(S)) \to \underset{\sim}{\mathbf{M}}^S$ is an isomorphism (by Lemma 2.2.1) and $e_{\mathbf{F}} : \mathbf{F}_{\underline{\mathbf{M}}}(S) \to ED(\mathbf{F}_{\underline{\mathbf{M}}}(S))$ is an embedding, the composite

$$u := E(\rho_S^{-1}) \circ e_{\mathbf{F}} : \mathbf{F}_{\underline{\mathbf{M}}}(S) \to E(\underset{\sim}{\mathbf{M}}^S) = \mathcal{X}(\underset{\sim}{\mathbf{M}}^S, \underset{\sim}{\mathbf{M}})$$

is an embedding. For $s \in S$ and $x \in M^S$, we compute

$$\begin{aligned} u(\pi_s)(x) &= E(\rho_S^{-1})(e_{\mathbf{F}}(\pi_s))(x) = (e_{\mathbf{F}}(\pi_s) \circ \rho_S^{-1})(x) \\ &= e_{\mathbf{F}}(\pi_s)(\rho_S^{-1}(x)) = \rho_S^{-1}(x)(\pi_s) = x(s) = \pi_s(x). \end{aligned}$$

Since the projections generate $\mathbf{F}$, we conclude that u is the inclusion map. Thus $(\mathrm{CLO})_S$ says precisely that u is surjective, which is true if and only if $e_{\mathbf{F}}$ is surjective. This proves (i).

To prove (ii), observe that the equivalences (a) $\Leftrightarrow$ (c) and (b) $\Leftrightarrow$ (d) follow from (i) and the implication (b) $\Rightarrow$ (a) is trivial. It remains to prove (a) $\Rightarrow$ (b), that is, that (CLO) implies $(\mathrm{CLO})_S$ for each infinite set S. Assume that (CLO) holds and let $t\colon \underset{\sim}{\mathbf{M}}^S \to \underset{\sim}{\mathbf{M}}$ be a morphism. By Lemma B.6(iv), there exist a finite subset T of S and a map $t'\colon M^T \to M$ which satisfies $t = t' \circ \pi_T$. Since π_T is given by restriction to T and $t' \circ \pi_T = t$ is a morphism, it follows that t' is a morphism (see Exercise 1.21). Hence, by (CLO), t' is a T-ary term function on $\underline{\mathbf{M}}$, whence $t = t' \circ \pi_T$ is an S-ary term function on $\underline{\mathbf{M}}$. Hence $(\mathrm{CLO})_S$ holds. This completes the proof of (ii). ■

We now return to the proof of the First Duality Theorem. Assume (i) to prove (iii). By Proposition 2.2.3(ii) it remains only to verify (INJ). Let $u\colon \mathbf{B} \to \mathbf{A}$ be surjective and let $\alpha\colon D(\mathbf{A}) \to \underset{\sim}{\mathbf{M}}$ be a morphism. Since, by (i), $\underset{\sim}{\mathbf{M}}$ yields a duality on $\mathbf{A}$, there exists $a \in A$ such that $e_{\mathbf{A}}(a) = \alpha$. As u is surjective, there exists $b \in B$ with $u(b) = a$. Let $\beta := e_{\mathbf{B}}(b)$. Then, for all $x \in D(\mathbf{B})$,

$$(\beta \circ D(u))(x) = e_{\mathbf{B}}(b)(x \circ u) = x(u(b)) = e_{\mathbf{A}}(a)(x) = \alpha(x),$$

whence $\beta \circ D(u) = \alpha$, as required to prove (iii).

Assume that (iii) holds and that $\alpha\colon D(\mathbf{A}) \to \underset{\sim}{\mathbf{M}}$ is a morphism for some $\mathbf{A} \in \mathcal{A}$. Let $u\colon \mathbf{F}_{\underline{\mathbf{M}}}(A) \to \mathbf{A}$ be the unique extension of $\mathrm{id}_A\colon A \to A$ to a homomorphism from $\mathbf{F}_{\underline{\mathbf{M}}}(A)$ onto $\mathbf{A}$. By the Dual Adjunction Theorem 1.5.3(iii), the dual map $D(u)\colon D(\mathbf{A}) \to D(\mathbf{F}_{\underline{\mathbf{M}}}(A))$ is an embedding and hence, by (INJ), there exists a morphism $\beta\colon D(\mathbf{F}_{\underline{\mathbf{M}}}(A)) \to \underset{\sim}{\mathbf{M}}$ satisfying $\beta \circ D(u) = \alpha$. By $(\mathrm{CLO})_A$, the morphism $t := \beta \circ \rho_A^{-1}\colon \underset{\sim}{\mathbf{M}}^A \to \mathbf{M}$ is an A-ary term function on $\underline{\mathbf{M}}$. Denote the inclusion of $D(\mathbf{A})$ into $\underset{\sim}{\mathbf{M}}^A$ by $\iota_{D(\mathbf{A})}$. To show that t extends α we must show that $t \circ \iota_{D(\mathbf{A})} = \alpha$. For all $x \in D(\mathbf{A})$, we have

$$\rho_A(D(u)(x)) = (x \circ u){\restriction}_A = x = \iota_{D(\mathbf{A})}(x)$$

whence $\rho_A \circ D(u) = \iota_{D(\mathbf{A})}$. Thus the three inner triangles of Figure 2.1 commute and a simple diagram chase shows that the outer triangle also commutes, that is, $t \circ \iota_{D(\mathbf{A})} = \alpha$. Hence (iii) implies (ii).

Finally, assume (ii) and let $\mathbf{A} \in \mathcal{A}$. To verify (i), we must prove that $e_{\mathbf{A}}\colon \mathbf{A} \to ED(\mathbf{A})$ is surjective. Let $\alpha \in ED(\mathbf{A})$, that is, $\alpha\colon D(\mathbf{A}) \to \underset{\sim}{\mathbf{M}}$ is a morphism. By (ii), there is an A-ary term function $t\colon M^A \to \underset{\sim}{M}$ which extends α. By the definition of an A-ary term function, there exist $a_1, \dots, a_n \in A$ and an n-ary term function $t'\colon M^n \to M$ such that $t(x) =$

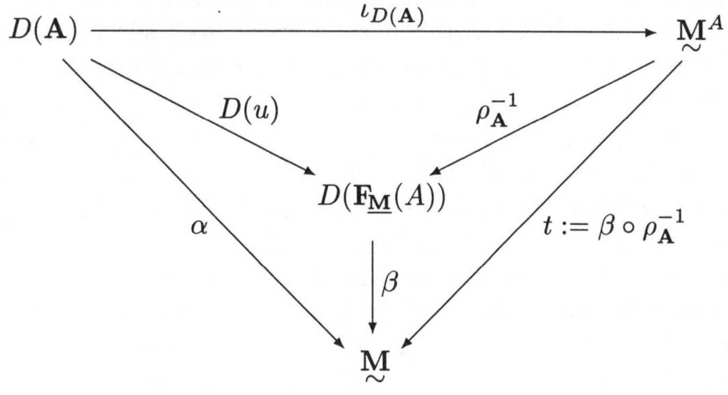

Figure 2.1 t extends α

$t'(x(a_1), \dots, x(a_n))$ for all *maps* $x \colon A \to M$. Define $a := t'(a_1, \dots, a_n)$. Then for all *homomorphisms* $x \colon \mathbf{A} \to \underline{\mathbf{M}}$ we find

$$\begin{aligned} e_{\mathbf{A}}(a)(x) = x(a) &= x(t'(a_1, \dots, a_n)) \\ &= t'(x(a_1), \dots, x(a_n)) = t(x) = \alpha(x) \end{aligned}$$

since t extends α. Thus $e_{\mathbf{A}}(a) = \alpha$, whence $e_{\mathbf{A}}$ is surjective. ■

As an immediate consequence of Lemma 2.2.1 and Proposition 2.2.3 we obtain the following corollary, which says that powers of $\underset{\sim}{\mathbf{M}}$ are dual to free algebras in $\mathcal{A}$ under very weak circumstances.

2.4 Corollary *If S is a non-empty set and $\underset{\sim}{\mathbf{M}}$ is algebraic over $\underline{\mathbf{M}}$, then*

(i) *$D(\mathbf{F}_{\underline{\mathbf{M}}}(S)) \cong \underset{\sim}{\mathbf{M}}^S$ and*

(ii) *$E(\underset{\sim}{\mathbf{M}}^S) = \mathbf{F}_{\underline{\mathbf{M}}}(S)$ if* (CLO) *holds.*

By the Preduality Theorem 1.5.2, every term function of $\underline{\mathbf{M}}$ is a morphism in $\mathcal{X}$. The condition (CLO) adds the converse: duality requires that the (finitary) term functions on $\underline{\mathbf{M}}$ must be exactly the morphisms from finite powers of $\underset{\sim}{\mathbf{M}}$ into $\underset{\sim}{\mathbf{M}}$. Thus (CLO) says precisely that $G \cup H \cup R$ determines the clone of term functions on $\underline{\mathbf{M}}$. It is important to note that (CLO) is not sufficient to guarantee duality. There is a two-element set R of relations which determines the clone of the three-element Kleene algebra but does not yield a duality. Nevertheless, the addition of one further relation to R does yield a duality. (See (CLO) versus (IC) in 4.3.12.)

The condition (INJ) guarantees that if $\underset{\sim}{\mathbf{M}}$ yields a duality on $\mathbf{B} \in \mathcal{A}$ then $\underset{\sim}{\mathbf{M}}$ yields a duality on every homomorphic image of $\mathbf{B}$ in $\mathcal{A}$ (Exercise 2.2). Thus the conditions (CLO) and (INJ) lead us along a natural algebraic path: we first show that $\underset{\sim}{\mathrm{M}}$ yields a duality on the free algebras by proving (CLO) and then, since every algebra is a homomorphic image of a free algebra, show that the duality extends to arbitrary algebras by proving (INJ).

Of course, (INJ) is a special case of the injectivity of $\underset{\sim}{\mathrm{M}}$ in $\mathcal{X}$. In general it is difficult to verify (INJ) directly because we have no practical way to identify those embeddings with respect to which $\underset{\sim}{\mathrm{M}}$ is supposed to be injective. As a result we find that, in almost every example to date, (INJ) has been verified by showing that $\underset{\sim}{\mathrm{M}}$ is injective with respect to *all* embeddings in $\mathcal{X}$; that is, $\underset{\sim}{\mathrm{M}}$ is injective in $\mathcal{X}$. Nevertheless, the injectivity of $\underset{\sim}{\mathrm{M}}$ in $\mathcal{X}$ is not a necessary condition for $\underset{\sim}{\mathrm{M}}$ to yield a duality on $\mathcal{A}$. In Exercise 4.11, we show that it is possible to start with a structure $\underset{\sim}{\mathrm{M}}'$ which yields a duality on $\mathcal{A}$ and is injective in $\mathcal{X}$ and, by reinterpreting its operations as relations, convert it into a structure $\underset{\sim}{\mathrm{M}}$ which yields a duality but is not injective. In Lemma 7.8.6, we give a more natural example of a structure which yields a duality but is not injective in $\mathcal{X}$. The theme of injectivity of $\underset{\sim}{\mathrm{M}}$ will be central in the next chapter.

In order to prove (INJ), we will begin by verifying a more accessible special case of the injectivity of $\underset{\sim}{\mathrm{M}}$ in $\mathcal{X}$. To see how to do this, let $\mathcal{A}_{\text{fin}}$ and $\mathcal{X}_{\text{fin}}$ be the categories of finite members of $\mathcal{A}$ and $\mathcal{X}$, respectively, and let D_{fin}, E_{fin}, e_{fin} and ε_{fin} be the respective restrictions of D, E, e and ε to finite structures. Since $D(\mathbf{A})$ and $E(\mathbf{X})$ are finite when $\mathbf{A} \in \mathcal{A}$ and $\mathbf{X} \in \mathcal{X}$ are finite, $\langle D_{\text{fin}}, E_{\text{fin}}, e_{\text{fin}}, \varepsilon_{\text{fin}} \rangle$ is a dual adjunction between $\mathcal{A}_{\text{fin}}$ and $\mathcal{X}_{\text{fin}}$. If $\underset{\sim}{\mathrm{M}}$ yields a duality on $\mathcal{A}_{\text{fin}}$, that is, $e_{\mathbf{A}} : \mathbf{A} \to ED(\mathbf{A})$ is surjective and therefore an isomorphism for all $\mathbf{A} \in \mathcal{A}_{\text{fin}}$, then $\langle D_{\text{fin}}, E_{\text{fin}}, e_{\text{fin}}, \varepsilon_{\text{fin}} \rangle$ is a dual representation.

The condition (CLO) together with the assertion that $\underset{\sim}{\mathrm{M}}$ is injective in $\mathcal{X}_{\text{fin}}$ are equivalent to the following simple **interpolation condition**:

(IC) *for each $n \in \mathbb{N}$ and each substructure $\mathbf{X}$ of $\underset{\sim}{\mathrm{M}}^n$, every morphism $\alpha \colon \mathbf{X} \to \underset{\sim}{\mathrm{M}}$ extends to a term function $t \colon M^n \to M$ of the algebra $\underline{\mathrm{M}}$.*

They are also equivalent to a statement about duality.

2.5 Lemma *The following are equivalent:*

(i) (IC) *holds;*

(ii) (CLO) *holds and $\underset{\sim}{\mathrm{M}}$ is injective in $\mathcal{X}_{\text{fin}}$;*

(iii) $\underset{\sim}{\mathrm{M}}$ *yields a duality on $\mathcal{A}_{\text{fin}}$ and is injective in $\mathcal{X}_{\text{fin}}$.*

Proof The equivalence (i) $\Leftrightarrow$ (ii) is immediate, and (iii) $\Rightarrow$ (ii) is true by Proposition 2.2.3(ii) since the finitely generated free algebras in $\mathcal{A}$ are finite (Lemma 1.3.2).

We assume (ii) to prove (iii). Let $\mathbf{A} \in \mathcal{A}_{\text{fin}}$ and let $\varphi \colon D(\mathbf{A}) \to \underset{\sim}{\mathrm{M}}$. Since $\underset{\sim}{\mathrm{M}}$ is injective in $\mathcal{X}_{\text{fin}}$, there is a morphism $\psi \colon \underset{\sim}{\mathrm{M}}^A \to \underset{\sim}{\mathrm{M}}$ which extends φ. By (CLO), there is an A-ary term t such that $t^{\mathbf{M}} = \psi$. Let $A = \{a_1, a_2, \dots, a_m\}$ and let $x \in D(\mathbf{A})$. Then

$$\begin{aligned}\varphi(x) = \psi(x) = t^{\mathbf{M}}(x) &= t^{\mathbf{M}}(x(a_1), x(a_2), \dots, x(a_m)) \\ &= x(t^{\mathbf{A}}(a_1, a_2, \dots, a_m)) = e_{\mathbf{A}}(t^{\mathbf{A}}(a_1, a_2, \dots, a_m))(x).\end{aligned}$$

Thus $e_{\mathbf{A}}$ is surjective. ■

We would like to obtain a duality for $\mathcal{A}$ in two steps: first show that $\underset{\sim}{\mathbf{M}}$ yields a duality on $\mathcal{A}_{\text{fin}}$, for example by verifying that (IC) holds, and then apply some general theory to show that the duality lifts automatically to a duality on the whole of $\mathcal{A}$. Our next two theorems shows that this is achievable provided $\underset{\sim}{\mathbf{M}}$ enjoys some degree of finiteness.

If $T \subseteq S$, then $\pi_T \colon \underset{\sim}{\mathbf{M}}^S \to \underset{\sim}{\mathbf{M}}^T$ denotes the natural restriction map given by $\pi_T(x) := x{\restriction}_T$ for all $x \in M^S$. Clearly, π_T is a morphism for all choices of $\underset{\sim}{\mathbf{M}}$.

2.6 Lemma *Assume that $\underset{\sim}{\mathbf{M}} = \langle M; R, \mathcal{T}\rangle$ where R is finite. Let $\mathbf{X}$ be a closed substructure of $\underset{\sim}{\mathbf{M}}^S$, for some non-empty set S, and let $\alpha \colon \mathbf{X} \to \underset{\sim}{\mathbf{M}}$ be a morphism. Then there exist a finite subset T of S and a unique morphism $\alpha' \colon \pi_T(\mathbf{X}) \to \underset{\sim}{\mathbf{M}}$ such that $\alpha = \alpha' \circ \pi_T$.*

Proof Let $\mathbf{X}$ be a closed substructure of $\underset{\sim}{\mathbf{M}}^S$ and let $\alpha \colon \mathbf{X} \to \underset{\sim}{\mathbf{M}}$ be a morphism. Let $\Delta_M := \{(a, a) \mid a \in M\}$ be the diagonal relation on M and define $R^* := R \cup \{\Delta_M\}$. To simplify the notation, let $\boldsymbol{x}$ denote the n-tuple $(x_1, \dots, x_n)$ where each x_i belongs to M^S, let $\boldsymbol{x}(s)$ denote the n-tuple $(x_1(s), \dots, x_n(s))$ and let $\alpha(\boldsymbol{x})$ denote the n-tuple $(\alpha(x_1), \dots, \alpha(x_n))$. For each ($n$-ary) relation $r \in R^*$, define

$$C_r := \{\, \boldsymbol{x} \in X^n \mid \alpha(\boldsymbol{x}) \notin r \,\} \subseteq (\underset{\sim}{\mathbf{M}}^S)^n.$$

Since α is continuous and r is open in (the discrete topology on) M^n, it follows that C_r is closed in $\mathbf{X}^n$. Let $\boldsymbol{x} \in C_r$. Since α preserves r we have $\boldsymbol{x} \notin r^{\mathbf{X}}$ and hence $\boldsymbol{x}(s) \notin r$ for some $s \in S$. Consequently, C_r is covered by the clopen sets

$$C_r(\boldsymbol{x}, s) := \{\, \boldsymbol{y} \in (\underset{\sim}{\mathbf{M}}^S)^n \mid \boldsymbol{y}(s) = \boldsymbol{x}(s) \notin r \,\}$$

and by compactness finitely many of these sets suffice. Thus, there is a finite subset T_r of S such that

$$(\forall \boldsymbol{x} \in C_r)(\exists s \in T_r)\ \boldsymbol{x}(s) \notin r.$$

Let T be the union of the sets T_r, where $r \in R^*$, and let $\pi_T \colon \underset{\sim}{\mathbf{M}}^S \to \underset{\sim}{\mathbf{M}}^T$ be the natural restriction map. Since R^* is finite, so is T. As Δ_M is in R^*, we have

$$\alpha(x) \neq \alpha(y) \implies \pi_T(x) \neq \pi_T(y), \text{ for all } x, y \in X,$$

that is $\ker \pi_T{\restriction}_X \subseteq \ker \alpha$. Thus there is a unique map $\alpha' \colon \pi_T(X) \to M$ satisfying $\alpha = \alpha' \circ \pi_T$.

We must now show that α' preserves the relations in R. Let $r \in R$ be n-ary and let $\boldsymbol{y} \in \pi_T(X)^n$ with $\alpha'(\boldsymbol{y}) \notin r$. Choose $\boldsymbol{x} \in X^n$ so that $\boldsymbol{y} = \pi_T(\boldsymbol{x})$. Then

$$\alpha(\boldsymbol{x}) = (\alpha' \circ \pi_T)(\boldsymbol{x}) = \alpha'(\boldsymbol{y}) \notin r$$

and consequently $\boldsymbol{x} \in C_r$. It follows that there is an $s \in T$ such that $\boldsymbol{y}(s) = \boldsymbol{x}(s) \notin r$, whence $\boldsymbol{y} \notin r^{\underset{\sim}{\mathbf{M}}^T}$. Hence, α' preserves r. ∎

If $\underset{\sim}{\mathbf{M}} = \langle M; G, R, \mathcal{T}\rangle$, that is, every operation of $\underset{\sim}{\mathbf{M}}$ is total rather than partial, then we call $\underset{\sim}{\mathbf{M}}$ a **total structure**.

2.7 Second Duality Theorem *Assume that $\underset{\sim}{\mathbf{M}}$ is a total structure with R finite. If* (IC) *holds, then $\underset{\sim}{\mathbf{M}}$ yields a duality on $\mathcal{A}$ and is injective in $\mathfrak{X}$.*

Proof Assume that $\underset{\sim}{\mathbf{M}} = \langle M; G, R, \mathcal{T}\rangle$ is a total structure with R finite and that (IC) holds. By Lemma 2.2.5 and the First Duality Theorem, it suffices to show that if $\underset{\sim}{\mathbf{M}}$ is injective in $\mathfrak{X}_{\text{fin}}$ then it is injective in $\mathfrak{X}$.

Let $\mathbf{X}$ be a closed substructure of $\underset{\sim}{\mathbf{M}}^S$ and let $\alpha \colon \mathbf{X} \to \underset{\sim}{\mathbf{M}}$ be a morphism. By Lemma 2.2.6, there exist a finite subset T of S and a unique R-preserving map $\alpha' \colon \pi_T(\mathbf{X}) \to \underset{\sim}{\mathbf{M}}$ such that $\alpha = \alpha' \circ \pi_T$. (See Figure 2.2.) Since π_T is a morphism, $\pi_T(X)$ is closed under the (total) operations in G. (It is at this point that we need $H = \varnothing$ as the image $\pi_T(X)$ of the morphism π_T need not be a substructure of $\underset{\sim}{\mathbf{M}}^T$ in the presence of partial operations!) A simple calculation, using the fact that both α and π_T preserve the operations in G, shows that α' also preserves the operations in G. Thus α' is a morphism.

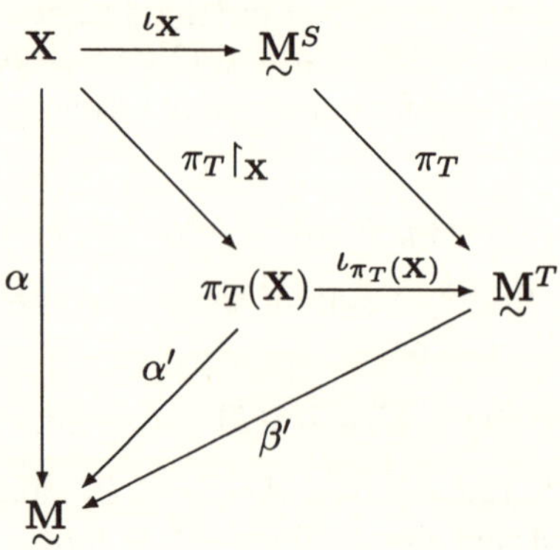

Figure 2.2 $\beta' \circ \pi_T$ extends α

If $\underset{\sim}{\mathbf{M}}$ is injective in $\mathfrak{X}_{\text{fin}}$, there exists a morphism $\beta'\colon \underset{\sim}{\mathbf{M}}^T \to \underset{\sim}{\mathbf{M}}$ which extends α' and then $\beta := \beta' \circ \pi_T$ is the required extension of α, whence $\underset{\sim}{\mathbf{M}}$ is injective in $\mathfrak{X}$. ■

This result is rather surprising. It gives us simple finitary conditions which yield both a dual adjunction between the categories $\mathcal{A}$ and $\mathfrak{X}$ and a topological representation of each algebra in $\mathcal{A}$, but it requires us to do no category theory and no topology! While (IC) is sufficient, under certain circumstances, to show that $\underset{\sim}{\mathbf{M}}$ yields a duality on $\mathcal{A}$, it is not necessary (see Exercise 4.11 and Lemma 7.8.6).

The final theorem in this section gives a reduction to the finite case without establishing (IC) and therefore yields no information on the injectivity of $\underset{\sim}{\mathbf{M}}$ in $\mathfrak{X}$. Its proof requires two preliminary lemmas. Recall from Chapter 1 that if $x_1, \dots, x_n\colon \mathbf{A} \to \underline{\mathbf{M}}$ are homomorphisms, then the homomorphism $x_1 \sqcap \cdots \sqcap x_n\colon \mathbf{A} \to \underline{\mathbf{M}}^n$ is defined by

$$(x_1 \sqcap \cdots \sqcap x_n)(a) := (x_1(a), \dots, x_n(a)) \text{ for all } a \in A.$$

If r is an n-ary relation in $\mathcal{B}$, then we denote r *qua* subalgebra of $\underline{\mathbf{M}}^n$ by $\mathbf{r}$ and $\rho_i\colon \mathbf{r} \to \underline{\mathbf{M}}$ denotes the ith coordinate projection given by $\rho_i(c_1, \dots, c_n) := c_i$ for all $(c_1, \dots, c_n) \in r$. We shall need the following simple but important observation.

2.8 Lemma *Let $\mathbf{A} \in \mathcal{A}$ and let $r \in \mathcal{B}$ be n-ary.*

(i) *Let $x_1, \dots, x_n \in D(\mathbf{A})$. Then $(x_1, \dots, x_n) \in r^{D(\mathbf{A})}$ if and only if the image of $x_1 \sqcap \cdots \sqcap x_n\colon \mathbf{A} \to \underline{\mathbf{M}}^n$ is a subset of r (whence $x_1 \sqcap \cdots \sqcap x_n\colon \mathbf{A} \to \mathbf{r}$ is a well-defined homomorphism).*

(ii) *If $x\colon \mathbf{A} \to \mathbf{r}$ is a homomorphism, then $(x_1, \dots, x_n) \in r^{D(\mathbf{A})}$ where $x_i := \rho_i \circ x$ for all i.*

Proof (i) is an immediate consequence of the fact that $r^{D(\mathbf{A})}$ is defined pointwise while (ii) follows from the fact that $\rho_1 \circ x \sqcap \cdots \sqcap \rho_n \circ x = x$ for every homomorphism $x\colon \mathbf{A} \to \mathbf{r}$. ■

To prove our next lemma we will again use the Inverse Limit Theorem 1.3.3. An alternative proof using the Compactness Theorem of First Order Logic is outlined in Exercise 2.8.

2.9 Lemma *Let $\mathbf{B}$ be a subalgebra of a locally finite algebra $\mathbf{A}$, let $\mathbf{D}$ be a finite algebra and let $h\colon \mathbf{B} \to \mathbf{D}$ be a homomorphism. If for every finite subalgebra $\mathbf{F}$ of $\mathbf{A}$ which intersects $\mathbf{B}$ there is a homomorphism $k\colon \mathbf{F} \to \mathbf{D}$ which agrees with h on $B \cap F$, then there is a homomorphism $g\colon \mathbf{A} \to \mathbf{D}$ which extends h.*

Proof Since $\mathbf{A}$ is locally finite, the set $\mathcal{F}$ of all finite subalgebras of $\mathbf{A}$ which intersect $\mathbf{B}$ is directed by inclusion. Our hypothesis says that, for each $\mathbf{F} \in \mathcal{F}$, the set

$$\mathcal{A}_{\mathbf{F}} := \{\, k\colon \mathbf{F} \to \mathbf{D} \mid k \text{ is a homomorphism which extends } h{\restriction}_{B\cap F} \,\}$$

is non-empty. For $\mathbf{F}, \mathbf{G} \in \mathcal{F}$ with $G \supseteq F$ define $\Phi_{G,F}(k) := k{\restriction}_F$ for all $k \in \mathcal{A}_{\mathbf{G}}$. This defines an inverse limit system of finite sets $\mathcal{A}_{\mathbf{F}}$ and maps $\Phi_{G,F} : \mathcal{A}_{\mathbf{G}} \to \mathcal{A}_{\mathbf{F}}$ between them. By the Inverse Limit Theorem 1.3.3 there is an element $(k_{\mathbf{F}})_{\mathbf{F}\in\mathcal{F}}$ of the inverse limit. For each $a \in A$ 'define' $g(a) := k_{\mathbf{F}}(a)$ where $\mathbf{F}$ is any member of $\mathcal{F}$ such that $a \in F$. Now it is easy to check that $g\colon \mathbf{A} \to \mathbf{D}$ is a well-defined homomorphism which extends h. ■

This lemma has an interesting corollary.

2.10 Corollary *If $\mathbf{D} \in \mathcal{A}_{\text{fin}}$ is injective in $\mathcal{A}_{\text{fin}}$, then $\mathbf{D}$ is injective in $\mathcal{A}$.*

The following important result was proved by R. Willard (private communication) and independently by L. Zádori [Zad95]. Our proof is a modified version of Willard's proof. Zádori's proof is given in Chapter 10 (see 10.6.4). If $\underset{\sim}{\mathbf{M}} = \langle M; G, H, R, \mathcal{T}\rangle$ and G, H and R are finite, then we say that $\underset{\sim}{\mathbf{M}}$ is of **finite type**.

2.11 Duality Compactness Theorem *If $\underset{\sim}{\mathbf{M}}$ is of finite type and yields a duality on $\mathcal{A}_{\text{fin}}$, then $\underset{\sim}{\mathbf{M}}$ yields a duality on $\mathcal{A}$.*

Proof Assume that $\underset{\sim}{\mathbf{M}}$ is of finite type and yields a duality on $\mathcal{A}_{\text{fin}}$. By Lemma 2.1.2, without loss of generality we may replace each (partial) operation by its graph and assume that $\underset{\sim}{\mathbf{M}} = \langle M; R, \mathcal{T}\rangle$ with R finite.

Let $\mathbf{A} \in \mathcal{A}$. We must show that $e_{\mathbf{A}}\colon \mathbf{A} \to ED(\mathbf{A})$ is surjective, that is, every morphism $\alpha\colon D(\mathbf{A}) \to \underset{\sim}{\mathbf{M}}$ is of the form $e_{\mathbf{A}}(a)$ for some $a \in A$. Let $\alpha\colon D(\mathbf{A}) \to \underset{\sim}{\mathbf{M}}$ be a morphism. Since $D(\mathbf{A})$ is a closed subset of M^A and since α is continuous, by Lemma B.6(iv) there is a finite subset S of A such that, for all $x_1, x_2 \in D(\mathbf{A})$, we have $\alpha(x_1) = \alpha(x_2)$ whenever $x_1{\restriction}_S = x_2{\restriction}_S$. Let $\mathbf{B}$ be the subalgebra of $\mathbf{A}$ generated by S. By Lemma 1.3.2, $\mathbf{B}$ is finite.

Let $r \in R$ be n-ary and let $y_1, \dots, y_n \in D(\mathbf{B})$ with $(y_1, \dots, y_n) \in r^{D(\mathbf{B})}$. Then, by Lemma 2.2.8, $y := y_1 \sqcap \dots \sqcap y_n\colon \mathbf{B} \to \mathbf{r}$ is a well-defined homomorphism. If $\mathbf{B}$ is a subalgebra of $\mathbf{B}'$ and $z_i \in D(\mathbf{B}')$ is an extension of y_i (for $i = 1, \dots, n$) with $(z_1, \dots, z_n) \in r^{D(\mathbf{B}')}$, then, by Lemma 2.2.8 again, $z := z_1 \sqcap \dots \sqcap z_n\colon \mathbf{B}' \to \mathbf{r}$ is a well-defined extension of y. Conversely, if $z\colon \mathbf{B}' \to \mathbf{r}$ is an extension of y, then $z_i := \rho_i \circ z \in D(\mathbf{B}')$ is an extension of y_i for each i and $(z_1, \dots, z_n) \in r^{D(\mathbf{B}')}$.

Now assume that the $(n+1)$-tuple $ry_1 \dots y_n$ satisfies the following non-extendability condition:

(NE) there exist no homomorphisms $x_1, \dots, x_n \in D(\mathbf{A})$ which satisfy $x_i{\restriction}_B = y_i$ for all i and also satisfy $(x_1, \dots, x_n) \in r^{D(\mathbf{A})}$.

Thus, by the remarks in the paragraph above with $\mathbf{B}' = \mathbf{A}$ and $z = x$, the homomorphism $y\colon \mathbf{B} \to \mathbf{r}$ has no extension to a homomorphism $x\colon \mathbf{A} \to \mathbf{r}$. By Lemma 2.2.9 (with $\mathbf{D} = \mathbf{r}$, $h = y$ and $g = x$), there is a finite subalgebra $\mathbf{F}$ of $\mathbf{A}$ such that there is no homomorphism $z\colon \mathbf{F} \to \mathbf{r}$ which agrees with y on $B \cap F$. Since B is finite, we may assume that $B \subseteq F$ (simply replace $\mathbf{F}$ by the subalgebra generated by $B \cup F$), hence that the homomorphism $y\colon \mathbf{B} \to \mathbf{r}$ has no extension to a homomorphism $z\colon \mathbf{F} \to \mathbf{r}$. Consequently, by the remarks in the paragraph above, there exist no homomorphisms $z_1, \dots, z_n \in D(\mathbf{F})$ such that z_i extends y_i for all i and such that $(z_1, \dots, z_n) \in r^{D(\mathbf{F})}$. Denote this subalgebra $\mathbf{F}$ of $\mathbf{A}$ by $\mathbf{C}_{ry_1\dots y_n}$.

Define $\mathbf{C}$ to be the subalgebra of $\mathbf{A}$ generated by

$$\bigcup\{\, C_{ry_1\dots y_n} \mid r \in R,\ y_1, \dots, y_n \in D(\mathbf{B}), \text{ and } ry_1 \dots y_n \text{ satisfies (NE)}\}.$$

Since R and $D(\mathbf{B})$ are finite, $\mathbf{C}$ is finitely generated and hence is finite. We claim that the map $\alpha'\colon D(\mathbf{C}) \to \underset{\sim}{\mathbf{M}}$ given by

$$\alpha'(z) := \begin{cases} \alpha(x) & \text{if there exists } x \in D(\mathbf{A}) \text{ with } x{\restriction}_B = z{\restriction}_B, \\ \text{arbitrary} & \text{otherwise,} \end{cases}$$

is a well-defined morphism. If $x_1, x_2 \in D(\mathbf{A})$ with $x_1{\restriction}_B = x_2{\restriction}_B$, then $\alpha(x_1) = \alpha(x_2)$ by the choice of the subalgebra $\mathbf{B}$ of $\mathbf{A}$; thus α' is well defined. In order to show that α' preserves an n-ary relation $r \in R$, let $z_1, \dots, z_n \in D(\mathbf{C})$ with $(z_1, \dots, z_n) \in r^{D(\mathbf{C})}$. Define $y_i := z_i{\restriction}_B \in D(\mathbf{B})$. Then $(y_1, \dots, y_n) \in r^{D(\mathbf{B})}$ and z_i is an extension of y_i to $\mathbf{C}$, for each i, such that $(z_1, \dots, z_n) \in r^{D(\mathbf{C})}$. The construction of $\mathbf{C}$ guarantees that each y_i has an extension to a homomorphism $x_i \in D(\mathbf{A})$ with $(x_1, \dots, x_n) \in r^{D(\mathbf{A})}$. Since $x_i{\restriction}_B = y_i = z_i{\restriction}_B$ and since α preserves r, we conclude that

$$(\alpha'(z_1), \dots, \alpha'(z_n)) = (\alpha(x_1), \dots, \alpha(x_n)) \in r,$$

whence α' preserves r. Thus $\alpha'\colon D(\mathbf{C}) \to \underset{\sim}{\mathbf{M}}$ is a morphism. Since $\underset{\sim}{\mathbf{M}}$ yields a duality on $\mathcal{A}_{\text{fin}}$ and since $\mathbf{C} \in \mathcal{A}_{\text{fin}}$, there is some $a \in C$ such that $\alpha' = e_{\mathbf{C}}(a)$. Let $x \in D(\mathbf{A})$ and define $z := x{\restriction}_C$. Then

$$e_{\mathbf{A}}(a)(x) = x(a) = z(a) = e_{\mathbf{C}}(a)(z) = \alpha'(z) = \alpha(x) \text{ as } x{\restriction}_B = z{\restriction}_B.$$

Thus $\alpha = e_{\mathbf{A}}(a)$, as required. Hence $\underset{\sim}{\mathbf{M}}$ yields a duality on $\mathcal{A}$. ■

Combined with Lemma 2.2.5, the Duality Compactness Theorem produces a variant of the Second Duality Theorem.

2.12 Corollary *Assume that* $\underset{\sim}{\mathrm{M}}$ *is of finite type. If* (IC) *holds, then* $\underset{\sim}{\mathrm{M}}$ *yields a duality on* $\mathcal{A}$ *and is injective in* $\mathfrak{X}_{\mathrm{fin}}$.

In every known example of a natural duality we have been able to use a structure $\underset{\sim}{\mathrm{M}}$ of finite type. Hence, we close this section by posing a fundamental question.

2.13 Finite Type Problem *Is there a finite algebra* $\underline{\mathrm{M}}$ *for which some choice of* $\underset{\sim}{\mathrm{M}}$ *(and therefore* $\underset{\sim}{\mathrm{M}} = \langle M; \mathcal{B}, \mathcal{T}\rangle$*) yields a duality on* $\mathcal{A}$ *but no choice of* $\underset{\sim}{\mathrm{M}}$ *of finite type yields a duality on* $\mathcal{A}$*?*

2.3 Taming brute force with near-unanimity

In the light of the Second Duality Theorem, the natural question now is, *'How can we force* (IC) *to hold?'* The answer is, *'Use brute force!'*

3.1 Brute Force Duality Theorem *Brute force yields a duality on* $\mathcal{A}_{\mathrm{fin}}$. *Indeed, if* $\underset{\sim}{\mathrm{M}} = \langle M; \mathcal{B}, \mathcal{T}\rangle$, *then* (IC) *holds and therefore* $\underset{\sim}{\mathrm{M}}$ *yields a duality on* $\mathcal{A}_{\mathrm{fin}}$ *and* $\underset{\sim}{\mathrm{M}}$ *is injective in* $\mathfrak{X}_{\mathrm{fin}}$.

Proof Let $X \subseteq M^n$ and let $\alpha\colon \mathbf{X} \to \underset{\sim}{\mathrm{M}}$ be a morphism. Since M is finite, we have $X = \{x_1, \dots, x_k\}$. Denote the k-tuples $(x_1, \dots, x_k)$ and $(x_1(i), \dots, x_k(i))$ by $\boldsymbol{x}$ and $\boldsymbol{x}(i)$ respectively. Define r to be the subalgebra of $\underline{\mathrm{M}}^k$ generated by $\{\boldsymbol{x}(1), \dots, \boldsymbol{x}(n)\}$. By construction, we have $(x_1(i), \dots, x_k(i)) = \boldsymbol{x}(i) \in r$ for all i whence $\boldsymbol{x} = (x_1, \dots, x_k) \in r^{\mathbf{X}}$. Since α preserves r, it follows that $\alpha(\boldsymbol{x}) := (\alpha(x_1), \dots, \alpha(x_k)) \in r$. Since r is generated by $\{\boldsymbol{x}(1), \dots, \boldsymbol{x}(n)\}$, there is an n-ary term function t on $\underline{\mathrm{M}}$ such that $\alpha(\boldsymbol{x}) = t(\boldsymbol{x}(1), \dots, \boldsymbol{x}(n))$. But this says precisely that

$$\alpha(x_i) = t(x_i(1), \dots, x_i(n)) = t(x_i) \text{ for all } i = 1, \dots, k,$$

and thus t extends α and the proof of (IC) is complete. ■

Even though brute force always yields a duality on the finite algebras in $\mathcal{A}$, it does not always yield a duality on $\mathcal{A}$. We shall see in Chapter 10 that not even brute force will yield a duality on the variety of implication algebras $\mathcal{A} := \mathbb{ISP}\underline{\mathrm{I}}$, where $\underline{\mathrm{I}} := \langle\{0,1\}; \to, 1\rangle$ and $\to$ is the usual Boolean implication. If we are willing to use a proper *class* of possibly *infinitary* relations, we can obtain a duality for $\mathcal{A}$ without the use of any topology. Let $\mathcal{B}_\infty$ denote the class consisting of all subalgebras of $\underline{\mathrm{M}}^\kappa$ where κ ranges over all non-zero cardinals. Then a simple modification of the proof of 2.3.1 shows that $\underset{\sim}{\mathrm{M}}_\infty := \langle M; \mathcal{B}_\infty\rangle$ yields a duality on $\mathcal{A}$. This is the **ultimate brute force** construction. (See also Exercise 2.5, where we check that the omission of topology from $\underset{\sim}{\mathrm{M}}_\infty$ is not a typo!)

While brute force may not always yield a duality on $\mathcal{A}$, 2.2.3, 2.2.5 and 2.3.1 combine to show that brute force always yields a duality on the free algebras in $\mathcal{A}$. This gives a standard result of clone theory as an immediate corollary.

3.2 Corollary *If $\underline{\mathbf{M}}$ is a finite algebra, then the set $\mathcal{B}$ of all finitary algebraic relations on $\underline{\mathbf{M}}$ determines the clone of $\underline{\mathbf{M}}$, that is, for all $n \in \mathbb{N}$, a map $t\colon M^n \to M$ is a term function of $\underline{\mathbf{M}}$ if and only if it preserves all finitary algebraic relations on $\underline{\mathbf{M}}$.*

In general, a finite set R of relations will not suffice to determine the clone of $\underline{\mathbf{M}}$, for example, the two-element implication algebra $\underline{\mathbf{I}}$ requires an infinite set R to determine its clone (see Exercise 10.13). In this context it is worth reiterating that in every known example of a natural duality, we have been able to use a structure $\underset{\sim}{\mathbf{M}}$ of finite type.

The Second Duality Theorem and the Brute Force Duality Theorem are in a tug-of-war. The former says that (IC) will give us duality if we use only finitely many (partial) operations and relations. The latter tells us that we get (IC) if we are willing to put in all possible relations. In order to bridge the gap, we need a condition which will ensure that preservation of certain finite sets of relations guarantees preservation of all algebraic relations. Such a condition was discovered by K. Baker and A. Pixley [BP75], quite independently of duality theory. They call a $(k{+}1)$-ary term $n(x_1, \dots, x_{k+1})$, with k at least 2, a **near-unanimity term** on $\underline{\mathbf{M}}$ if $\underline{\mathbf{M}}$ satisfies the identities

$$n(x, \dots, x, y) \approx n(x, \dots, x, y, x) \approx \cdots \approx n(y, x, \dots, x) \approx x.$$

A ternary near-unanimity term on $\underline{\mathbf{M}}$ is called a **majority term**. For example, on any algebra with an underlying lattice structure, the median,

$$m(x, y, z) := (x \wedge y) \vee (y \wedge z) \vee (z \wedge x)$$

is a majority term since it satisfies the identities

$$m(x, x, y) \approx m(x, y, x) \approx m(y, x, x) \approx x.$$

The significance of a near-unanimity term from our perspective is that its presence gives us the full benefit of brute force at a very small price.

3.3 NU Lemma (K. Baker and A. Pixley [BP75]) *Let $k \geqslant 2$ and assume that $\underline{\mathbf{M}}$ has a $(k{+}1)$-ary near-unanimity term. Let X be a subset of M^m and let $\alpha\colon X \to M$ be a map that preserves every k-ary relation in $\mathcal{B}$. Then α preserves every relation in $\mathcal{B}$.*

Proof Assume that α preserves every k-ary relation in $\mathcal{B}$. Certainly, α preserves every relation in $\mathcal{B}$ of arity not more than k. We prove by induction that α preserves every $(l+1)$-ary relation r in $\mathcal{B}$, where $l \geq k$. Choose $\boldsymbol{x} = (x_0, x_1, \dots, x_l) \in X^{l+1}$ such that, for $i = 1, 2, \dots, m$, we have

$$\boldsymbol{x}(i) = (x_0(i), x_1(i), \dots, x_l(i)) \in r.$$

By induction, α preserves the induced, one-factor-deleted l-ary relations

$$r_j := (\pi_0 \times \cdots \times \pi_{j-1} \times \pi_{j+1} \times \cdots \times \pi_l)(r) \subseteq M^l$$

for $0 \leq j \leq l$. Consequently

$$(\alpha(x_0), \ldots, \alpha(x_{j-1}), \alpha(x_{j+1}), \ldots, \alpha(x_l)) \in r_j$$

for $0 \leq j \leq l$. Thus there is an $a_j \in M$ so that

$$\boldsymbol{y}_j := (\alpha(x_0), \ldots, \alpha(x_{j-1}), a_j, \alpha(x_{j+1}), \ldots, \alpha(x_l)) \in r.$$

Let n be a $(k+1)$-ary near-unanimity term for $\underline{\mathbf{M}}$. Since r is a subalgebra of $\underline{\mathbf{M}}^{l+1}$, we obtain

$$\alpha(\boldsymbol{x}) = (\alpha(x_0), \alpha(x_1), \ldots, \alpha(x_l)) = n(\boldsymbol{y}_0, \boldsymbol{y}_1, \ldots, \boldsymbol{y}_k) \in r,$$

and hence α preserves r. ■

The theorem below is extremely useful. We shall see in Chapter 3 that it always can be extended to a full duality and in Chapter 10 that it has a very strong converse.

3.4 NU Duality Theorem *Let $k \geqslant 2$ and assume that $\underline{\mathbf{M}}$ has a $(k+1)$-ary near-unanimity term. Then $\underset{\sim}{\mathbf{M}} = \langle M; \mathbb{S}(\underline{\mathbf{M}}^k), \mathcal{T}\rangle$ satisfies* (IC), *yields a duality on $\boldsymbol{\mathcal{A}}$ and is injective in $\boldsymbol{\mathfrak{X}}$.*

Proof Let X be a subset of M^m and let $\alpha\colon X \to M$ preserve every k-ary algebraic relation on $\underline{\mathbf{M}}$. By the NU Lemma α preserves every finitary algebraic relation on $\underline{\mathbf{M}}$. By the Brute Force Duality Theorem, α extends to a term function. Thus (IC) holds for $\underset{\sim}{\mathbf{M}}$ and, by the Second Duality Theorem, $\underset{\sim}{\mathbf{M}}$ yields a duality on $\boldsymbol{\mathcal{A}}$ and is injective in $\boldsymbol{\mathfrak{X}}$. ■

3.5 Distributive Lattices Revisited We saw in Chapter 1 that $\underset{\sim}{\mathbf{D}} := \langle\{0,1\}; 0, 1, \leqslant, \mathcal{T}\rangle$ yields a full duality on the class $\boldsymbol{\mathcal{D}} = \mathbb{ISP}\underline{\mathbf{D}}$ of distributive lattices. The fact that $\underset{\sim}{\mathbf{D}}$ yields a duality is a very simple consequence of the theory we have developed so far. Since $\underline{\mathbf{D}}$ is a lattice, it has a majority term and thus the NU Duality Theorem applies with $k = 2$. By inspection, we can see that the only binary algebraic relations on $\underline{\mathbf{D}}^2$ are

$$\{(0,0)\},\ \{(0,1)\},\ \{(1,0)\},\ \{(1,1)\},\ \{0\} \times D,$$
$$\{1\} \times D,\ D \times \{0\},\ D \times \{1\},\ \Delta_D,\ \leqslant,\ \geqslant,\ D^2.$$

Let $X \subseteq D^n$ and let $\alpha\colon X \to D$ be a map. Then α must preserve the trivial relations Δ_D and D^2. Moreover, it is easily seen that if α preserves the constants 0 and 1 then it will preserve the first eight relations in the list, and, of course, if α preserves $\leqslant$ then it preserves the converse relation $\geqslant$. Consequently, if α preserves 0, 1, and $\leqslant$, then it preserves every subalgebra of $\underline{\mathbf{D}}^2$. Since $\mathbb{S}(\underline{\mathbf{D}}^2)$ yields a duality on $\mathcal{D}$, we see at once that the structure $\underset{\sim}{\mathrm{D}} = \langle D; 0, 1, \leqslant, \mathcal{T}\rangle$ yields a duality on $\mathcal{D}$. This establishes the algebraic half of Priestley's duality.

2.4 Refining a duality via entailment

The NU Duality Theorem is a very powerful tool. It tells us, in particular, that if $\underline{\mathbf{M}}$ is a finite algebra, which has an underlying lattice structure, then $\underset{\sim}{\mathrm{M}} := \langle M; \mathbb{S}(\underline{\mathbf{M}}^2), \mathcal{T}\rangle$ yields a natural duality on $\mathcal{A} := \mathbb{ISP}\underline{\mathbf{M}}$. Unfortunately, it has one drawback: even when M is small $\mathbb{S}(\underline{\mathbf{M}}^2)$ can be monstrous and so yield a duality of no practical use. We must look for ways to delete relations from $\mathbb{S}(\underline{\mathbf{M}}^2)$, without destroying the duality, until a minimal subset is obtained. The analysis of the distributive lattice case in 2.3.5, while far simpler than most examples, is typical of what is involved. Starting from the 12 sublattices of $\underline{\mathbf{D}}^2$ along with the 3 sublattices of $\underline{\mathbf{D}}$, we deleted all relations bar 0, 1, and $\leqslant$ without destroying the duality given by the NU Duality Theorem. It is straightforward to show that this is an optimal duality, that is, if any further relation were deleted, the duality would be destroyed. In fact, much more is true. For example, amongst subsets of $\mathbb{S}(\underline{\mathbf{D}}) \cup \mathbb{S}(\underline{\mathbf{D}}^2)$, any set R which yields a duality on the class of distributive lattices must include either $\leqslant$ or $\geqslant$. (See Exercise 8.7.) In the remainder of this section we shall introduce just enough of the theory to facilitate our discussion of examples in Chapter 4.

4.1 Entailment Let $\underset{\sim}{\mathrm{M}} = \langle M; G, H, R, \mathcal{T}\rangle$ and let s be a fixed finitary algebraic relation or (partial) operation on $\underline{\mathbf{M}}$. Given $\mathbf{A} \in \mathcal{A}$, we say that $G \cup H \cup R$, or simply $\underset{\sim}{\mathrm{M}}$, **entails** s **on** $D(\mathbf{A})$ if every continuous map α from $D(\mathbf{A})$ to M which preserves the relations and (partial) operations in $G \cup H \cup R$ also preserves s. The set $G \cup H \cup R$ is said to **entail** s, in symbols, $G \cup H \cup R \vdash s$, if $G \cup H \cup R$ entails s on $D(\mathbf{A})$ for all $\mathbf{A} \in \mathcal{A}$. Let $\underset{\sim}{\mathrm{M}} = \langle M; G, H, R, \mathcal{T}\rangle$ and $\underset{\sim}{\mathrm{M}}' = \langle M; G', H', R', \mathcal{T}\rangle$. If $G \cup H \cup R \vdash s$ for all $s \in G' \cup H' \cup R'$, then we say that $G \cup H \cup R$ **entails** $G' \cup H' \cup R'$, or $\underset{\sim}{\mathrm{M}}$ **entails** $\underset{\sim}{\mathrm{M}}'$, and write $(G \cup H \cup R) \vdash (G' \cup H' \cup R')$. The following lemma is a simple but frequently used consequence of the definition.

4.2 $\underset{\sim}{\mathrm{M}}$-Shift Duality Lemma *If $\underset{\sim}{\mathrm{M}}$ entails $\underset{\sim}{\mathrm{M}}'$ and $\underset{\sim}{\mathrm{M}}'$ yields a duality on some subclass $\mathcal{C}$ of $\mathcal{A}$, then $\underset{\sim}{\mathrm{M}}$ also yields a duality on $\mathcal{C}$.*

Proof Define $\mathcal{X} = \mathbb{IS}_c\mathbb{P}^+ \underset{\sim}{\mathrm{M}}$ and $\mathcal{X}' = \mathbb{IS}_c\mathbb{P}^+ \underset{\sim}{\mathrm{M}}'$ and let $D\colon \mathcal{A} \to \mathcal{X}$, $E\colon \mathcal{X} \to \mathcal{A}$, $D'\colon \mathcal{A} \to \mathcal{X}'$, $E'\colon \mathcal{X}' \to \mathcal{A}$ be the hom-functors obtained from $\underset{\sim}{\mathrm{M}}$ and $\underset{\sim}{\mathrm{M}}'$ respectively. Assume that $\underset{\sim}{\mathrm{M}}$ entails $\underset{\sim}{\mathrm{M}}'$ and that $\underset{\sim}{\mathrm{M}}'$ yields a duality on $\mathcal{C}$. Let $\mathbf{A} \in \mathcal{C}$. Then, by the definition of entailment,

$\mathfrak{X}(D(\mathbf{A}),\underset{\sim}{\mathbf{M}}) \subseteq \mathfrak{X}'(D'(\mathbf{A}),\underset{\sim}{\mathbf{M}}')$. Since $\underset{\sim}{\mathbf{M}}'$ yields a duality on $\mathbf{A}$, every morphism from $D'(\mathbf{A})$ to $\underset{\sim}{\mathbf{M}}'$ is an evaluation. Hence

$$ED(\mathbf{A}) = \mathfrak{X}(D(\mathbf{A}),\underset{\sim}{\mathbf{M}}) \subseteq \mathfrak{X}'(D'(\mathbf{A}),\underset{\sim}{\mathbf{M}}') = E'D'(\mathbf{A}) = e_{\mathbf{A}}(\mathbf{A}).$$

Consequently every morphism from $D(\mathbf{A})$ to $\underset{\sim}{\mathbf{M}}$ is also an evaluation, whence $\underset{\sim}{\mathbf{M}}$ yields a duality on $\mathbf{A}$. Hence, $\underset{\sim}{\mathbf{M}}$ yields a duality on $\mathcal{C}$. ∎

This lemma allows us to combine the Brute Force Duality Theorem and the Duality Compactness Theorem 2.2.11.

4.3 Duality and Entailment Theorem *Let* $\underset{\sim}{\mathbf{M}} = \langle M; G, H, R, \mathcal{T}\rangle$.

(i) *If* $\underset{\sim}{\mathbf{M}}$ *yields a duality on* $\mathcal{A}$, *then* $G \cup H \cup R$ *entails every finitary algebraic relation and every finitary algebraic (partial) operation on* $\underline{\mathbf{M}}$.

(ii) *If* $\underset{\sim}{\mathbf{M}}$ *is of finite type, then the following are equivalent:*

(a) $\underset{\sim}{\mathbf{M}}$ *yields a duality on* $\mathcal{A}$;

(b) $G \cup H \cup R$ *entails every finitary algebraic relation on* $\underline{\mathbf{M}}$;

(c) $G \cup H \cup R$ *entails every finitary algebraic partial operation on* $\underline{\mathbf{M}}$.

Proof If $\underset{\sim}{\mathbf{M}}$ yields a duality on $\mathcal{A}$, then, for all $\mathbf{A} \in \mathcal{A}$, the only morphisms from $D(\mathbf{A})$ to $\underset{\sim}{\mathbf{M}}$ are the evaluation maps. Since the evaluation maps preserve every finitary algebraic relation and (partial) operation, it follows at once that $G \cup H \cup R$ entails every finitary algebraic relation and (partial) operation. Thus (i) holds.

Note that (i) says that the implications (a) ⇒ (b) and (a) ⇒ (c) of (ii) hold. By Lemma 2.1.1, a map $\alpha \colon D(\mathbf{A}) \to M$ preserves a (partial) operation if and only if it preserves its graph. Thus (b) ⇒ (c) of (ii) holds.

Assume (c) to prove (b). Let s be an n-ary algebraic relation on $\underline{\mathbf{M}}$ and let $\rho_1 \colon \mathbf{s} \to \underline{\mathbf{M}}$ be the first projection. By (c), $G \cup H \cup R$ entails ρ_1. An easy calculation shows that a map $\alpha \colon D(\mathbf{A}) \to M$ preserves s if and only if it preserves ρ_1 and consequently $G \cup H \cup R$ entails s. Thus (c) ⇒ (b) holds.

Finally, assume (b). Since $G \cup H \cup R$ entails every relation $s \in \mathcal{B}$, it follows that $\underset{\sim}{\mathbf{M}}$ entails $\underset{\sim}{\mathbf{M}}'$ where $\underset{\sim}{\mathbf{M}}' := \langle M; \mathcal{B}, \mathcal{T}\rangle$. But, by the Brute Force Duality Theorem, $\underset{\sim}{\mathbf{M}}'$ yields a duality on $\mathcal{A}_{\text{fin}}$, whence the $\underset{\sim}{\mathbf{M}}$-Shift Duality Lemma implies that $\underset{\sim}{\mathbf{M}}$ yields a duality on $\mathcal{A}_{\text{fin}}$. As $\underset{\sim}{\mathbf{M}}$ is, by assumption, of finite type, Theorem 2.2.11 implies that $\underset{\sim}{\mathbf{M}}$ yields a duality on $\mathcal{A}$. Hence (b) ⇒ (a), which concludes the proof of (ii). ∎

In practice, when we prove that $G \cup H \cup R$ entails s it is usual to apply the following observation (which is a trivial consequence of the fact that $D(\mathbf{A})$ is a closed substructure of $\underset{\sim}{\mathbf{M}}^A$) and establish a slightly stronger but simpler condition:

4.4 Lemma *Let $\underset{\sim}{\mathbf{M}} = \langle M; G, H, R, \mathcal{T}\rangle$. Then $G \cup H \cup R$ entails s provided that, for each non-empty set S and each closed substructure $\mathbf{X}$ of $\underset{\sim}{\mathbf{M}}^S$, every morphism α from $\mathbf{X}$ to $\underset{\sim}{\mathbf{M}}$ preserves s.*

If we have some finite family Ω of relations which yields a duality on $\mathcal{A}$, typically $\Omega = \mathbb{S}(\underline{\mathbf{M}}^2)$, we would like to delete relations one at a time from Ω until we obtain a minimal subset R such that $R \vdash s$ for all $s \in \Omega \backslash R$. In this way, we obtain minimal subsets of Ω which yield a duality on $\mathcal{A}$. Such dualities will be referred to as **optimal dualities** and will be studied in detail in Chapter 8.

4.5 Constructs for Entailment The definition of 'entailment' does not provide any obvious way to determine if a relation s is entailed by even a finite set R since it quantifies over all $\mathbf{A} \in \mathcal{A}$. In practice we would like to know that if $G \cup H \cup R \vdash s$ then s can be obtained from a finite subset of $G \cup H \cup R$ by applying a finite number of constructs like intersection, product, converse and composition of (partial) functions.

By an **admissible construct** of type $\langle G, H, R\rangle$ we mean a rule which associates each structure $\underset{\sim}{\mathbf{M}} = \langle M; G, H, R, \mathcal{T}\rangle$ with an algebraic operation, partial operation or relation s that is entailed by $\underset{\sim}{\mathbf{M}}$. A set $\mathcal{C}$ of admissible constructs will be called **complete** if s can be constructed from a finite subset of $G \cup H \cup R$, by applying a finite number of the constructs in $\mathcal{C}$, whenever $G \cup H \cup R \vdash s$. For the moment we will give an extensive list of constructs that is sufficient to establish the examples of Chapter 4. While this list of constructs is not complete in general, it can be made complete by the addition of three further constructs (see the Completeness of Entailment Theorem 9.2.6). Many of the examples $\underline{\mathbf{M}}$ considered in Chapter 4 have the property that every non-trivial subalgebra of $\underline{\mathbf{M}}$ is subdirectly irreducible and $\underline{\mathbf{M}}$ generates a congruence distributive variety—for such algebras the constructs listed below are complete (see Theorem 9.3.4). In fact, each of our examples where we need to use these constructs has a ternary near-unanimity term and consequently, by the NU Duality Theorem, we can obtain a duality by using the set Ω which consists of all unary algebraic relations, all binary algebraic relations and all (partial) endomorphisms. Several of the constructs below become extremely simple when restricted to this set Ω. Note that we have listed the purely relational constructs first followed by those which involve (partial) operations.

(1) **Trivial relations** If θ is an equivalence relation on $\{1, \dots, n\}$ then any set of relations entails the relation

$$\Delta^\theta := \{(c_1, \dots, c_n) \in M^n \mid i\,\theta\, j \Rightarrow c_i = c_j\}.$$

This is an example of a nullary construct and includes as special cases the relation M^2 and the relation $\Delta_M := \{(c,c) \mid c \in M\}$, which may be viewed as the equality relation on M or as the graph of the endomorphism $\mathrm{id}_{\underline{\mathbf{M}}}$.

(2) **Subscript manipulation** If $\varepsilon\colon \{1,\dots,m\} \to \{1,\dots,n\}$, then the n-ary relation $r^\varepsilon := \{(c_1,\dots,c_n) \in M^n \mid (c_{\varepsilon(1)},\dots,c_{\varepsilon(m)}) \in r\}$ is entailed by the m-ary relation r (provided that r^ε is non-empty). In most applications we shall be concerned only with relations which are at most binary. Apart from the two identity maps, $\mathrm{id}_{\{1\}}$ and $\mathrm{id}_{\{1,2\}}$, there are six choices for ε when we restrict to $m, n \in \{1,2\}$: if r is unary, we can construct the binary relations $r \times M$ and $M \times r$; if r is binary we can construct the unary relation $r^1 := \pi_1(r \cap \Delta)$ and the binary relations $r^1 \times M$, $M \times r^1$, and, finally, the **converse** $r^{\smile} := \{(c_2,c_1) \mid (c_1,c_2) \in r\}$ of r.

(3) **Trivial expansion** This is the special case of subscript manipulation in which ε is one-to-one and order-preserving. For example, when $\varepsilon\colon \{1,2\} \to \{1,2,3\}$ is given by $\varepsilon(1) = 1$, $\varepsilon(2) = 3$, we obtain the ternary relation $r^\varepsilon = \{(c_1,c_2,c_3) \mid (c_1,c_3) \in r\}$.

(4) **Permutation** This is the special case of subscript manipulation in which $m = n$ and ε is a permutation. The converse, $r^{\smile}$, of a binary relation r is obtained when $m = n = 2$ and ε is the 2-cycle (1 2).

(5) **Repetition removal** If r is an n-ary relation and, for fixed i and j, we have $c_i = c_j$ for all $(c_1,\dots,c_n) \in r$, then r entails $r'_j := \{(c_1,\dots,c_{j-1},c_{j+1},\dots,c_n) \in M^{n-1} \mid (c_1,\dots,c_n) \in r\}$. This is also a special case of subscript manipulation.

(6) **Intersection** If r and s are n-ary relations, then $r \cap s$ is entailed by $\{r, s\}$ (provided $r \cap s$ is non-empty).

(7) **Product** Let r be n-ary and s be m-ary. For $\mathbf{c} = (c_1,\dots,c_n)$ and $\mathbf{d} = (d_1,\dots,d_n)$, let $(\mathbf{c},\mathbf{d})$ denote the corresponding $(n+m)$-tuple. Then $\{r, s\}$ entails $r \times s := \{(\mathbf{c},\mathbf{d}) \in M^{n+m} \mid \mathbf{c} \in r \text{ and } \mathbf{d} \in s\}$.

(8) **Coordinate projections** Any set of relations entails the ith coordinate projection $\pi_i\colon M^n \to M$.

(9) **Restriction of domain** If $h\colon r \to M$ is an n-ary partial operation and $s \subseteq r$, then $\{h, s\}$ entails $h{\restriction_s}\colon s \to M$.

(10) **Graph** It follows immediately from Lemma 2.1.1 that a (partial) operation, h, entails its graph, $\mathrm{graph}(h)$, and conversely, if r is the graph of an n-ary partial operation, $\mathrm{graph}^{-1}(r)$, then r entails $\mathrm{graph}^{-1}(r)$. In particular, a nullary operation c entails the unary relation $\{c\}$ and conversely.

(11) **Composition** If g is a k-ary partial operation and $h_1,\dots,h_k$ are n-ary partial operations, then $\{g, h_1,\dots,h_k\}$ entails the composite $g(h_1,\dots,h_k)$. Recall that if g is a nullary operation, then the com-

position of g with the empty set of n-ary operations is the n-ary constant operation corresponding to g. Thus $\underset{\sim}{\mathbf{M}} = \langle M; G, H, R, \mathcal{T}\rangle$ entails every member of its enriched partial clone $[G \cup H]$. Note that this is exactly the set of partial operations which are interpretations in $\underset{\sim}{\mathbf{M}}$ of the terms of type $\langle G, H, R\rangle$.

(12) **Domain** If h is a partial operation, then h entails $\mathrm{dom}(h)$.

(13) **Equaliser** If h_1 and h_2 are n-ary partial operations, then $\{h_1, h_2\}$ entails $\mathrm{eq}(h_1, h_2) := \{\mathbf{c} \in \mathrm{dom}(h_1) \cap \mathrm{dom}(h_2) \mid h_1(\mathbf{c}) = h_2(\mathbf{c})\}$ (provided the latter is non-empty). A useful special case is the **fixpointset**, $\mathrm{fix}(e) := \{c \in N \mid e(c) = c\} = \mathrm{eq}(e, \mathrm{id}_{\mathbf{M}})$, of a **partial endomorphism** $e\colon \mathbf{N} \to \underline{\mathbf{M}}$, where $\mathbf{N}$ is a subalgebra of $\underline{\mathbf{M}}$.

(14) **Joint kernel** If h_1 and h_2 are respectively n-ary and m-ary partial operations, then the set $\{h_1, h_2\}$ entails the $(n+m)$-ary relation $\ker(h_1, h_2) := \{(\mathbf{c}, \mathbf{d}) \in \mathrm{dom}(h_1) \times \mathrm{dom}(h_2) \mid h_1(\mathbf{c}) = h_2(\mathbf{d})\}$ (provided this set is non-empty).

(15) **Action by a partial endomorphism** If r is an n-ary relation and e is a partial endomorphism of $\underline{\mathbf{M}}$, then the set $\{e, r\}$ entails $e \cdot r := \{(c_1, \dots, c_n) \in M^n \mid c_1 \in \mathrm{dom}(e) \;\&\; (e(c_1), c_2, \dots, c_n) \in r\}$ (provided this set is non-empty). If r is a unary relation, this is simply $e^{-1}(r) := \{c \in \mathrm{dom}(e) \mid e(c) \in r\}$. We then have $\{e, r\} \vdash e^{-1}(r)$.

In Exercise 2.6 we ask the reader to check the soundness of these constructs.

4.6 Soundness Theorem *Each of the constructs from* (1) *to* (15) *is admissible.*

One construct which is notable by its absence is the **relational product** of an n-ary relation r and an m-ary relation s:

$$r \cdot s := \{(c_1, \dots, c_{n+m-2}) \in M^{n+m-2} \mid$$
$$(\exists c \in M)(c_1, \dots, c_{n-1}, c) \in r \text{ and } (c, c_n, \dots, c_{n+m-2}) \in s\}.$$

In general, relational product is not an admissible construct (see 9.1.3). The action, $e \cdot r$, of a partial endomorphism e on a relation r is an instance of an admissible relational product. Indeed, $e \cdot r$ is an alternative notation for the relational product $\mathrm{graph}(e) \cdot r$. If r happens to be the graph of a partial endomorphism f, then $e \cdot r$ is the graph of the composition of the maps e and f with e applied first: thus

$$e \cdot \mathrm{graph}(f) = \mathrm{graph}(e) \cdot \mathrm{graph}(f) = \mathrm{graph}(f \circ e).$$

There are various interconnections between these constructs. For example, repetition removal can be achieved by taking the domain of the function whose graph is obtained from r via the permutation $\sigma := (n\ n-1\ \dots\ j)$ which moves the jth coordinate to the end: in symbols, $r'_j = \mathrm{dom}(\mathrm{graph}^{-1}(r^\sigma))$. See Exercise 2.7 for further examples.

Exercises 2

Exercise from the text. Complete the proof of Lemma 2.1.1.

2.1 (Our first five exercises describe conditions under which $\underset{\sim}{\mathbf{M}}$ yields a duality on a single algebra $\mathbf{A}$. This one shows that local dualisability is equivalent to a purely topological condition—see [DHM95].) For each topological space $\mathbf{X}$, denote the algebra of all continuous maps from $\mathbf{X}$ into $\underline{\mathbf{M}}$, with the discrete topology, by $C(\mathbf{X}, \underline{\mathbf{M}})$. The topology which $C(\mathbf{X}, \underline{\mathbf{M}})$ inherits as a subspace of $\mathbf{M}^X$ is called the **pointwise topology**. For each algebra $\mathbf{A} \in \mathcal{A}$, let

$$A^\# := \{\, e_{\mathbf{A}}(a) \colon D(\mathbf{A}) \to M \mid a \in A \,\}$$

denote the set of all evaluation maps from $D(\mathbf{A})$ to M. Note that $\mathbf{A}^\# \leqslant ED(\mathbf{A}) \leqslant C(D(\mathbf{A}), \underline{\mathbf{M}})$ and $\mathbf{A}^\# = ED(\mathbf{A})$ precisely when $\underset{\sim}{\mathbf{M}}$ yields a duality on $\mathbf{A}$ (by Lemma 2.1.3). Prove the following claims.

(i) For all choices of $\underset{\sim}{\mathbf{M}}$, the second dual $ED(\mathbf{A})$ of $\mathbf{A}$ is closed in the pointwise topology on $C(D(\mathbf{A}), \underline{\mathbf{M}})$.

(ii) Under the brute force construction, $ED(\mathbf{A})$ is the closure of $A^\#$ in the pointwise topology on $C(D(\mathbf{A}), \underline{\mathbf{M}})$.

(iii) There is some structure $\underset{\sim}{\mathbf{M}}$ which yields a duality on $\mathbf{A}$ if and only if $A^\#$ is closed in the pointwise topology on $C(D(\mathbf{A}), \underline{\mathbf{M}})$.

2.2 Assume that (INJ) holds. Prove that if $\underset{\sim}{\mathbf{M}}$ yields a duality on $\mathbf{B} \in \mathcal{A}$, then $\underset{\sim}{\mathbf{M}}$ yields a duality on every homomorphic image of $\mathbf{B}$ in $\mathcal{A}$.

2.3 Show that if $\underset{\sim}{\mathbf{M}}$ yields a duality on $\mathbf{A}$, then $\underset{\sim}{\mathbf{M}}$ yields a duality on every retract of $\mathbf{A}$.

2.4 (This exercise shows the advantages of a schizophrenic world view.)

(i) Show that the set $\mathrm{End}(\underline{\mathbf{M}}) = D(\underline{\mathbf{M}})$ of endomorphisms of $\underline{\mathbf{M}}$ yields a duality on $\underline{\mathbf{M}}$.

(ii) More generally, show that if $\mathbf{A}$ is a subalgebra of $\underline{\mathbf{M}}$, then the set $D(\mathbf{A}) = \mathcal{A}(\mathbf{A}, \underline{\mathbf{M}})$ of partial endomorphisms of $\underline{\mathbf{M}}$ which have domain $\mathbf{A}$ yields a duality on $\mathbf{A}$.

2.5 (In this exercise we give a localised version of the Ultimate Brute Force construction. We show that for each $\mathbf{A} \in \mathcal{A}$, there is a single, possibly infinitary, algebraic relation r such that $R = \{r\}$ yields a duality on $\mathbf{A}$ in the natural sense. Moreover, r can be chosen so that the corresponding algebra $\mathbf{r}$ is isomorphic to $\mathbf{A}$!) Write the dual of $\mathbf{A}$ as $D(\mathbf{A}) = \{\, x_\lambda \mid \lambda < \kappa \,\}$ for some cardinal κ and let

$$r := \{(e_{\mathbf{A}}(a)(x_\lambda))_{\lambda<\kappa} \mid a \in A\} \leqslant \underline{\mathbf{M}}^\kappa$$

be the κ-ary relation corresponding to the subalgebra $\mathbf{A}^\#$ of $\underline{\mathbf{M}}^{D(\mathbf{A})}$ (see Exercise 2.1 above). Prove that

(i) $\underset{\sim}{\mathbf{M}} := \langle M, r \rangle$ yields a duality on $\mathbf{A}$ (without the aid of topology);
(ii) $\mathbf{r}$ is isomorphic to $\mathbf{A}$.

2.6 (i) Let e be a partial endomorphism of $\underline{\mathbf{M}}$ and let $r \in \mathcal{B}$. Show that if $X \subseteq M^S$ is closed under e and $\alpha\colon X \to M$ preserves both e and r, then α preserves $e \cdot r$. (By Lemma 2.4.4 it follows that $\{e, r\} \vdash e \cdot r$, whence Construct (15) is admissible.)
(ii) Show, in a similar manner, that the remaining 14 constructs listed in 2.4.5 are admissible.

2.7 Prove each of the following claims concerning the constructs listed in 2.4.5.
(i) Product can be obtained from trivial expansion and intersection.
(ii) Trivial expansion can be obtained from trivial relations, repetition removal, intersection and product.
(iii) Permutation can be obtained from trivial relations, repetition removal, intersection and product.
(iv) Subscript manipulation can be obtained from trivial relations, repetition removal, intersection and product.
(v) Restriction of domain can be obtained from trivial relations, intersection, product, graph and graph^{-1}.
(vi) Domain can be obtained from equaliser.

2.8 (This is for those familiar with mathematical logic.) Use the Compactness Theorem of First-Order Logic to give an alternative proof of Lemma 2.2.9. Specifically, write down a set Σ of first-order sentences which assert that $\{\, a^* \mid a \in A \,\}$ forms an algebra $\mathbf{A}^*$ isomorphic to $\mathbf{A}$, that $\{\, d^* \mid d \in D \,\}$ forms an algebra $\mathbf{D}^*$ isomorphic to $\mathbf{D}$, and that g^* is a homomorphism from $\mathbf{A}^*$ into $\mathbf{D}^*$ which agrees with h on $\mathbf{B}^*$. Show that every finite subset of Σ has a model, and use the Compactness Theorem to conclude that Σ has a model.

2.9 (Much of our theory carries over to infinite topological algebras $\underline{\mathbf{M}}$. This exercise is the infinite version of the First Duality Theorem.) Let $\underline{\mathbf{M}} = \langle M; F \rangle$ be a (possibly infinite) algebra, let $\mathcal{T}$ be a compact Hausdorff topology on M with respect to which every operation in F is continuous and let $\underset{\sim}{\mathbf{M}} = \langle M; G, H, R, \mathcal{T} \rangle$ be algebraic over $\underline{\mathbf{M}}$. Define $\mathcal{A} := \mathbb{ISP}\underline{\mathbf{M}}$, $\mathcal{X} := \mathbb{IS}_c\mathbb{P}^+ \underset{\sim}{\mathbf{M}}$, D, E, e and ε just as they were in the finite case (see Exercise 1.19). Show that $\underset{\sim}{\mathbf{M}}$ yields a duality on $\mathcal{A}$ if and only if (INJ) and (CLO) hold and for every non-empty set S and each morphism $\alpha\colon \underset{\sim}{\mathbf{M}}^S \to \underset{\sim}{\mathbf{M}}$ there exists a finite subset T of S such that α factors through $\pi_T\colon \underset{\sim}{\mathbf{M}}^S \to \underset{\sim}{\mathbf{M}}^T$ (that is, there exists a morphism $\beta\colon \underset{\sim}{\mathbf{M}}^T \to \underset{\sim}{\mathbf{M}}$ such that $\alpha = \beta \circ \pi_T$).

3

Strong Dualities

In this chapter we will look for methods to extend a natural duality to a dual equivalence between the categories $\mathcal{A}$ and $\mathcal{X}$. To do so we continue to consider the context in which $\underset{\sim}{\mathbf{M}} = \langle M; G, H, R, \mathcal{T} \rangle$ carries the discrete topology and is algebraic over the finite algebra $\underline{\mathbf{M}}$. In the last chapter we discovered conditions under which the embeddings

$$e_{\mathbf{A}} : \mathbf{A} \to ED(\mathbf{A})$$

would all be surjective, that is, that $\underset{\sim}{\mathbf{M}}$ would yield a duality on $\mathcal{A}$ and consequently determine a dual representation between $\mathcal{A}$ and $\mathcal{X}$. Following this terminology, we say that $\underset{\sim}{\mathbf{M}}$ yields a **full duality** on $\mathcal{A}$ if it yields a duality on $\mathcal{A}$ and, for each $\mathbf{X} \in \mathcal{X}$,

$$\varepsilon_{\mathbf{X}} : \mathbf{X} \to DE(\mathbf{X})$$

is also surjective. Thus $\underset{\sim}{\mathbf{M}}$ yields a full duality on $\mathcal{A}$ exactly when it yields a duality on $\mathcal{A}$ which is a dual equivalence.

It turns out that there are a number of additional features which may be present under a full duality, but do not follow from dual equivalence alone.

(a) Embeddings in $\mathcal{X}$ may be associated with surjections in $\mathcal{A}$.

(b) $\underset{\sim}{\mathbf{M}}$ may be injective in $\mathcal{X}$.

(c) Embeddings in $\mathcal{A}$ may be associated with surjections in $\mathcal{X}$.

(d) $\underline{\mathbf{M}}$ may be injective in $\mathcal{A}$.

In the presence of full duality it will not be hard to show that (a) and (b) are equivalent and are implied by (c) and (d), which are also equivalent. The ideas in this chapter arise from the observation that all existing proofs of full duality actually establish a condition which is stronger, more natural and considerably more useful than full duality. We will say that $\underset{\sim}{\mathbf{M}}$ yields a **strong duality** on $\mathcal{A}$ when this additional condition is met. In the second section we will define strong duality and see exactly when a full duality is strong. Serendipitously, strong duality always gives us (a) and (b) as well. Because strong duality provides our only apparent access to full duality and is required for almost all interesting full duality theorems and applications, we have made it the central focus of this chapter. We will show in Chapter 6 that conditions (c) and (d) hold exactly when the strong duality can be obtained without the use of partial operations in the dual category.

Our goal in this chapter will be to develop a number of methods for modifying a choice of $\underset{\sim}{\mathbf{M}}$ which is already known to yield a duality on $\mathcal{A}$, to obtain one which yields a strong duality on $\mathcal{A}$. This will always be done by adding operations as well as partial operations to its structure. In particular, we will see just why *partial* operations are in general necessary to do this. While it is not known if every duality can be upgraded to a full duality or to a strong duality, the efficacy of the methods presented here will be demonstrated in the next chapter when we use them to upgrade a broad range of dualities to strong dualities by adding finitely many (partial) operations to $\underset{\sim}{\mathbf{M}}$.

We saw in the last chapter that (IC) easily implies that $\underset{\sim}{\mathbf{M}}$ yields a duality on $\mathcal{A}_{\text{fin}}$, and then found conditions in the Second Duality Theorem under which this duality extends to all of $\mathcal{A}$. Many of our results in this chapter will follow a similar pattern. We will identify finitary conditions that give what we want for finite structures and then prove that, under the right circumstances, they imply what we want for all structures. The first section is devoted to a very careful examination of the connection between dualities and full dualities. In the next section we will see how the most natural path to full duality leads to the notion of strong duality, and we will prove three general theorems telling how strong dualities can be obtained. These efforts culminate in the last section where we prove the Two-for-One Strong Duality Theorem and the NU Strong Duality Theorem, each of which gives finitary conditions for a strong duality to exist. Some readers may wish to go directly to these theorems to obtain a strong duality for their favourite quasi-variety. In Chapter 4 we will exhibit many examples of strong dualities obtained from these theorems.

The results presented in this chapter arose in the foundational papers of B. Davey and H. Werner [DW83] and D. Clark and P. Krauss [CK84], and were subsequently refined and extended by the present authors in [CD95].

3.1 Full duality and the dual category

A duality is, in particular, a representation theorem: for every algebra $\mathbf{A} \in \mathcal{A}$ there are a structure $\mathbf{X} \in \mathcal{X}$ and an isomorphism $u \colon \mathbf{A} \to E(\mathbf{X})$, in symbols, $\mathcal{A} = \mathbb{I}E(\mathcal{X})$, where we take the canonical choices $\mathbf{X} = D(\mathbf{A})$ and $u = e_{\mathbf{A}}$. If $\underset{\sim}{\mathbf{M}}$ yields a full duality on $\mathcal{A}$, then we also have a representation on the topological side: for every structure $\mathbf{X} \in \mathcal{X}$ there are an algebra $\mathbf{A} \in \mathcal{A}$ and an isomorphism $\varphi \colon \mathbf{X} \to D(\mathbf{A})$, in symbols, $\mathcal{X} = \mathbb{I}D(\mathcal{A})$, where again we take the canonical choices $\mathbf{A} = E(X)$ and $\varphi = \varepsilon_{\mathbf{X}}$. Our first lemma shows that if both $\mathcal{A} = \mathbb{I}E(\mathcal{X})$ and $\mathcal{X} = \mathbb{I}D(\mathcal{A})$ hold and one can be done with the canonical choices, then the other can as well. We also find that while a duality between $\mathcal{A}$ and $\mathcal{X}$ need not be full, it always yields a dual equivalence between $\mathcal{A}$ and a natural subcategory of $\mathcal{X}$.

1.1 Full Duality Lemma

(i) *Assume that* $\underset{\sim}{\mathbf{M}}$ *yields a duality on* $\mathcal{A}$. *Let* $\mathcal{X}' = \mathbb{I}D(\mathcal{A})$ *and let* E' *and* ε' *be the restrictions of* E *and* ε *to* $\mathcal{X}'$. *Then* $\langle D, E', e, \varepsilon' \rangle$ *is a dual equivalence between* $\mathcal{A}$ *and* $\mathcal{X}'$.

(ii) *The following are equivalent:*

(a) $\underset{\sim}{\mathbf{M}}$ *yields a full duality on* $\mathcal{A}$;

(b) $e_{\mathbf{A}}$ *is an isomorphism for each* $\mathbf{A}$ *in* $\mathcal{A}$ *(that is,* $\underset{\sim}{\mathbf{M}}$ *yields a duality on* $\mathcal{A}$*) and* $\mathcal{X} = \mathbb{I}D(\mathcal{A})$;

(c) $\varepsilon_{\mathbf{X}}$ *is an isomorphism for each* $\mathbf{X}$ *in* $\mathcal{X}$ *and* $\mathcal{A} = \mathbb{I}E(\mathcal{X})$.

Moreover, (i) *and* (ii) *are also true if* $\mathcal{A}$ *and* $\mathcal{X}$ *are replaced by the subcategories* $\mathcal{A}_{\text{fin}}$ *and* $\mathcal{X}_{\text{fin}}$ *consisting of their respective finite members.*

Proof Assume that $\underset{\sim}{\mathbf{M}}$ yields a duality on $\mathcal{A}$ and let $\mathbf{X} \in \mathcal{X}'$. Then there are an $\mathbf{A} \in \mathcal{A}$ and an isomorphism $\varphi\colon \mathbf{X} \to D(\mathbf{A})$. Let $u\colon \mathbf{A} \to E(\mathbf{X})$ be the corresponding homomorphism given in the definition of a dual adjunction (see Figure 1.2). By assumption, $e_{\mathbf{A}}$ is an isomorphism, and $E(\varphi)$ is an isomorphism since φ is. Consequently u is an isomorphism, whence $D(u)$ is an isomorphism and therefore $\varepsilon'_{\mathbf{X}} = D(u)^{-1} \circ \varphi$ is an isomorphism. Thus (i) holds.

Part (ii) follows almost at once from (i). It is trivial that (a) implies both (b) and (c); that (b) implies (a) is a consequence of (i); and (c) implies (a) by symmetry. It remains only to check that these arguments apply equally well to the finitary version of the lemma, and we leave this as an exercise. ■

In the context of full duality a special role is played by the (possibly empty) set K of elements of M which determine one-element subalgebras of $\underline{\mathbf{M}}$. If $\underset{\sim}{\mathbf{M}}$ is algebraic over $\underline{\mathbf{M}}$, then the Preduality Theorem 1.5.2(iv) assures us that K will determine a substructure $\mathbf{K}$ of $\underset{\sim}{\mathbf{M}}$. Under a full duality we can say much more about $\mathbf{K}$. Let $\underline{\mathbf{1}}$ denote the one-element algebra $\underline{\mathbf{M}}^{\varnothing}$ in $\mathcal{A}$ and recall that $[G \cup H]$ denotes the enriched partial clone generated by $G \cup H$, that is, the enriched partial clone of $\underset{\sim}{\mathbf{M}}$.

1.2 Lemma *Assume that* $\underset{\sim}{\mathbf{M}}$ *yields a full duality on* $\mathcal{A}$.

(i) $\mathbf{K}$ *and* $\underline{\mathbf{1}}$ *are dual to one another:* $E(\mathbf{K}) \cong \underline{\mathbf{1}}$ *and* $D(\underline{\mathbf{1}}) \cong \mathbf{K}$.

(ii) $\mathbf{K}$ *is the substructure of* $\underset{\sim}{\mathbf{M}}$ *generated by its distinguished elements. Indeed, the following are equivalent:*

(a) c *determines a one-element subalgebra of* $\underline{\mathbf{M}}$, *that is,* $c \in K$;

(b) c *is the value of a constant unary function in* $[G \cup H]$;

(c) *c is the value of a constant unary total function in $[G \cup H]$;*
(d) *c is the value of a nullary term in $[G \cup H]$;*
(e) *c is in the substructure of $\underset{\sim}{\mathbf{M}}$ generated by its distinguished elements.*

(iii) *For every $\mathbf{X} \in \mathcal{X}$ there is a unique embedding of $\mathbf{K}$ into $\mathbf{X}$.*

(iv) *$\mathbf{K}$ is an initial object (free object on the empty set) in $\mathcal{X}$ while $\underline{\mathbf{1}}$ is a final object in $\mathcal{A}$.*

(v) *$\underset{\sim}{\varnothing} \in \mathcal{X}$ if and only if $\mathbf{K} = \underset{\sim}{\varnothing}$ if and only if $\underset{\sim}{\mathbf{M}}$ has no nullary operations if and only if $\underline{\mathbf{1}} \notin \mathbb{IS}\underline{\mathbf{M}}$.*

Proof $D(\underline{\mathbf{1}}) \cong \mathbf{K}$ since the embeddings of $\underline{\mathbf{1}}$ into $\underline{\mathbf{M}}$ correspond exactly to the elements of K. Since we have a duality, $E(\mathbf{K}) \cong ED(\underline{\mathbf{1}}) \cong \underline{\mathbf{1}}$, that is, the only morphism from $\mathbf{K}$ into $\underset{\sim}{\mathbf{M}}$ is the inclusion. Hence, (i) holds. Now consider (ii). Since the substructure of $\underset{\sim}{\mathbf{M}}$ generated by its distinguished elements is exactly the set of all values of nullary terms in $[G \cup H]$, we have (d) $\Leftrightarrow$ (e). The implications (d) $\Rightarrow$ (c) $\Rightarrow$ (b) hold by the definition of $[G \cup H]$, and (b) $\Rightarrow$ (a) since every element of $[G \cup H]$ is algebraic over $\underline{\mathbf{M}}$. Let $\mathbf{C}$ denote the (possibly empty) substructure of $\underset{\sim}{\mathbf{M}}$ generated by the distinguished elements of $\underset{\sim}{\mathbf{M}}$. Then, clearly, $E(\mathbf{C}) \cong \underline{\mathbf{1}} \cong E(\mathbf{K})$ and, since the duality is full, $\mathbf{C} \cong DE(\mathbf{C}) \cong DE(\mathbf{K}) \cong \mathbf{K}$. From the definition of $\mathbf{C}$, this isomorphism must be the identity, that is, $\mathbf{C} = \mathbf{K}$, whence (a) $\Leftrightarrow$ (e). This completes the proof of (ii). Now (iii), (iv) and (v) follow immediately from (ii). ■

In view of (ii) above, in order to achieve full or strong duality, we are obliged to include as distinguished elements of $\underset{\sim}{\mathbf{M}}$ enough members of $\mathbf{K}$ to generate all of $\mathbf{K}$ in $\underset{\sim}{\mathbf{M}}$. For $\mathbf{X} \in \mathcal{X}$ we will denote by $\mathbf{K}_{\mathbf{X}}$ the image of $\mathbf{K}$ under the unique embedding of (iii) above.

The Full Duality Lemma 3.1.1 suggests a general strategy for extending dualities to full dualities. Suppose that we begin with a choice of $\underset{\sim}{\mathbf{M}}$ that yields a duality on $\mathcal{A}$. According to part (i) of the Full Duality Lemma 3.1.1, D and E give us a dual equivalence between $\mathcal{A}$ and the subcategory $\mathbb{I}D(\mathcal{A})$ of $\mathcal{X}$. By itself this fact is of rather limited value since in practice we have no way to identify members of $\mathbb{I}D(\mathcal{A})$ within $\mathcal{X}$. Now suppose that we augment $G \cup H$ with additional algebraic (partial) operations. The new structure $\underset{\sim}{\mathbf{M}}'$ will then entail $\underset{\sim}{\mathbf{M}}$, and therefore will also yield a duality on $\mathcal{A}$. Again D and E give us a dual equivalence between $\mathcal{A}$ and the subcategory $\mathbb{I}D(\mathcal{A})$ of $\mathcal{X}' = \mathbb{IS}_c\mathbb{P}^+ \underset{\sim}{\mathbf{M}}'$. But now $\mathcal{X}'$ will be a smaller extension of $\mathbb{I}D(\mathcal{A})$ since we have eliminated the substructures of powers of $\underset{\sim}{\mathbf{M}}$ that are not closed under the new (partial) operations. By adding enough operational structure to $\underset{\sim}{\mathbf{M}}$ we would like to squeeze the category $\mathcal{X}$ down until it coincides with the category $\mathbb{I}D(\mathcal{A})$, thus giving us a full duality.

It turns out that this strategy can be further extended, as the members of $D(\mathcal{A})$ are closed under more than just the *finitary* algebraic (partial) operations. Let I be an arbitrary set, $\mathbf{B}$ a subalgebra of $\underline{\mathbf{M}}^I$ and $h\colon \mathbf{B} \to \underline{\mathbf{M}}$ a homomorphism, that is, h is an **algebraic I-ary partial operation** on $\underline{\mathbf{M}}$. Just as we do in the finitary case, we may extend the map h pointwise to an I-ary partial operation $\boldsymbol{h}$ on any power M^S of M. For each $s \in S$, let $\pi_s\colon M^S \to M$ denote the sth **projection** given by $\pi_s(y) = y(s)$ for each $y \in M^S$. Then the domain of the extension $\boldsymbol{h}$ is

$$\mathrm{dom}(\boldsymbol{h}) = \{\, \boldsymbol{x} \in (M^S)^I \mid \pi_s \circ \boldsymbol{x} \in B \text{ for all } s \in S \,\} \subseteq (M^S)^I$$

and $\boldsymbol{h}\colon \mathrm{dom}(\boldsymbol{h}) \to \underline{\mathbf{M}}^S$ is defined by $(\boldsymbol{h}(\boldsymbol{x}))(s) = h(\pi_s \circ \boldsymbol{x})$ for $\boldsymbol{x} \in \mathrm{dom}(\boldsymbol{h})$. As is customary, we say that a subset X of M^S is **closed under** h provided $\boldsymbol{h}(\boldsymbol{x}) \in X$ whenever $\boldsymbol{x} \in \mathrm{dom}(\boldsymbol{h})$ and $\boldsymbol{x}(i) \in X$ for each $i \in I$. We shall say that X is **hom-closed (in M^S)** if, for each set I, the set X is closed under every algebraic I-ary partial operation h on $\underline{\mathbf{M}}$. In particular, taking $I = \varnothing$, if X is hom-closed, then it contains the constant map $\boldsymbol{a}\colon S \to M$ onto $\{a\}$ whenever $\{a\}$ forms a one-element subalgebra of $\underline{\mathbf{M}}$. Note that $\varnothing$ is hom-closed in $\underline{\mathbf{M}}^S$ if and only if $\underset{\sim}{\varnothing}$ is in $\mathfrak{X}$.

We will see in Corollary 3.1.5 that the members of $D(\mathcal{A})$ can be characterised up to isomorphism as the hom-closed subsets of powers of M. This will be more easily accomplished by first giving an alternative and equally important description of $\mathbb{I}D(\mathcal{A})$ in terms of the $\mathcal{A}$-free algebras. For an arbitrary non-empty set S, the subalgebra of $\underline{\mathbf{M}}^{M^S}$ ($\mathcal{A}$-freely) generated by the projections $\{\, \pi_s \mid s \in S \,\}$ is called the **free algebra of S-ary term functions** over $\underline{\mathbf{M}}$ and is denoted by $\mathbf{F}_{\underline{\mathbf{M}}}(S)$. We say that $y \in M^S$ is in the **term closure** of a set $X \subseteq M^S$ if any two S-ary term functions that agree on X also agree at y. (See Appendix A for a review of free term algebras and closure operations.) Thus a subset X of M^S is **term-closed (in M^S)** if for all $y \in M^S \backslash X$ there exist S-ary term functions $\sigma, \tau\colon M^S \to M$ on $\underline{\mathbf{M}}$ that agree on X but not at y. Alternatively, X is term-closed in M^S provided it is an intersection of equaliser sets

$$\mathrm{eq}(\sigma, \tau) = \{\, x \in M^S \mid \sigma(x) = \tau(x)\}$$

of pairs (σ, τ) of S-ary term functions from $\mathbf{F}_{\underline{\mathbf{M}}}(S)$. Note that $\varnothing$ is term-closed in M^S if and only if the conditions of Lemma 3.1.2(v) hold.

Despite their disparate origins, these two notions of closure prove to be equivalent. The proof of this fact requires a modification of the embedding $\varepsilon_{\mathbf{X}}$. Let $S \neq \varnothing$, let $\mathbf{X} \leqslant \underset{\sim}{\mathbf{M}}^S$ and let $\mathbf{F}_{\underline{\mathbf{M}}}(S){\restriction}_X \leqslant \underline{\mathbf{M}}^X$ be the restriction of the algebra of S-ary term functions to X:

$$\mathbf{F}_{\underline{\mathbf{M}}}(S){\restriction}_X = \{\, \tau{\restriction}_X\colon X \to M \mid \tau \in \mathbf{F}_{\underline{\mathbf{M}}}(S)\}.$$

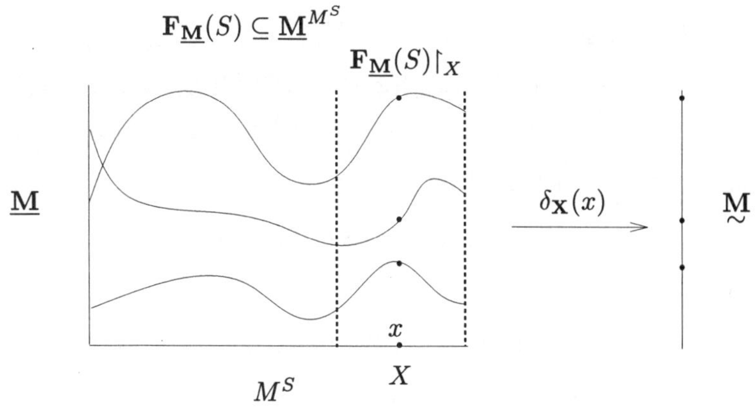

Figure 3.1 the map $\delta_{\mathbf{X}}$

Define the evaluation map $\delta_{\mathbf{X}}\colon X \to D(\mathbf{F}_{\underline{\mathbf{M}}}(S)\restriction_X)$ by $\delta_{\mathbf{X}}(x)(\tau\restriction_X) = \tau(x)$ for $x \in X$ and $\tau \in \mathbf{F}_{\underline{\mathbf{M}}}(S)$. (See Figure 3.1.)

In the proof of (ii) $\Rightarrow$ (iii) below, notice the schizophrenic nature of X, which acts simultaneously as a substructure of a power of $\underset{\sim}{\mathbf{M}}$ and as the exponent of a power of $\underline{\mathbf{M}}$.

1.3 Closure Theorem *If $S \neq \varnothing$ and $X \subseteq M^S$, then the following are equivalent:*

(i) *X is term-closed in M^S;*
(ii) *X is hom-closed in M^S;*
(iii) *every homomorphism $u\colon \mathbf{F}_{\underline{\mathbf{M}}}(S)\restriction_X \to \underline{\mathbf{M}}$ is given by evaluation at some point $x \in X$;*
(iv) *$\mathbf{X} \in \mathfrak{X}$ (that is, $\mathbf{X}$ is a closed substructure of $\underset{\sim}{\mathbf{M}}^S$) and the map $\delta_{\mathbf{X}}\colon \mathbf{X} \to D(\mathbf{F}_{\underline{\mathbf{M}}}(S)\restriction_X)$ is an isomorphism.*

Proof If $X = \varnothing$, then the conditions are equivalent as each is equivalent to the statement that $\underline{\mathbf{M}}$ has no one-element subalgebras. (Note that $\mathbf{F}_{\underline{\mathbf{M}}}(S)\restriction_\varnothing$ is a one-element algebra.) Hence we may assume that X is non-empty. Suppose that X is term-closed in M^S. Let $\mathbf{B}$ be a subalgebra of $\underline{\mathbf{M}}^I$, let $h\colon \mathbf{B} \to \underline{\mathbf{M}}$ and let $\boldsymbol{x} \in X^I$ be in the domain of $\boldsymbol{h}$. We will show that $\boldsymbol{h}(\boldsymbol{x})$ is in the term closure of X. Since $\mathbf{F}_{\underline{\mathbf{M}}}(S)$ is generated by $\{\, \pi_s \mid s \in S \,\}$, each member of $\mathbf{F}_{\underline{\mathbf{M}}}(S)$ is of the form $\tau^{(\mathbf{F}_{\underline{\mathbf{M}}}(S))}(\pi_{s_1}, \dots, \pi_{s_n})$ where τ is an n-ary term. Assume, then, that

$$\tau^{\mathbf{F}_{\underline{\mathbf{M}}}(S)}(\pi_{s_1}, \dots, \pi_{s_n}) \quad \text{and} \quad \sigma^{\mathbf{F}_{\underline{\mathbf{M}}}(S)}(\pi_{s_1}, \dots, \pi_{s_n})$$

agree on X. Observe that

$$\tau^{\mathbf{B}}(\pi_{s_1} \circ \boldsymbol{x}, \dots, \pi_{s_n} \circ \boldsymbol{x}), \sigma^{\mathbf{B}}(\pi_{s_1} \circ \boldsymbol{x}, \dots, \pi_{s_n} \circ \boldsymbol{x}) \in M^I$$

and that, for any $i \in I$,

$$\begin{aligned}\tau^{\mathbf{B}}(\pi_{s_1} \circ \boldsymbol{x}, \ldots, \pi_{s_n} \circ \boldsymbol{x})(i) &= \tau^{\underline{\mathbf{M}}}(\pi_{s_1}\boldsymbol{x}(i), \ldots, \pi_{s_n}\boldsymbol{x}(i))\\ &= \tau^{\mathbf{F}_{\underline{\mathbf{M}}}(S)}(\pi_{s_1}, \ldots, \pi_{s_n})(\boldsymbol{x}(i)) = \sigma^{\mathbf{F}_{\underline{\mathbf{M}}}(S)}(\pi_{s_1}, \ldots, \pi_{s_n})(\boldsymbol{x}(i))\\ &= \sigma^{\underline{\mathbf{M}}}(\pi_{s_1}\boldsymbol{x}(i), \ldots, \pi_{s_n}\boldsymbol{x}(i)) = \sigma^{\mathbf{B}}(\pi_{s_1} \circ \boldsymbol{x}, \ldots, \pi_{s_n} \circ \boldsymbol{x})(i)\end{aligned}$$

from which we conclude that

$$\tau^{\mathbf{B}}(\pi_{s_1} \circ \boldsymbol{x}, \ldots, \pi_{s_n} \circ \boldsymbol{x}) = \sigma^{\mathbf{B}}(\pi_{s_1} \circ \boldsymbol{x}, \ldots, \pi_{s_n} \circ \boldsymbol{x}).$$

Thus

$$\begin{aligned}\tau^{\mathbf{F}_{\underline{\mathbf{M}}}(S)}(\pi_{s_1}, \ldots, \pi_{s_n})(\boldsymbol{h}(\boldsymbol{x})) &= \tau^{\underline{\mathbf{M}}}(\boldsymbol{h}(\boldsymbol{x})(s_1), \ldots, \boldsymbol{h}(\boldsymbol{x})(s_n))\\ &=\tau^{\underline{\mathbf{M}}}(h(\pi_{s_1} \circ \boldsymbol{x}), \ldots, h(\pi_{s_n} \circ \boldsymbol{x})) = h\tau^{\mathbf{B}}((\pi_{s_1} \circ \boldsymbol{x}), \ldots, (\pi_{s_n} \circ \boldsymbol{x}))\\ &=h\sigma^{\mathbf{B}}((\pi_{s_1} \circ \boldsymbol{x}), \ldots, (\pi_{s_n} \circ \boldsymbol{x})) = \sigma^{\underline{\mathbf{M}}}(h(\pi_{s_1} \circ \boldsymbol{x}), \ldots, h(\pi_{s_n} \circ \boldsymbol{x}))\\ &=\sigma^{\underline{\mathbf{M}}}(\boldsymbol{h}(\boldsymbol{x})(s_1), \ldots, \boldsymbol{h}(\boldsymbol{x})(s_n)) = \sigma^{\mathbf{F}_{\underline{\mathbf{M}}}(S)}(\pi_{s_1}, \ldots, \pi_{s_n})(\boldsymbol{h}(\mathbf{x})).\end{aligned}$$

Since $\boldsymbol{h}(\boldsymbol{x})$ is in the term closure of X, it is in X. Thus (i) implies (ii).

Assume that (ii) holds. In order to prove (iii), we apply (ii) to the closure of X under u. Let $\mathbf{B} = \mathbf{F}_{\underline{\mathbf{M}}}(S){\restriction}_X \leqslant \underline{\mathbf{M}}^X$ and let $\boldsymbol{i} \in (M^S)^X$ be the inclusion map $\boldsymbol{i}(x) = x$. Then $\boldsymbol{i} \in \mathrm{dom}(\boldsymbol{u})$ since $\pi_s \circ \boldsymbol{i} = \pi_s{\restriction}_X$, for each $s \in S$. By hom-closure, $\boldsymbol{u}(\boldsymbol{i}) \in X$. We check that u is exactly the evaluation $\delta_{\mathbf{X}}(\boldsymbol{u}(\boldsymbol{i}))$. Since both are homomorphisms, it is sufficient to see that they agree on each generator $\pi_s{\restriction}_X$ where $s \in S$:

$$\delta_{\mathbf{X}}(\boldsymbol{u}(\boldsymbol{i}))(\pi_s{\restriction}_X) = \pi_s(\boldsymbol{u}(\boldsymbol{i})) = \boldsymbol{u}(\boldsymbol{i})(s) = u(\pi_s \circ \boldsymbol{i}) = u(\pi_s{\restriction}_X).$$

We now show that (iii) implies (i). Suppose $z \in M^S$ and any two S-ary term functions which agree on X also agree at z. Then the map $u\colon \mathbf{F}_{\underline{\mathbf{M}}}(S){\restriction}_X \to M$ defined by $u(\tau{\restriction}_X) = \tau(z)$ is a well-defined homomorphism. By (iii) there is an $x \in X$ such that u is evaluation at x. Thus for all $s \in S$ we have

$$z(s) = \pi_s(z) = u(\pi_s{\restriction}_X) = \pi_s(x) = x(s)$$

and consequently $z = x \in X$. Hence X is term-closed.

Since it is trivial that (iv) implies (iii), to complete the proof it suffices to prove that (i) and (ii) together imply (iv). Topological closure is a simple consequence of term closure, and closure under algebraic (partial) operations is an immediate consequence of hom-closure. Thus $\mathbf{X} \in \mathfrak{X}$. The proof that $\delta_{\mathbf{X}}$ is an embedding is similar to the proof that $\varepsilon_{\mathbf{X}}$ is an embedding (part (i) of the Dual Adjunction Theorem 1.5.3), and we leave its verification to the reader. ■

The fact that condition (i) of the Preduality Theorem 1.5.2 extends to hom-closure is more easily verified by looking at term closure.

1.4 Lemma *$D(\mathbf{A}) = \mathcal{A}(\mathbf{A}, \underline{\mathbf{M}})$ is term-closed (= hom-closed) in M^A for all $\mathbf{A} \in \mathcal{A}$.*

Proof The assertion is true in the case $D(\mathbf{A}) = \underset{\sim}{\varnothing}$, so assume that $D(\mathbf{A}) \neq \underset{\sim}{\varnothing}$. Let $y \in M^A$ and assume that any two term functions in $\mathbf{F}_{\underline{\mathbf{M}}}(A)$ that agree on $D(\mathbf{A})$ also agree at y. Let f be an n-ary operation symbol in the type of $\mathcal{A}$ and let $a_1, \dots, a_n \in A$. For any $x \in D(\mathbf{A})$, the maps $\sigma = \pi_{f(a_1,\dots,a_n)}$ and $\tau = f(\pi_{a_1}, \dots, \pi_{a_n})$ agree at x:

$$\begin{aligned}\sigma(x) &= \pi_{f(a_1,\dots,a_n)}(x) = x(f(a_1, \dots, a_n)) = f(x(a_1), \dots, x(a_n)) \\ &= f(\pi_{a_1}(x), \dots, \pi_{a_n}(x)) = f(\pi_{a_1}, \dots, \pi_{a_n})(x) = \tau(x).\end{aligned}$$

Consequently, σ and τ also agree at y. Thus

$$\begin{aligned}y(f(a_1, \dots, a_n)) &= \pi_{f(a_1,\dots,a_n)}(y) = \sigma(y) = \tau(y) \\ &= f(\pi_{a_1}, \dots, \pi_{a_n})(y) = f(\pi_{a_1}(y), \dots, \pi_{a_n}(y)) = f(y(a_1), \dots, y(a_n)),\end{aligned}$$

whence $y \in D(\mathbf{A})$. ∎

If the structure $\underset{\sim}{\mathbf{M}}$ yields a duality on $\mathcal{A}$, then part (i) of the Full Duality Lemma 3.1.1 tells us that D and E (restricted to $\mathbb{I}D(\mathcal{A})$) give us a dual equivalence between the categories $\mathcal{A}$ and $\mathbb{I}D(\mathcal{A})$. By part (iv) of the Closure Theorem 3.1.3 we know that every term-closed $\mathbf{X}$ is isomorphic to $D(\mathbf{A})$ for some $\mathbf{A} \in \mathcal{A}$. Combining this fact with Lemma 3.1.4 gives us two fundamental descriptions of the dual category $\mathbb{I}D(\mathbf{A})$ that will be used throughout the rest of this book.

1.5 Corollary *Let $\mathbf{X}$ be in $\mathfrak{X}$. Then $\mathbf{X} \in \mathbb{I}D(\mathcal{A})$ if and only if $\mathbf{X}$ is isomorphic to a term-closed (= hom-closed) subset of $\underset{\sim}{\mathbf{M}}^S$ for some non-empty set S.*

The content of the last three results is summarised in Figure 3.2.

We can now prove a variant of the First Duality Theorem 2.2.2 which is more exactly suited to our needs. Here we have replaced (INJ) with a different collection (INJ)′ of instances of injectivity of $\underset{\sim}{\mathbf{M}}$ in $\mathfrak{X}$ expressed fully in terms of the category $\mathfrak{X}$.

1.6 Third Duality Theorem *The following are equivalent:*

(i) *$\underset{\sim}{\mathbf{M}}$ yields a duality on $\mathcal{A}$;*

(ii) *if $\mathbf{X}$ is a term-closed substructure of $\underset{\sim}{\mathbf{M}}^S$ for some $S \neq \varnothing$, then $E(\mathbf{X}) = \mathbf{F}_{\underline{\mathbf{M}}}(S){\restriction}_X$, that is, every morphism $\alpha \colon \mathbf{X} \to \underset{\sim}{\mathbf{M}}$ extends to an S-ary term function $\tau \colon \underset{\sim}{\mathbf{M}}^S \to \underset{\sim}{\mathbf{M}}$;*

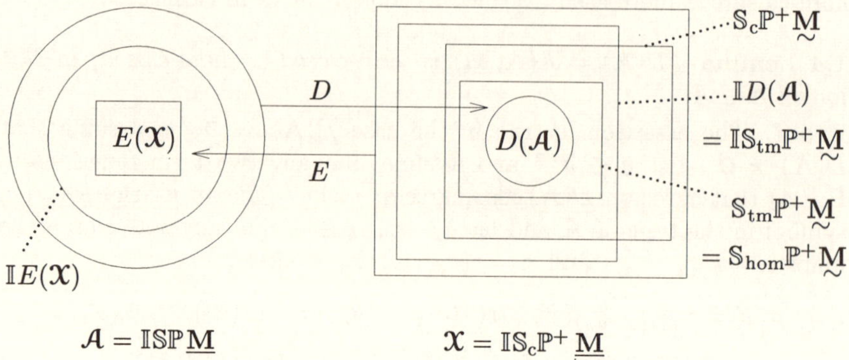

Figure 3.2 subcategories of $\mathcal{X}$ under a dual adjunction

(iii) *the following two conditions hold—*

(INJ)′ $\underset{\sim}{\mathrm{M}}$ *is injective with respect to term-closed sets* $\mathbf{X} \subseteq \underset{\sim}{\mathrm{M}}^S$ *(with* $S \neq \varnothing$*) and their inclusion maps, that is, each morphism* $\alpha\colon \mathbf{X} \to \underset{\sim}{\mathrm{M}}$ *extends to a morphism* $\beta\colon \underset{\sim}{\mathrm{M}}^S \to \underset{\sim}{\mathrm{M}}$,

(CLO) *for each* $n \in \mathbb{N}$*, every morphism* $\tau\colon \underset{\sim}{\mathrm{M}}^n \to \underset{\sim}{\mathrm{M}}$ *is an* n*-ary term function on* $\underline{\mathbf{M}}$.

Moreover, the same is true if $\mathcal{A}$ *is replaced by* $\mathcal{A}_{\mathrm{fin}}$ *in* (i) *and* S *is restricted to finite sets in* (ii) *and* (iii).

Proof Assume that $\underset{\sim}{\mathrm{M}}$ yields a duality on $\mathcal{A}$ and that $\mathbf{X}$ is term-closed in $\underset{\sim}{\mathrm{M}}^S$. The resulting dual adjunction yields the triangles in Figure 3.3, where $i\colon \mathbf{F}_{\underline{\mathbf{M}}}(S){\restriction}_X \to E(\mathbf{X})$ is the inclusion map. Since $\underset{\sim}{\mathrm{M}}$ yields a duality on $\mathcal{A}$, the map $e_{(\mathbf{F}_{\underline{\mathbf{M}}}(S){\restriction}_X)}$ is surjective. By the Closure Theorem 3.1.3, $\delta_{\mathbf{X}}$ is an isomorphism and hence $E(\delta_{\mathbf{X}})$ is an isomorphism. By the left-hand triangle, the inclusion map i is surjective and thus we have $E(\mathbf{X}) = \mathbf{F}_{\underline{\mathbf{M}}}(S){\restriction}_X$. Hence, (i) implies (ii).

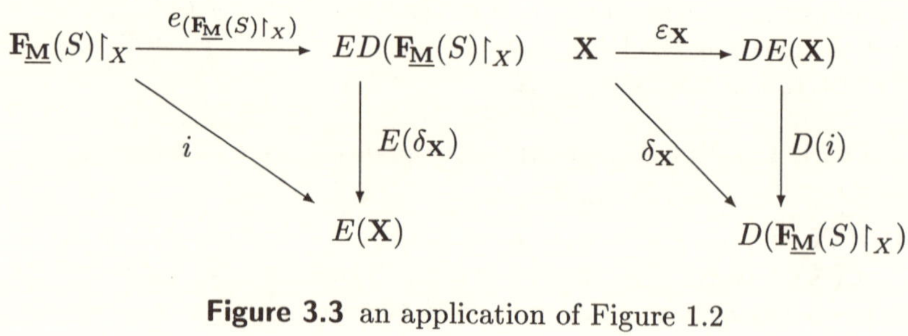

Figure 3.3 an application of Figure 1.2

As (ii) implies (iii) trivially, it remains to prove that (iii) implies (i). We verify part (ii) of the First Duality Theorem 2.2.2. Let $\mathbf{A} \in \mathcal{A}$ and let $\alpha\colon D(\mathbf{A}) \to \underset{\sim}{\mathbf{M}}$. By Lemma 3.1.4, $D(\mathbf{A})$ is term-closed and therefore, by (INJ)$'$, it follows that α extends to a morphism $\beta\colon \underset{\sim}{\mathbf{M}}^A \to \underset{\sim}{\mathbf{M}}$ which, since (CLO) implies $(\mathrm{CLO})_A$, is an A-ary term function on $\underline{\mathbf{M}}$.

To prove the finitary version ('Moreover, ... '), first observe that the same argument again establishes that (i) implies (ii), and that now (ii) and (iii) are identical. To prove that (ii) implies (i), consider $n \in \mathrm{N}$ and $\eta\colon D(\mathbf{A}) \to \underset{\sim}{\mathbf{M}}$ where $\mathbf{A} \leqslant \underline{\mathbf{M}}^n$. By Lemma 3.1.4, the set $D(\mathbf{A})$ is term-closed and consequently β extends to an A-ary term function τ. Thus there are $a, b, c, \dots \in A$ such that for all $x \in D(\mathbf{A})$, we have that

$$\begin{aligned}\eta(x) &= \tau(\pi_a, \pi_b, \pi_c, \dots)(x) = \tau(x(a), x(b), x(c), \dots)\\ &= x(\tau(a, b, c, \dots)) = e_{\mathbf{A}}(\tau(a, b, c, \dots))(x).\end{aligned}$$

Thus $e_{\mathbf{A}}$ is surjective and we have (i). ■

In practice we will normally begin with a duality which we seek to upgrade to a full duality. Our characterisation of the dual category tells us just what will be required to do this.

1.7 Full Duality Theorem *If $\underset{\sim}{\mathbf{M}}$ yields a duality on $\mathcal{A}$, then the following are equivalent:*

(i) *$\underset{\sim}{\mathbf{M}}$ yields a full duality on $\mathcal{A}$;*

(ii) *$\mathcal{X} = \mathbb{I}D(\mathcal{A})$;*

(iii) *every closed substructure of a power of $\underset{\sim}{\mathbf{M}}$ is isomorphic to a term-closed (= hom-closed) substructure of a power of $\underset{\sim}{\mathbf{M}}$.*

Moreover, the same is true if $\mathcal{A}$ and $\mathcal{X}$ are replaced by $\mathcal{A}_{\mathrm{fin}}$ and $\mathcal{X}_{\mathrm{fin}}$ in (i) *and* (ii), *and* (iii) *is restricted to finite powers of $\underset{\sim}{\mathbf{M}}$.*

Proof Use the Full Duality Lemma 3.1.1(ii) and Corollary 3.1.5. ■

This theorem tells us that the duality which $\underset{\sim}{\mathbf{M}}$ yields on $\mathcal{A}$ will be full exactly when we have the cleaner arrangement of subcategories of $\mathcal{X}$ illustrated in Figure 3.4. We turn next to the question of how and when this will be the case.

3.2 Strong duality and the role of injectivity

We can at last reveal the exact goal of this chapter. Assume that $\underset{\sim}{\mathbf{M}}$ yields a duality on $\mathcal{A}$. To establish that $\underset{\sim}{\mathbf{M}}$ yields a full duality on $\mathcal{A}$, the Full Duality Theorem 3.1.7 says that we must find, for each non-empty set S and each closed substructure $\mathbf{X}$ of $\underset{\sim}{\mathbf{M}}^S$, a non-empty set T and term-closed (=hom-closed) substructure $\underset{\sim}{\mathbf{Y}}$ of $\underset{\sim}{\mathbf{M}}^T$ such that $\mathbf{X}$ is isomorphic

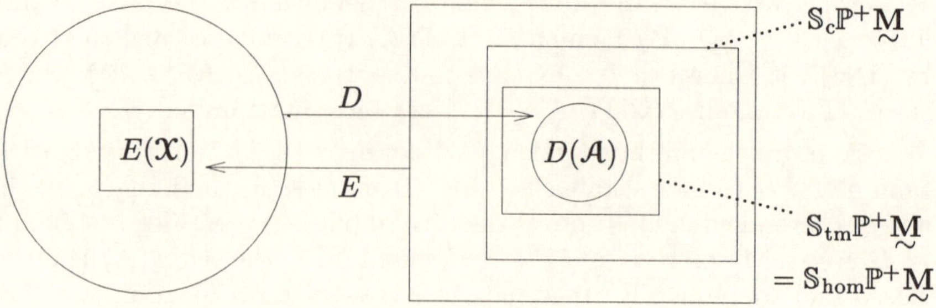

$\mathcal{A} = \mathbb{ISP}\underline{\mathbf{M}} = \mathbb{I}E(\mathcal{X})$ $\qquad$ $\mathcal{X} = \mathbb{IS}_c\mathbb{P}^+\,\underset{\sim}{\mathbf{M}} = \mathbb{I}D(\mathcal{A})$

Figure 3.4 Subcategories of $\mathcal{X}$ under a full duality

to $\mathbf{Y}$. But where, given $\mathbf{X}$, are we to find the structure $\mathbf{Y}$? This seemingly daunting task is actually carried out in every known example of full duality by establishing that the natural candidate $\mathbf{Y} = \mathbf{X}$ will suffice, that is, that every closed substructure of a power of $\underset{\sim}{\mathbf{M}}$ *is* term-closed (or, equivalently, that it *is* hom-closed). When $\underset{\sim}{\mathbf{M}}$ yields a duality on $\mathcal{A}$ and every closed substructure of a power of $\underset{\sim}{\mathbf{M}}$ is term-closed, we say that $\underset{\sim}{\mathbf{M}}$ **yields a strong duality on** $\mathcal{A}$. Note that in Figure 3.4 the duality is strong when the two inner boxes on the right are really one.

While proving strong duality appears to be the most natural way to establish that a duality is full, it also provides a valuable enhancement of full duality which will prove useful in many ways. In this section we will see that strong duality implies that $\underset{\sim}{\mathbf{M}}$ is injective in $\mathcal{X}$ and, conversely, that a full duality for which $\underset{\sim}{\mathbf{M}}$ is injective in $\mathcal{X}$ is necessarily strong. Strong dualities also give us a nice representation of $\mathcal{A}$-congruences, as the following application of term closure shows. For each closed substructure $\mathbf{Z} \leqslant \mathbf{X}$ we define the **projection congruence**

$$\theta_Z := \{(\alpha, \beta) \mid \alpha(z) = \beta(z) \text{ for all } z \in Z\}$$

on $E(\mathbf{X})$. Note that θ_Z is an $\mathcal{A}$-congruence. (See Theorem 1.3.5.)

2.1 Theorem (Davey and Priestley [DP87]) *Assume that $\underset{\sim}{\mathbf{M}}$ yields a strong duality on $\mathcal{A}$ and that $\mathbf{A} = E(\mathbf{X})$ where $\mathbf{X} \in \mathcal{X}$. Then the lattice of closed substructures of $\mathbf{X}$ is dually isomorphic to the lattice*

$$\mathrm{Con}_{\mathcal{A}}(\mathbf{A}) := \{\theta \in \mathrm{Con}(\mathbf{A}) \mid \mathbf{A}/\theta \in \mathcal{A}\}$$

of $\mathcal{A}$-congruences on $\mathbf{A}$ under the mapping $\mathbf{Z} \mapsto \theta_Z$.

Proof To see that the mapping is one-to-one, suppose $\mathbf{Y} \neq \mathbf{Z}$ are distinct closed substructures of $\mathbf{X}$. Choose an element y in one but not in the other, say, $y \in Y \backslash Z$. Since $\mathbf{X} \in \mathcal{X}$ we may assume that $\mathbf{X} \leqslant \underset{\sim}{\mathbf{M}}^S$ for some non-empty set S. Since Z is term-closed, there are S-ary term functions $\alpha, \beta\colon \underset{\sim}{\mathbf{M}}^S \to \underset{\sim}{\mathbf{M}}$ that agree on Z but not at y. Thus (α, β) is in θ_Z but not in θ_Y.

Next we check that it is onto. Let $\psi \in \mathrm{Con}_{\mathcal{A}}(\mathbf{A})$, and let $Z := \{z \in X \mid \psi \subseteq \theta_z\}$ where $\theta_z := \{(\alpha, \beta) \mid \alpha(z) = \beta(z)\}$. Then $\psi \subseteq \theta_Z$. Now suppose that $(\alpha, \beta) \notin \psi$ where $\alpha, \beta \in A$. Let $n\colon \mathbf{A} \to \mathbf{A}/\psi$ be the natural map. Since $\psi \in \mathrm{Con}_{\mathcal{A}}(\mathbf{A})$ we can apply the Algebraic Separation Theorem 1.3.1 to obtain a homomorphism $u\colon \mathbf{A}/\psi \to \underset{\sim}{\mathbf{M}}$ such that $u \circ n(\alpha) \neq u \circ n(\beta)$. Since $u \circ n \in DE(\mathbf{X})$ and the duality is full, we can find an x in $\mathbf{X}$ such that $u \circ n = \varepsilon_{\mathbf{X}}(x)$. Thus $\alpha(x) \neq \beta(x)$. Since $\psi \subseteq \ker(u \circ n) = \ker(\varepsilon_{\mathbf{X}}(x)) = \theta_x$, we conclude that $x \in Z$ and therefore $(\alpha, \beta) \notin \theta_Z$. This shows that $\psi = \theta_Z$.

It remains to verify that the order is reversed. Let $\mathbf{Y}$ and $\mathbf{Z}$ be closed substructures of $\mathbf{X}$, where again we assume that $\mathbf{X} \leqslant \underset{\sim}{\mathbf{M}}^S$. Clearly $\mathbf{Y} \leqslant \mathbf{Z}$ implies that $\theta_Z \subseteq \theta_Y$. To prove the converse, assume that $\theta_Z \subseteq \theta_Y$ and let $y \in Y$. We claim that y is in the term closure of Z. Let ρ and σ be S-ary term functions that agree on Z. Then $\rho, \sigma \in E(\mathbf{X})$ where $(\rho, \sigma) \in \theta_Z \subseteq \theta_Y \subseteq \theta_y$. Consequently ρ and σ agree at y. Since the duality is strong, $\mathbf{Z}$ is term-closed and therefore $y \in Z$. Thus $\mathbf{Y} \leqslant \mathbf{Z}$. ■

In order to see how strong dualities are to be obtained, we need an analogue of the notion of entailment. Let

$$\mathcal{P} := \{\, h \mid h\colon \mathbf{A} \to \underline{\mathbf{M}} \text{ for some } n \geq 0 \text{ and some } \mathbf{A} \leqslant \underline{\mathbf{M}}^n \,\}$$

denote the set of all finitary, algebraic, partial or total operations on $\underline{\mathbf{M}}$. The set $\mathcal{P}$, with the empty map $\eta\colon \varnothing \to M$ added if $\underline{\mathbf{M}}$ has no constant unary term functions (see Exercise 3.9), is closed under composition and includes the coordinate projections, and is therefore an example of an enriched partial clone on M. We will refer to it as the **enriched partial hom-clone** of $\underline{\mathbf{M}}$. The set $\mathcal{P}$ is contained in

$$\mathcal{P}_\infty := \{\, h \mid h\colon \mathbf{A} \to \underline{\mathbf{M}} \text{ for some set } S \text{ and some } \mathbf{A} \leqslant \underline{\mathbf{M}}^S \,\},$$

the class of all algebraic, partial or total operations on $\underline{\mathbf{M}}$ of arbitrary arity. If $P \subseteq \mathcal{P}$ and $h \in \mathcal{P}_\infty$, we say that P **hom-entails** h provided that every subset of a power of M which is closed under the (partial) operations of P is also closed under h. We say that P **hom-entails** a subclass of $\mathcal{P}_\infty$ if it hom-entails each member of that subclass. In this language the hom-closure criterion for strong duality simply says that $\underset{\sim}{\mathbf{M}} = \langle M; G, H, R, \mathcal{T}\rangle$ yields a strong duality on $\mathcal{A}$ provided that it yields a duality on $\mathcal{A}$ and that $G \cup H$ hom-entails $\mathcal{P}_\infty$.

This perspective explains why operations, and particularly *partial* operations, are generally required for a strong duality. In order to obtain strong duality from a structure $\underset{\sim}{\mathrm{M}}$ which yields a duality on $\mathcal{A}$, we must add enough members of $\mathcal{P}$ to hom-entail all members of $\mathcal{P}_\infty$. If this is possible at all, then it can certainly be achieved by adding *all* of $\mathcal{P}$. We refer to this as the **strong brute force** construction.

2.2 Brute Force Strong Duality Theorem *Suppose that some structure $\underset{\sim}{\mathrm{M}}'$ yields a strong duality on $\mathcal{A}$. Then $\underset{\sim}{\mathrm{M}} = \langle M; G, H, R, \mathcal{T}\rangle$ yields a strong duality on $\mathcal{A}$ if and only if $\underset{\sim}{\mathrm{M}}$ yields a duality on $\mathcal{A}$ and $G \cup H$ hom-entails $\mathcal{P}$. In particular, if some structure $\underset{\sim}{\mathrm{M}}'$ yields a strong duality on $\mathcal{A}$, then the brute force structure $\underset{\sim}{\mathrm{M}} = \langle M; \mathcal{P}, \mathcal{T}\rangle$ yields a strong duality on $\mathcal{A}$.*

Proof The first assertion is immediately verifiable. Suppose that $\underset{\sim}{\mathrm{M}}' = \langle M; G', H', R', \mathcal{T}\rangle$ yields a strong duality on $\mathcal{A}$ and let $\underset{\sim}{\mathrm{M}} = \langle M; \mathcal{P}, \mathcal{T}\rangle$. If $r \in R'$, then $\rho_1 = \pi_1{\restriction}_r$, the restriction of the projection on the first coordinate to the algebra $\mathbf{r}$, is in $\mathcal{P}$. Clearly ρ_1 entails r and, consequently, $\underset{\sim}{\mathrm{M}}$ entails $\underset{\sim}{\mathrm{M}}'$. Thus $\underset{\sim}{\mathrm{M}}$ yields a duality on $\mathcal{A}$. Since $G' \cup H'$ hom-entails $\mathcal{P}_\infty$, we see that $\mathcal{P}$ certainly does so as well. ∎

We will return to a more detailed study of hom-entailment in Chapter 9. Pressing the concept of 'brute force' a bit further, we can consider the structure $\underset{\sim}{\mathrm{M}}^\infty := \langle M; \mathcal{P}_\infty\rangle$ with a proper class of operations and partial operations and *no topology*. An appropriate modification of the above proof shows that $\underset{\sim}{\mathrm{M}}^\infty$ yields a strong duality on $\mathcal{A}$. (See Exercise 3.2.) This might be thought of as the **ultimate strong brute force** construction.

It will often be important to be able to modify a structure $\underset{\sim}{\mathrm{M}}$ yielding a strong duality on $\mathcal{A}$ without losing that strong duality. Our next lemma tells how this can be done. We say that $G \cup H \cup R$ **strongly entails** $G' \cup H' \cup R'$ if it entails $G' \cup H' \cup R'$ and $G \cup H$ hom-entails $G' \cup H'$. In Chapter 9 (Theorem 9.4.2) we will see that every partial operation hom-entailed by $G \cup H$ can be obtained from $G \cup H$ using the constructs described in the next lemma.

2.3 $\underset{\sim}{\mathrm{M}}$-Shift Strong Duality Lemma *Consider the structure $\underset{\sim}{\mathrm{M}}' = \langle M; G', H', R', \mathcal{T}\rangle$.*

(i) *If $\underset{\sim}{\mathrm{M}}$ strongly entails $\underset{\sim}{\mathrm{M}}'$ and $\underset{\sim}{\mathrm{M}}'$ yields a strong duality on $\mathcal{A}$, then $\underset{\sim}{\mathrm{M}}$ also yields a strong duality on $\mathcal{A}$.*

(ii) *$\underset{\sim}{\mathrm{M}}$ strongly entails $\underset{\sim}{\mathrm{M}}'$ if it is obtained from $\underset{\sim}{\mathrm{M}}'$ by*
   - (a) *enlarging G', H' or R',*
   - (b) *deleting members of G' or H' which can be obtained as compositions of the remaining members of G' and H' and the projection mappings,*
   - (c) *deleting a member h of H' which has an extension among the remaining members of $G' \cup H'$ and adding $\mathrm{dom}(h)$ to R'.*

(iii) $\underset{\sim}{\mathbf{M}}$ *strongly entails* $\underset{\sim}{\mathbf{M}}'$ *if* $\underset{\sim}{\mathbf{M}}$ *entails* $\underset{\sim}{\mathbf{M}}'$ *and is obtained from* $\underset{\sim}{\mathbf{M}}'$ *by*
  (a) *deleting members of* R', *or*
  (b) *deleting members of* H' *which have an extension in* G' *or* H'.

Proof Each of these facts follows directly from the $\underset{\sim}{\mathbf{M}}$-Shift Duality Lemma 2.4.2 and the definition of strong duality, as we ask the reader to verify. ∎

Recall that $\underset{\sim}{\mathbf{M}}$ yields a duality on $\mathcal{A}_{\text{fin}}$ if $e_{\mathbf{A}}\colon \mathbf{A} \to ED(\mathbf{A})$ is an isomorphism for every finite algebra $\mathbf{A}$ in $\mathcal{A}$. Following this terminology, we will say that $\underset{\sim}{\mathbf{M}}$ **yields a strong duality on** $\mathcal{A}_{\text{fin}}$ if it yields a duality on $\mathcal{A}_{\text{fin}}$ and every (closed) substructure of a finite power of $\underset{\sim}{\mathbf{M}}$ is term-closed (= hom-closed).

As we have said, every known full duality can be shown to be full by showing that it is in fact strong. In every known full duality, (INJ) has been established by proving that $\underset{\sim}{\mathbf{M}}$ is in fact injective in $\mathcal{X}$. It turns out that these two enhancements of full duality are equivalent. A more category-theoretic proof of the following result is presented in Exercises 3.13 and 3.14.

2.4 First Strong Duality Theorem $\underset{\sim}{\mathbf{M}}$ *yields a strong duality on* $\mathcal{A}$ *if and only if* $\underset{\sim}{\mathbf{M}}$ *yields a full duality on* $\mathcal{A}$ *and is injective in* $\mathcal{X}$. *The corresponding result holds at the finite level, that is,* $\underset{\sim}{\mathbf{M}}$ *yields a strong duality on* $\mathcal{A}_{\text{fin}}$ *if and only if* $\underset{\sim}{\mathbf{M}}$ *yields a full duality on* $\mathcal{A}_{\text{fin}}$ *and is injective in* $\mathcal{X}_{\text{fin}}$.

Proof Assume $\underset{\sim}{\mathbf{M}}$ yields a strong duality on $\mathcal{A}$. Then $\underset{\sim}{\mathbf{M}}$ yields a full duality on $\mathcal{A}$ by the Full Duality Theorem 3.1.7. To prove injectivity, let $\varphi\colon \mathbf{Y} \to \underset{\sim}{\mathbf{M}}^S$ be an embedding and let $\alpha\colon \mathbf{Y} \to \underset{\sim}{\mathbf{M}}$ be a morphism in $\mathcal{X}$. Then the image of $\mathbf{Y}$ under φ is a closed substructure $\mathbf{X}$ of $\underset{\sim}{\mathbf{M}}^S$ and we have $\alpha \circ \varphi^{-1}\colon \mathbf{X} \to \underset{\sim}{\mathbf{M}}$. By strong duality $\mathbf{X}$ is term-closed in $\underset{\sim}{\mathbf{M}}^S$, and by the Third Duality Theorem 3.1.6, the map $\alpha \circ \varphi^{-1}$ extends to an S-ary term function τ on $\underline{\mathbf{M}}$. Thus $\tau\colon \underset{\sim}{\mathbf{M}}^S \to \underset{\sim}{\mathbf{M}}$ is the required extension of α.

Conversely, suppose $\underset{\sim}{\mathbf{M}}$ yields a full duality on $\mathcal{A}$ and is injective in $\mathcal{X}$. Let $S \neq \varnothing$, let $\mathbf{X} \leqslant \underset{\sim}{\mathbf{M}}^S$ and let $y \in M^S \backslash X$. Denote by $\mathbf{Y}$ the closed substructure of $\underset{\sim}{\mathbf{M}}^S$ generated by $X \cup \{y\}$, that is, $\mathbf{Y}$ is the intersection of all closed substructures of $\underset{\sim}{\mathbf{M}}^S$ which contain $X \cup \{y\}$ (Exercise 3.4). Let $\iota\colon \mathbf{X} \to \mathbf{Y}$ be the inclusion map. By injectivity, each member of $E(\mathbf{X})$ has an extension in $E(\mathbf{Y})$, that is to say, $E(\iota)\colon E(\mathbf{Y}) \to E(\mathbf{X})$ is surjective. Since $\mathbf{X} \neq \mathbf{Y}$, the inclusion ι is not an isomorphism and hence $E(\iota)$ is not an isomorphism. Thus $E(\iota)$ is not an embedding and we must have $\sigma, \tau \in E(\mathbf{Y})$ with $\sigma \neq \tau$ but $\sigma{\restriction}_X = \tau{\restriction}_X$. Suppose that $\sigma(y) = \tau(y)$. Then $X \cup \{y\} \subseteq \text{eq}(\sigma, \tau)$, whence $Y \subseteq \text{eq}(\sigma, \tau)$, by the definition of $\mathbf{Y}$, since $\text{eq}(\sigma, \tau)$ is a closed substructure of $\underset{\sim}{\mathbf{M}}^S$. Thus $Y = \text{eq}(\sigma, \tau)$, which contradicts the fact that $\sigma \neq \tau$. Hence, $\sigma(y) \neq \tau(y)$. Finally, σ and τ extend to morphisms on $\underset{\sim}{\mathbf{M}}^S$ by injectivity, and these are S-ary term functions by part (ii) of the Third Duality Theorem 3.1.6.

The same proof, restricted to finite powers of $\underset{\sim}{\mathbf{M}}$, yields the finite version. ■

As an immediate consequence, we can spell out the properties of $\underset{\sim}{\mathbf{M}}$ required to establish a strong duality.

2.5 Strong Duality Corollary *$\underset{\sim}{\mathbf{M}}$ yields a strong duality on $\mathcal{A}$ if and only if*

(i) *$\underset{\sim}{\mathbf{M}}$ satisfies* (CLO),

(ii) *$\underset{\sim}{\mathbf{M}}$ is injective in $\mathfrak{X}$ and*

(iii) *every closed substructure of a power of $\underset{\sim}{\mathbf{M}}$ is term-closed.*

In the presence of full duality, the injectivity of $\underset{\sim}{\mathbf{M}}$ in $\mathfrak{X}$ has immediate algebraic consequences. It says precisely that embeddings go to surjections under E. But in this setting it gives us a further strengthening of the Dual Adjunction Theorem 1.5.3.

2.6 Lemma *Assume that $\underset{\sim}{\mathbf{M}}$ yields a full duality on $\mathcal{A}$. Then the following are equivalent:*

(i) *$\underset{\sim}{\mathbf{M}}$ is injective in $\mathfrak{X}$, that is, $\underset{\sim}{\mathbf{M}}$ yields a strong duality on on $\mathcal{A}$;*

(ii) *$\varphi\colon \mathbf{X} \to \mathbf{Y}$ is an embedding in $\mathfrak{X}$ (if and) only if $E(\varphi)\colon E(\mathbf{Y}) \to E(\mathbf{X})$ is a surjection in $\mathcal{A}$;*

(iii) *$u\colon \mathbf{A} \to \mathbf{B}$ is a surjection in $\mathcal{A}$ if (and only if) $D(u)\colon D(\mathbf{B}) \to D(\mathbf{A})$ is an embedding in $\mathfrak{X}$.*

Proof The equivalence of (i) and (ii) is an immediate consequence of the definitions.

To prove that (ii) implies (iii), we use the fact that each $e_{\mathbf{A}}$ is an isomorphism. Let $D(u)$ be an embedding. By (ii), the map $ED(u)$ is surjective. From Figure 1.2 (since D and E are dually adjoint) we have that $u = e_{\mathbf{B}}^{-1} \circ ED(u) \circ e_{\mathbf{A}}$ is also surjective.

To prove that (iii) implies (ii), we use the fact that each $\varepsilon_{\mathbf{X}}$ is an isomorphism. Let φ be an embedding. From Figure 1.2 we have that $DE(\varphi) = \varepsilon_{\mathbf{Y}} \circ \varphi \circ \varepsilon_{\mathbf{X}}^{-1}$ is also an embedding. By (iii), we conclude that $E(\varphi)$ is surjective. ■

The First Duality Theorem 2.2.2 and the Third Duality Theorem 3.1.6 say that many instances of the injectivity of $\underset{\sim}{\mathbf{M}}$ in $\mathfrak{X}$ must hold once we have a duality. If duality is to be established by means of the Second Duality Theorem 2.2.7, then $\underset{\sim}{\mathbf{M}}$ must be injective in $\mathfrak{X}$. Nevertheless, it is not known whether or not the injectivity of $\underset{\sim}{\mathbf{M}}$ in $\mathcal{X}$ is actually required for $\underset{\sim}{\mathbf{M}}$ to yield a full duality, and this remains one of the oldest and most tantalising open questions in the foundations of duality theory.

2.7 Full vs Strong Problem *Do there exist a finite algebra $\underline{\mathbf{M}}$ and a choice of $\underset{\sim}{\mathbf{M}}$ such that $\underset{\sim}{\mathbf{M}}$ yields a duality on $\mathcal{A} = \mathbb{ISP}\underline{\mathbf{M}}$ which is full but not strong, or equivalently, such that $\underset{\sim}{\mathbf{M}}$ yields a full duality on $\mathcal{A}$ with $\underset{\sim}{\mathbf{M}}$ non-injective in $\mathcal{X}$?*

In every known full duality, $\underset{\sim}{\mathbf{M}}$ is injective in $\mathcal{X}$. The corresponding statement about $\underline{\mathbf{M}}$ is false: the four-element Heyting algebra chain has an internal isomorphism which does not extend to an endomorphism (Theorem 4.2.3(iii)). Often we come to duality with prior information about the injectivity of $\underline{\mathbf{M}}$ in $\mathcal{A}$. The injectivity of $\underline{\mathbf{M}}$ in $\mathcal{A}$ says precisely that D maps embeddings to surjections, and from this we obtain the following analogue of Lemma 3.2.6 for $\underline{\mathbf{M}}$ by the exact dual argument.

2.8 Lemma *Assume that $\underset{\sim}{\mathbf{M}}$ yields a full duality on $\mathcal{A}$. Then the following are equivalent:*

(i) *$\underline{\mathbf{M}}$ is injective in $\mathcal{A}$;*

(ii) *a homomorphism $u\colon \mathbf{A} \to \mathbf{B}$ is an embedding in $\mathcal{A}$ (if and) only if the morphism $D(u)\colon D(\mathbf{B}) \to D(\mathbf{A})$ is a surjection in $\mathcal{X}$;*

(iii) *a morphism $\varphi\colon \mathbf{X} \to \mathbf{Y}$ is a surjection in $\mathcal{X}$ if (and only if) the homomorphism $E(\varphi)\colon E(\mathbf{Y}) \to E(\mathbf{X})$ is an embedding in $\mathcal{A}$.*

Proof The proof is the same as that of Lemma 3.2.6. The equivalence of (i) and (ii) is an immediate consequence of the definitions. To prove that (ii) implies (iii), use the fact that each $\varepsilon_{\mathbf{X}}$ is an isomorphism. To prove that (iii) implies (ii), use the fact that each $e_{\mathbf{A}}$ is an isomorphism. ■

Recall that $\underset{\sim}{\mathbf{M}}$ is a **total structure** if $H = \varnothing$, that is, if every operation of $\underset{\sim}{\mathbf{M}}$ is total rather than partial. In this case we find that we can establish strong duality by checking term closure for just the substructures of finite powers of $\underset{\sim}{\mathbf{M}}$. We state this condition explicitly in the next theorem. On a first reading, it is easy to miss the point where this argument depends on the absence of partial operations in $\underset{\sim}{\mathbf{M}}$.

2.9 Second Strong Duality Theorem *If $\underset{\sim}{\mathbf{M}}$ is a total structure which yields a duality on $\mathcal{A}$, then the following are equivalent:*

(i) *$\underset{\sim}{\mathbf{M}}$ yields a strong duality on $\mathcal{A}$;*

(ii) *$\underset{\sim}{\mathbf{M}}$ yields a strong duality on $\mathcal{A}_{\text{fin}}$;*

(iii) *$\underset{\sim}{\mathbf{M}}$ satisfies the **Finite Term Closure** condition—*

(FTC) *If $\mathbf{X}$ is a substructure of $\underset{\sim}{\mathbf{M}}^n$ for some $n \in \mathrm{N}$ and $y \in M^n \backslash X$, then there exist term functions $\sigma, \tau\colon \underset{\sim}{\mathbf{M}}^n \to \underset{\sim}{\mathbf{M}}$ on $\underline{\mathbf{M}}$ (that is, morphisms) which agree on X but not at y.*

Proof It is trivial that (i) implies (ii), and (ii) is equivalent to (iii) since (FTC) is precisely the condition for a duality on $\mathcal{A}_{\mathrm{fin}}$ to be a strong duality on $\mathcal{A}_{\mathrm{fin}}$.

Finally, to prove that (iii) implies (i), assume (FTC). Let $\mathbf{X}$ be a closed substructure of $\underset{\sim}{\mathbf{M}}^S$, and let $z \in M^S \backslash X$. Then there is a basic clopen set U containing z and not intersecting X given as

$$U = \{\, x \in M^S \mid \pi_F(x) = \pi_F(z) \,\}$$

where $F \subseteq S$ is finite and $\pi_F \colon \underset{\sim}{\mathbf{M}}^S \to \underset{\sim}{\mathbf{M}}^F$ is the projection morphism. Then we have that $\pi_F(z) \notin \pi_F(X)$ and that $\pi_F(\mathbf{X})$ is a closed substructure of $\underset{\sim}{\mathbf{M}}^F$. By (FTC) there are F-ary term functions σ and τ that agree on $\pi_F(X)$ but disagree at $\pi_F(z)$. Consequently the S-ary term functions $\sigma \circ \pi_F$ and $\tau \circ \pi_F$ agree on X but not at z. ■

The careful reader will note that the image of $\mathbf{X}$ under the morphism π_F need not be a substructure of $\underset{\sim}{\mathbf{M}}^F$ if partial operations are present. (See Warning 1.4.1.) This point will play a significant role in the remainder of this chapter. While Lemmas 3.2.6 and 3.2.8 highlight the symmetry between $\underline{\mathbf{M}}$ in $\mathcal{A}$ and $\underset{\sim}{\mathbf{M}}$ in $\mathcal{X}$ which has characterised our development of duality theory, this symmetry begins to break down in the presence of partial operations in $\underset{\sim}{\mathbf{M}}$. For example, partial operations preclude a full converse to our next lemma.

2.10 Injectivity Lemma *Assume that $\underset{\sim}{\mathbf{M}}$ yields a full duality on $\mathcal{A}$. If $\underline{\mathbf{M}}$ is injective in $\mathcal{A}$, then $\underset{\sim}{\mathbf{M}}$ is injective in $\mathcal{X}$. The converse is true provided that for every morphism $\varphi \colon \mathbf{X} \to \mathbf{Y}$ in $\mathcal{X}$, the image of $\mathbf{X}$ under φ is a substructure of $\mathbf{Y}$. In particular this is the case when $\underset{\sim}{\mathbf{M}}$ is a total structure.*

Proof We will verify the third condition of Lemma 3.2.6. Let $u \colon \mathbf{A} \to \mathbf{B}$ in $\mathcal{A}$ and assume that $D(u)$ is an embedding. Because $\mathcal{A}$ is a quasi-variety of (total) algebras, we can factor $u = w \circ v$ where $v \colon \mathbf{A} \to \mathbf{C}$ is a surjection and $w \colon \mathbf{C} \to \mathbf{B}$ is an embedding. Then $D(u) = D(v) \circ D(w)$. Focusing on $D(w)$ we notice that it is an embedding since $D(u)$ is an embedding, and it is a surjection by Lemma 3.2.8. Consequently $D(w)$ is an isomorphism, whence $ED(w)$ is an isomorphism as well. By Figure 1.2, $w = e_{\mathbf{B}}^{-1} \circ ED(w) \circ e_{\mathbf{C}}$ is an isomorphism, and from this it follows that u is a surjection. This establishes (iii) of Lemma 3.2.6 which gives us the injectivity of $\underset{\sim}{\mathbf{M}}$.

The converse is established by the same argument applied dually, where we observe that the factoring of u into a surjection followed by an embedding can be mimicked when its image is a substructure. ■

We can now give purely finite conditions for the existence of a strong duality when $\underset{\sim}{\mathrm{M}}$ is a total structure and R is finite by combining the Second Duality Theorem 2.2.7 and the Second Strong Duality Theorem 3.2.9.

2.11 Third Strong Duality Theorem *Assume that the structure* $\underset{\sim}{\mathrm{M}} = \langle M; G, R, \mathcal{T}\rangle$ *is total and that* R *is finite. Then the following are equivalent:*

(i) $\underset{\sim}{\mathrm{M}}$ *yields a strong duality on* $\mathcal{A}$*;*

(ii) $\underset{\sim}{\mathrm{M}}$ *yields a strong duality on* $\mathcal{A}_{\mathrm{fin}}$*;*

(iii) (IC) *and* (FTC) *hold.*

Proof (i) implies (ii) trivially, so now assume (ii). To prove (IC), let $n \in \mathbb{N}$ and let $\mathbf{X} \leqslant \underset{\sim}{\mathrm{M}}^n$. By (ii) we have that $\mathbf{X}$ is term-closed, and from the finitary version of the Third Duality Theorem 3.1.6 we conclude that each morphism from $\mathbf{X}$ into $\underset{\sim}{\mathrm{M}}$ extends to an n-ary term function on $\underline{\mathbf{M}}$. Since (FTC) is given directly by (ii), we have (iii). Finally, (iii) implies (i) since duality follows from (IC) by the Second Duality Theorem 2.2.7 and strong duality follows from (FTC) by the Second Strong Duality Theorem 3.2.9. ■

3.3 Producing strong dualities

In this section we will gather together the theory that we have developed so far and use it to prove two theorems: the Two-for-One Strong Duality Theorem and the NU Strong Duality Theorem, together with several useful corollaries of each. Until very recently these two results, augmented by the $\underset{\sim}{\mathrm{M}}$-Shift Strong Duality Lemma, could be used to produce all known full dualities, all of which are in fact strong dualities. The Two-for-One Strong Duality Theorem is derived from our description of the dual category $\mathbb{I}D(\mathcal{A})$ as the isomorphs of term-closed sets, while the NU Strong Duality Theorem derives from our characterisation of $\mathbb{I}D(\mathcal{A})$ as the isomorphs of hom-closed sets. It is here that each of these two divergent approaches comes to fruition.

We first address the case where both partial operations *and* relations can be avoided so that our strongest results will apply. These results reduce the discussion to properties of $\mathcal{A}_{\mathrm{fin}}$ and $\mathcal{X}_{\mathrm{fin}}$ where topology vanishes. Thus there arises a complete symmetry between the roles of $\underline{\mathbf{M}}$ and $\underset{\sim}{\mathrm{M}}$ which allows us to get two strong dualities for the price of one. If $\underset{\sim}{\mathrm{M}} = \langle M; G, \mathcal{T}\rangle$, that is, the sets H of partial operations and R of relations are empty, then we say that $\underset{\sim}{\mathrm{M}}$ is a **total algebra**. We say that an algebra $\mathbf{A}$ **has named constants** if the value of every constant unary term function on $\mathbf{A}$ is the value of a nullary term. The following is an immediate consequence of Lemma 3.1.2(ii).

3.1 Lemma *If a total algebra* $\underset{\sim}{\mathbf{M}} = \langle M; G, \mathcal{T}\rangle$ *yields a full duality on* $\mathcal{A}$, *then the algebra* $\underline{\mathbf{M}}' = \langle M; G\rangle$ *has named constants.*

In the following theorem, we include a new proof of the equivalence of (v), a fact which was recently discovered by Daniels [Dan97].

3.2 Two-for-One Strong Duality Theorem *Assume that both* $\underline{\mathbf{M}} = \langle M; F\rangle$ *and* $\underline{\mathbf{M}}' = \langle M; G\rangle$ *have named constants. Consider the total algebras* $\underset{\sim}{\mathbf{M}} = \langle M; G, \mathcal{T}\rangle$ *and* $\underset{\sim}{\mathbf{M}}' = \langle M; F, \mathcal{T}\rangle$ *and define* $\mathcal{A}' = \mathbb{ISP}\underline{\mathbf{M}}'$ *and* $\mathcal{X}' = \mathbb{IS}_{\mathrm{c}}\mathbb{P}^{+}\underset{\sim}{\mathbf{M}}'$. *Then the following are equivalent:*

(i) $\underset{\sim}{\mathbf{M}}$ *yields a strong duality on* $\mathcal{A}$;

(ii) $\underset{\sim}{\mathbf{M}}$ *and* $\underset{\sim}{\mathbf{M}}'$ *yield strong dualities on* $\mathcal{A}$ *and* $\mathcal{A}'$ *respectively;*

(iii) (IC) *and* (FTC) *hold with respect to* $\underset{\sim}{\mathbf{M}}$;

(iv) (IC) *and* (FTC) *hold with respect to* $\underset{\sim}{\mathbf{M}}'$;

(v) (FTC) *holds with respect to both* $\underset{\sim}{\mathbf{M}}$ *and* $\underset{\sim}{\mathbf{M}}'$;

(vi) (IC) *holds with respect to both* $\underset{\sim}{\mathbf{M}}$ *and* $\underset{\sim}{\mathbf{M}}'$;

(vii) *the algebras* $\underline{\mathbf{M}}$ *and* $\underline{\mathbf{M}}'$ *satisfy the following symmetric conditions–*

(a) *every homomorphism from a subalgebra of* $\underline{\mathbf{M}}^n$ *into* $\underline{\mathbf{M}}$ *extends to an* n*-ary term function on* $\underline{\mathbf{M}}'$, *and*

(b) *every homomorphism from a subalgebra of* $(\underline{\mathbf{M}}')^n$ *into* $\underline{\mathbf{M}}'$ *extends to an* n*-ary term function on* $\underline{\mathbf{M}}$.

Proof For the purposes of this argument, we say that $\underline{\mathbf{M}}$ **yields a duality on** $\mathcal{X}$ (on $\mathcal{X}_{\mathrm{fin}}$), or that $\underset{\sim}{\mathbf{M}}$ **yields a coduality on** $\mathcal{A}$ (on $\mathcal{A}_{\mathrm{fin}}$), if $\varepsilon_{\mathbf{X}}$ is an isomorphism for each $\mathbf{X}$ in $\mathcal{X}$ (in $\mathcal{X}_{\mathrm{fin}}$). We first establish the equivalence of (i), (ii), (iii) and (iv). Observe that $\underset{\sim}{\mathbf{M}}$ yields a duality (a coduality) on $\mathcal{A}_{\mathrm{fin}}$ if and only if $\underset{\sim}{\mathbf{M}}'$ yields a coduality (a duality) on $\mathcal{A}'_{\mathrm{fin}}$, since exactly the same sets and maps are involved. (This may not be the case if constants are not named: see Exercise 3.8.) Thus, $\underset{\sim}{\mathbf{M}}$ yields a full duality on $\mathcal{A}_{\mathrm{fin}}$ if and only if $\underset{\sim}{\mathbf{M}}'$ yields a full duality on $\mathcal{A}'_{\mathrm{fin}}$.

We now claim that the same is true of strong dualities. To prove this, assume that $\underset{\sim}{\mathbf{M}}'$ yields a strong duality on $\mathcal{A}'$. By the First Strong Duality Theorem 3.2.4, $\underset{\sim}{\mathbf{M}}'$ is injective in $\mathcal{X}'$. Applying the Injectivity Lemma 3.2.10 we conclude that $\underline{\mathbf{M}}'$ is injective in $\mathcal{A}'$ and therefore $\underset{\sim}{\mathbf{M}}$ is injective in $\mathcal{X}_{\mathrm{fin}}$. Since $\underset{\sim}{\mathbf{M}}'$ yields a strong duality on $\mathcal{A}'_{\mathrm{fin}}$, it also yields a full duality on $\mathcal{A}'_{\mathrm{fin}}$ and therefore $\underset{\sim}{\mathbf{M}}$ yields a full duality on $\mathcal{A}_{\mathrm{fin}}$. By the First Strong Duality Theorem 3.2.4, we conclude that $\underset{\sim}{\mathbf{M}}$ yields a strong duality on $\mathcal{A}_{\mathrm{fin}}$. From the Third Strong Duality Theorem 3.2.11 it follows that $\underset{\sim}{\mathbf{M}}$ yields a strong duality on $\mathcal{A}$. By symmetry, we have

$\underset{\sim}{\mathbf{M}}$ *yields a strong duality on* $\mathcal{A}$
if and only if
$\underset{\sim}{\mathbf{M}}'$ *yields a strong duality on* $\mathcal{A}'$.

By the Third Strong Duality Theorem 3.2.11, the former statement is equivalent to (iii) and the latter to (iv). Their conjunction is exactly (ii). Since the equivalence of (i) and (iii) is given by the Third Strong Duality Theorem 3.2.11, we see that (i), (ii), (iii) and (iv) are equivalent.

Condition (vii) is simply an explicit version of (vi) expressed as (finitary) properties of the algebras alone. Since (iii) and (iv) trivially imply (vi), and (ii) trivially implies (v), we will complete the proof by showing that (vi) implies (i) and that (v) implies (iii).

Assuming (vi), we apply Lemma 2.2.5 to conclude that $\underset{\sim}{\mathbf{M}}$ yields a duality on $\mathcal{A}_{\text{fin}}$ and that $\underset{\sim}{\mathbf{M}}$ is injective in $\mathfrak{X}_{\text{fin}}$. Similarly, $(\widetilde{\text{IC}})$ holding with respect to $\underset{\sim}{\mathbf{M}}'$ tells us that $\underset{\sim}{\mathbf{M}}'$ yields a duality on $\mathcal{A}'_{\text{fin}}$, that is, $\underset{\sim}{\mathbf{M}}$ yields a coduality on $\mathcal{A}_{\text{fin}}$. Thus $\underset{\sim}{\mathbf{M}}$ yields a strong duality on $\mathcal{A}_{\text{fin}}$. Now (i) follows by the Third Strong Duality Theorem 3.2.11.

To prove that (v) implies (iii), we assume that (FTC) holds with respect to $\underset{\sim}{\mathbf{M}}'$ and show that (IC) must hold with respect to $\underset{\sim}{\mathbf{M}}$. Let $\mathbf{X} \leqslant \underset{\sim}{\mathbf{M}}^n$ and let $\varphi\colon \mathbf{X} \to \underset{\sim}{\mathbf{M}}$. We take $\mathbf{F} \leqslant \underline{\mathbf{M}}^{M^n}$ to be the algebra of n-ary term functions of $\underset{\sim}{\mathbf{M}}$ and take $\mathbf{F}_X \leqslant \underset{\sim}{\mathbf{M}}^X$ to be the restriction of $\mathbf{F}$ to X. Let u and v be X-ary term functions of $\underline{\mathbf{M}}'$ that agree on $\mathbf{F}_X$. We will show that u and v agree at φ. Since $\mathbf{F}_X \leqslant (\underset{\sim}{\mathbf{M}}')^X$ and (FTC) holds with respect to $\underset{\sim}{\mathbf{M}}'$, we will then be able to conclude that $\varphi \in F_X$, that is, φ extends to an n-ary term function of $\underline{\mathbf{M}}$.

To see that $u(\varphi) = v(\varphi)$, let $X = \{x_1, \dots, x_k\}$ and let $x_i = (x_{i1}, \dots, x_{in}) \in M^n$. Then

$$\begin{aligned} u(\varphi) &= u(\varphi(x_1), \dots, \varphi(x_k)) \\ &= u(\varphi(x_{11}, \dots, x_{1n}), \dots, \varphi(x_{k1}, \dots, x_{kn})) \\ &= \varphi(u(x_{11}, \dots, x_{k1}), \dots, u(x_{1n}, \dots, x_{kn})) \end{aligned}$$

(since $\mathbf{X} \leqslant (\underline{\mathbf{M}}')^n$ is a subalgebra, $\varphi\colon \mathbf{X} \to \underline{\mathbf{M}}'$ is a homomorphism and u is a term function of $\underline{\mathbf{M}}'$)

$$\begin{aligned} &= \varphi(u(\pi_1{\restriction}_X), \dots, u(\pi_k{\restriction}_X)) \\ &= \varphi(v(\pi_1{\restriction}_X), \dots, v(\pi_k{\restriction}_X)) \end{aligned}$$

since each $\pi_i{\restriction}_X$ is in $\mathbf{F}_X$. By the reverse computation, we see that this is equal to $v(\varphi)$. ■

Given $\underline{\mathbf{M}}$, where do we find the algebra $\underline{\mathbf{M}}'$? The following corollary tells us when we can use the first candidate that comes to mind.

3.3 Strong Self-Duality Theorem *Assume that $\underline{\mathbf{M}}$ has named constants and let $\underset{\sim}{\mathbf{M}} = \langle M; F, \mathcal{T}\rangle$ be obtained by augmenting $\underline{\mathbf{M}} = \langle M; F\rangle$ with the discrete topology. Then $\underset{\sim}{\mathbf{M}}$ yields a strong duality on $\mathcal{A}$ (and $\underline{\mathbf{M}}$ is injective in $\mathcal{A}$) if and only if $(\widetilde{\text{IC}})$ holds.*

Underlying our quest for strong dualities is a fundamental question: can every duality be upgraded to a strong duality? We state this question more precisely below.

3.4 Strong Upgrade Problem *Assume that $\underset{\sim}{\mathbf{M}}$ yields a duality on $\mathcal{A}$. Can we extend the structure $\underset{\sim}{\mathbf{M}}$ by adding (finitely many) partial or total operations and so obtain a structure $\underset{\sim}{\mathbf{M}}'$ which yields a strong duality (or a full duality) on $\mathcal{A}$?*

Consider, for example, a finite algebra $\underline{\mathbf{M}}$ which is a lattice under some pair of binary term functions. The NU Duality Theorem 2.3.4 gives us a duality on $\mathcal{A} = \mathbb{ISP}\underline{\mathbf{M}}$ by taking for R (a set that entails all of) the binary algebraic relations on $\underline{\mathbf{M}}$. Can finite G and H be chosen to make this into a full duality? A strong duality? We will show that *every* duality which arises from the NU Duality Theorem 2.3.4 can be upgraded to a strong duality by adding finitely many (partial) operations to $\underset{\sim}{\mathbf{M}}$. Our approach will be to use the Brute Force Strong Duality Theorem 3.2.2 and look for partial operations which will ensure that substructures of powers of $\underset{\sim}{\mathbf{M}}$ are hom-closed. The key is to observe that, in a congruence distributive variety, Bjarni Jónsson has already done all but a finite amount of the work for us, and that a handful of partial operations will do the rest!

3.5 Lemma *Let M be a finite set, let U be an ultrafilter on a set I, and let $f\colon M^I \to M$ be defined by*

$$f(\beta) = a \quad \textit{if and only if} \quad \beta^{-1}(a) \in U.$$

Then every topologically closed subspace X of a product space M^S is closed under f.

Proof Let $\boldsymbol{x} \in X^I$. To show that $\boldsymbol{f}(\boldsymbol{x}) \in X$ we let $F \subseteq S$ be finite and check that $\boldsymbol{f}(\boldsymbol{x})$ agrees with a member of X on F. For each $s \in F$, denote $f(\pi_s \circ \boldsymbol{x}) \in M$ by a_s. Thus $(\pi_s \circ \boldsymbol{x})^{-1}(a_s) \in U$. Since F is finite,

$$J = \bigcap \{\, (\pi_s \circ \boldsymbol{x})^{-1}(a_s) \mid s \in F \,\} \in U.$$

Then, for any $j \in J$, we find that $\boldsymbol{f}(\boldsymbol{x})$ agrees with $\mathbf{x}(j)$ on F since, for all $s \in F$,

$$\boldsymbol{f}(\boldsymbol{x})(s) = f(\pi_s \circ \boldsymbol{x}) = a_s = (\pi_s \circ \boldsymbol{x})(j) = \boldsymbol{x}(j)(s). \qquad \blacksquare$$

Every congruence on a finite algebra $\mathbf{Q}$ is a meet of meet-irreducible congruences on $\mathbf{Q}$ (Lemma A.9). Let $\operatorname{irr}(\mathbf{Q})$ be the least n such that the zero congruence $\mathbf{0}^{\mathbf{Q}}$ on $\mathbf{Q}$ is a meet of n meet-irreducible congruences. Then $\mathbf{Q}$ is subdirectly irreducible if and only if $\operatorname{irr}(\mathbf{Q}) = 1$. We define the **irreducibility index** of a finite algebra $\underline{\mathbf{M}}$ by

$$\operatorname{Irr}(\underline{\mathbf{M}}) = \max\{\, \operatorname{irr}(\mathbf{Q}) \mid \mathbf{Q} \text{ is a subalgebra of } \underline{\mathbf{M}} \,\}.$$

Our next lemma shows how to apply Jónsson's Lemma A.15 in a congruence distributive variety to obtain strong brute force.

3.6 Lemma *Assume that* $\underline{\mathbf{M}}$ *generates a congruence distributive variety and that the enriched partial clone of* $\underset{\sim}{\mathbf{M}}$ *includes all* n*-ary algebraic partial operations on* $\underline{\mathbf{M}}$ *for* $0 \leqslant n \leqslant \operatorname{Irr}(\underline{\mathbf{M}})$. *Then every closed substructure of a power of* $\underset{\sim}{\mathbf{M}}$ *is hom-closed.*

Proof Let $\mathbf{X} \leqslant \underset{\sim}{\mathbf{M}}^S$ be a closed substructure, $\mathbf{B} \leqslant \underline{\mathbf{M}}^I$ a subalgebra and $g\colon \mathbf{B} \to \underline{\mathbf{M}}$ a homomorphism. If $g(\mathbf{B})$ is a trivial subalgebra of $\underline{\mathbf{M}}$, then we may take g to be unary and conclude that $\mathbf{X}$ is closed under g. If $g(\mathbf{B})$ is non-trivial, then there are $k \leqslant \operatorname{Irr}(\underline{\mathbf{M}})$ completely meet-irreducible congruences $\psi_1, \psi_2, \dots, \psi_k$ on $\mathbf{B}$ with $\bigcap \psi_i = \ker(g)$. Since $\underline{\mathbf{M}}$ generates a congruence distributive variety we can apply Jónsson's Lemma A.15 to get ultrafilters $U_1, U_2, \dots, U_k$ on I determining congruences $\theta_1 \subseteq \psi_1$, $\theta_2 \subseteq \psi_2, \dots, \theta_k \subseteq \psi_k$ on $\mathbf{B}$. For $i = 1, 2, \dots, k$, let $f_i\colon \underline{\mathbf{M}}^I \to \underline{\mathbf{M}}$ be the homomorphism defined by

$$f_i(\beta) = a \quad \text{if and only if} \quad \beta^{-1}(a) \in U_i$$

and let

$$f = f_1 \sqcap \cdots \sqcap f_k\colon \underline{\mathbf{M}}^I \to \underline{\mathbf{M}}^k,$$

so that

$$f(\beta) = (f_1(\beta), \dots, f_k(\beta))$$

for $\beta \in M^I$. Let $\mathbf{A} \leqslant \underline{\mathbf{M}}^k$ be the image of $\mathbf{B}$ under f, and let $f{\restriction_B}\colon \mathbf{B} \to \mathbf{A}$ be the restriction of f. We claim that g can be factored through $f{\restriction_B}$, that is, that there is a homomorphism $h\colon \mathbf{A} \to \underline{\mathbf{M}}$ such that $g = h \circ f{\restriction_B}$. To prove this we must show that $\ker(f{\restriction_B}) \subseteq \ker(g)$. Let $\beta, \gamma \in B$ with $f(\beta) = f(\gamma)$. Then, for $i = 1, 2, \dots, k$, we have

$$f_i(\beta) = f_i(\gamma) = a_i \quad \text{for some} \quad a_i \in M.$$

Consequently, $\beta^{-1}(a_i)$ and $\gamma^{-1}(a_i)$ are in U_i and therefore

$$\beta^{-1}(a_i) \cap \gamma^{-1}(a_i) \in U_i.$$

Moreover, β and γ have the same value, a_i, on $\beta^{-1}(a_i) \cap \gamma^{-1}(a_i)$ which means that $(\beta, \gamma) \in \theta_i$. Since this is true for each i, we have

$$(\beta, \gamma) \in \bigcap \{\, \theta_i \mid i = 1, \dots, k \,\} \subseteq \bigcap \{\, \psi_i \mid i = 1, \dots, k \,\} = \ker(g).$$

From this we conclude that $\ker(f{\restriction_B}) \subseteq \ker(g)$.

The structure $\mathbf{X}$ is closed under each f_i, by Lemma 3.3.5, and is closed under the k-ary partial operation h by hypothesis. Assume that

$x \in X^I$ such that, for each $s \in S$, the composition $\pi_s \circ x$ is in B. Then

$$\begin{aligned} g(x)(s) = g(\pi_s \circ x) &= hf(\pi_s \circ x) \\ &= h(f_1(\pi_s \circ x), f_2(\pi_s \circ x), \dots, f_k(\pi_s \circ x)) \\ &= h(f_1(x), f_2(x), \dots, f_k(x))(s) \end{aligned}$$

whence $g(x) \in X$ since $\mathbf{X}$ is closed under h and each f_i. Thus $\mathbf{X}$ is hom-closed. ■

This result allows us to offer the following partial solution to The Strong Upgrade Problem 3.3.4.

3.7 Theorem *Assume that $\underline{\mathbf{M}}$ generates a congruence distributive variety and that $\underset{\sim}{\mathbf{M}}$ yields a duality on $\mathcal{A}$. If $\underset{\sim}{\mathbf{M}}'$ is obtained from $\underset{\sim}{\mathbf{M}}$ by adding to $G \cup H$ all n-ary algebraic partial operations where $0 \leqslant n \leqslant \mathrm{Irr}(\underline{\mathbf{M}})$, then $\underset{\sim}{\mathbf{M}}'$ yields a strong duality on $\mathcal{A}$.*

Since any algebra which has a near-unanimity term generates a congruence distributive variety (Mitschke's Theorem A.14), we can combine these results with the NU Duality Theorem 2.3.4 to obtain our central strong duality theorem. In order to state this result we define $\mathcal{P}_n$ and $\mathcal{B}_n$, for $n = 1, 2, 3, \dots$, to be the sets of all n-ary (partial) operations and relations, respectively, which are algebraic over $\underline{\mathbf{M}}$. Recall that K denotes the set of one-element subalgebras of $\underline{\mathbf{M}}$.

3.8 NU Strong Duality Theorem *Let $k \geqslant 2$ and assume that $\underline{\mathbf{M}}$ has a $(k{+}1)$-ary near-unanimity term. If*

$$\underset{\sim}{\mathbf{M}} = \langle M; K, H, \mathcal{B}_k, \mathcal{T} \rangle \text{ where } H = \bigcup \{ \mathcal{P}_n \mid 1 \leqslant n \leqslant \mathrm{Irr}(\underline{\mathbf{M}}) \},$$

then any structure that strongly entails $\underset{\sim}{\mathbf{M}}$ yields a strong duality on $\mathcal{A}$.

Most known strong dualities are obtained as applications of this theorem. It is a powerful result, as it gives us an exact recipe for constructing from a near-unanimity term for $\underline{\mathbf{M}}$ a structure $\underset{\sim}{\mathbf{M}}$ that will yield a strong duality on $\mathcal{A}$. With this result in hand, the focus of our quest shifts to that of applying the $\underset{\sim}{\mathbf{M}}$-Shift Strong Duality Lemma to obtain a simple and manageable choice of $\underset{\sim}{\mathbf{M}}$ which will still yield a strong duality on $\mathcal{A}$. This may be difficult if k and $\mathrm{Irr}(\underline{\mathbf{M}})$ are not both small. Even if k and $\mathrm{Irr}(\underline{\mathbf{M}})$ are small the structure on $\underset{\sim}{\mathbf{M}}$ may be huge. Given that a duality is intended to produce a category $\mathcal{X}$ which can be used to better understand the original category $\mathcal{A}$, we are led to the rather loose notion of a useful duality. A duality is most likely to be useful if, in some reasonable sense, the structures in $\mathcal{X}$ are simpler or more tractable than their duals in $\mathcal{A}$.

In Chapter 6 we will discover just when a strong duality is available with one of several different useful properties. Our present goal is to carve out a few special cases of the NU Strong Duality Theorem which will lead to dualities that we can use. Recall that a ternary near-unanimity term is called a **majority term**.

3.9 NU Strong Duality Corollary *Assume that $\underline{\mathbf{M}}$ has a majority term and that all of the non-trivial subalgebras of $\underline{\mathbf{M}}$ are subdirectly irreducible. Then any structure that strongly entails $\underset{\sim}{\mathbf{M}} = \langle M; K, \mathcal{P}_1, \mathcal{B}_2, \mathcal{T}\rangle$ yields a strong duality on $\mathcal{A}$.*

The NU Strong Duality Theorem remains a highly untapped resource, as almost all interesting applications of it to date follow from the NU Strong Duality Corollary where we have the strongest special case: $k = 2$ and $\mathrm{Irr}(\underline{\mathbf{M}}) = 1$. Even the NU Strong Duality Corollary can lead to complex dual categories, for $\underline{\mathbf{M}}^2$ may have many subalgebras. In Chapter 4 we will see a range of examples in which the methods of Chapter 2 can be used to reduce these binary relations to a small and convenient number. We return to this problem in Chapters 8 and 9 where a full theory for achieving such reductions will be developed.

Judging by the complexity of the type of $\underset{\sim}{\mathbf{M}}$, we can expect to get the best strong dualities from the NU Strong Duality Theorem when k is small. Since k must be at least 2, this never leaves us less than binary relations. In order to eliminate binary relations from the type of $\underset{\sim}{\mathbf{M}}$, we will need something more. The key turns out to be the presence of a term p on $\underline{\mathbf{M}}$ satisfying

$$p(x, y, y) = p(x, y, x) = p(y, y, x) = x.$$

Such a term is called a **Pixley arithmeticity term**, and its presence witnesses the fact that the algebras in the variety generated by $\underline{\mathbf{M}}$ all have a distributive lattice of permuting congruences. (See Pixley's Theorem A.17.) Simultaneously it gives us the ternary near-unanimity term, or majority term, $p(x, p(x, y, z), z)$ which leads to a strong duality with at most binary relations. While congruence distributivity is subsumed under near-unanimity (Mitschke's Theorem A.14), congruence permutability is new. Congruence permutability will allow us to use Fleischer's Theorem A.12 to recognise the binary relations of $\underset{\sim}{\mathbf{M}}$ (which are subalgebras of $\underline{\mathbf{M}} \times \underline{\mathbf{M}}$) and, under the right conditions, replace them with partial unary operations.

3.10 Lemma *Suppose congruences on subalgebras of $\underline{\mathbf{M}} \times \underline{\mathbf{M}}$ permute and each non-trivial homomorphic image of a subalgebra of $\underline{\mathbf{M}}$ is in $\mathcal{A}$. Then for every $m \geq 1$, the partial operations $\mathcal{P}_m$ entail $\mathcal{B}_2$.*

Proof Since $\mathcal{P}_m \vdash \mathcal{P}_1$, we need only show that $\mathcal{P}_1 \vdash \mathcal{B}_2$. Let $r \in \mathcal{B}_2$ be binary. Since congruences on the subalgebras of $\underline{\mathbf{M}} \times \underline{\mathbf{M}}$ permute, we can use Fleischer's characterisation of its subalgebras (Fleischer's Theorem A.12). Let $\mathbf{P}_1$ and $\mathbf{P}_2$ be subalgebras of $\underline{\mathbf{M}}$, let $\mathbf{C}$ be an algebra, and let $f_1: \mathbf{P}_1 \to \mathbf{C}$ and $f_2: \mathbf{P}_2 \to \mathbf{C}$ be surjections such that $r = \ker(f_1, f_2)$ is the set of $(p_1, p_2) \in P_1 \times P_2$ such that $f_1(p_1) = f_2(p_2)$. Since $\mathbf{C} \in \mathbb{ISP}\underline{\mathbf{M}}$ is finite, there is a finite set of homomorphisms $h_1, h_2, \dots, h_m: \mathbf{C} \to \underline{\mathbf{M}}$ that separate $\mathbf{C}$.

We leave it to the reader to check that

$$r = \bigcap_{j=1}^{m} \ker(h_j \circ f_1, h_j \circ f_2).$$

By Constructs for Entailment 2.4.5 (6) and (14) we conclude that the partial endomorphisms $h_j \circ f_1$ and $h_j \circ f_2$, for $j = 1, \dots, m$, entail r. Thus $\mathcal{P}_1$ entails r. ■

Combining this result with the NU Strong Duality Theorem gives us conditions under which $K \cup \mathcal{P}_m$ alone will give us a strong duality.

3.11 Arithmetic Strong Duality Theorem *Assume that* $\underline{\mathbf{M}}$ *generates an arithmetical variety and that every homomorphic image of a subalgebra of* $\underline{\mathbf{M}}$ *is in* $\mathcal{A}$. *Let* $m = \mathrm{Irr}(\underline{\mathbf{M}})$. *Then any structure that strongly entails* $\underset{\sim}{\mathbf{M}} = \langle M; K, \mathcal{P}_m, \mathcal{T}\rangle$ *yields a strong duality on* $\mathcal{A}$.

Proof Let p, as above, be a Pixley arithmeticity term for $\underline{\mathbf{M}}$ so that $p(x, p(x, y, z), z)$ is the resulting majority term. By the NU Strong Duality Theorem 3.3.8, we obtain a strong duality for the quasi-variety $\mathcal{A}$ if we define $\underset{\sim}{\mathbf{M}}' = \langle M; K, \mathcal{P}_m, \mathcal{B}_2, \mathcal{T}\rangle$, and $\mathcal{B}_2$ can be eliminated using Lemma 3.3.10 and the $\underset{\sim}{\mathbf{M}}$-Shift Strong Duality Lemma 3.2.3(iii). ■

In the early seventies A. Pixley [Pix70] and H. Werner [Wer70] identified a special class of algebras $\underline{\mathbf{M}}$ which satisfy all of the hypotheses of the Arithmetic Strong Duality Theorem. They called $\underline{\mathbf{M}}$ **quasi-primal** if it satisfied the equivalent conditions of the theorem below. By a **partial automorphism** of an algebra we mean an isomorphism between two of its subalgebras.

3.12 Quasi-primality Theorem *For a finite algebra* $\underline{\mathbf{M}}$, *the following are equivalent:*

(i) *the* **ternary discriminator** *operation*

$$t^M(x, y, z) := \begin{cases} x & \text{if } x \neq y, \\ z & \text{if } x = y \end{cases}$$

is a term function of $\underline{\mathbf{M}}$;

(ii) *every finitary operation on M that preserves the subalgebras of $\underline{\mathbf{M}}$ and the partial automorphisms of $\underline{\mathbf{M}}$ is a term function of $\underline{\mathbf{M}}$;*

(iii) *$\underline{\mathbf{M}}$ generates an arithmetical variety and every non-trivial subalgebra of $\underline{\mathbf{M}}$ is simple;*

(iv) *$\underline{\mathbf{M}}$ has a ternary majority term and every subalgebra of $\underline{\mathbf{M}}^2$ is either a product of two subalgebras of $\underline{\mathbf{M}}$ or the graph of a partial automorphism of $\underline{\mathbf{M}}$.*

Proof Trivially (ii) implies (i), so assume (i) to prove (iii). Since a ternary discriminator term is also an arithmeticity term, it remains to show that every subalgebra $\mathbf{N}$ of $\underline{\mathbf{M}}$ with more than one element is simple. Accordingly, let θ be a congruence on $\mathbf{N}$ and assume that $a\theta b$ where $a \neq b$. Choose any $c \in M$. Then

$$a = t^{\mathbf{M}}(a, b, c)\,\theta\, t^{\mathbf{M}}(b, b, c) = c,$$

showing that $a\theta c$ and consequently θ is the universal congruence. Next assume (iii) to prove (iv). If p is a Pixley arithmeticity term for $\underline{\mathbf{M}}$, then $m(x, y, z) := p(x, p(x, y, z), z)$ is a majority term for $\underline{\mathbf{M}}$. Since the non-trivial subalgebras of $\underline{\mathbf{M}}$ are simple, the description in (iv) of the subalgebras of $\underline{\mathbf{M}}^2$ follows immediately from Fleischer's Theorem A.12. Finally assume (iv) to prove (ii). From the NU Duality Theorem 2.3.4 we see that we will obtain a duality by taking for H the partial automorphisms of $\underline{\mathbf{M}}$ and for R the subalgebras of $\underline{\mathbf{M}}$. This duality satisfies condition (CLO), which gives us (ii). ■

3.13 Quasi-primal Strong Duality Theorem *Assume that $\underline{\mathbf{M}}$ is quasi-primal. Then $\mathcal{A}$ is a variety and*

$$\underset{\sim}{\mathrm{M}} = \langle M; K \cup \mathrm{Aut}(\underline{\mathbf{M}}), \mathrm{Iso}(\underline{\mathbf{M}}), \mathcal{T}\rangle$$

yields a strong duality on $\mathcal{A}$, where K is the set of one-element subalgebras of $\underline{\mathbf{M}}$ (viewed as distinguished elements of $\underset{\sim}{\mathrm{M}}$), where $\mathrm{Aut}(\underline{\mathbf{M}})$ is the set of automorphisms of $\underline{\mathbf{M}}$ (viewed as unary operations of $\underset{\sim}{\mathrm{M}}$) and $\mathrm{Iso}(\underline{\mathbf{M}})$ is the set of proper partial automorphisms of $\underline{\mathbf{M}}$ (viewed as unary partial operations on $\underset{\sim}{\mathrm{M}}$).

Proof The fact that $\mathcal{A}$ is a variety follows from Theorem 1.3.6(iii), and the duality result is an immediate consequence of the Arithmetic Strong Duality Theorem and the Quasi-primality Theorem (iii). ■

Even more specialised are the **semi-primal** algebras, that is, quasi-primal algebras having no isomorphisms between subalgebras with more than one element other than the identity maps. If $\underline{\mathbf{M}}$ is semi-primal, then its only partial automorphisms are the identity maps on its subalgebras and isomorphisms between one-element subalgebras. In this case a better duality is obtained by shifting from partial operations to relations, as it leads to a very nice description of the dual category.

3.14 Semi-primal Strong Duality Theorem *Let* $\underline{\mathbf{M}}$ *be a semi-primal algebra, where* K *is the set of one-element subalgebras of* $\underline{\mathbf{M}}$ *(viewed as distinguished elements) and* $\mathcal{B}_1$ *is the set of all subalgebras of* $\underline{\mathbf{M}}$ *(viewed as unary relations).*

(i) $\underset{\sim}{\mathbf{M}} = \langle M; K, \mathcal{B}_1, \mathcal{T}\rangle$ *yields a strong duality on the variety* $\mathcal{A}$.

(ii) *The non-trivial members of the dual category* $\mathcal{X} = \mathbb{IS}_c\mathbb{P}^+\underset{\sim}{\mathbf{M}}$ *are the non-trivial structured Boolean spaces* $\mathbf{X} = \langle X; K^{\mathbf{X}}, \mathcal{B}_1^{\mathbf{X}}, \mathcal{T}\rangle$ *such that, for all* $e \in K$ *and all subalgebras* $\mathbf{A}$ *and* $\mathbf{B}$ *of* $\underline{\mathbf{M}}$,

  (a) $A^{\mathbf{X}}$ *is a closed subspace of* $\mathbf{X}$ *and* $M^{\mathbf{X}} = X$,

  (b) $A^{\mathbf{X}} = \{e^{\mathbf{X}}\}$ *if* $A = \{e\}$,

  (c) *if* $A \cap B \neq \varnothing$, *then* $(A \cap B)^{\mathbf{X}} = A^{\mathbf{X}} \cap B^{\mathbf{X}}$,

  (d) *if* $A \cap B = \varnothing$, *then* $A^{\mathbf{X}} \cap B^{\mathbf{X}} = \varnothing$.

Proof (i) Use part (ii)(c) of the $\underset{\sim}{\mathbf{M}}$-Shift Strong Duality Lemma 3.2.3 to replace the partial identity maps with their domains and the identity map on M, and use part (i) to eliminate the identity on M. Isomorphisms between trivial subalgebras can be eliminated since K has been included.

(ii) We ask the reader to check that conditions (a)–(d) can be expressed as quasi-atomic formulæ and therefore, by the Preservation Theorem 1.4.3, hold if $\mathbf{X} \in \mathcal{X}$. Conversely, assume (a)–(d). To verify the conditions of the Separation Theorem 1.4.4, let $x_1 \in X$ and let $\mathbf{B}$ be a subalgebra of $\underline{\mathbf{M}}$ such that $x \notin B^{\mathbf{X}}$. For each $x \in X$, use (a) to choose a clopen set $Y(x)$ such that $x \in Y(x)$ and $Y(x) \cap A^{\mathbf{X}} = \varnothing$ for each $\mathbf{A} \in \mathcal{B}_1$ where $x \notin A^{\mathbf{X}}$. By compactness we can find a finite set $\{x_1, x_2, \dots, x_n\} \subseteq X$ such that $Y(x_1) \cup Y(x_2) \cup \dots \cup Y(x_n) = X$. For $i = 1, 2, \dots, n$, we construct disjoint clopen subsets $X_i \subseteq Y(x_i)$ which still cover X by taking $X_1 = Y(x_1)$ and $X_{k+1} = Y(x_{k+1})\backslash(Y(x_1) \cup \dots \cup Y(x_k))$. By (d) we can choose, for $i = 1, 2, \dots, n$, an element $a_i \in C_i := \bigcap\{\, A \in \mathcal{B}_1 \mid x_i \in A^{\mathbf{X}} \,\}$. If $i = 1$, then we further constrain the choice of a_1 by using (c) and (d) to observe that $x_1 \in \bigcap\{\, A^{\mathbf{X}} \mid x_1 \in A^{\mathbf{X}} \,\} = C_1^{\mathbf{X}} \nsubseteq B^{\mathbf{X}}$. By (d), we have $C_1 \nsubseteq B$ so that we can choose $a_1 \in C_1\backslash B$.

We leave it to the reader to check that φ is a morphism from $\mathbf{X}$ to $\underset{\sim}{\mathbf{M}}$, where $\varphi(x) = a_i$ if $x \in X_i$. However, we chose $\varphi(x_1) = a_1 \notin B$. A similar argument shows that distinct elements x_1 and x_2 can be separated by a morphism into $\mathbf{X}$: simply choose $Y(x_1)$ and $Y(x_2)$ to be disjoint and use (b) to ensure that $a_1 \neq a_2$. ■

Exercises 3

Exercises from the text. Review the proof of Lemma 3.1.1 and verify the final assertion, that (i) and (ii) hold when $\mathcal{A}$ and $\mathcal{X}$ are replaced by $\mathcal{A}_{\text{fin}}$ and $\mathcal{X}_{\text{fin}}$, respectively. Prove that the map $\delta_{\mathbf{X}}$ in part (iv) of the Closure Theorem 3.1.3 is an embedding. Fill in the proof of the $\underset{\sim}{\mathbf{M}}$-Shift

Strong Duality Lemma 3.2.3. Review the proof of the First Strong Duality Theorem 3.2.4 and verify that the same argument holds in the finite case. Prove Lemma 3.2.8. Verify the details of the proof of the Semi-primal Strong Duality Theorem 3.3.14.

3.1 Let $X \subseteq M^n$ for some $n \geqslant 1$. Show that X is term-closed if and only if it is closed under every map $h \in \mathcal{P}$.

3.2 Consider the ultimate brute force structure $\underset{\sim}{\mathbf{M}}^\infty := \langle M; \mathcal{P}_\infty \rangle$. State the extensions to this context of the relevant notions, and show that $\underset{\sim}{\mathbf{M}}^\infty$ always yields a strong duality on $\mathcal{A}$ without the aid of topology.

3.3 For a finite algebra $\underline{\mathbf{M}}$ and a non-empty set S, the **term closure** X^{tm} of a subset $X \subseteq M^S$ is the set of all y in M^S such that any two S-ary term functions which agree on X also agree at y. As a generalisation of the Closure Theorem 3.1.3, prove that the following are equivalent:

(i) $y \in X^{\text{tm}}$;

(ii) there exists a homomorphism $u\colon \mathbf{F}_{\underline{\mathbf{M}}}(S){\restriction}_X \to \underline{\mathbf{M}}$ with $u(\pi_s{\restriction}_X) = y(s)$ for all $s \in S$;

(iii) there exists a homomorphism $u\colon \mathbf{F}_{\underline{\mathbf{M}}}(S){\restriction}_X \to \underline{\mathbf{M}}$ with $u(t{\restriction}_X) = t(y)$ for all $t \in \mathbf{F}_{\underline{\mathbf{M}}}(S)$.

3.4 Let $\mathbf{X} \in \mathcal{X}$ and let $T \subseteq X$. The intersection of all closed substructures of $\mathbf{X}$ containing T is denoted by $[T]$ and is called the **closed substructure of X generated by** T. This terminology is justified by the following alternative method of obtaining $[T]$. Define a non-decreasing ordinal sequence $T_0, T_1, \dots, T_\alpha, \dots$ of subsets of X by

$T_0 = T$,

$T_{\alpha+1}$ is the topological closure of the algebraic closure of T_α,

$T_\lambda = \bigcup\{\, T_\alpha \mid \alpha < \lambda \,\}$ for each limit ordinal λ.

Prove each of the following.

(i) There is an ordinal α such that $T_\alpha = T_\beta$ for all $\beta > \alpha$.

(ii) If $H = \varnothing$, then $T_1 = T_\beta$ for all $\beta > 1$.

(iii) If $T_\alpha = T_\beta$ for all $\beta > \alpha$, then $T_\alpha = [T]$.

(iv) For every $\mathbf{Y} \in \mathcal{X}$, every pair of morphisms $\varphi, \psi\colon \mathbf{X} \to \mathbf{Y}$ that agree on T also agree on $[T]$.

(v) Suppose $\underset{\sim}{\mathbf{M}}$ yields a strong duality on $\mathcal{A}$. Then $[T]$ is the only closed substructure of $\mathbf{X}$ which contains T and satisfies (iv).

3.5 Let S be a non-empty set and let $\Pi_S = \{\, \pi_s\colon M^S \to M \mid s \in S \,\}$ be the corresponding set of projections. Notice that (CLO) says precisely that $E(\underset{\sim}{\mathbf{M}}^S)$ is freely generated in $\mathcal{A}$ by Π_S. Assume that $\underset{\sim}{\mathbf{M}}$ yields a strong duality on $\mathcal{A}$ and prove the dual fact,

$(\text{CLO})^\partial_S$ *the structure* $D(\underline{\mathbf{M}}^S)$ *is freely generated in* $\mathcal{X}$ *by* Π_S, in three steps.

(i) If $\mathbf{X} \in \mathcal{X}$ and $x_s \in X$ for all $s \in S$, then there is a morphism $\varphi\colon D(\underline{\mathbf{M}}^S) \to \mathbf{X}$ satisfying $\varphi(\pi_s) = x_s$ for all $s \in S$. [Hint: Let $\varphi = \varepsilon_{\mathbf{X}}^{-1} \circ D(u)$ where $u(\alpha) = \alpha \circ x$.]

(ii) The morphism φ in part (i) is unique. [Hint: Suppose that $\varphi, \psi\colon D(\underline{\mathbf{M}}^S) \to \mathbf{X}$ agree on Π_S but not at $x \in D(\underline{\mathbf{M}}^S)$. Apply the Separation Theorem 1.4.4 to $\varphi(x)$ and $\psi(x)$ and the definition of duality to reach a contradiction.]

(iii) Use Exercise 3.4(v) to conclude that $[\Pi_S] = D(\underline{\mathbf{M}}^S)$.

3.6 We now give a partial converse to Exercise 3.5.

(i) Assume that $\underset{\sim}{\mathbf{M}}$ yields a duality on $\mathcal{A}$. Show that if $\underline{\mathbf{M}}$ is injective in $\mathcal{A}$ and $(\mathrm{CLO})^{\partial}_{S}$ holds for every non-empty set S, then $\underset{\sim}{\mathbf{M}}$ yields a strong duality on $\mathcal{A}$.

(ii) Assume that $\underset{\sim}{\mathbf{M}}$ is a total structure and that R is finite. Then $\underset{\sim}{\mathbf{M}}$ yields a strong duality on $\mathcal{A}$ and $\underline{\mathbf{M}}$ is injective in $\mathcal{A}$ provided that the following conditions hold:

(a) $\underset{\sim}{\mathbf{M}}$ is injective in $\mathcal{X}_{\text{fin}}$,

(b) $\underline{\mathbf{M}}$ is injective in $\mathcal{A}_{\text{fin}}$,

(c) $\mathcal{A}(\underline{\mathbf{M}}^n, \underline{\mathbf{M}})$ is freely generated in $\mathcal{X}_{\text{fin}}$ by the projections,

(d) $\mathcal{X}(\underset{\sim}{\mathbf{M}}^n, \underset{\sim}{\mathbf{M}})$ is freely generated in $\mathcal{A}_{\text{fin}}$ by the projections.

3.7 Show that if L is any subset of M, then the multi-pointed set $\underline{\mathbf{M}}^L := \langle M; L\rangle$, obtained by making every element of L the value of a nullary operation, is strongly dualisable. [Hint: Let F be the clone of all total operations on M which preserve each singleton set $\{a\}$ where $a \in L$. Use the Semi-primal Strong Duality Theorem 3.3.14 to prove that $\underset{\sim}{\mathbf{M}} := \langle M; L, \mathcal{T}\rangle$ yields a strong duality on the variety generated by $\underline{\mathbf{M}} := \langle M; F\rangle$, and then use the Two-for-One Strong Duality Theorem 3.3.2 to reverse the roles of $\underset{\sim}{\mathbf{M}}$ and $\underline{\mathbf{M}}$.]

3.8 Show that the Two-for-One Strong Duality Theorem fails if we do not require named constants. [Check that the total algebra $\underset{\sim}{\mathbf{M}} = \langle \{0,1\}; \varnothing, \mathcal{T}\rangle$ yields a strong duality on the quasi-variety generated by the Sheffer-stroke algebra $\underline{\mathbf{M}} = \langle \{0,1\}; |\rangle$, and then show that $\underset{\sim}{\mathbf{M}}' = \langle \{0,1\}; |, \mathcal{T}\rangle$ does not yield a strong duality on the quasi-variety generated by $\underline{\mathbf{M}}' = \langle \{0,1\}; \varnothing\rangle$ as $E(\underset{\sim}{\varnothing}) \cong E(\underset{\sim}{\mathbf{M}}')$.]

3.9 Show that the following are equivalent:

(i) the set $\mathcal{P}$ is closed under composition, that is, if g is a k-ary algebraic partial operation and $h_1, \dots, h_k$ are n-ary algebraic partial operations on $\underline{\mathbf{M}}$, then the composition $g(h_1, \dots, h_k)$ has a non-empty domain;

(ii) for all $n \in \mathbb{N}$, if $\mathbf{A}$ and $\mathbf{B}$ are subalgebras of $\underline{\mathbf{M}}^n$, then $A \cap B$ is non-empty;

(iii) $\underline{\mathbf{M}}$ has a constant unary term function.

[Hint: Consider the composition $g(h_1, \dots, h_k)$ with $g = \rho_1^A$ and $h_i = \rho_i^B$ for each i, where ρ_i^C denotes the restriction to $C \subseteq M^n$ of the projection π_i.]

3.10 Let $\underset{\sim}{\mathbf{M}}$ be algebraic over $\underline{\mathbf{M}}$ and let $a \in M$.

(i) Assume $\underset{\sim}{\mathbf{M}}$ yields a duality on $\mathcal{A}$. Show that a is the value of a constant unary term function of $\underline{\mathbf{M}}$ if and only if it forms a complete one-element substructure of $\underset{\sim}{\mathbf{M}}$.

(ii) Assume $\underset{\sim}{\mathbf{M}}$ yields a full duality on $\mathcal{A}$ and that $\underline{\mathbf{M}}$ has *no* constant unary term function. Let $\underset{\sim}{\mathbf{1}}$ denote the complete one-element structure with underlying set $\{\varnothing\}$ and, for $\mathbf{X} \in \mathcal{X}$, define $\mathbf{X}^{\varnothing}$ to be $\underset{\sim}{\mathbf{1}}$. Let $\mathcal{X}'$ consist of $\mathcal{X}$ together with all complete one-element structures, let $\mathcal{A}'$ consist of $\mathcal{A}$ together with the **empty algebra** $\underline{\varnothing}$ and let $\langle D', E', e', \varepsilon' \rangle$ be the natural extension of $\langle D, E, e, \varepsilon \rangle$ to $\mathcal{A}'$ and $\mathcal{X}'$. Show that $\langle D', E', e', \varepsilon' \rangle$ is a dual equivalence where $D'(\underline{\varnothing}) = \underset{\sim}{\mathbf{1}}$ and $E'(\underset{\sim}{\mathbf{1}}) = \underline{\varnothing}$.

In the remaining exercises we assume, as usual, that $\underset{\sim}{\mathbf{M}}$ is algebraic over $\underline{\mathbf{M}}$ and we consider the categories $\mathcal{A} := \mathbb{ISP}\underline{\mathbf{M}}$ and $\mathcal{X} := \mathbb{IS}_c\mathbb{P}^+\underset{\sim}{\mathbf{M}}$. Nevertheless, some of the category-theoretic results concerning $\mathcal{A}$ and $\mathcal{X}$ are true more generally. Our aim is to give an alternative proof of the First Strong Duality Theorem using more category-theoretic methods.

3.11 A morphism $u\colon \mathbf{A} \to \mathbf{B}$ in a category $\mathcal{C}$ is called **epic** if it is right cancellative, that is, for all morphisms $s, t\colon \mathbf{B} \to \mathbf{C}$ in $\mathcal{C}$ we have $s = t$ whenever $s \circ u = t \circ u$. By reversing all the arrows we obtain the dual concept of a **monic** morphism $u\colon \mathbf{B} \to \mathbf{A}$ in $\mathcal{C}$.

(i) Show that a homomorphism $u\colon \mathbf{B} \to \mathbf{A}$ in $\mathcal{A}$ is monic if and only if it is an embedding. [Hint: Take $\mathbf{C}$ to be the free algebra in $\mathcal{A}$ on one generator.]

(ii) (a) Show that every embedding in $\mathcal{X}$ is monic.

(b) Show that in the category of bounded Priestley spaces a morphism u is monic if and only if it is one-to-one and hence give an example of a monic morphism which is not an embedding.

(iii) Show that surjective morphisms in both $\mathcal{A}$ and $\mathcal{X}$ are epic.

(iv) Show that an embedding of a three-element chain into $\underline{\mathbf{D}}^2$ is a non-surjective epic in the category of distributive lattices.

(v) Show that a morphism is epic in the category of bounded Priestley spaces if and only if it is surjective.

3.12 (i) Show that a morphism φ is epic in $\mathcal{X}$ if and only if $E(\varphi)$ is one-to-one. [Similarly, a morphism u is epic in $\mathcal{A}$ if and only if $D(u)$ is one to-one.]

(ii) Generalising Exercise 3.11(v), show that if $\underset{\sim}{\mathrm{M}}$ yields a full duality on $\mathcal{A}$ and $\underline{\mathbf{M}}$ is injective in $\mathcal{A}$, then a morphism in $\mathcal{X}$ is epic if and only if it is surjective.

(Refinements of this result and its relationship to the problem of whether every full duality is strong are presented in the exercises for Chapter 6.)

3.13 Let $u, v\colon \mathbf{A} \to \mathbf{B}$ be morphisms in a category $\mathcal{C}$. A morphism $e\colon \mathbf{B} \to \mathbf{E}$ in $\mathcal{C}$ is called a **coequaliser** of u and v if

(a) $e \circ u = e \circ v$, and

(b) for every morphism $e'\colon \mathbf{B} \to \mathbf{E}'$ in $\mathcal{C}$ such that $e' \circ u = e' \circ v$ there exists a unique morphism $w\colon \mathbf{E} \to \mathbf{E}'$ satisfying $w \circ e = e'$.

The dual concept, the **equaliser** of a pair of morphisms $u, v\colon \mathbf{B} \to \mathbf{A}$ in $\mathcal{C}$, is obtained by reversing all the morphisms.

(i) Prove that if $e\colon \mathbf{B} \to \mathbf{E}$ and $e'\colon \mathbf{B} \to \mathbf{E}'$ are both coequalisers of a pair $u, v\colon \mathbf{A} \to \mathbf{B}$ of morphisms then the morphism $w\colon \mathbf{E} \to \mathbf{E}'$, guaranteed by the definition of coequaliser, is an isomorphism.

(ii) Show that every coequaliser is epic and, dually, that every equaliser is monic.

(iii) Let $u, v\colon \mathbf{A} \to \mathbf{B}$ be homomorphisms in $\mathcal{A}$. Let θ be the intersection of all $\mathcal{A}$-congruences on $\mathbf{B}$ which contain the set $\{\, (u(a), v(a)) \mid a \in A \,\}$. Show that the natural homomorphism from $\mathbf{B}$ to $\mathbf{B}/\theta$ is a coequaliser of u and v in $\mathcal{A}$.

(iv) Let $\varphi, \psi\colon \mathbf{X} \to \mathbf{Y}$ be morphisms in $\mathcal{X}$.

(a) Show that $\mathrm{eq}(\varphi, \psi) := \{x \in X \mid \varphi(x) = \psi(x)\}$ is a closed substructure of $\mathbf{X}$.

(b) Show that the inclusion map, $e\colon \mathrm{eq}(\varphi, \psi) \to X$, is an equaliser in $\mathcal{X}$ of the morphisms φ and ψ.

(A similar result holds in the category $\mathcal{A}$. Note that it is common to refer to the *object* $\mathrm{eq}(\varphi, \psi)$ as the equaliser whereas, strictly speaking, an equaliser should be a *morphism*—the inclusion of $\mathrm{eq}(\varphi, \psi)$ into X.)

3.14 A morphism $e\colon \mathbf{A} \to \mathbf{B}$ in a category $\mathcal{C}$ is called a **regular epic** if there exist morphisms $u, v\colon \mathbf{C} \to \mathbf{A}$ in $\mathcal{C}$ such that e is a coequaliser of u and v. Similarly, a morphism is a **regular monic** if it is an equaliser. By Exercise 3.13(ii), a regular epic is epic and a regular monic is monic.

(i) (a) Prove that a homomorphism in $\mathcal{A}$ is a regular epic if and only if it is surjective. [Hint: For the reverse implication, show that a surjective map e is the coequaliser of $\rho_1 := \pi_1{\restriction}_{\mathbf{C}}$ and $\rho_2 := \pi_2{\restriction}_{\mathbf{C}}$ where $\mathbf{C}$ is $\ker(e)$ regarded as a subalgebra of $\mathbf{A}^2$.]

(b) Show that the embedding described in Exercise 3.11(iv) is not a regular monic.

(c) Show that every regular monic in $\mathcal{X}$ is an embedding.

(ii) (a) Let $\mathbf{Y} \in \mathcal{X}$ and let $\mathbf{X}$ be a substructure of $\mathbf{Y}$. Show that the inclusion of $\mathbf{X}$ into $\mathbf{Y}$ is a regular monic in $\mathcal{X}$ if and only if for all $y \in Y \backslash X$ there exist morphisms $\alpha, \beta \colon \mathbf{Y} \to \underset{\sim}{\mathbf{M}}$ satisfying $\alpha{\restriction}_{\mathbf{X}} = \beta{\restriction}_{\mathbf{X}}$ and $\alpha(y) \neq \beta(y)$. [Hint: Given these morphisms, define morphisms $\sigma, \tau \colon \mathbf{Y} \to \underset{\sim}{\mathbf{M}}^{Y \backslash X}$ that have the inclusion map as their equaliser.]

(b) Let $\mathbf{X}$ be a substructure of $\underset{\sim}{\mathbf{M}}^S$. Show that if X is term-closed in M^S, then the inclusion of $\mathbf{X}$ into $\underset{\sim}{\mathbf{M}}^S$ is a regular monic in $\mathcal{X}$. Then show that the converse is true provided that (CLO) holds.

(iii) Use the Third Duality Theorem 3.1.6 to show that $\underset{\sim}{\mathbf{M}}$ yields a duality on $\mathcal{A}$ if and only if (CLO) holds and $\underset{\sim}{\mathbf{M}}$ is injective with respect to regular monics in $\mathcal{X}$.

(iv) Assume that $\underset{\sim}{\mathbf{M}}$ yields a full duality on $\mathcal{A}$. Use Exercise 3.5 and (i)(a) above to prove the following.

(a) Use (i)(a) above to prove that φ is a regular monic in $\mathcal{X}$ if and only if $E(\varphi)$ is a surjection in $\mathcal{A}$.

(b) Use Corollary 2.2.4 to prove that $E(\underset{\sim}{\mathbf{M}})$ is **projective** in $\mathcal{A}$, that is, if $u \colon \mathbf{A} \to \mathbf{B}$ is a surjection in $\mathcal{A}$ and $v \colon E(\underset{\sim}{\mathbf{M}}) \to \mathbf{B}$ is a homomorphism, then there exists a homomorphism $v' \colon E(\underset{\sim}{\mathbf{M}}) \to \mathbf{A}$ such that $u \circ v' = v$.

(v) Assume that $\underset{\sim}{\mathbf{M}}$ yields a duality on $\mathcal{A}$. By using the hints below, prove that the following are equivalent. (This gives an alternative proof of the First Strong Duality Theorem.)

(a) $\underset{\sim}{\mathbf{M}}$ yields a strong duality on $\mathcal{A}$;

(b) a morphism in $\mathcal{X}$ is an embedding if and only if it is a regular monic;

(c) $\underset{\sim}{\mathbf{M}}$ yields a full duality on $\mathcal{A}$ and $\underset{\sim}{\mathbf{M}}$ is injective in $\mathcal{X}$.

[Hint: Prove (a) $\Rightarrow$ (b) $\Rightarrow$ (c) $\Rightarrow$ (b) $\Rightarrow$ (a). When proving that (b) $\Rightarrow$ (c), use (ii)(a) above and Exercise 1.4 to show that the duality is full, and then use part (iv) above to show that $\underset{\sim}{\mathbf{M}}$ is injective in $\mathcal{X}$.]

4

Examples of Strong Dualities

The theory we have put together will now reward us with a rich harvest of examples. This chapter is organised into five sections, the first four based on the major strong duality theorems obtained in the last section of the previous chapter. Each example will appear as an application of the strongest theorem that applies to the appropriate generating algebra $\underline{\mathbf{M}}$, and our theorems will immediately give us a choice of $\underset{\sim}{\mathbf{M}}$ that yields a strong duality on $\mathcal{A} = \mathbb{ISP}\underline{\mathbf{M}}$. Two further steps will remain:

(1) reduce and simplify the structure on $\underset{\sim}{\mathbf{M}}$ to obtain an optimal strong duality using the Constructs for Entailment 2.4.5, and

(2) give a useful description of the category $\mathcal{X} = \mathbb{IS}_c\mathbb{P}^+\underset{\sim}{\mathbf{M}}$ using the Preservation Theorem 1.4.3 and the Separation Theorem 1.4.4.

Our treatments of these two issues will vary. In most cases we resolve them both in complete detail, while in some cases we outline the argument and leave the details as an exercise. In those cases where the reduction process or the description requires a substantial theoretical development, we simply give appropriate references and report the results.

Since most strong dualities come from the various offshoots of the NU Strong Duality Theorem 3.3.8, we present them in the first three sections. Sample consequences of the Two-for-One Strong Duality Theorem 3.3.2 are given in the fourth section. In the last section we exhibit without proof a collection of strong dualities which are not consequences of either of these two general theorems. The existence of such vagrant examples demonstrates that the theory we have presented, powerful as it is, is still not complete. In many of our examples Theorem 1.3.6 can be applied to show that the quasi-variety $\mathcal{A}$ is actually the *variety* generated by $\underline{\mathbf{M}}$. We will often mention this fact without explicit reference, as it is not really a problem of duality theory *per se*.

4.1 Strong dualities from primality and its variants

In 1936 Stone [Sto36] published the seminal work on duality theory, exhibiting a dual equivalence between the category of all Boolean algebras and the category of all Boolean spaces. Seventeen years later, Foster [Fos53] made another major step forward, recognising the significance of the fact that every Boolean algebra is a special type of subdirect power of the two-element Boolean algebra. This two-element algebra is an example of what Foster called a functionally strictly complete algebra and later shortened to **primal algebra**, that is, a finite algebra in which every finitary operation on its underlying set is a term function. By the Quasi-primality Theorem 3.3.12, a finite algebra $\underline{\mathbf{M}}$ is primal if and only if it is quasi-primal and has no proper subalgebras and no non-trivial automorphisms. Consequently, $\underline{\mathbf{M}}$ is primal provided only that the ternary discriminator and the constant unary maps are term functions, and, in this case, $\mathcal{A}$ is a variety. Exploiting this idea, Hu demonstrated that Stone's theorem applies equally well to the variety generated by any primal algebra.

1.1 Theorem [Hu71] *Let $\underline{\mathbf{P}}$ be a primal algebra, and let $\underset{\sim}{\mathbf{P}} := \langle P, \mathcal{T} \rangle$.*

(i) *$\underset{\sim}{\mathbf{P}}$ yields a strong duality on the variety $\mathbb{ISP}\underline{\mathbf{P}}$ generated by $\underline{\mathbf{P}}$.*

(ii) *The dual $\mathbb{IS}_c\mathbb{P}^+ \underset{\sim}{\mathbf{P}}$ is exactly the category of all Boolean spaces.*

Proof We leave it to the reader to check that (i) is a consequence of both the Semi-primal Strong Duality Theorem 3.3.14 and the Two-for-One Strong Duality Theorem 3.3.2, while (ii) follows immediately from the Separation Theorem 1.4.4. ∎

Taking the two-element **Boolean algebra** and **Boolean space**,

$$\underline{\mathbf{2}} = \langle \{0,1\}; \vee, \wedge, ', 0, 1 \rangle \quad \text{and} \quad \underset{\sim}{\mathbf{2}} = \langle \{0,1\}; \mathcal{T} \rangle,$$

we immediately get a formulation of Stone's theorem. As the original and normally quoted version of Stone's theorem does not mention hom-functors at all, we will give it here explicitly. A maximal, proper, non-empty filter x of a Boolean algebra $\mathbf{B} = \langle B; \vee, \wedge, ', 0, 1 \rangle$ is called an **ultrafilter**. By Exercise 4.1(ii), this is equivalent to x being a (proper, non-empty) prime filter. As in Theorem 1.2.4, we can check that the set $\mathbf{B}^*$ of all ultrafilters of $\mathbf{B}$ forms a Boolean space whose clopen subsets are exactly the sets $e_{\mathbf{B}}(a) := \{\, x \in \mathbf{B}^* \mid a \in x \,\}$ where a ranges over B. Dually, if $\mathbf{X}$ is a Boolean space, then the set of all clopen subsets of $\mathbf{X}$ forms a Boolean algebra $\mathbf{X}^*$ under union, intersection and complement. Using the compactness of $\mathbf{X}$, we can check that the ultrafilters of this Boolean algebra are exactly the sets $\varepsilon_{\mathbf{X}}(x) := \{\alpha \in \mathbf{X}^* \mid x \in \alpha\}$ where x ranges over X. This gives us an isomorphism

$$e_{\mathbf{B}} : \mathbf{B} \to \mathbf{B}^{**} \quad \text{and} \quad \varepsilon_{\mathbf{X}} : \mathbf{X} \to \mathbf{X}^{**}$$

for each Boolean algebra $\mathbf{B}$ and each Boolean space $\mathbf{X}$.

1.2 Stone's Duality Theorem [Sto36] *The quadruple $\langle {}^*, {}^*, e, \varepsilon \rangle$ is a dual equivalence between the category $\mathcal{B} = \mathbb{ISP}\underline{\mathbf{2}}$ of Boolean algebras and the category $\mathcal{Z} = \mathbb{IS}_c\mathbb{P}^+ \underset{\sim}{\mathbf{2}}$ of Boolean spaces.*

We leave it to the reader either to prove this theorem directly or to use Exercise 4.1 in order to extract it from Theorem 4.1.1.

Other familiar primal algebras are the **Sheffer-stroke algebra**

$$\underline{\mathbf{S}} = \langle \{0,1\}; \mid \rangle \quad \text{where} \quad x \mid y = \neg x \wedge \neg y$$

and the **Boolean ring**

$$\underline{\mathbf{R}} = \langle \{0,1\}; +, \cdot, 0, 1 \rangle$$

(Exercise 4.2). An important primal algebra associated with n-valued logic is the n-element **Post algebra**

$$\underline{\mathbf{P}}_n = \langle \{0, \ldots, n-1\}; \vee, \wedge, {}^*, D_1, \ldots, D_{n-1}, 0, \ldots, n-1 \rangle$$

obtained by augmenting the n-element chain with unary operations *, $D_1, \ldots, D_{n-1}$. If we write $\mathbf{1} := n-1$, then the additional operations are given by

$$a^* = \begin{cases} \mathbf{1} & \text{if } a = 0, \\ 0 & \text{if } a \neq 0, \end{cases} \quad \text{and} \quad D_i(a) = \begin{cases} \mathbf{1} & \text{if } i \leqslant a, \\ 0 & \text{if } i > a. \end{cases}$$

The ternary discriminator operation on $\underline{\mathbf{P}}_n$ can be obtained as follows. Let $\mathrm{equal}(x, y)$ be the meet of the $n-1$ terms $D_i(x \wedge y) \vee D_i(x \vee y)^*$ where $i = 1, 2, \ldots, n-1$. For $a, b \in P_n$, we have

$$\mathrm{equal}(a, b) = \begin{cases} \mathbf{1} & \text{if } a = b, \\ 0 & \text{if } a \neq b. \end{cases}$$

This gives us the ternary discriminator term

$$t(x, y, z) = (\mathrm{equal}(x, y) \wedge z) \vee (\mathrm{equal}(x, y)^* \wedge x).$$

We refer to Rasiowa [Ras:NL] for a careful discussion of Post algebras.

Turning to a slightly less restrictive class of quasi-primal algebras, we find many new examples among the semi-primal quasi-varieties. Recall from the last chapter that a finite algebra is **semi-primal** if it has the ternary discriminator as a term function but has no isomorphisms between its non-trivial subalgebras other than the identity maps.

1.3 Theorem *Let $\underline{\mathbf{GF}}(p) = \langle \{0, \ldots, p-1\}; +, \cdot, 0 \rangle$ be the field of prime order p, and let $\underset{\sim}{\mathbf{GF}}(p) = \langle \{0, \ldots, p-1\}; 0, \mathcal{T} \rangle$ have a single distinguished element 0.*

(i) *$\underset{\sim}{\mathbf{GF}}(p)$ yields a strong duality on the variety $\mathbb{ISP}(\underline{\mathbf{GF}}(p))$ of commutative rings satisfying $px \approx 0$ and $x^p \approx x$.*

(ii) *The dual $\mathbb{IS}_c\mathbb{P}^+(\underset{\sim}{\mathbf{GF}}(p))$ is exactly the category of pointed Boolean spaces $\mathbf{X} = \langle X; 0, \mathcal{T} \rangle$.*

Proof (i) In Exercise 4.3 we ask the reader to show that the variety of commutative rings defined by these two identities is exactly $\mathbb{ISP}(\underline{\mathbf{GF}}(p))$.

Observe that $\underline{\mathbf{GF}}(p)$ is semi-primal as $\{0\}$ is its only proper subalgebra, it has no proper automorphisms, and it has the ternary discriminator term

$$t(x, y, z) = (x - y)^{p-1}(x - z) + z.$$

The conclusion follows immediately from the Semi-primal Strong Duality Theorem 3.3.14. Part (ii) follows from the Separation Theorem 1.4.4. ∎

In Chapter 3 we saw that, from the point of view of the theory of strong dualities, we might normally expect to encounter partial operations in our dual structures. The need for proper partial operations arises among quasi-primal algebras, as we see in the following example.

Monk [Mon70] introduced the notion of monadic algebras to model the logic of statements in one variable where, in addition to propositional connectives, we allow quantification over that variable. A **monadic algebra** is an algebra $\mathbf{A} = \langle A; \vee, \wedge, ', C, 0, 1\rangle$ where $\langle A; \vee, \wedge, ', 0, 1\rangle$ is a Boolean algebra and C is a unary operation (corresponding to existential quantification) satisfying the (redundant) axioms

$$C0 \approx 0, \quad x \leqslant Cx, \quad x \leqslant y \Rightarrow Cx \leqslant Cy, \quad C(x \wedge Cy) \approx Cx \wedge Cy.$$

The proper subvarieties of the variety of monadic algebras form an ω-chain, the nth member of which is generated by the simple monadic algebra

$$\underline{\mathbf{M}}_n = \langle 2^n; \vee, \wedge, ', C, 0, 1\rangle$$

formed from the 2^n-element Boolean algebra, where

$$Ca = \begin{cases} 1 & \text{if} \quad a \neq 0, \\ 0 & \text{if} \quad a = 0. \end{cases}$$

These algebras are quasi-primal, as we see from the discriminator term

$$t(x, y, z) = [\text{notequal}(x, y) \wedge x] \vee [\text{notequal}(x, y)' \wedge z],$$

where

$$\text{notequal}(x, y) := C[(x \wedge y') \vee (x' \wedge y)] = \begin{cases} 1 & \text{if } x \neq y, \\ 0 & \text{if } x = y. \end{cases}$$

From the Quasi-primal Strong Duality Theorem 3.3.13 we see that $\mathbb{ISP}\underline{\mathbf{M}}_n$ is exactly the variety of monadic algebras generated by $\underline{\mathbf{M}}_n$, and that we obtain a strong duality for it by taking

$$\underset{\sim}{\mathbf{M}}_n = \langle 2^n; K \cup \mathrm{Aut}(\underline{\mathbf{M}}_n), \mathrm{Iso}(\underline{\mathbf{M}}_n), \mathcal{T}\rangle.$$

Here $\mathcal{P}_1 = \mathrm{Aut}(\underline{\mathbf{M}}_n) \cup \mathrm{Iso}(\underline{\mathbf{M}}_n)$ forms an inverse semigroup of partial bijections on M_n. In general, $\mathcal{P}_1$ contains many unary partial operations which are not restrictions of automorphisms. In Chapter 6 we will see that these partial operations are unavoidable if we want a strong duality. The dual structures form a category of Boolean spaces acted on by an inverse semigroup of partial homeomorphisms. We refer the reader to Davey and Werner [DW83] for a detailed description of these spaces.

A thorough survey of primality is given in Quackenbush's appendix to Grätzer [Grä:UA], and many additional examples of interesting primal, semi-primal and quasi-primal algebras may be found in Davey, Schumann and Werner [DSW91] or Werner [Wer:DA]. A finite simple non-abelian group with each element distinguished is primal (Maurer and Rhodes [MR65]); the finite simple members of the varieties of Łukasiewicz algebras and complementary semigroups are semi-primal while every finite simple cylindric algebra and every finite field are quasi-primal (Werner [Wer:DA]). For the benefit of any reader who might still harbour a suspicion that algebras of this kind are rare and obscure, we quote an amusing result of Murskiǐ [Mur75]: for any fixed type which includes at least one more-than-unary operation, the proportion of semi-primal algebras among all algebras of a fixed size approaches 1 !

4.2 Arithmetical strong dualities

The third condition of the Quasi-primality Theorem 3.3.12 leads to a wider class of strong dualities: a finite algebra is quasi-primal if and only if it generates an arithmetical variety and all its non-trivial subalgebras are simple. As we have seen, each of these conditions is important for duality: arithmeticity gives us a ternary near-unanimity term ensuring that binary relations will yield a duality, congruence permutability, via Fleischer's Theorem, allows us to eliminate relations in favour of unary partial operations, and simple subalgebras tells us that unary partial operations are sufficient for strong duality. If we give up simplicity of subalgebras, we still have strong duality without relations but we must now consider partial operations of higher arity. We now look at several examples obtained as applications of the resulting Arithmetic Strong Duality Theorem 3.3.11.

2.1 Heyting Algebras Heyting algebras were developed to model intuitionistic logic by replacing the unary negation operation with a binary operation $\Rightarrow$ to represent implication (see Rasiowa [Ras:NL] or Balbes and Dwinger [BD:DL]). In a Boolean algebra, $a \Rightarrow b := \neg a \vee b$ can be described as the largest element whose meet with a is less than or equal to b. A **Heyting algebra** is an algebra $\mathbf{A} = \langle A; \vee, \wedge, \Rightarrow, 0, 1\rangle$ which is a bounded distributive lattice with a **relative pseudocomplementation** operation $\Rightarrow$

such that

$$a \Rightarrow b = \max\{\, c \in A \mid c \wedge a \leqslant b \,\}.$$

Many nice properties of Heyting algebras result from the fact that they have a Pixley arithmeticity term

$$p(x, y, z) = (y \Rightarrow x) \wedge (x \vee z) \wedge (y \Rightarrow z).$$

For example, the Arithmetic Strong Duality Theorem 3.3.11 gives us a multitude of strong dualities for quasi-varieties of Heyting algebras.

2.2 Heyting Strong Duality Theorem *Assume that $\underline{\mathbf{M}}$ is a finite subdirectly irreducible Heyting algebra such that $\mathbb{HS}\underline{\mathbf{M}} \subseteq \mathbb{ISP}\underline{\mathbf{M}}$. Then $\underset{\sim}{\mathbf{M}} = \langle M; G, H, \mathcal{T}\rangle$ yields a strong duality on $\mathcal{A}$, provided that $G \cup H$ is a generating set for the monoid of partial endomorphisms of $\underline{\mathbf{M}}$.*

Proof A Heyting algebra $\underline{\mathbf{M}}$ is subdirectly irreducible if and only if it is obtained as $\underline{\mathbf{M}} = \mathbf{L} \oplus \mathbf{1}$ by adding a new top to a finite distributive lattice $\mathbf{L}$, and any subalgebra of such an algebra has the same form. ∎

There is an abundance of finite Heyting algebras satisfying these conditions, although at present no easy method is known to identify them. (See Lemma 7.3.2(ii) for a convenient way to generate examples.) Among them are the finite Heyting algebra chains and Heyting algebras $\underline{\mathbf{2}}^n \oplus \underline{\mathbf{1}}$ obtained from finite Boolean lattices.

Now assume that $\underline{\mathbf{M}}$ is a subdirectly irreducible Heyting algebra such that $\mathbb{HS}\underline{\mathbf{M}} \subseteq \mathbb{ISP}\underline{\mathbf{M}}$. It remains to find a (small) generating set for the monoid of partial endomorphisms of $\underline{\mathbf{M}}$. When we are lucky, the endomorphisms alone will be sufficient. We say that $\underline{\mathbf{M}}$ is (**strongly**) **endodualisable** if $\underset{\sim}{\mathbf{M}} = \langle M; K \cup \mathrm{End}(\underline{\mathbf{M}}), \mathcal{T}\rangle$ yields a (strong) duality on $\mathcal{A}$, where, as usual, K denotes the set of elements which determine one-element subalgebras of $\underline{\mathbf{M}}$. (See Theorem 6.4.5 for a characterisation of strong endodualisability and Section 8 of Chapter 7 for a discussion of endodualisability in general.) Referring to the $\underset{\sim}{\mathbf{M}}$-Shift Strong Duality Lemma 3.2.3 and the Constructs for Entailment 2.4.5, we recognise a simple strategy to show that $\underline{\mathbf{M}}$ is strongly endodualisable:

- *show that $\underline{\mathbf{M}}$ is* **self injective**, *that is, that each partial endomorphism extends to an endomorphism, and*
- *show that each subalgebra of $\underline{\mathbf{M}}$ is the intersection of equalisers of endomorphisms.*

In Chapter 6 we will prove the Total Structure Theorem 6.1.2 which will tell us exactly when partial operations can be eliminated in general. As one application of this result, we will prove the Unary Total Algebra Theorem 6.4.5 which will show that $\underline{\mathbf{M}}$ is strongly endodualisable only if the above strategy is successful. We conclude with several examples.

2.3 Theorem

(i) *The Heyting algebra* $\underline{\mathbf{M}} = \underline{\mathbf{2}}^2 \oplus \underline{\mathbf{1}}$ *is strongly endodualisable.*

(ii) *The n-element Heyting algebra chain* $\underline{\mathbf{C}}_n$ *is strongly endodualisable if and only if* $n < 4$.

(iii) *Let h be the partial automorphism of the four-element Heyting algebra chain* $\underline{\mathbf{C}}_4$*, between its two three-element subalgebras, which moves the third element down to the second. Then the structure* $\underset{\sim}{C}_4 = \langle C_4; \mathrm{End}(\underline{\mathbf{C}}_4), h, \mathcal{T}\rangle$ *yields a strong duality on* $\mathbb{ISP}\underline{\mathbf{C}}_4$.

Proof We outline the proof, leaving the reader to verify the details in each example.

(i) Check that $\underline{\mathbf{M}}$ has exactly four endomorphisms which include extensions of each partial endomorphism, and that each subalgebra is an equaliser of endomorphisms.

(ii) Show that $\underset{\sim}{C}_2 = \langle\{0,1\}, \mathcal{T}\rangle$ yields a strong duality on $\mathbb{ISP}\underline{\mathbf{C}}_2$ and that $\underset{\sim}{C}_3 = \langle\{0,a,1\}; d, \mathcal{T}\rangle$, where d is the endomorphism taking a to 1, yields a strong duality on $\mathbb{ISP}\underline{\mathbf{C}}_3$. Let $n \geqslant 4$ and let

$$C_n = \{0, a_1, \dots, a_{n-2}, 1\} \quad \text{with} \quad 0 < a_1 < \dots < a_{n-2} < 1.$$

It is easily seen that $X = C_n \backslash \{a_1\}$ is closed under the endomorphisms of $\underline{\mathbf{C}}_n$ but is not closed under the partial map $h\colon X \to C_n$, given by $h(a_2) = a_1$ and $h(a) = a$ for $a \neq a_2$. Thus X is not hom-closed in C_n and consequently $\underline{\mathbf{C}}_n$ is not strongly endodualisable.

(iii) Observe that every partial endomorphism, other than h, extends to an endomorphism and that every subalgebra of $\underline{\mathbf{M}}_4$, other than $\mathrm{dom}(h) = \{0, a_2, 1\}$, is an equaliser of endomorphisms. ∎

In the spirit of (iii), we can find generating sets for the partial endomorphisms of longer finite Heyting algebra chains in order to simplify their strong dualities. We return to the Heyting Strong Duality Theorem, though from a different perspective, in Section 3 of Chapter 7. In Chapter 8 we will see that all non-trivial finite Heyting algebra chains, though not necessarily *strongly* endodualisable, are still endodualisable.

2.4 The Ring $\underline{\mathbf{Z}}_6$ Our next example uses the NU Strong Duality Theorem to obtain a strong duality with non-unary operations. As we noted earlier, a finite field of order $n > 1$ is quasi-primal; indeed, the discriminator is given by

$$p(x,y,z) = (x-y)^{n-1}(x-z) + z.$$

Exploiting this idea further, we find that for any ring satisfying the identity $x^n \approx x$ this same term, although not the discriminator, is still a Pixley arithmeticity term. We consider here the ring with identity

$$\underline{\mathbf{Z}}_6 = \langle\{0,1,2,3,4,5\}; +, \cdot, 0, 1\rangle,$$

where we have included the identity, 1, in the type. This ring is not quasi-primal, but does satisfy the identity $x^3 \approx x$. It has no proper subalgebras, no endomorphisms other than id and exactly two non-trivial congruences, $\equiv_2$ and $\equiv_3$, whose meet is $\mathbf{0}$. Consequently, since $\mathcal{P}_2$ entails $\mathcal{B}_2$, we obtain a strong duality from (partial) operations alone by taking $H = \mathcal{P}_2$.

It remains to find a generating set for these partial operations and to describe the dual category. It turns out that only a single binary operation is required, and that the dual category is a familiar one.

2.5 Theorem *Let* $\underset{\sim}{\mathbf{Z}}_6 = \langle\{0,1,2,3,4,5\}; *, \mathcal{T}\rangle$ *where* $a * b$ *is the unique solution to the equations* $x \equiv_2 a$ *and* $x \equiv_3 b$. *Then* $\underset{\sim}{\mathbf{Z}}_6$ *yields a strong duality on* $\mathcal{A} = \mathbb{ISP}\underline{\mathbf{Z}}_6$.

Proof In order to identify the members of $\mathcal{P}_2$, we must first determine the subalgebras $\mathbf{r}$ of $\underline{\mathbf{Z}}_6 \times \underline{\mathbf{Z}}_6$. By Fleischer's Theorem A.12, a subalgebra of the square must have the form $\ker(g, h)$, where g and h are homomorphisms from $\underline{\mathbf{Z}}_6$ onto $\mathbf{S}$ for some ring with unit $\mathbf{S}$ (not necessarily in $\mathcal{A}$). There are four possible choices for $\mathbf{S}$, namely, $\underline{\mathbf{Z}}_6$, $\mathbf{Z}_3$, $\mathbf{Z}_2$ and $\{0\}$, which correspond to the four congruences $\mathbf{0}$, $\equiv_2$, $\equiv_3$ and $\mathbf{1}$, respectively. Thus the subalgebras of the square are exactly the four congruences.

Now consider any non-zero subalgebra $\mathbf{r}$ of $\underline{\mathbf{Z}}_6 \times \underline{\mathbf{Z}}_6$ and a homomorphism $f\colon \mathbf{r} \to \underline{\mathbf{Z}}_6$. Since the unit $(1,1)$ has additive order 6, the kernel $\ker(f)$ is an ideal of index 6 whose intersection with the diagonal $\{(0,0),(1,1),(2,2),(3,3),(4,4),(5,5)\}$ is $\{(0,0)\}$. We refer to such an ideal as a $\underline{\mathbf{Z}}_6$**-ideal** of $\mathbf{r}$. Observe that f is uniquely determined by $\ker(f)$ as $f(\ker(f) + (a,a)) = \{a\}$.

We must therefore determine the $\underline{\mathbf{Z}}_6$-ideals of each subalgebra of the square. It is easy to check that the only $\underline{\mathbf{Z}}_6$-ideals of $\mathbf{0}$, $\equiv_2$ and $\equiv_3$ are those corresponding to the projection mappings. Since $\underline{\mathbf{Z}}_6$ is a ring with unit, every ideal of its square is a product of two of its own ideals. It has exactly four ideals, $\{0\}$, $\{0,3\}$, $\{0,2,4\}$ and itself. Taking direct products of pairs of these gives us exactly four $\underline{\mathbf{Z}}_6$-ideals, $\{0\} \times Z_6$ and $Z_6 \times \{0\}$, $\{0,2,4\} \times \{0,3\}$ and $\{0,3\} \times \{0,2,4\}$. The first two are the kernels of the two projections, the third is the kernel of the operation $*$, and the last is the kernel of the opposite operation $\circ$ where $a \circ b$ is the unique solution to the equations $x \equiv_3 a$ and $x \equiv_2 b$. Thus $\mathcal{P}_2$ consists of the seven pairwise distinct projection maps together with $*$ and $\circ$.

We can now argue that the single operation $*$ strongly entails $\mathcal{P}_2$. Applying the Constructs for Entailment 2.4.5, we conclude that $*$ entails the binary relations $\mathbf{0} = \Delta_{Z_6}$ and $\mathbf{1} = Z_6 \times Z_6$, by (1), and $\equiv_2 = \mathrm{eq}(*, \pi_2)$ and $\equiv_3 = \mathrm{eq}(\pi_1, *)$, by (8) and (13). Using (8) and (11) we also see that $*$ strongly entails the composition $\circ = *(\pi_2, \pi_1)$. ■

Next we describe the dual category $\mathcal{X} = \mathbb{IS}_c\mathbb{P}^+ \underset{\sim}{\mathbf{Z}}_6$. A **(Boolean) rectangular band** is a (topological) semigroup (over a Boolean space) satisfying the **anti-commutative law**

$$(xy \approx yx) \Rightarrow (x \approx y).$$

Note that $\underset{\sim}{\mathbf{Z}}_6$ is an example of a Boolean rectangular band. A **left Boolean semigroup** is a Boolean space with left trivial multiplication defined by the identity $xy \approx x$, and a **right Boolean semigroup** is defined analogously.

2.6 Theorem *If $\mathbf{X} = \langle X; *, \mathcal{T}\rangle$, then the following are equivalent:*

(i) $\mathbf{X} \in \mathcal{X} = \mathbb{IS}_c\mathbb{P}^+ \underset{\sim}{\mathbf{Z}}_6$;

(ii) $\mathbf{X}$ *is a Boolean rectangular band;*

(iii) $\mathbf{X}$ *is isomorphic to the direct product of a left Boolean semigroup and a right Boolean semigroup.*

Proof The implication (i) ⇒ (ii) is true since $\mathbf{X}$ is a Boolean space by Lemma B.5 and $\mathbf{X}$ is a rectangular band by the Preservation Theorem 1.4.3. Since the two projection maps reduce the arguments to the special cases where $\mathbf{X}$ is a Boolean left or right semigroup, (iii) ⇒ (i) is an easy consequence of the Separation Theorem 1.4.4. We outline the proof of (ii) ⇒ (iii), leaving the details as an exercise. Let $\mathbf{X}$ be a Boolean rectangular band. Show that $\mathbf{X}$ satisfies $xyx \approx x$ and therefore also $xyz \approx xz$. Choose any element $a \in X$ and check that $\mathbf{L} = \mathbf{X}a$ is a left Boolean semigroup, that $\mathbf{R} = a\mathbf{X}$ is a right Boolean semigroup, and that $\varphi\colon \mathbf{L} \times \mathbf{R} \to \mathbf{X}$ is an isomorphism where $\varphi(xa, ay) = xy$. ■

4.3 NU strong dualities

In this section we will exhibit a few of the many non-arithmetical applications of the NU Strong Duality Corollary 3.3.9. In all of these examples we have the **median operation**

$$m(x, y, z) = (x \wedge y) \vee (x \wedge z) \vee (y \wedge z)$$

generated by operations $\vee$ and $\wedge$ which are either lattice operations or close enough to lattice operations to ensure that m is a majority term. But the absence of congruence permutability now means that we must sift through the subalgebras of $\underline{\mathbf{M}} \times \underline{\mathbf{M}}$ in order to eliminate redundant relations in the type of $\underset{\sim}{\mathbf{M}}$.

3.1 Distributive Lattices The variety $\mathcal{D}$ of **distributive lattices** is normally defined as the models $\mathbf{L} = \langle L; \vee, \wedge\rangle$ of the equations (DL) of Chapter 1.2. In Lemma 1.2.2 we found that $\mathcal{D} = \mathbb{ISP}\underline{\mathbf{D}}$ where

$$\underline{\mathbf{D}} = \langle\{0, 1\}; \vee, \wedge\rangle.$$

Applying the NU Strong Duality Corollary we obtain a strong duality by taking $G = \{0,1\}$, $H = \varnothing$ and $R = \mathbb{S}(\underline{\mathbf{D}} \times \underline{\mathbf{D}})$. In the example of 2.3.5 we saw that

$$\underset{\sim}{\mathrm{D}} = \langle \{0,1\}; 0, 1, \leqslant, \mathcal{T} \rangle$$

entails all binary algebraic relations on $\underline{\mathbf{D}}$. Using the $\underset{\sim}{\mathrm{M}}$-Shift Strong Duality Lemma 3.2.3(iii)(a) and recalling Lemma 1.2.8, we obtain a strengthened version of the Priestley Dual Equivalence Theorem 1.2.5.

3.2 Theorem (Priestley [Pri70, Pri72]) $\underset{\sim}{\mathrm{D}}$ *yields a strong duality between the variety* $\mathcal{D}$ *of distributive lattices and the category* $\mathbb{IS}_c\mathbb{P}^+ \underset{\sim}{\mathrm{D}}$ *of bounded Priestley spaces.* ■

In Exercises 4.4 and 4.5 we will give two modifications of this result for **bounded distributive lattices**.

3.3 Median Algebras An algebra with a single ternary operation satisfying the equations

$$m(x,y,z) \approx m(y,x,z) \approx m(y,z,x), \quad m(x,x,y) \approx x,$$
$$m(m(x,y,z),u,v) \approx m(x, m(y,u,v), m(z,u,v))$$

is called a **median algebra**. These equations uniquely define the two-element median algebra

$$\underline{\mathbf{M}} = \langle \{0,1\}; m \rangle,$$

which is clearly the simplest possible near-unanimity algebra. The quasi-variety $\mathcal{M} = \mathbb{ISP}\underline{\mathbf{M}}$ that it generates turns out to be the variety of all median algebras. (See Isbell [Isb80] for a proof of this fact and a survey of other work on median algebras.) Applying the NU Strong Duality Corollary 3.3.9 we get a strong duality for median algebras by taking G to be $\{0,1,{}^*\}$, where * is the automorphism reversing 0 and 1, and R to be all subalgebras of $\underline{\mathbf{M}} \times \underline{\mathbf{M}}$. Again the usual order $\leqslant$ on $\{0,1\}$ plays a key role. Let

$$\underset{\sim}{\mathrm{M}} = \langle \{0,1\}; {}^*, 0, 1, \leqslant, \mathcal{T} \rangle.$$

3.4 Theorem (Isbell [Isb80], Werner [Wer81])

(i) $\underset{\sim}{\mathrm{M}}$ *yields a strong duality on the variety* $\mathcal{M}$ *of median algebras.*

(ii) *The dual* $\mathbb{IS}_c\mathbb{P}^+ \underset{\sim}{\mathrm{M}}$ *is exactly the category of bounded Priestley spaces with a continuous homeomorphism that is order-reversing, interchanges 0 and 1 and satisfies* $x^{**} \approx x$ *and* $x \leqslant x^* \Rightarrow x \approx 0$.

Proof (i) Using the $\underset{\sim}{\mathrm{M}}$-Shift Strong Duality Lemma 3.2.3(iii)(a), we must show that $\underset{\sim}{\mathrm{M}}$ entails all binary algebraic relations on $\underline{\mathbf{M}}$. Since all non-empty subsets of $M \times M$ determine subalgebras of the square, there are 15 cases to consider. Of these, 9 are obtainable as direct products of $\{0\}$, $\{1\}$ and $\{0,1\}$, and the remaining 6 are

$$\Delta_{\{0,1\}}, \quad \leqslant, \quad \text{graph}(^*), \quad \leqslant^{\smile} = \geqslant, \quad ^* \cdot \leqslant, \quad ^* \cdot \geqslant$$

which are entailed by $\underset{\sim}{\mathbf{M}}$ according to the Constructs for Entailment 2.4.5 (1), (4) and (15).

(ii) By the Preservation Theorem 1.4.3 and the Separation Theorem 1.4.4, each member of $\mathbb{IS}_c\mathbb{P}^+ \underset{\sim}{\mathbf{M}}$ has the listed properties. Conversely, let $\mathbf{X} = \langle X; ^*, 0, 1, \leqslant, \mathcal{T}\rangle$ be such a space. By a **disjoint ideal** of $\mathbf{X}$ we mean a non-empty clopen decreasing set which is disjoint from its image under *. For example, if W is a proper clopen decreasing set, then $W \backslash W^*$ is a disjoint ideal. If $z \neq 1$, then $z^* \nleqslant z$, so there is a clopen decreasing set W containing z but not z^*. Replacing W with $W \backslash W^*$, we conclude that

- *if $z \neq 1$, then there is a disjoint ideal containing z but not z^*.*

Now choose $x, y \in X$ where $x \nleqslant y$. Then $x \neq 0$ so that $x \nleqslant x^*$, and consequently there is a clopen decreasing set U containing y and x^* but not x. Similarly, there is a clopen decreasing set V containing y and x^* but not y^*. Now $Z = (U \cap V)\backslash(U \cap V)^*$ is a disjoint ideal containing y and x^*, that is, $x \in Z^*$ and $y \in Z$.

Since X is compact, we can find a collection $V_1, V_2, \ldots, V_n, Z$ of $n+1$ disjoint ideals such that

$$\{V_1, V_2, \ldots, V_n, Z, V_1^*, V_2^*, \ldots, V_n^*, Z^*\}$$

covers X and Z contains y and x^*. Removing V_n and redefining Z to be $(V_n \backslash Z^*) \cup Z$, we obtain a collection $V_1, V_2, \ldots, V_{n-1}, Z$ of n disjoint ideals which, together with their images under *, cover X, and where the new Z contains y and x^*. Repeating this process n times will leave us with a single disjoint ideal Z containing y and x^* such that $X = Z \cup Z^*$. Now the morphism $Z \mapsto 0$ and $Z^* \mapsto 1$ separates x and y. ∎

As we found in the Representation Theorem 1.2.4, the representation part of Priestley duality is often more easily viewed as saying that every distributive lattice is isomorphic to the lattice of clopen decreasing subsets of a Priestley space. A similar construction applies here, the verification of which we leave to the reader.

3.5 Proposition *Let $\mathbf{X} \in \mathbb{IS}_c\mathbb{P}^+ \underset{\sim}{\mathbf{M}}$ and let $\mathbf{A}$ be the algebra of disjoint ideals of $\mathbf{X}$ with the operation*

$$m(U, V, W) = (U \cap V) \cup (U \cap W) \cup (V \cap W).$$

Then $\mathbf{A}$ is a median algebra and $\alpha \mapsto \alpha^{-1}(0)$ is an isomorphism from $E(\mathbf{X})$ onto $\mathbf{A}$.

3.6 Stone Algebras A three-valued logic named after Stone offers a generalisation of classical propositional logic using a weakened negation * in which $x \vee x^*$ is not a tautology but $x^* \vee x^{**}$ is. The associated algebras $\underline{\mathbf{S}} = \langle S; \vee, \wedge, {}^*, 0, 1\rangle$, called **Stone algebras**, are bounded distributive lattices satisfying the additional axioms

$$(x \vee y)^* \approx x^* \wedge y^*, \quad (x \wedge y)^* \approx x^* \vee y^*, \quad x \wedge x^* \approx 0, \quad x^* \vee x^{**} \approx 1.$$

The element x^*, the **pseudocomplement** of x, is the greatest element that meets x at 0. Models of these axioms form a variety $\mathcal{S} = \mathbb{ISP}\underline{\mathbf{S}}$ generated by the three-element chain

$$\underline{\mathbf{S}} = \langle \{0, a, 1\}; \vee, \wedge, {}^*, 0, 1\rangle$$

where $0 < a < 1$, and $*$ is defined by $1^* = a^* = 0$ and $0^* = 1$ (see Balbes and Dwinger [BD:DL]). The only non-identity homomorphism between subalgebras of $\underline{\mathbf{S}}$ is the endomorphism $d\colon \underline{\mathbf{S}} \to \underline{\mathbf{S}}$ given by $0 \mapsto 0$, $a \mapsto 1$ and $1 \mapsto 1$. Applying the NU Strong Duality Corollary we obtain a strong duality by taking $G = \{d\}$, $H = \varnothing$ and $R = \mathcal{B}_2$. In order to simplify our dual category, we choose from R the order

$$\preccurlyeq = \{(0,0), (a,a), (1,1), (1,a)\}$$

of Figure 4.1 and define the dual structure

$$\underset{\sim}{\mathbf{S}} = \langle \{0, a, 1\}; d, \preccurlyeq, \mathcal{T}\rangle.$$

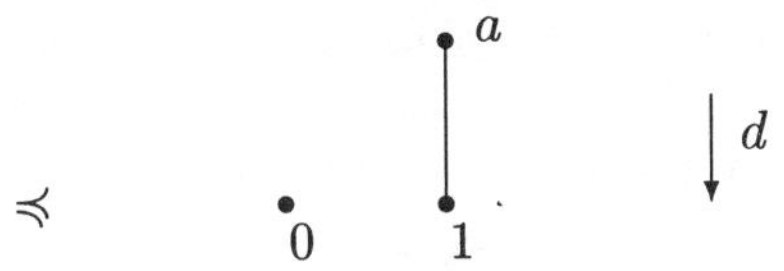

Figure 4.1 the order $\preccurlyeq$ and retraction d on $\underset{\sim}{\mathbf{S}}$

3.7 Theorem (Davey [Dav78, Dav82])

(i) $\underset{\sim}{\mathbf{S}}$ *yields a strong duality on the variety* $\mathcal{S}$ *of Stone algebras.*

(ii) *The dual category* $\mathbb{IS}_c\mathbb{P}^+ \underset{\sim}{\mathbf{S}}$ *is exactly the category of Priestley spaces* $\mathbf{X} = \langle X; d, \preccurlyeq, \mathcal{T}\rangle$ *with a continuous retraction* $d = d \circ d$ *mapping each element (down) to the unique minimal element below it.*

Proof Part (ii) is Example 1.4.7. To prove (i) we must, according to the $\underset{\sim}{\mathbf{M}}$-Shift Strong Duality Lemma 3.2.3(iii)(a), show that d and $\preccurlyeq$ together entail all algebraic binary relations. Using the Constructs for Entailment 2.4.5 we obtain the eleven algebraic binary relations as follows:

$$\Delta_S, \quad \mathrm{graph}(d), \quad \Delta_S \cap \mathrm{graph}(d), \quad \mathrm{graph}(d)^{\smile}, \quad \preccurlyeq, \quad \preccurlyeq^{\smile} = \succcurlyeq, \quad \ker(d),$$
$$\mathrm{fix}(d) \times \mathrm{fix}(d), \quad \mathrm{fix}(d) \times S, \quad S \times \mathrm{fix}(d), \quad S \times S,$$

where $\mathrm{fix}(d) = \{0, 1\}$ denotes the fixpointset of d. ∎

See Exercise 8.8 for further information. We leave to the reader the verification of the following representation for Stone algebras.

3.8 Proposition *Let* $\mathbf{X} \in \mathbb{IS}_c\mathbb{P}^+ \underset{\sim}{\mathbf{S}}$ *and let* $\mathbf{A}$ *be the lattice of clopen decreasing subsets of* $\mathbf{X}$ *with the additional operation* $I^* = X \backslash d^{-1}(I)$ *for* $I \in A$. *Then* $\mathbf{A}$ *is a Stone algebra and* $\alpha \mapsto \alpha^{-1}(1)$ *is an isomorphism from* $E(\mathbf{X})$ *onto* $\mathbf{A}$.

3.9 Kleene Algebras For a different approach to three-valued logic, an algebra $\underline{\mathbf{K}} = \langle K; \vee, \wedge, \neg, 0, 1\rangle$ is called a **Kleene algebra** if it is a bounded distributive lattice satisfying the axioms

$$\neg\neg x \approx x, \quad \neg 0 \approx 1, \quad \neg(x \wedge y) \approx \neg x \vee \neg y, \quad x \wedge \neg x \leqslant y \vee \neg y.$$

The models of these axioms form a variety $\mathcal{K} = \mathbb{ISP}\underline{\mathbf{K}}$ generated by the three-element chain

$$\underline{\mathbf{K}} = \langle \{0, a, 1\}; \vee, \wedge, \neg, 0, 1\rangle$$

where

$$0 < a < 1, \quad \neg 0 = 1, \quad \neg 1 = 0 \quad \text{and} \quad \neg a = a$$

(see Balbes and Dwinger [BD:DL]). In addition to the Boolean truth values of 0 (*false*) and 1 (*true*), we include the value a for *don't know*. The fact that $\neg a = a$ reflects the observation that, if we don't know if a statement is true, then we also don't know if its negation is true.

Because $\underline{\mathbf{K}}$ and its only subalgebra $\mathbf{K}_0 = \langle \{0, 1\}; \vee, \wedge, \neg, 0, 1\rangle$ are both simple and have no non-identity endomorphisms, we obtain a strong duality by taking $G = H = \varnothing$ and $R = \mathbb{S}(\underline{\mathbf{K}} \times \underline{\mathbf{K}})$. Among the relations of R we single out the order $\preccurlyeq$, illustrated in Figure 4.2, together with the unary relation K_0 and the reflexive, symmetric relation

$$\sim\, = \{(0,0), (a,a), (1,1), (0,a), (1,a), (a,0), (a,1)\}$$

relating all pairs except 0 and 1. Let

$$\underset{\sim}{\mathbf{K}} = \langle \{0, a, 1\}; \preccurlyeq, \sim, K_0, \mathcal{T}\rangle.$$

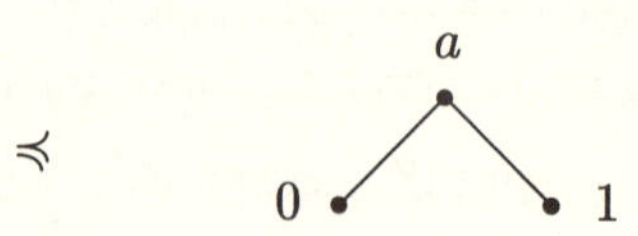

Figure 4.2 the order $\preccurlyeq$ on $\underset{\sim}{\mathbf{K}}$

3.10 Theorem (Davey and Werner [DW83])

(i) $\underset{\sim}{\mathbf{K}}$ *yields a strong duality on the variety* $\mathcal{K}$ *of Kleene algebras.*

(ii) $\mathbf{X} = \langle X; \preccurlyeq, \sim, X_0, \mathcal{T}\rangle$ *belongs to the dual category* $\mathbb{IS}_c\mathbb{P}^+ \underset{\sim}{\mathbf{K}}$ *if and only if* $\langle X; \preccurlyeq\rangle$ *is a Priestley space,* $\sim$ *is a closed binary relation,* X_0 *is a closed subspace and the following universal axioms are satisfied:*

(a) $x \sim x$,

(b) $x \sim y$ *and* $x \in X_0 \implies x \preccurlyeq y$,

(c) $x \sim y$ *and* $y \preccurlyeq z \implies z \sim x$.

Proof (i) As in the previous examples, we prove the first statement by showing that $\underset{\sim}{\mathbf{K}}$ entails all algebraic binary relations. Observe that every subalgebra $\mathbf{r}$ of $\underline{\mathbf{K}} \times \underline{\mathbf{K}}$ is generated by a set L which is either empty or a sublattice of $\langle\{(0a), (aa), (01), (a1)\}; \vee, \wedge\rangle$ as $r = L \cup \neg L \cup \{(00), (11)\}$. There are exactly 13 such choices of L yielding 11 distinct subalgebras $\mathbf{r}$. We obtain these from $\underset{\sim}{\mathbf{K}}$ using the Constructs for Entailment 2.4.5 as

$$\Delta_K,\quad \Delta_{K_0},\quad \preccurlyeq,\quad \preccurlyeq^{\smile} = \succcurlyeq,\quad \sim,\quad K \times K,\quad K \times K_0,\quad K_0 \times K,$$
$$K_0 \times K_0,\quad \preccurlyeq \cap\, (K_0 \times K) \quad\text{and}\quad \preccurlyeq^{\smile} \cap\, (K \times K_0).$$

(ii) The Preservation Theorem 1.4.3 assures us that every member of $\mathbb{IS}_c\mathbb{P}^+ \underset{\sim}{\mathbf{K}}$ has the listed properties. Conversely, let us assume that $\mathbf{X} = \langle X; \preccurlyeq, \sim, X_0, \mathcal{T}\rangle$ has these properties. The quasi-equations

(d) $x \sim y \implies y \sim x$,

(e) $y \preccurlyeq z \implies z \sim y$,

(f) $y \preccurlyeq x$ and $x \in X_0 \implies y \approx x$,

(g) $x \preccurlyeq y$ and $x \preccurlyeq z \implies y \sim z$,

(h) $x \in X_0 \implies (x \sim y \Leftrightarrow x \preccurlyeq y)$

follow easily from (a), (b) and (c). In order to apply the Separation Theorem 1.4.4, we notice that a map $\alpha\colon X \to K$ is a continuous morphism if and only if $I = \alpha^{-1}(0)$ and $J = \alpha^{-1}(1)$ are clopen decreasing sets such that $(I \times J) \cap \sim\, = \varnothing$ and $X_0 \subseteq I \cup J$.

Choose $x, y \in X$ so that $x \not\sim y$. We must find I, J as above with $x \in I$ and $y \in J$. Since $\sim$ is closed, the set $\{z \mid z \sim x\}$ is closed, and by (c) it is $\preccurlyeq$-increasing. Thus $\{z \mid z \not\sim x\}$ is an open $\preccurlyeq$-decreasing set containing y. Since $\mathbf{X}$ is a Priestley space, this is a union of clopen decreasing sets. Let J be a clopen decreasing set containing y such that $z \not\sim x$ for every $z \in J$. Let L be the set of all $w \in X$ that are not $\sim$ to any member of J. Then L is open since $\sim$ is closed and J is compact; L is decreasing by (c); each member of $X_0 \backslash J$ is in L by (h) and $x \in L$. Since $\mathbf{X}$ is a Priestley space, L is a union of clopen decreasing sets. Consequently, there is a clopen decreasing subset I of L containing the closed set $\{x\} \cup (X_0 \backslash J)$. This choice of I and J gives us the separating morphism α.

Choose $x, y \in X$ so that $x \not\preccurlyeq y$. If $x \nsim y$, we can use the α obtained above. Assume that $x \sim y$. By (h) we know that $x \notin X_0$ and by (f) we know that $x \not\preccurlyeq z$ for each $z \in X_0$. Thus we can find a clopen decreasing set J containing the compact set $\{y\} \cup X_0$ which does not contain x. Let $\beta \colon \mathbf{X} \to \underset{\sim}{\mathbf{K}}$ take J to 0 and $X \backslash J$ to a.

Finally, suppose $x \notin X_0$. If $X_0 = \varnothing$, let γ map X to a. Otherwise, choose any $y \in X_0$. By (f) we have $x \not\preccurlyeq y$ and we again use β. ∎

We leave to the reader the verification of the following representation for Kleene algebras.

3.11 Proposition *Let $\mathbf{X} \in \mathbb{IS}_c\mathbb{P}^+ \underset{\sim}{\mathbf{K}}$, let C be the set of clopen decreasing subsets of $\mathbf{X}$, and let*

$$A = \{(I, J) \mid I, J \in C;\ (I \times J) \cap \sim = \varnothing \text{ and } X_0 \subseteq I \cup J\}.$$

Then

$$\mathbf{A} = \langle A; \vee, \wedge, \neg, (X, \varnothing), (\varnothing, X)\rangle$$

is a Kleene algebra where

$$(I, J) \vee (I', J') = (I \cap I', J \cup J'),$$

$$(I, J) \wedge (I', J') = (I \cup I', J \cap J') \quad \text{and} \quad \neg(I, J) = (J, I).$$

Moreover, $\alpha \mapsto (\alpha^{-1}(0), \alpha^{-1}(1))$ is an isomorphism from $E(\mathbf{X})$ onto $\mathbf{A}$.

3.12 (CLO) versus (IC) From the example of Kleene algebras we obtain a simple illustration of the fact that a choice of relations for $\underset{\sim}{\mathbf{M}}$ which determine the clone of $\underline{\mathbf{M}}$ may not be enough to give us the duality condition (IC). Consider the structure

$$\underset{\sim}{\mathbf{K}}' = \langle\{0, a, 1\}; \preccurlyeq, K_0, \mathcal{T}\rangle.$$

If φ is a *total* operation on K which preserves $\preccurlyeq$ and K_0, it will also preserve $\succcurlyeq$ as well as the relational product $\sim = \succcurlyeq \cdot \preccurlyeq$. Since $\underset{\sim}{\mathbf{K}}$ satisfies (CLO), φ must be a term function. Thus $\underset{\sim}{\mathbf{K}}'$ also satisfies (CLO). But $\underset{\sim}{\mathbf{K}}'$ does not satisfy (IC), as we see by taking $X = \{(0, a), (a, 0)\}$ and $\gamma \colon (0, a) \mapsto 0;\ (a, 0) \mapsto 1$. Then $\gamma \colon \mathbf{X} \to \underset{\sim}{\mathbf{K}}'$ preserves $\preccurlyeq$ and K_0 (vacuously), but it does not preserve $\sim$ and therefore does not extend to a term function. In particular, $\underset{\sim}{\mathbf{K}}'$ determines the clone of $\underline{\mathbf{K}}$ but does not yield a duality on the quasi-variety it generates.

Kleene algebras have played an important role in the development of natural duality theory. They occur as seminal examples several times later in this text: see Section 5 of Chapter 7 and Section 4 of Chapter 8.

3.13 Double Stone Algebras An algebra $\mathbf{A} = \langle A; \vee, \wedge, {}^*, {}^\circ, 0, 1\rangle$ is a **double Stone algebra** if both $\langle A; \vee, \wedge, {}^*, 0, 1\rangle$ and $\langle A; \wedge, \vee, {}^\circ, 1, 0\rangle$ are Stone algebras. The dual operation $^\circ$ is called **dual pseudocomplementation**. The variety $\mathcal{DS} = \mathbb{ISP}\underline{\mathbf{DS}}$ of double Stone algebras is generated by the algebra

$$\underline{\mathbf{DS}} = \langle \{0, a, b, 1\}; \vee, \wedge, {}^*, {}^\circ, 0, 1\rangle$$

where $0 < a < b < 1$ and

$$1^* = b^* = a^* = 0, \quad 0^* = 1, \quad 0^\circ = a^\circ = b^\circ = 1, \quad 1^\circ = 0$$

(see Balbes and Dwinger [BD:DL]). Let $\mathbf{D}_a$ and $\mathbf{D}_b$ be the subalgebras of $\underline{\mathbf{DS}}$ determined by the subsets $\{0, a, 1\}$ and $\{0, b, 1\}$, respectively, and let

$$f\colon \mathbf{D}_a \to \mathbf{D}_b, \quad d\colon \underline{\mathbf{DS}} \to \mathbf{D}_a \quad \text{and} \quad u\colon \underline{\mathbf{DS}} \to \mathbf{D}_b$$

be the unique surjections. From the NU Strong Duality Corollary we obtain a strong duality for $\mathcal{DS}$ by taking

$$G = \{u, d\}, \quad H = \{f, f^{-1}\} \quad \text{and} \quad R = \mathcal{B}_2.$$

Again we find that $G \cup H \cup R$ can be vastly reduced. Consider the algebraic order $\preccurlyeq$ shown in Figure 4.3 and let

$$\underset{\sim}{\mathbf{DS}} = \langle \{0, a, b, 1\}; d, u, \preccurlyeq, \mathcal{T}\rangle.$$

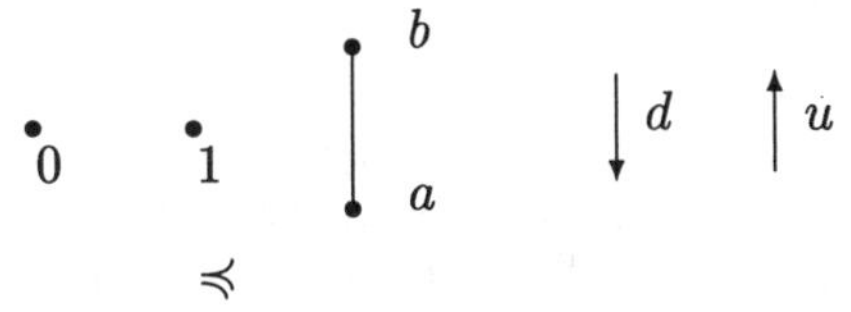

Figure 4.3 the order $\preccurlyeq$ and retractions d and u on $\underset{\sim}{\mathbf{DS}}$

3.14 Theorem (Davey [Dav82])

(i) $\underset{\sim}{\mathbf{DS}}$ *yields a strong duality on the variety* $\mathcal{DS}$ *of double Stone algebras.*

(ii) *The dual category* $\mathbb{IS}_c\mathbb{P}^+\underset{\sim}{\mathbf{DS}}$ *is exactly the category of Priestley spaces* $\mathbf{X} = \langle X; d, u, \preccurlyeq, \mathcal{T}\rangle$ *with continuous retractions* $d = d \circ d$ *mapping each element (down) to the unique minimal element below it and* $u = u \circ u$ *mapping each element (up) to the unique maximal element above it.*

Proof (i) The only non-identity homomorphisms between subalgebras of $\underline{\mathbf{DS}}$ are d, u, f and f^{-1}. Since $\mathrm{graph}(u) \cap \mathrm{graph}(d)^\smile = \mathrm{graph}(f)$, applying parts (i) and (iii)(b) of the $\underset{\sim}{\mathbf{M}}$-Shift Strong Duality Lemma 3.2.3, we see that strong duality is preserved if we replace G with $\{d, u\}$ and H with $\varnothing$. By (iii)(a) it remains to show that $\{d, u, \preccurlyeq\}$ entails the rest of R. We leave it to the reader to check that R has exactly 29 members, 13 subalgebras of ker(d) and 16 direct products of subalgebras of $\underline{\mathbf{DS}}$. Since $\underline{\mathbf{DS}}$ has exactly 4 subalgebras, itself, fix(d), fix(u) and their intersection, we see that $\underset{\sim}{\mathbf{DS}}$ entails all 16 direct products of subalgebras and all 4 identity maps. The remaining relations of R are ker(d) together with graph(d), graph(f) = graph(u) $\cap$ graph(d)$\breve{}$, graph(u), $\preccurlyeq$, and their converses.

(ii) This proof is similar to that of Example 1.4.7, and we ask the reader to either construct it or find it in [Dav82], [DW83] or [Cla89]. ∎

3.15 De Morgan Algebras A different generalisation of propositional logic is modelled by **de Morgan algebras** $\mathbf{A} = \langle A; \vee, \wedge, \neg, 0, 1\rangle$ consisting of a bounded distributive lattice with a negation operation satisfying

$$\neg\neg x \approx x, \quad \neg 0 \approx 1, \quad \neg(x \wedge y) \approx \neg x \vee \neg y, \quad \neg(x \vee y) \approx \neg x \wedge \neg y.$$

The variety of de Morgan algebras is given as $d\mathcal{M} = \mathbb{ISP}\underline{\mathbf{dM}}$ where

$$\underline{\mathbf{dM}} = \langle\{0, a, b, 1\}; \vee, \wedge, \neg, 0, 1\rangle,$$

is the bounded four-element non-chain lattice in which $\neg$ interchanges 0 and 1 and fixes both a and b (see Balbes and Dwinger [BD:DL]). The algebra $\underline{\mathbf{dM}}$ has four subalgebras determined by the subsets

$$dM = \{0, a, b, 1\}, \quad A = \{0, a, 1\}, \quad B = \{0, b, 1\}, \quad C = \{0, 1\}.$$

Since each of these is simple we obtain a strong duality for $d\mathcal{M}$ by taking for G the automorphism f which interchanges a and b, for H the isomorphisms between $\mathbf{A}$ and $\mathbf{B}$, and for $\mathcal{B}_2$ the subalgebras of $\underline{\mathbf{dM}} \times \underline{\mathbf{dM}}$. Unfortunately the number of relations in $\mathcal{B}_2$ is now up to 55, requiring some additional effort to prune it down to a manageable size. Surprisingly, there are two simple and natural members of $\mathcal{B}_2$ that strongly entail the rest. Let $\preccurlyeq \in \mathcal{B}_2$ be the lattice order shown in Figure 4.4 and let

$$\underset{\sim}{\mathbf{dM}} = \langle\{0, a, b, 1\}; f, \preccurlyeq, \mathcal{T}\rangle.$$

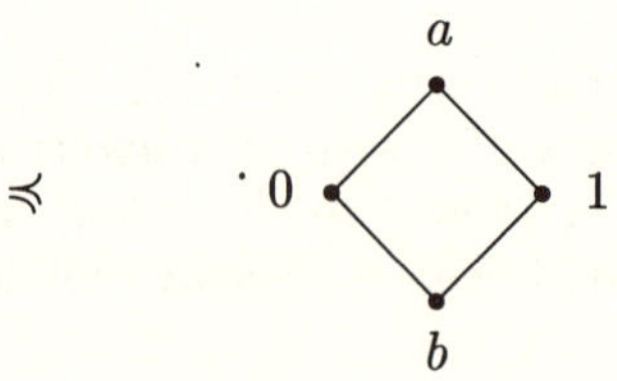

Figure 4.4 the order $\preccurlyeq$ on $\underset{\sim}{\mathbf{dM}}$

3.16 Theorem (Cornish and Fowler [CF77])

(i) $\underset{\sim}{\mathbf{dM}}$ *yields a strong duality on the variety* $d\mathcal{M}$ *of de Morgan algebras.*

(ii) *The dual category* $\mathbb{IS}_c\mathbb{P}^+ \underset{\sim}{\mathbf{dM}}$ *is exactly the category of Priestley spaces* $\mathbf{X} = \langle X; f, \preccurlyeq, \mathcal{T}\rangle$ *with an order-reversing homeomorphism* f *of order* 2.

Proof (i) We give a brief outline of the argument, leaving the tedious but character building details to the reader. By the $\underset{\sim}{\mathbf{M}}$-Shift Strong Duality Lemma 3.2.3(ii)(c),

$$\langle \{0, a, b, 1\}; f, \mathcal{B}_2 \cup \{A, B, C\}, \mathcal{T}\rangle$$

also yields a strong duality on $d\mathcal{M}$. First show that $\mathcal{B}_2$ does indeed contain exactly 55 subalgebras. Then check that each of these can be obtained by first taking direct products of **A**, **B**, **C** and $\underline{\mathbf{dM}}$, and then taking intersections of these with graph(f), $\preccurlyeq$, $\succcurlyeq$, $f \cdot \preccurlyeq$, $f \cdot \succcurlyeq$ and Δ_{dM}. Finally, eliminate $C = \text{fix}(f)$, and eliminate A and B which can be obtained from $f \cdot \preccurlyeq$ and $f \cdot \succcurlyeq$, respectively, using repetition removal.

(ii) Every member of $\mathbb{IS}_c\mathbb{P}^+ \underline{\mathbf{dM}}$ is a Priestley space with an order-reversing homeomorphism by the Preservation Theorem 1.4.3. To prove the converse using the Separation Theorem 1.4.4, we consider $x, y \in X$ where $\mathbf{X}$ is such a space and $x \not\preccurlyeq y$. Let U be a clopen decreasing set containing y but not x. It is immediate that α, given by

$$\alpha(z) = \begin{cases} 1 & \text{if } z \in U \cap f(U), \\ b & \text{if } z \in U \backslash f(U), \\ a & \text{if } z \in f(U) \backslash U, \\ 0 & \text{otherwise}, \end{cases}$$

is the required separating morphism. ∎

3.17 Distributive p-Algebras A **distributive p-algebra** is defined to be a bounded distributive lattice $\mathbf{A} = \langle A; \wedge, \vee, {}^*, 0, 1\rangle$ equipped with an operation * of **pseudocomplementation** where

$$a^* = \max\{b \in A \mid a \wedge b = 0\}.$$

It was shown by Lee [Lee70] that the subvarieties of the variety $\mathcal{B}_\omega$ of distributive p-algebras form an $(\omega{+}1)$-chain

$$\mathcal{B}_{-1} \subset \mathcal{B}_0 \subset \mathcal{B}_1 \subset \cdots \subset \mathcal{B}_n \subset \cdots \subset \mathcal{B}_\omega,$$

where $\mathcal{B}_{-1}$ is the trivial variety, $\mathcal{B}_0$ is the variety of Boolean algebras, $\mathcal{B}_1 = \mathcal{S}$ is the variety of Stone algebras and, for $n \geqslant 1$, the variety $\mathcal{B}_n = \mathbb{ISP}\mathbf{B}_n$ is also the quasi-variety generated by the subdirectly irreducible

p-algebra $\mathbf{B}_n$ constructed by adding a new top element 1 to the n-atom Boolean lattice (whose top and bottom we denote by a and 0).

For each $\mathcal{B}_n$ we obtain a strong duality by taking for $\underset{\sim}{\mathbf{B}}_n$ all endomorphisms and partial endomorphisms together with all algebraic binary relations. When $n = 0$ this reduces to Stone's duality for Boolean algebras, and when $n = 1$ we obtain the above strong duality for Stone algebras. When $n > 1$, the problem of reducing this structure becomes decidedly more difficult. Whereas the example of de Morgan algebras (where the square had 16 elements) pressed the limits of what we might be willing to do with a straightforward, bare-hands approach, these limits are exceeded by $\underset{\sim}{\mathbf{B}}_n$ as soon as $n > 1$ (and consequently the square has at least 25 elements). We describe some of the results here without proof, referring the reader to the literature or to later chapters for more details.

A strong duality for $\mathcal{B}_2$ due to Davey and Werner [DW85] has proven tractable and useful (see Clark and Schmid [CS96]). Label the generator

$$\underline{\mathbf{B}}_2 = \langle\{0, a, b, c, 1\}; \vee, \wedge, *, 0, 1\rangle$$

where b and c meet to 0 and join to a, its dual atom. From among the 40 subalgebras of $\underline{\mathbf{B}}_2 \times \underline{\mathbf{B}}_2$ we pick out the two shown in Figure 4.5. Here $\leqslant$ is an order while $\preccurlyeq$ is an antisymmetric, transitive relation that is reflexive only on $S = \{0, a, 1\}$. Let f be the automorphism which fixes each member of S and interchanges b and c, and let d be the endomorphism which maps each element to the $\preccurlyeq$-minimal element below it.

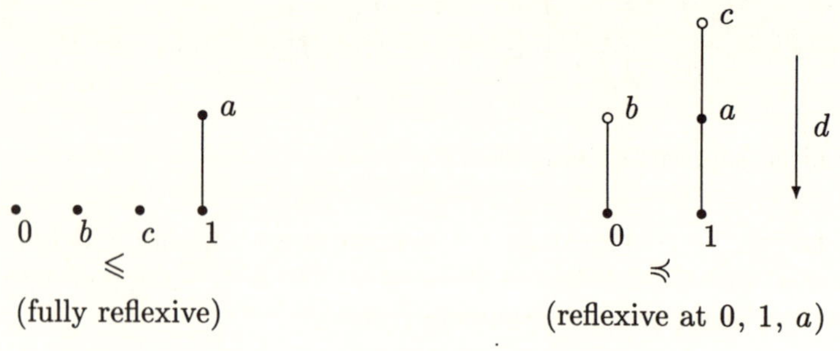

Figure 4.5 the relations $\leqslant$ and $\preccurlyeq$ of $\underset{\sim}{\mathbf{B}}_2$

3.18 Theorem (Davey and Werner [DW85, DW86]) *The structure*

$$\underset{\sim}{\mathbf{B}}_2 = \langle\{0, a, b, c, 1\}; f, d, \leqslant, \preccurlyeq, \mathcal{T}\rangle$$

yields a strong duality on $\mathcal{B}_2$.

We refer the reader to Clark and Schmid [CS96] for a complete set of axioms for the dual category $\mathbb{IS}_c\mathbb{P}^+ \underset{\sim}{\mathbf{B}}_2$.

If $n > 2$, the number of subalgebras of $\underline{\mathbf{B}}_n \times \underline{\mathbf{B}}_n$ is exceedingly large. Although many relations can be eliminated, a substantial number is still necessary to guarantee a duality. In this context the question arises as to when a set of relations is optimal, that is, minimal with respect to entailing the rest. Optimal dualities for the variety $\mathcal{B}_n$ are discussed in detail in Section 6 of Chapter 8.

4.4 Two-for-one strong dualities

Unlike the NU Strong Duality Theorem and its corollaries, the Two-for-One Strong Duality Theorem 3.3.2 gives us no clue as to how to determine $\underset{\sim}{\mathbf{M}}$ when we are given $\underline{\mathbf{M}}$. Successful applications of the Two-for-One Theorem have arisen by circumventing this problem in one of two ways. If the operations of the algebra $\underline{\mathbf{M}}$ happen to be algebraic over $\underline{\mathbf{M}}$ itself, then $\underline{\mathbf{M}}$ becomes a candidate for strong self-duality by choosing $\underset{\sim}{\mathbf{M}}$ to be $\underline{\mathbf{M}}$ with the discrete topology. We will exhibit several applications of the Strong Self-Duality Theorem 3.3.3. The other approach is to appeal to the true spirit of the Two-for-One Strong Duality Theorem: if we can by any means whatsoever establish that a *total algebra* $\underset{\sim}{\mathbf{M}}$ yields a strong duality, then we will be awarded a second strong duality absolutely free!

4.1 Abelian Groups One of the significant prototypes for the theory of natural dualities was the duality of Pontryagin for abelian groups. Let $\underline{\mathbf{A}}$ denote the **circle group** and let $\underset{\sim}{\mathbf{A}}$ be the **compact circle group**, that is, the circle group with the usual compact topology. Since every abelian group is a subdirect product of subgroups of $\underline{\mathbf{A}}$, the class $\mathcal{A} = \mathbb{ISP}\underline{\mathbf{A}}$ is the variety of all abelian groups. Similarly, $\mathfrak{X} = \mathbb{IS}_c\mathbb{P}^+\underset{\sim}{\mathbf{A}}$ is the category of all compact abelian groups (though this fact is quite difficult to prove). Pontryagin's theorem [Pon:TG] says that here as well D and E are contravariant hom-functors which produce a dual equivalence between the categories $\mathcal{A}$ and $\mathfrak{X}$ (see Exercise 2.9). We refer the reader to Davey [Dav78] for an alternative proof given in the spirit of our approach.

The proper subvarieties of $\mathcal{A}$ fall into a one-to-one correspondence with the positive integers. For a positive integer m, let $\mathcal{A}_m$ be the variety of abelian groups satisfying the identity $mx \approx 0$. Such a group is a direct sum of cyclic groups each having order dividing m, and is therefore embedded in a power of the cyclic group

$$\underline{\mathbf{A}}_m = \langle\{0, 1, \ldots, m-1\}; +, 0\rangle,$$

that is to say, $\mathcal{A}_m = \mathbb{ISP}\underline{\mathbf{A}}_m$. The group $\underline{\mathbf{A}}_m$ is an example of an algebra that is algebraic over itself, suggesting that we choose

$$\underset{\sim}{\mathbf{A}}_m = \langle\{0, 1, \ldots, m-1\}; +, 0, \mathcal{T}\rangle.$$

This choice does yield a strong duality which is exactly the restriction of the Pontryagin duality to the subcategory $\mathcal{A}_m$.

4.2 Theorem (Davey and Werner [DW83])

(i) $\underset{\sim}{\mathbf{A}}_m$ *yields a strong duality on the variety* $\mathcal{A}_m$ *of abelian groups of exponent* m.

(ii) *The dual* $\mathbb{IS}_c\mathbb{P}^+ \underset{\sim}{\mathbf{A}}_m$ *is exactly the category of Boolean spaces with a continuous abelian group operation satisfying* $mx \approx 0$.

Proof Claim (ii) is Example 1.4.8. To prove (i), we use the Strong Self-Duality Theorem 3.3.3 and verify (IC). This we do in two steps, using Lemma 2.2.5(ii). Let $\mathfrak{X} = \mathbb{IS}_c\mathbb{P}^+ \underset{\sim}{\mathbf{A}}_m$. We must therefore prove (CLO) and prove that $\underset{\sim}{\mathbf{A}}_m$ is injective in $\mathfrak{X}_{\text{fin}}$.

To prove (CLO), let $f\colon (\underline{\mathbf{A}}_m)^n \to \underline{\mathbf{A}}_m$ be a group homomorphism. Define

$$a_1 = f(1,0,\ldots,0),\ a_2 = f(0,1,\ldots,0),\ \ldots,\ a_n = f(0,0,\ldots,1).$$

Then, for $x = (x_1, x_2, \ldots, x_n) \in (A_m)^n$ we have

$$\begin{aligned} f(x) &= f(x_1,0,\ldots,0) + f(0,x_2,\ldots,0) + \cdots + f(0,0,\ldots,x_n) \\ &= a_1x_1 + a_2x_2 + \cdots + a_nx_n, \end{aligned}$$

showing that f is a term function.

Next we need to show that $\underset{\sim}{\mathbf{A}}_m$ is injective in $\mathfrak{X}_{\text{fin}}$. Let $\mathbf{X} \leqslant (\underset{\sim}{\mathbf{A}}_m)^n$ and let $\varphi\colon \mathbf{X} \to \underset{\sim}{\mathbf{A}}_m$. It suffices to prove that for any $a \in (A_m)^n$, we can extend φ to a homomorphism from the subgroup $\langle a \rangle + X$ into $\underset{\sim}{\mathbf{A}}_m$. Let k be the least positive integer such that ka is a generator of the cyclic group $\langle a \rangle \cap X$. Thus $ta \in X$ if and only if k divides t. Since $ma = 0$, there is an integer j such that $m = jk$. Thus $0 = \varphi(ma) = j\varphi(ka)$ from which we conclude that there is an integer c such that $\varphi(ka) = ck$. We now leave it to the reader to show that ψ is a well-defined extension of φ if we define $\psi(ta + x) = tc + \varphi(x)$. ■

4.3 Vector Spaces Let $\underline{\mathbf{V}}$ be the one-dimensional vector space over a finite field F. Since $\underline{\mathbf{V}}$ is also algebraic over itself, we choose $\underset{\sim}{\mathbf{V}}$ to be $\underline{\mathbf{V}}$ augmented with the discrete topology.

4.4 Theorem (Davey and Werner [DW83])

(i) $\underset{\sim}{\mathbf{V}}$ *yields a strong duality on the variety of vector spaces over* F.

(ii) *The dual* $\mathbb{IS}_c\mathbb{P}^+ \underset{\sim}{\mathbf{V}}$ *is exactly the category of Boolean topological vector spaces over* F.

Proof The argument uses the Strong Self-Duality Theorem 3.3.3 as in the abelian group case, but is simpler, and is left as an exercise. ∎

If we use instead of $\underline{\mathbf{V}}$ a different finite non-trivial vector space $\underline{\mathbf{M}}$ over F as generator of the variety, we obtain an application of the full strength of the Two-for-One Strong Duality Theorem 3.3.2. Again $\underline{\mathbf{M}}$ is algebraic over itself, but this time it does not carry enough structure to yield a duality. This situation can be rectified by adding further algebraic operations which dissect the structure on $\underline{\mathbf{M}}$. Fix a basis for $\underline{\mathbf{M}}$ over F, let p_i be the projection onto the ith summand and let $g_{i,j}$ interchange the ith and jth summands. Let λ_a denote scalar multiplication by a.

4.5 Theorem (Clark and Davey [CD95]) *The structure* $\underset{\sim}{\mathbf{M}} = \langle M; G, \mathcal{T}\rangle$, *where*

$$G = \{+, 0\} \cup \{\lambda_a\}_{a\in F} \cup \{p_i\} \cup \{g_{i,j}\},$$

yields a strong duality on the variety of vector spaces over F with generating algebra $\underline{\mathbf{M}}$.

Proof We refer the reader to [CD95] for the full proof which uses the Two-for-One Strong Duality Theorem 3.3.2. Let $\underline{\mathbf{M}}' = \langle M; G\rangle$ and let $\underset{\sim}{\mathbf{M}}' = \langle M; +, 0, \{\lambda_a\}_{a\in F}, \mathcal{T}\rangle$. In [CD95] it is shown that (IC) holds with respect to both $\underset{\sim}{\mathbf{M}}$ and $\underset{\sim}{\mathbf{M}}'$. ∎

See Examples 7.7.6 and 7.8.5 for an alternative approach to the dualisability and endodualisability of finite vector spaces.

4.6 Meet Semilattices with 1 An early duality was discovered by Hofmann, Mislove and Stralka [HMS74] which also follows from the Strong Self-Duality Theorem. We describe the result, referring the reader to Davey and Werner [DW83] or Exercise 4.10 for a complete proof. A **meet semilattice with one** is a commutative idempotent monoid. The variety of meet semilattices with 1 is generated as a quasi-variety by

$$\underline{\mathbf{SL}} = \langle \{0, 1\}; \wedge, 1\rangle.$$

Let $\underset{\sim}{\mathbf{SL}}$ be obtained from $\underline{\mathbf{SL}}$ by adding the discrete topology.

4.7 Theorem (Hofmann, Mislove and Stralka [HMS74])

(i) $\underset{\sim}{\mathbf{SL}}$ *yields a strong duality on the variety of meet semilattices with* 1.

(ii) *The dual* $\mathbb{IS}_c\mathbb{P}^+\,\underset{\sim}{\mathbf{SL}}$ *is exactly the category of all Boolean topological meet semilattices with* 1.

4.8 Rectangular Bands We end this section with an example where we truly get two dualities for the price of one by using the Two-for-One Strong Duality Theorem to reverse the strong duality of Theorem 4.2.5. Recall that a **rectangular band** is a semigroup satisfying the anti-commutative law $(xy \approx yx) \implies (x \approx y)$. We ask the reader to prove the following simplified analogue of Theorem 4.2.6.

4.9 Theorem *The quasi-variety $\mathcal{RB}$ of rectangular bands is generated as $\mathcal{RB} = \mathbb{ISP}\underline{\mathbf{R}}_6$ by the semigroup*

$$\underline{\mathbf{R}}_6 = \langle\{0, 1, 2, 3, 4, 5\}; *\rangle$$

where $*$ is uniquely defined by $a \equiv_2 a * b \equiv_3 b$.

4.10 Theorem *The ring of integers modulo* 6 *endowed with the discrete topology,*

$$\underset{\sim}{\mathbf{R}}_6 = \langle\{0, 1, 2, 3, 4, 5\}; +, \cdot, 0, 1, \mathcal{T}\rangle,$$

yields a strong duality on $\mathcal{RB}$.

Proof Apply parts (i) and (ii) of the Two-for-One Strong Duality Theorem 3.3.2, using the strong duality of Theorem 4.2.5. ∎

4.5 Vagrant dualities

In this chapter we have seen that a vast range of strong dualities can be obtained as immediate consequences of the culminating results of Chapter 3, the NU Strong Duality Theorem and the Two-for-One Strong Duality Theorem. When these results were first established in [CD95] they were proclaimed to encompass all known strong dualities. But as of this writing the subject of natural dualities is active and expanding, and new strong dualities are being discovered which do not have a home among the consequences of our general theorems. In this section we present three such examples without giving their proofs. Each was obtained by directly establishing the conditions of the Strong Duality Corollary 3.2.5. The existence of such examples surely suggests that there are many others, and that there should be a broader theory that will predict them as well. We hope our readers will become engaged in the search for these examples and in the development of that broader theory, joining our effort to stem the rising tide of vagrancy ....

5.1 Dihedral Groups Let $\underline{\mathbf{D}}_m = \langle D_m; \cdot, ^{-1}, 1\rangle$ be the dihedral group of order $2m$ generated by the two elements a and b where $a^m = b^2 = 1$ and $ba = a^{m-1}b$. In the case that m is odd, Davey and Quackenbush [DQ96] found a strong duality for the (quasi-)variety $\mathcal{D}_m = \mathbb{ISP}\underline{\mathbf{D}}_m$ generated by

$\underline{\mathbf{D}}_m$. We will quote their result here, referring the reader to [DQ96] for the proof. The dual category is generated by the structure

$$\underset{\sim}{\mathbf{D}}_m = \langle D_m; f, 1, +, *, \mathcal{T} \rangle$$

whose total operations f and 1 and partial operations $+$ and $*$ are defined as follows. We take f to be the automorphism given by $f(a) = a$ and $f(b) = ab$, and we distinguish the element 1. The partial operation $+$ is defined to be the group operation restricted to the subgroup $\{1, b\}$, that is, $b^i + b^j = b^{i+j}$. Let $N = \{1, a, a^2, \dots, a^{m-1}\}$ and let $\theta = (N \times N) \cup (Nb \times Nb)$ be the congruence on $\underline{\mathbf{D}}_m$ corresponding to the normal subgroup N. We define the other partial operation $*\colon \theta \to D_m$ by $a^i b^k * a^j b^k = a^{i+j} b^k$.

5.2 Theorem (Davey and Quackenbush [DQ96]) *If m is odd, then $\underset{\sim}{\mathbf{D}}_m$ yields a strong duality on the variety $\mathfrak{D}_m$ generated by the dihedral group $\underline{\mathbf{D}}_m$.*

5.3 Paraprimal Algebras A finite non-trivial algebra $\underline{\mathbf{P}}$ is **paraprimal** if its non-trivial subalgebras are simple and it generates a congruence permutable variety. If every subalgebra of $\underline{\mathbf{P}}$ contains a one-element subalgebra, then the quasi-variety $\mathcal{A} = \mathbb{ISP}\underline{\mathbf{P}}$ is in fact a variety (see Clark and Krauss [CK76, CK77a] or Quackenbush's Appendix 5 to [Grä:UA]). Quasi-primal algebras and one-dimensional vector spaces over a finite field are two quite different examples of paraprimal algebras. Theorem 3.3.13 derives a strong duality for the former from the NU Strong Duality Theorem 3.3.8 while Theorems 4.4.4 and 4.4.5 derive strong dualities for the latter from the Two-for-One Strong Duality Theorem 3.3.2. Combining the outputs of both of these approaches, Davey and Quackenbush [DQ95] showed that *every* paraprimal algebra is strongly dualisable. Since the dual structures $\underset{\sim}{\mathbf{M}}$ are in general quite complex, we exhibit one particular example which illustrates the method.

Let $\underline{\mathbf{P}}$ denote the seven-point, seven-line projective plane. We view $\underline{\mathbf{P}} = \langle P, \circledast \rangle$ as an algebra by defining $a \circledast b$ to be a if $a = b$, and otherwise to be the third point on the unique line containing a and b. The relevant facts about $\underline{\mathbf{P}}$ are easy to check: each of the seven points forms a one-element subalgebra and the only other subalgebras are the seven three-point lines and the entire plane, all of which are simple. Since $p(x, y, z) = y \circledast (x \circledast z)$ is a Mal'cev term for $\underline{\mathbf{P}}$, we see that $\underline{\mathbf{P}}$ is paraprimal.

Let $\underline{\mathbf{L}}$ be any fixed line of $\underline{\mathbf{P}}$ and label its points as $L = \{0, 1, 2\}$. Then the restriction $*$ of $\circledast$ to L is given by $x*y = 2x+2y \bmod 3$. From this fact we see that the partial operation $*$ is algebraic over $\underline{\mathbf{P}}$ (although $\circledast$ is not), and it is not difficult to check that $\underset{\sim}{\mathbf{L}} = \langle L; 0, 1, 2, *, \mathcal{T} \rangle$ yields a strong two-for-one duality on $\mathbb{ISP}\underline{\mathbf{L}}$. Now let $\underline{\mathbf{P}}_t$ be the quasi-primal algebra obtained by adding the ternary discriminator operation to $\underline{\mathbf{P}}$. The Quasi-primal Strong Duality Theorem 3.3.13 tells us that $\underset{\sim}{\mathbf{P}}_t = \langle P; \mathrm{Aut}(\underline{\mathbf{P}}), 0, \mathcal{T}\rangle$ yields a strong duality on $\mathbb{ISP}\underline{\mathbf{P}}_t$. Combining these two strong dualities leads to the following result, which is proven by directly establishing the three conditions of Corollary 3.2.5.

5.4 Theorem (Davey and Quackenbush [DQ95]) *The structure* $\underset{\sim}{\mathbf{P}} = \langle P; \mathrm{Aut}(\underline{\mathbf{P}}), 0, *, \mathcal{T}\rangle$ *yields a strong duality on the variety* $\mathbb{ISP}\underline{\mathbf{P}}$ *generated by the paraprimal algebra* $\underline{\mathbf{P}}$.

5.5 Commutative Rings with 1 Let $\underline{\mathbf{M}} = \langle M; +, \cdot, 0, 1\rangle$ be a finite commutative ring with 1 as a named constant. The **Jacobson radical** of $\underline{\mathbf{M}}$ is the set J of all nilpotent elements of $\underline{\mathbf{M}}$ or, equivalently, the intersection of the maximal ideals of $\underline{\mathbf{M}}$. We say that J is **self-annihilating** if $ab = 0$ for every $a, b \in J$. Recently a full characterisation of dualisability for these rings was discovered. Let $\underset{\sim}{\mathbf{M}} = \langle M; H, R, \mathcal{T}\rangle$ where $R = \mathcal{B}_3$ and H consists of $\mathcal{P}_1 \cup \mathcal{P}_2$ together with the ternary partial operation $r(x, y, z) = x - y + z$ whose domain is all triples (x, y, z) such that x, y and z are in the same coset of J.

5.6 Theorem (Clark, Idziak, Sabourin, Szabó, Willard [CISSW]) *Let* $\underline{\mathbf{M}} = \langle M; +, \cdot, 0, 1\rangle$ *be a finite commutative ring with* 1. *Then the following are equivalent:*

(i) J *is self-annihilating;*

(ii) $\underline{\mathbf{M}}$ *is dualisable;*

(iii) $\underline{\mathbf{M}}$ *is strongly dualisable;*

(iv) $\underset{\sim}{\mathbf{M}}$ *strongly dualises* $\mathcal{A} = \mathbb{ISP}\underline{\mathbf{M}}$.

We note that none of the strong dualities presented in this section arises from the NU Strong Duality Theorem since, in each case, $\mathcal{A}$ is not congruence distributive, nor do they arise from the Two-for-One Strong Duality Theorem since each uses partial operations.

Exercises 4

Exercises from the text. Prove Theorems 4.1.1, 4.1.2 (directly) and 4.1.3(ii). Prove each of the concrete representations, Propositions 4.3.5, 4.3.8 and 4.3.11. Complete the proofs of Theorems 4.2.3, 4.2.6, 4.3.2, 4.3.14, 4.3.16, 4.4.2, 4.4.4 and 4.4.9.

4.1 In this exercise we show how to obtain Stone's Duality Theorem 4.1.2 from Theorem 4.1.1. If D and D' are contravariant functors from $\mathcal{A}$

to $\mathcal{X}$, we say that $\eta\colon D \to D'$ is a **natural isomorphism** provided that, for each $\mathbf{B} \in \mathcal{A}$, there is an isomorphism $\eta_{\mathbf{B}}\colon D(\mathbf{B}) \to D'(\mathbf{B})$ such that for every morphism $f\colon \mathbf{A} \to \mathbf{B}$ in $\mathcal{A}$, we have the commuting diagram $\eta_{\mathbf{A}} \circ D(f) = D'(f) \circ \eta_{\mathbf{B}}$ similar to the squares of Figure 1.2.

(i) Let $\langle D, E, e, \varepsilon\rangle$ be a dual equivalence between categories $\mathcal{A}$ and $\mathcal{X}$, assume $\eta\colon D \to D'$ and $u\colon E \to E'$ are natural isomorphisms and, for all $\mathbf{A} \in \mathcal{B}$ and $\mathbf{X} \in \mathcal{X}$, define

$$e'_{\mathbf{A}} = u_{D'(\mathbf{A})} \circ E(\eta_{\mathbf{A}})^{-1} \circ e_{\mathbf{A}} \quad \text{and} \quad \varepsilon'_{\mathbf{X}} = \eta_{E'(\mathbf{X})} \circ D(u_{\mathbf{X}})^{-1} \circ \varepsilon_{\mathbf{X}}.$$

Show that $\langle D', E', e', \varepsilon'\rangle$ is a dual equivalence.

(ii) Let x be a non-empty, proper filter of a Boolean algebra B. Show that the following are equivalent:
  (a) x is an ultrafilter;
  (b) x is a prime filter;
  (c) $B\backslash x$ is a (proper, non-empty) prime ideal;
  (d) for all $b \in B$ either b or b' is in x.

(iii) Let $\langle D, E, e, \varepsilon\rangle$ be the strong duality between the variety $\mathcal{B}$ of Boolean algebras and the category $\mathcal{Z}$ of Boolean spaces given by Theorem 4.1.1, and let ${}^*\colon \mathcal{B} \to \mathcal{Z}$ and ${}^*\colon \mathcal{Z} \to \mathcal{B}$ be as defined in Stone's Duality Theorem. Show that $\eta\colon D \to {}^*$ is a natural isomorphism where, for each $\mathbf{B} \in \mathcal{B}$, we define $\eta_{\mathbf{B}}\colon D(\mathbf{B}) \to \mathbf{B}^*$ by $\eta(x) = x^{-1}(1)$ for each $x \in D(\mathbf{B})$. Similarly, show that $u\colon E \to {}^*$ is a natural isomorphism where, for each $\mathbf{X} \in \mathcal{Z}$, we define $u_{\mathbf{X}}\colon E(\mathbf{X}) \to \mathbf{X}^*$ by $u(\alpha) = \alpha^{-1}(1)$ for each $\alpha \in E(\mathbf{X})$.

(iv) Use (i), (iii) and Theorem 4.1.1 to prove Stone's Duality Theorem 4.1.2.

4.2 Prove that the Boolean algebra $\underline{\mathbf{B}}$ is primal. Show that the Sheffer-stroke algebra $\underline{\mathbf{S}}$ and the Boolean ring $\underline{\mathbf{R}}$ are term-equivalent to $\underline{\mathbf{B}}$ and so are primal also.

4.3 Prove that a ring $\mathbf{R} = \langle R; +, \cdot, 0\rangle$ is in $\mathbb{ISP}\underline{\mathbf{GF}}(p)$ if and only if it satisfies the equations $xy \approx yx$, $px \approx 0$ and $x^p \approx x$ by filling in the following argument. By the Algebraic Separation Theorem 1.3.1 it suffices to show that, for $0 \neq a \in R$, there is an ideal I not containing a such that $\mathbf{R}/I \cong \underline{\mathbf{GF}}(p)$. Consider the idempotent $e = a^{p-1}$ and the ideal $J = \{x - ex \mid x \in R\}$. Let I be an ideal that is maximal with respect to containing J but neither a nor e. Since $\mathbf{R}/I$ has identity $e + I$, the ideal I must be maximal and therefore $\mathbf{R}/I$ is a field (of characteristic p). Since $x^p - x$ has at most p roots in $\mathbf{R}/I$, we conclude that $\mathbf{R}/I \cong \underline{\mathbf{GF}}(p)$.

4.4 Prove that $\langle\{0,1\};0,\leqslant,\mathcal{T}\rangle$ yields a strong duality between the variety $\mathbb{ISP}\langle\{0,1\};\vee,\wedge,0\rangle$ of distributive lattices with least element 0 and the category $\mathbb{IS}_c\mathbb{P}^+\langle\{0,1\};0,\leqslant,\mathcal{T}\rangle$ of Priestley spaces with least element 0.

4.5 Prove similarly that $\langle\{0,1\};\leqslant,\mathcal{T}\rangle$ yields a strong duality between the variety $\mathbb{ISP}\langle\{0,1\};\vee,\wedge,0,1\rangle$ of bounded distributive lattices and the category $\mathbb{IS}_c\mathbb{P}^+\langle\{0,1\};\leqslant,\mathcal{T}\rangle$ of (all) Priestley spaces.

4.6 In example 4.3.12, let $\sim$ denote the subalgebra of $\underline{\mathbf{K}}\times\underline{\mathbf{K}}$ determined by $\sim$ and use $\underset{\sim}{\mathbf{K}}'$ to obtain a dual adjunction. Show that $D(\sim)$ contains only the two projection maps ρ_1, ρ_2, and that $\gamma\colon \pi_1\mapsto 0;\ \pi_2\mapsto 1$ is an extension of φ in $ED(\sim)$ which is not an evaluation map. (Show that it does not preserve $\sim$ on $D(\sim)$.) This shows directly that $\underset{\sim}{\mathbf{K}}'$ does not yield a duality on $\mathcal{K}$. (In 8.4.2 we will see that our duality for Kleene algebras is in fact optimal.)

4.7 Let $\underline{\mathbf{W}}_n = \langle\{0,1,\ldots,n-1\};\wedge,\vee\rangle$ where $\wedge$ and $\vee$ are defined in terms of the reflexive, non-transitive order $0<2,3,\ldots,n-1<1<0$ in which $2,3,\ldots,n-1$ are incomparable. Show that the median operation m is nevertheless a near-unanimity term. The resulting duality, due to Werner [Wer81], is described in [DW83].

4.8 Consider the strong dualities for the Heyting algebra chains $\underline{\mathbf{C}}_2$ and $\underline{\mathbf{C}}_3$ described in Theorem 4.2.3(ii).

(i) Show that the dual $\mathbb{IS}_c\mathbb{P}^+\,\underset{\sim}{\mathbf{C}}_2$ is exactly the category of all Boolean spaces.

(ii) Show that the dual $\mathbb{IS}_c\mathbb{P}^+\,\underset{\sim}{\mathbf{C}}_3$ is exactly the category of all Boolean spaces with a continuous retraction d.

(iii) Find a concise description of the dual category for the Heyting algebra $\underline{\mathbf{M}} = \underline{\mathbf{2}}^2\oplus\underline{\mathbf{1}}$.

4.9 [Ban71, DW83] Let $\underline{\mathbf{Z}} = \langle\{0,1\};\varnothing\rangle$ be a two-element set and $\underset{\sim}{\mathbf{Z}} = \langle\{0,1\};\vee,\wedge,',0,1,\mathcal{T}\rangle$ a two-element topological Boolean algebra.

(i) Show that $\mathcal{A} := \mathbb{ISP}\underline{\mathbf{Z}}$ is the class of all non-empty sets.

(ii) Show that $\mathcal{X} := \mathbb{IS}_c\mathbb{P}^+\,\underset{\sim}{\mathbf{Z}}$ is the class of all non-trivial topological Boolean algebras whose underlying topology is Boolean.

(iii) Use the Two-for-One Strong Duality Theorem and Stone's Duality Theorem to show that $\underset{\sim}{\mathbf{Z}}$ yields a strong duality on $\mathcal{A}$.

4.10 Let $\mathcal{A}$ be the variety of meet semilattices with 1.

(i) Use the Algebraic Separation Theorem 1.3.1 to prove that $\mathcal{A} = \mathbb{ISP}\underline{\mathbf{SL}}$ as follows. If $\mathbf{A}\in\mathcal{A}$, show that $\leqslant$ is an order on $\mathbf{A}$ where $a\leqslant b$ whenever $a\wedge b = a$. Now choose $a\nleqslant b$ and let I be an order ideal that is maximal with respect to containing b but not a. Show that the map taking I to 0 and its complement to 1 is a homomorphism into $\underline{\mathbf{SL}}$ separating a and b.

(ii) Use the Strong Self-Duality Theorem 3.3.3 to prove part (i) of Theorem 4.4.7.

(iii) Use the Separation Theorem 1.4.4 to prove part (ii) of Theorem 4.4.7.

4.11 Following the steps below, exhibit an example of a quasi-variety $\mathcal{A}$ and a structure $\underset{\sim}{\mathbf{M}}$ which yields a duality on $\mathcal{A}$ but fails to satisfy (IC) and is therefore not injective in $\mathfrak{X}$. Let $\underline{\mathbf{M}}$ be the abelian group $\underline{\mathbf{A}}_4 = \langle\{0, 1, 2, 3\}; +, 0\rangle$ and let $\underset{\sim}{\mathbf{M}}$ be the purely relational structure $\underset{\sim}{\mathbf{A}}_4 = \langle\{0, 1, 2, 4\}; \mathrm{graph}(+), \{0\}, \mathcal{T}\rangle$.

(i) Use Theorem 4.4.2 and the $\underset{\sim}{\mathbf{M}}$-Shift Duality Lemma 2.4.2 to show that $\underset{\sim}{\mathbf{A}}_4$ yields a duality on $\mathcal{A} = \mathbb{ISP}\underline{\mathbf{A}}_4$.

(ii) Show that the set $\{2\}$ determines a singleton substructure $\mathbf{X}$ of $\underset{\sim}{\mathbf{A}}_4$, and that the map φ taking 2 to 1 is a morphism from $\mathbf{X}$ to $\underset{\sim}{\mathbf{A}}_4$ which does not extend to a morphism from $\underset{\sim}{\mathbf{A}}_4$ to $\underset{\sim}{\mathbf{A}}_4$.

(iii) Conclude that $\underset{\sim}{\mathbf{A}}_4$ is not injective in $\mathfrak{X}$ and that, by Lemma 2.2.5, (IC) does not hold.

In the following exercises we give a method of generating many interesting quasi-primal algebras that are strongly dualised by various combinations of their subalgebras and partial automorphisms. We say that an algebra $\underline{\mathbf{L}} = \langle L; \vee, \wedge, ', 0, 1\rangle$ is a **pseudo-logic** if

(i) $\langle L; \vee, \wedge, 0, 1\rangle$ is a bounded lattice with smallest element 0 and largest element 1, and

(ii) $'\colon L \to L$ is a unary operation satisfying $0' = 1$ and $1' = 0$.

For the remainder of these exercises we shall assume that $\underline{\mathbf{L}}$ is a finite pseudo-logic and let $\mathcal{A} = \mathbb{ISP}\underline{\mathbf{L}}$. We think of $\vee, \wedge, ', 0, 1$ as OR, AND, NOT, FALSE and TRUE respectively. Note that the median term is a majority term for $\underline{\mathbf{L}}$. According to the characterisation (iv) of the Quasi-primality Theorem 3.3.12, $\underline{\mathbf{L}}$ is quasi-primal provided that the only subalgebras of its square are the graphs of its partial automorphisms and the direct products of its subalgebras.

4.12 Let $\mathbf{r}$ be a subalgebra of $\underline{\mathbf{L}}^2$.

(i) Show that $(1, 0), (a, b), (c, d) \in r$ imply that $(a, d) \in r$.

(ii) Show that r is a product of subalgebras of $\underline{\mathbf{L}}$ if and only if $(1, 0) \in r$ if and only if $(0, 1) \in r$.

For $u \in L$, define $u^{(1)} := u'$ and for $k \geqslant 1$ define $u^{(k+1)} := (u^{(k)})'$. We identify two conditions that $\underline{\mathbf{L}}$ may satisfy.

($*$) If $u \neq 1$, then there is an $n \geq 1$ such that $u \wedge u^{(2n)} = 0$.

($**$) If $u \neq 0$, then $u' \neq 1$.

4.13 Suppose that $\underline{\mathbf{L}}$ satisfies $(*)$ and $(**)$. Let $\mathbf{r}$ be a subalgebra of $\underline{\mathbf{L}}^2$ containing a pair (u, v) where $u \neq v$ and either u or v is in $\{0, 1\}$. Show that $\mathbf{r}$ must be a product of subalgebras.

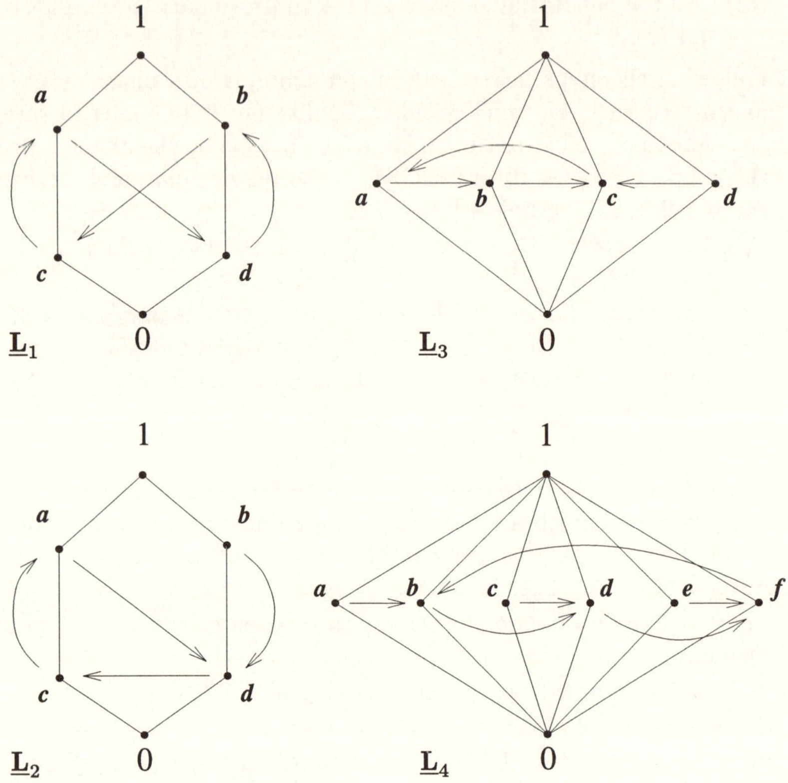

Figure 4.6 Four quasi-primal pseudo-logics.

4.14 Let $\underline{\mathbf{L}}$ be the pseudo-logic $\underline{\mathbf{L}}_1$ in Figure 4.6. Use the steps below to show that $\mathcal{A}$ is **strongly automorphism-dualisable**.

(i) Show that $\underline{\mathbf{L}}$ satisfies both $(*)$ and $(**)$, and then use Theorem 3.3.12(iv) to verify that $\underline{\mathbf{L}}$ is quasi-primal.

(ii) Show that $A_1 = \{0, 1\}$ is the only proper subalgebra of $\underline{\mathbf{L}}$.

(iii) Show that $\underline{\mathbf{L}}$ has exactly one partial automorphism σ other than the identity maps, where σ is the automorphism $0 \leftrightarrow 0$, $1 \leftrightarrow 1$, $a \leftrightarrow b$ and $c \leftrightarrow d$ of $\underline{\mathbf{L}}$.

(iv) Use the Quasi-primal Strong Duality Theorem 3.3.13 and the $\underset{\sim}{\mathbf{M}}$-Shift Strong Duality Lemma 3.2.3 to show that the structure

$\underset{\sim}{\mathrm{L}} = \langle\{0,1,a,b,c,d\}; \sigma, \mathcal{T}\rangle$ yields a strong duality on $\mathcal{A}$.

4.15 Let $\underline{\mathbf{L}}$ be the pseudo-logic $\underline{\mathbf{L}}_2$ in Figure 4.6. Use the steps below to show that $\mathcal{A}$ is **strongly subalgebra-dualisable**.

(i) Show that $\underline{\mathbf{L}}$ satisfies both $(*)$ and $(**)$, and then use Theorem 3.3.12(iv) to verify that $\underline{\mathbf{L}}$ is semi-primal.

(ii) Show that $\underline{\mathbf{L}}$ has exactly three subalgebras, determined by the sets $A_2 = \{0,1\}$, $B_2 = \{0,1,a,c,d\}$ and $L = \{0,1,a,b,c,d\}$.

(iii) Use the Semi-primal Strong Duality Theorem to show that $\underset{\sim}{\mathrm{L}} = \langle\{0,1,a,b,c,d\}; A_2, B_2, \mathcal{T}\rangle$ yields a strong duality on $\mathcal{A}$.

4.16 Let $\underline{\mathbf{L}}$ be the pseudo-logic $\underline{\mathbf{L}}_3$ in Figure 4.6. Use the steps below to show that $\mathcal{A}$ is **strongly proper-partial-automorphism-dualisable**.

(i) Show that $\underline{\mathbf{L}}$ satisfies both $(*)$ and $(**)$, and then use Theorem 3.3.12(iv) to verify that $\underline{\mathbf{L}}$ is quasi-primal.

(ii) Show that $\underline{\mathbf{L}}$ has exactly three subalgebras, determined by the sets $A_3 = \{0,1\}$, $B_3 = \{0,1,a,b,c\}$ and $L = \{0,1,a,b,c,d\}$.

(iii) Show that $\underline{\mathbf{L}}$ has exactly two partial automorphisms other than the identity maps, namely, σ and σ^2 where σ is the automorphism $0 \leftrightarrow 0$, $1 \leftrightarrow 1$, $a \mapsto b \mapsto c \mapsto a$ of $\mathbf{B}_3$.

(iv) Use the Quasi-primal Strong Duality Theorem 3.3.13 and the $\underset{\sim}{\mathrm{M}}$-Shift Strong Duality Lemma 3.2.3 to show that the structure $\underset{\sim}{\mathrm{L}} = \langle\{0,1,a,b,c,d\}; \sigma, \mathcal{T}\rangle$ yields a strong duality on $\mathcal{A}$.

4.17 Let $\underline{\mathbf{L}}$ be the pseudo-logic $\underline{\mathbf{L}}_4$ in Figure 4.6. Use the steps below to show that $\mathcal{A}$ is **strongly subalgebra-automorphism-dualisable**.

(i) Show that $\underline{\mathbf{L}}$ satisfies both $(*)$ and $(**)$, and then use Theorem 3.3.12(iv) to verify that $\underline{\mathbf{L}}$ is quasi-primal.

(ii) Show that $\underline{\mathbf{L}}$ has exactly nine subalgebras, determined by the sets $A_4 = \{0,1\}$, $B_X = \{0,1,b,d,f\} \cup X$ where X ranges over the eight subsets of $\{a,c,e\}$.

(iii) Show that, other than the identity map, the partial automorphisms of $\underline{\mathbf{L}}$ are exactly the restrictions to its subalgebras of σ and σ^2 where σ is the automorphism that takes $0,1,a,b,c,d,e,f$, respectively, to $0,1,c,d,e,f,a,b$.

(iv) Use the Quasi-primal Strong Duality Theorem 3.3.13, the Constructs for Entailment 2.4.5 and the $\underset{\sim}{\mathrm{M}}$-Shift Strong Duality Lemma 3.2.3 to show that $\underset{\sim}{\mathrm{L}} = \langle\{0,1,a,b,c,d,e,f\}; \sigma, B_Y, \mathcal{T}\rangle$ yields a strong duality on $\mathcal{A}$ where $Y = \{0,1,b,c,d,e,f\}$.

5

Sample Applications

From the outset we have billed natural dualities as a tool capable of helping to solve real problems that arise in algebraic quasi-varieties. The purpose of the present chapter is to substantiate this claim by exhibiting several samples of algebraic problems which have been successfully solved using dualities. Each sample application will require some digression into the particulars of the quasi-variety at hand and of the problem that is to be solved.

5.1 Compactness of propositional logic

Our first application is a simple one, but one of historical significance since it contributed to the motivation for the original duality between Boolean algebras and Boolean spaces (Theorem 4.1.2).

Boolean algebra was invented to give an algebraic representation of propositional logic. We think of the operation symbols $\vee$, $\wedge$, $\neg$, 0 and 1 as representing OR, AND, NOT, FALSE and TRUE, respectively. If we view each variable as representing an arbitrary sentence, then each term in the language of Boolean algebra becomes a compound sentence whose truth or falsity is completely determined by the truth or falsity of its constituent variables. More precisely, a map f from the set C of all compound sentences into $\{0, 1\}$ is a **truth assignment** if it respects the usual truth table rules:

$f(a \wedge b) = 1$ exactly when both $f(a) = 1$ and $f(b) = 1$;

$f(a \vee b) = 1$ exactly when either $f(a) = 1$ or $f(b) = 1$;

$f(\neg a) = 1$ exactly when $f(a) = 0$.

Using these rules we can compute $f(a)$ for any compound sentence a as soon as we know the value of $f(v)$ for each variable v.

This situation gives rise to a natural consequence relation. We say that a compound sentence s is a (**truth table**) **consequence** of a set Σ of compound sentences if every truth assignment which assigns 1 to each member of Σ also assigns 1 to s. Expressed in symbols, we write $\Sigma \models s$. Much of propositional logic is concerned with the understanding of the consequence relation. Typically, we are interested in finding a set of rules which allow us, by purely syntactic means, to extract consequences of Σ from finite subsets of Σ. Such systems of rules normally involve either **natural deduction**, **resolution** or generation of **tautologies**. We write

$\Sigma \vdash s$ to indicate that the compound sentence s is derivable from Σ using a particular set of rules.

Given any set of derivation rules, two questions immediately arise. First, we would like to know if the rules are **sound**, that is, does $\Sigma \vdash s$ imply $\Sigma \models s$? Secondly, we would like to know if the rules are **complete**, that is, does $\Sigma \models s$ imply $\Sigma \vdash s$? Soundness simply says that the rules are reasonable; they only allow us to derive valid consequences. This is usually easy to verify by examining the rules individually. Completeness says that the rules are powerful enough to derive all valid consequences. The verification of completeness is usually more involved, and typically proceeds in two steps.

(i) Show that if $\Sigma \models s$, then $\Gamma \models s$ for some finite $\Gamma \subseteq \Sigma$.

(ii) Show that if Γ is finite and $\Gamma \models s$, then $\Gamma \vdash s$.

Observe that (i) is totally general, and depends in no way on the particular rules being used. Only (ii) depends on the nature of the particular set of rules we are using. We will do (i) by interpreting the compound sentences as elements of a Boolean algebra and then using the fact that the dual space is compact.

1.1 Compactness Theorem *If Σ is a set of compound sentences, s is a compound sentence and $\Sigma \models s$, then $\Gamma \models s$ for some finite subset Γ of Σ.*

Proof We consider two compound sentences to be **equivalent** if they have the same truth table, and we identify each compound sentence with its equivalence class. Under this identification, the compound sentences form a Boolean algebra $\underline{\mathbf{L}}$ (called the **Lindenbaum algebra**). Let $\underline{\mathbf{B}} = \langle \{0,1\}; \vee, \wedge, ', 0, 1\rangle$ be the primal generator of the variety of Boolean algebras. Notice that the truth assignments are exactly the homomorphisms from $\underline{\mathbf{L}}$ into $\underline{\mathbf{B}}$, that is, the points of the dual space $\mathbf{X} = D(\underline{\mathbf{L}})$. Consider the clopen subsets

$$U_1 = \{x \in X \mid x(s) = 1\} \quad \text{and} \quad U_t = \{x \in X \mid x(t) = 0\}$$

of $\mathbf{X}$ where t ranges over Σ. Since $\Sigma \models s$, the sets $\{U_1\} \cup \{U_t \mid t \in \Sigma\}$ cover $\mathbf{X}$. Since $\mathbf{X}$ is compact, there is a finite subset $\Gamma \subseteq \Sigma$ such that $\{U_1\} \cup \{U_t \mid t \in \Gamma\}$ cover $\mathbf{X}$. Thus $\Gamma \models s$. ∎

5.2 Endomorphism monoids and upside down lattices

In this section we present in detail one instance of a class of applications which can be obtained from strong dualities, leaving it to our readers to discover other instances where the same method is successful. Given an object $\mathbf{A}$ in a category $\mathfrak{C}$, we denote by $\mathrm{End}(\mathbf{A})$ the **endomorphism**

monoid of $\mathbf{A}$, that is, the set $\mathcal{C}(\mathbf{A}, \mathbf{A})$ viewed as a monoid under composition. We address the following question: to what extent is $\mathbf{A}$ determined within $\mathcal{C}$ by the monoid $\mathrm{End}(\mathbf{A})$? A survey of the literature reveals a spectrum of possible answers which focus at two extremes. We begin by citing some of these examples.

Independently, Magill [Mag72], Maxson [Max72] and Schein [Sch70] showed that non-trivial Boolean algebras are determined up to isomorphism by their endomorphism monoids. (See Exercise 5.2.) It follows from Theorem 4.1.1 that the same is true of the algebras in the variety generated by any primal algebra. This is also the case for median algebras (Bandelt [Ban81]), distribute lattices with 0 and distributive lattices with 1 (Schein [Sch70]). The conclusion is slightly weaker for distributive lattices (without bounds): Schein [Sch70] showed that if $\mathrm{End}(\mathbf{L}_1) \cong \mathrm{End}(\mathbf{L}_2)$, then $\mathbf{L}_1$ is isomorphic either to $\mathbf{L}_2$ or to the lattice obtained by reversing the order of $\mathbf{L}_2$. The same result for bounded distributed lattices was later proven by McKenzie and Tsinakis [MT81]. Using the duality-theoretic methods we will present here, Adams and Clark [AC90] showed that, in the variety generated by almost any quasi-primal algebra $\mathbf{Q}$, there are at most k non-isomorphic algebras having a given endomorphism monoid, where k is the index of the weak automorphism group of $\mathbf{Q}$ in the automorphism group of $\mathbf{Q}$.

To put these positive results in perspective we note that, at the other extreme, a number of authors have found finitely generated varieties for which there are monoids that are isomorphic to the endomorphism monoid of a proper class of non-isomorphic algebras (see, for example, the text of Pultr and Trnková [PT:GSC]). A striking case of this dichotomy was discovered by Adams, Koubek and Sichler [AKS83, AKS84]. Recall from 4.3.17 that the subvarieties of distributive p-algebras form an $(\omega+1)$-chain. Adams, Koubek and Sichler found that non-trivial algebras in $\mathcal{B}_1$ (like $\mathcal{B}_{-1}$ and $\mathcal{B}_0$) are determined up to isomorphism, in $\mathcal{B}_2$ (like distributive lattices) they are determined to within two, while for $n \geqslant 3$ the variety $\mathcal{B}_n$ contains a proper class of non-isomorphic algebras with isomorphic endomorphism monoids.

We present here a general strategy for using strong dualities to obtain positive results like those cited above. Suppose that $\underline{\mathbf{M}}$ yields a strong duality on $\mathcal{A}$. We are given the endomorphism monoid $\boldsymbol{S}_{\mathbf{A}} = \mathrm{End}(\mathbf{A})$ of an algebra $\mathbf{A} \in \mathcal{A}$ *as an abstract monoid* and asked to reconstruct $\mathbf{A}$ from it. Let $\mathbf{X} = D(\mathbf{A})$ be the dual of $\mathbf{A}$. Since D is a full and faithful contravariant functor, we observe that the monoid $\boldsymbol{S}_{\mathbf{X}} = \mathrm{End}(\mathbf{X})$ is anti-isomorphic to $\boldsymbol{S}_{\mathbf{A}} = \mathrm{End}(\mathbf{A})$, and may therefore serve as our starting point. Beginning with $\boldsymbol{S}_{\mathbf{X}}$ we attempt to carry out some variation of the following steps. What is subtle but essential is that we find ways to identify the objects and sets that we need by referring only to the monoid $\boldsymbol{S}_{\mathbf{X}}$ and not to $\mathbf{X}$ itself.

(i) For each point x in X find a unique endomorphism $x^* \in \boldsymbol{S}_{\mathbf{X}}$ in such a way that the set $X^* = \{x^* \mid x \in X\}$ can be identified within $\boldsymbol{S}_{\mathbf{X}}$.

(ii) For each $h\colon \mathbf{X} \to \mathbf{X}$ in $\boldsymbol{S}_{\mathbf{X}}$, let $h^*\colon X^* \to X^*$ be the corresponding map $h^*(x^*) = h(x)^*$. Find a way to define h^* within $\boldsymbol{S}_{\mathbf{X}}$.

(iii) Find a subset $M^+ \subseteq X$ that determines a substructure $\underset{\sim}{\mathbf{M}}^+$ of $\mathbf{X}$ so that $\underset{\sim}{\mathbf{M}}^+ \cong \underset{\sim}{\mathbf{M}}$ and the set $M^* = \{x^* \mid x \in M^+\}$ can be identified from within $\boldsymbol{S}_{\mathbf{X}}$.

(iv) Identify the set $E(\mathbf{X})^+ := \{h \in \boldsymbol{S}_{\mathbf{X}} \mid h(\mathbf{X}) \subseteq M^+\}$ from within $\boldsymbol{S}_{\mathbf{X}}$.

(v) Identify the correct algebraic structure on the set M^* that makes it isomorphic to $\underline{\mathbf{M}}$ under $x \mapsto x^*$. This induces an algebraic structure on
$$E(\mathbf{X})^* := \{\, h^* \mid h \in \boldsymbol{S}_{\mathbf{X}} \text{ and } h(X) \subseteq M^+ \,\} \subseteq (M^*)^{X^*}$$
which makes it isomorphic to $E(\mathbf{X})$, and consequently also to $\mathbf{A}$.

The reader will hopefully appreciate the fact that this strategy constitutes a rather complex application of the duality. We do not attempt to construct $\mathbf{A}$ within $\boldsymbol{S}_{\mathbf{A}}$, but rather we reverse the operation in $\boldsymbol{S}_{\mathbf{A}}$ to obtain the anti-isomorphic monoid $\boldsymbol{S}_{\mathbf{X}}$. We then identify a subset X^* of $\boldsymbol{S}_{\mathbf{X}}$ that looks like $\mathbf{X}$. But instead of searching for the right algebraic and topological structure on X^* to recover $\mathbf{X}$ (and therefore also $\mathbf{A}$), we again exploit the natural duality construction and build from X^* a set $E(\mathbf{X})^*$ of maps that looks just like $E(\mathbf{X})$, which we know to be isomorphic to $\mathbf{A}$.

Once we have completed step (iv) it remains only to identify the correct algebraic structure on the set M^*, that is, the correct labelling of M^* with the members of $\underline{\mathbf{M}}$. If M has m elements, then we know that one of the $m!$ possible labellings must be correct. This means that completion of step (iv) assures us that the endomorphism monoid of $\mathbf{A}$ determines $\mathbf{A}$ to within at most $m!$ possibilities.

We will illustrate this strategy by giving a proof of the above mentioned result of McKenzie and Tsinakis [MT81]. The order-theoretic dual of a bounded lattice $\mathbf{L} = \langle L; \vee, \wedge, 0, 1\rangle$ is the bounded lattice $\mathbf{L}^\partial = \langle L; \wedge, \vee, 1, 0\rangle$ obtained by turning $\mathbf{L}$ upside down, that is, replacing the order $\leqslant$ with $\leqslant^{\smile} = \geqslant$. Although $\mathbf{L}$ and $\mathbf{L}^\partial$ are not in general isomorphic, exactly the same maps $h\colon L \to L$ are endomorphisms of each of them, and consequently they have the same endomorphism monoid.

2.1 Theorem *Let $\mathbf{L}_1$ and $\mathbf{L}_2$ be non-trivial bounded distributive lattices. Then* $\mathrm{End}(\mathbf{L}_1) \cong \mathrm{End}(\mathbf{L}_2)$ *if and only if either $\mathbf{L}_1 \cong \mathbf{L}_2$ or $\mathbf{L}_1 \cong \mathbf{L}_2^\partial$.*

Proof We apply Exercise 4.5 of Chapter 4, where we saw that the structure $\underset{\sim}{\mathbf{D}} = \langle \{0,1\}; \leqslant, \mathcal{T}\rangle$ yields a strong duality between the variety

$\mathcal{A}$ of all bounded distributive lattices $\mathbf{L} = \langle L; \vee, \wedge, 0, 1 \rangle$ and the category $\mathcal{X}$ of all Priestley spaces $\mathbf{X} = \langle X; \leqslant, \mathcal{T} \rangle$. Let $\mathbf{X} \in \mathcal{X}$ and let $\boldsymbol{S}_{\mathbf{X}}$ be its endomorphism monoid. We will follow steps (i) through (v) above to show how we can uniformly construct from $\boldsymbol{S}_{\mathbf{X}}$ a pair of bounded distributive lattice $\mathbf{L}_{\mathbf{X}}$ and $\mathbf{L}_{\mathbf{X}}^{\partial}$, one of which must be isomorphic to the lattice $\mathbf{L} = E(\mathbf{X})$.

(i) To carry out the first step, recall that an element $l \in \boldsymbol{S}_{\mathbf{X}}$ is called a **left zero** if $lh = l$ for all $h \in \boldsymbol{S}_{\mathbf{X}}$. For example, if $x \in X$, then the constant map $x^*: X \to \{x\}$ is a left zero of $\boldsymbol{S}_{\mathbf{X}}$. In fact, it is easy to check that we can identify the copy $X^* = \{x^* \mid x \in X\}$ of X within $\boldsymbol{S}_{\mathbf{X}}$ as being exactly the set of *all* left zeros of $\boldsymbol{S}_{\mathbf{X}}$.

(ii) We can also lift each endomorphism $h: \mathbf{X} \to \mathbf{X}$ to a corresponding map $h^*: X^* \to X^*$ by defining $h^*(x^*) = h(x)^*$. This map can be identified within the monoid $\boldsymbol{S}_{\mathbf{X}}$, as it can also be expressed as $h^*(x^*) = hx^*$, the product of h and x^* in $\boldsymbol{S}_{\mathbf{X}}$.

(iii) In this application some extra effort will be required since the space $\mathbf{X}$ only contains a copy of the generating structure $\underset{\sim}{\mathbf{D}}$ if it is not an antichain, that is, if $\mathbf{L} = E(\mathbf{X})$ is not a Boolean lattice. Fortunately, the required two-element chains—when they exist—are easily recognised from within X^*. Let u and v be two distinct elements of X. We ask the reader to check that u and v are incomparable in $\mathbf{X}$ if and only if there is an $h \in \boldsymbol{S}_{\mathbf{X}}$ such that $hu^* = v^*$ and $hv^* = u^*$; otherwise u and v form a 2-chain in $\mathbf{X}$.

Using this observation we fix a choice of two distinct left zeros a^* and b^* as follows. If $\mathbf{X}$ is not an antichain, choose a^* and b^* so that a and b form a 2-chain in $\mathbf{X}$. If $\mathbf{X}$ is an antichain, choose a^* and b^* to be distinct but otherwise arbitrary. Let $\leqslant_?$ be the restriction of the order on $\mathbf{X}$ to $\{a, b\}$ if $\{a, b\}$ forms a chain, and let it be either of the two 2-chain orders otherwise. Even if a and b are comparable, the order $\leqslant_?$ must remain a mystery since it will be different in the two spaces $\mathbf{X}$ and $\mathbf{X}^{\partial}$, both of which have $\boldsymbol{S}_{\mathbf{X}}$ as their endomorphism monoid. This order gives us a Priestley space and a bounded distributive lattice,

$$\underset{\sim}{\mathbf{D}}_? = \langle \{a, b\}; \leqslant_?, \mathcal{T} \rangle \quad \text{and} \quad \underline{\mathbf{D}}_? = \langle \{a, b\}; \vee_?, \wedge_?, 0_?, 1_? \rangle,$$

respectively, where $0_?$ and $1_?$ are a and b in one or the other order.

(iv) Next we choose any retraction $g \in \boldsymbol{S}_{\mathbf{X}}$ of $\mathbf{X}$ onto the set $\{a, b\}$. We ask the reader to verify that such retractions can be identified within $\boldsymbol{S}_{\mathbf{X}}$ as those $g \in \boldsymbol{S}_{\mathbf{X}}$ satisfying $ga^* = a^*$, $gb^* = b^*$ and, for every left zero x^*, either $gx^* = a^*$ or $gx^* = b^*$. Now let

$$E(\mathbf{X})^+ := \{\, h \in \boldsymbol{S}_{\mathbf{X}} \mid h(X) \subseteq \{a, b\} \,\} \subseteq \{a, b\}^X$$

and observe that in fact $E(\mathbf{X})^+ = \{h \in \boldsymbol{S}_\mathbf{X} \mid gh = h\}$.

(v) Let $\mathbf{X}^*$, $\underset{\sim}{\mathbf{D}}_?^*$ and $\underline{\mathbf{D}}_?^*$ be the unique structures such that ${}^*\colon \mathbf{X} \to \mathbf{X}^*$, ${}^*\colon \underset{\sim}{\mathbf{D}}_? \to \underset{\sim}{\mathbf{D}}_?^*$ and ${}^*\colon \underline{\mathbf{D}}_? \to \underline{\mathbf{D}}_?^*$ are isomorphisms. Let $e\colon \{0,1\} \to \{a,b\}$ where $e(0) = 0_?$ and $e(1) = 1_?$, that is, e is the unique isomorphism $e\colon \underset{\sim}{\mathbf{D}} \to \underset{\sim}{\mathbf{D}}_?$ and $e\colon \underline{\mathbf{D}} \to \underline{\mathbf{D}}_?$. These isomorphisms lift in a simple way to isomorphisms between the homset lattices:

$$E(\mathbf{X}) = \mathcal{X}(\mathbf{X}, \underset{\sim}{\mathbf{D}}) \underset{e\circ\_}{\to} \mathcal{X}(\mathbf{X}, \underset{\sim}{\mathbf{D}}_?) \underset{*}{\to} \mathcal{X}(\mathbf{X}^*, \underset{\sim}{\mathbf{D}}_?^*) \leqslant (\underline{\mathbf{D}}_?)^{X^*}$$

where $(e \circ \_)(f) = ef$ and $*$ is as defined in (ii) above. In order to retrieve the lattice structure on $\mathcal{X}(\mathbf{X}^*, \underset{\sim}{\mathbf{D}}_?^*)$ it remains only to impose a lattice structure on $\{a^*, b^*\}$. There are exactly two ways to do this which correspond to the two possible choices for the unknown order $\leqslant_?$ and which give us two lattices, $\mathbf{L}_\mathbf{X}$ and $\mathbf{L}_\mathbf{X}^\partial$, one of which must be isomorphic to $\mathbf{L} = E(\mathbf{X})$.

At first glance it may well look as if we are finished. If so, then look again, because this is the issue that makes these arguments subtle. In order to construct the lattice $\mathcal{X}(\mathbf{X}^*, \underset{\sim}{\mathbf{D}}_?^*)$, it still remains to identify the maps from X^* to $\{a^*, b^*\}$ which form its underlying set within the monoid $\boldsymbol{S}_\mathbf{X}$. This underlying set is exactly the image of the underlying set of $\mathcal{X}(\mathbf{X}, \underset{\sim}{\mathbf{D}}_?)$ under the isomorphism $*$. We claim that the underlying set of $\mathcal{X}(\mathbf{X}, \underset{\sim}{\mathbf{D}}_?)$ is exactly the set $E(\mathbf{X})^+$ in (iv). To see this, consider two cases. If $\underset{\sim}{\mathbf{X}}$ is not an antichain, then $\underset{\sim}{\mathbf{D}} \cong \underset{\sim}{\mathbf{D}}_?$ where $\underset{\sim}{\mathbf{D}}_?$ is a substructure of $\mathbf{X}$, and the underlying set of $\mathcal{X}(\mathbf{X}, \underset{\sim}{\mathbf{D}}_?)$ is clearly $E(\mathbf{X})^+$. Suppose that $\mathcal{X}$ is an antichain. Then, although $\underset{\sim}{\mathbf{D}}_?$ is not a substructure of $\mathbf{X}$, every member of $\mathcal{X}(\mathbf{X}, \underset{\sim}{\mathbf{D}}_?)$ is still in $\boldsymbol{S}_\mathbf{X}$. (Notice that this would not have been the case had we chosen a and b differently.) Thus, in either case, we conclude that the underlying set of the lattice $\mathcal{X}(\mathbf{X}, \underset{\sim}{\mathbf{D}}_?)$ is exactly $E(\mathbf{X})^+$. We now identify the underlying set of $\mathcal{X}(\mathbf{X}^*, \underset{\sim}{\mathbf{D}}_?^*)$ as

$$E(\mathbf{X})^* := \{\, h^* \mid h \in \boldsymbol{S}_\mathbf{X} \text{ and } gh = h \,\} \subseteq \{a^*, b^*\}^{X^*},$$

by the result of (iv). ∎

5.3 Theory of algebraic and existential closure

We would expect our search for applications of strong duality to focus on problems which can be phrased in the language of categories, and which consequently can benefit from a transfer of information between the categories $\mathcal{A}$ and $\mathcal{X}$. Recovery of an algebra from its endomorphism monoid is a typical example, as an algebra and its dual space always have anti-isomorphic endomorphism monoids. We will now look at a quite different kind of problem which, as it is given, is not at all phrased in terms of homomorphisms between structures. Upon closer scrutiny, however, we will see that under the right conditions it can be rephrased in terms of diagram completion properties which will be equivalent to dual diagram completion properties in the dual category. The goal of this section is to establish this fact in our final result, Theorem 5.3.5. In the next section we exhibit a series of examples to show that, in many instances, these dual properties can be much easier to recognise in the dual category and can lead to a more thorough understanding of the original problem.

The problem in question comes from mathematical logic and requires a few definitions. Although we are essentially concerned with algebras, the inclusion of relations does not alter the theory and allows many additional interesting examples. We therefore fix a set F of total operation symbols, to which we offer the option of adding a set R of relation symbols. These sets determine a type $\langle F, R\rangle$ of relational structures. We construct the **(first-order) formulæ (of type $\langle F, R\rangle$)** from atomic formulæ using the connectives $\vee$, $\wedge$ and $\neg$ together with the quantifiers $\forall$ and $\exists$. A formula λ is **quantifier-free** if it contains no quantifiers. A **sentence** is a formula containing no **free** (that is, unquantified) variables. Given a structure $\mathbf{A} = \langle A; F, R\rangle$ of type $\langle F, R\rangle$, we can extend this type by adding to F a set of distinct new constants $\{\mathbf{a} \mid a \in A\}$ corresponding to the elements of $\mathbf{A}$. A formula with **parameters from A** is a formula in this extended type. A formula is **satisfiable** in $\mathbf{A}$ if there is some assignment of members of $\mathbf{A}$ to its free variables which makes it true.

Let $\mathcal{C}$ be a category of structures of type $\langle F, R\rangle$. A structure $\mathbf{A}$ in $\mathcal{C}$ is **algebraically closed** if every conjunction of atomic formulæ with parameters in $\mathbf{A}$ that is satisfiable in some extension of $\mathbf{A}$ in $\mathcal{C}$ is also satisfiable in $\mathbf{A}$. We say that $\mathbf{A}$ is **existentially closed** if the same is true for every quantifier-free formula with parameters in $\mathbf{A}$. We denote by $\mathcal{C}^{\mathrm{ac}}$ and $\mathcal{C}^{\mathrm{ec}}$ the classes of algebraically and existentially closed members of $\mathcal{C}$, respectively. These notions originated in the classical theory of fields, where we learn that every field can be extended to an algebraically closed field. In the theory of fields $(r \approx 0) \vee (s \approx 0)$ is equivalent to $rs \approx 0$, and $r \neq 0$ is equivalent to $(\exists x)(rx \approx 1)$. It follows that every existential sentence in the language of fields is equivalent to a sentence stating that a conjunction of equations has a solution, and consequently every algebraically closed field is existentially closed as well.

In general there is no reason to expect the category $\mathcal{C}$ to contain any existentially closed members at all. If $\mathcal{C}$ is closed under unions of chains, then a very general argument shows that every member of $\mathcal{C}$ is contained in an existentially closed member of $\mathcal{C}$. This argument is given in Chang and Keisler [CK:MT], where the reader will find an exposition of the central role of existential closure in the study of model completeness as it was initiated by Abraham Robinson. Unfortunately the transfinite argument used to establish this general result gives us no indication as to how existentially closed structures can be constructed in specific cases. If $\mathcal{C} = \mathbb{ISP}\underline{\mathbf{M}}$ is a finitely generated quasi-variety, then Burris and Werner [BW79, Bur84] used sheaves to construct many existentially closed structures and to show that $\mathcal{C}^{\mathrm{ec}}$ is always a first-order definable subclass of $\mathcal{C}$.

We will be interested in finding concrete constructions of the members of $\mathcal{C}^{\mathrm{ac}}$ and $\mathcal{C}^{\mathrm{ec}}$ that will lead to simple sets of axioms for those classes. The contribution of natural dualities to this subject begins with the observation that, under the right conditions, we will be able to prove that the members $\mathbf{A}$ of $\mathcal{C}^{\mathrm{ac}}$ and $\mathcal{C}^{\mathrm{ec}}$ are characterised by the following diagram completion properties.

(FHP) **Finite Homomorphism Property** If $\mathbf{B}, \mathbf{C} \in \mathcal{C}$ are finite and $u\colon \mathbf{C} \to \mathbf{A}$ and $v\colon \mathbf{C} \to \mathbf{B}$ are embeddings, then there is a homomorphism $w\colon \mathbf{B} \to \mathbf{A}$ such that $u = wv$.

(FEP) **Finite Embedding Property** If $\mathbf{B}, \mathbf{C} \in \mathcal{C}$ are finite and $u\colon \mathbf{C} \to \mathbf{A}$ and $v\colon \mathbf{C} \to \mathbf{B}$ are embeddings, then there is an embedding $w\colon \mathbf{B} \to \mathbf{A}$ such that $u = wv$.

We will give a short proof of Burris and Werner's axiomatisability result under the assumptions which will apply later in this section. We say that $\mathcal{C}$ satisfies the **amalgamation property** (AP) if for any pair $u\colon \mathbf{C} \to \mathbf{A}$ and $v\colon \mathbf{C} \to \mathbf{B}$ of embeddings in $\mathcal{C}$, there are a $\mathbf{D} \in \mathcal{C}$ and embeddings $r\colon \mathbf{A} \to \mathbf{D}$ and $s\colon \mathbf{B} \to \mathbf{D}$ such that $ru = sv$. A **universal sentence** is a sentence of the form $(\forall x_1)(\forall x_2)\dots(\forall x_n)\mu$ where μ is quantifier-free, and the class of all models of a collection of universal sentences is called a **universal class**. Notice that equations and implications are examples of universal sentences, where the quantifiers were suppressed in Chapter 1. An $\forall\exists$-**sentence** is a sentence of the form $(\forall x_1)(\forall x_2)\dots(\forall x_n)(\exists y_1)(\exists y_2)\dots(\exists y_m)\mu$ where μ is quantifier-free. The class of all models of a collection of $\forall\exists$-sentences is called an $\forall\exists$-**class**.

3.1 Theorem *Let $\mathcal{C}$ be a locally finite universal class satisfying* (AP), *and let $\mathbf{A} \in \mathcal{C}$.*

(i) *$\mathbf{A}$ is algebraically closed if and only if it satisfies* (FHP).

(ii) *$\mathbf{A}$ is existentially closed if and only if it satisfies* (FEP).

In particular, both $\mathcal{C}^{\mathrm{ac}}$ and $\mathcal{C}^{\mathrm{ec}}$ are $\forall\exists$-classes.

Proof (i) Assume that $\mathbf{A}$ is algebraically closed, and let $u\colon \mathbf{C} \to \mathbf{A}$ and $v\colon \mathbf{C} \to \mathbf{B}$ be any embeddings where $\mathbf{B}, \mathbf{C} \in \mathcal{C}$ are finite. Applying (AP), we can find a structure $\mathbf{D} \in \mathcal{C}$ and embeddings $r\colon \mathbf{A} \to \mathbf{D}$ and $s\colon \mathbf{B} \to \mathbf{D}$ such that $ru = sv$. Suppose that $C = \{c_1, c_2, \dots, c_n\}$

and $B = \{v(c_1), v(c_2), \ldots, v(c_n), b_1, b_2, \ldots, b_m\}$. Then we can construct a quantifier-free formula $\mu(v(\mathbf{c}_1), v(\mathbf{c}_2), \ldots, v(\mathbf{c}_n), \mathbf{b}_1, \mathbf{b}_2, \ldots, \mathbf{b}_m)$ with parameters in **B** that completely defines the operation and relation tables of **B**. It follows that $\mu(sv(\mathbf{c}_1), sv(\mathbf{c}_2), \ldots, sv(\mathbf{c}_n), x_1, x_2, \ldots, x_m)$ is a formula with parameters in $sv(\mathbf{C}) \leqslant r(\mathbf{A})$ that is satisfied in **D** by choosing $s(b_i)$ for x_i.

Since $r(\mathbf{A})$ is isomorphic to **A**, it is also algebraically closed. This means that we can choose elements $d_1, d_2, \ldots, d_m \in r(A)$ so that the formula $\mu(sv(\mathbf{c}_1), sv(\mathbf{c}_2), \ldots, sv(\mathbf{c}_n), x_1, x_2, \ldots, x_m)$ is also satisfied by choosing d_i for x_i. Consequently $v(c_j) \mapsto sv(c_j)$ and $b_i \mapsto d_i$ defines a homomorphism $t\colon \mathbf{B} \to r(\mathbf{A})$ that agrees with s on $v(\mathbf{C})$. Taking $w = r^{-1}t\colon \mathbf{B} \to \mathbf{A}$, we have $wv = r^{-1}tv = r^{-1}sv = r^{-1}ru = u$.

Conversely, assume that **A** satisfies (FHP). Consider a conjunction $\mu(\mathbf{c}_1, \mathbf{c}_2, \ldots, \mathbf{c}_n, x_1, x_2, \ldots, x_m)$ of atomic formulæ with parameters from **A** that is satisfied in an extension $\mathbf{D} \in \mathfrak{C}$ of **A** by choosing $d_1, d_2, \ldots, d_m \in D$ for $x_1, x_2, \ldots, x_m$, respectively. Since **D** is locally finite, $\{c_1, c_2, \ldots, c_n\}$ generates a finite substructure **C** of **D** and $\{c_1, c_2, \ldots, c_n, d_1, d_2, \ldots, d_m\}$ generates a finite substructure **B** of **D**. By (FHP), there is a homomorphism $w\colon \mathbf{B} \to \mathbf{A}$ that fixes each member of **C**. Since w preserves μ, we conclude that μ is satisfied in **A** by choosing $w(d_i)$ for x_i.

(ii) Use exactly the same argument as in (i), but this time the formula μ describes **B** up to isomorphism in the first part and is preserved by embeddings in the second part.

The final statement is true since each instance of either (FHP) or (FEP) is expressible by an $\forall\exists$-sentence in the language of $\mathfrak{C}$. ■

Several interesting aspects of the subcategory $\mathfrak{C}^{\mathrm{ec}}$ are described in the next theorem, which is easy to prove under the conditions of Theorem 5.3.1. We say that $\mathfrak{C}$ has the **joint embedding property** (JEP) if any two non-trivial members of $\mathfrak{C}$ can be embedded in a third. We say that a structure **A** in $\mathfrak{C}$ is **$\mathfrak{C}$-universal** if every non-trivial member of $\mathfrak{C}$ of equal or smaller cardinality can be embedded in it. (These non-triviality stipulations here are not standard, but will exactly fit our context. For example, the variety of Boolean algebras satisfies (JEP) with this restriction but not without it, and the countable atomless Boolean algebra is universal with this restriction but not without it.) A countable structure **A** is **homogeneous** if every isomorphism between finite substructures of **A** extends to an automorphism of **A**. A class of structures is **$\aleph_0$-categorical** if it has, up to isomorphism, exactly one countably infinite member.

3.2 Lemma *Let $\mathfrak{C}$ be a locally finite universal class satisfying* (JEP) *and* (AP), *and let $\mathbf{A} \in \mathfrak{C}^{\mathrm{ec}}$ be non-trivial. Then every non-trivial finite member of $\mathfrak{C}$ can be embedded in* **A**.

Proof Let $\mathbf{F} \in \mathcal{C}$ be finite but non-trivial. By (JEP) we may assume that $\mathbf{A}$ and $\mathbf{F}$ are both substructures of $\mathbf{D} \in \mathcal{C}$. Choose any $a \in A$ and let $\mathbf{B}$ be the substructure of $\mathbf{D}$ generated by $F \cup \{a\}$. Since $\mathbf{D}$ is locally finite, $\mathbf{B}$ is finite. Since $\mathbf{B}$ and $\mathbf{A}$ have non-empty intersection, this intersection forms a substructure $\mathbf{C}$ of $\mathbf{A}$. We can now apply (FEP) from Theorem 5.3.1 to obtain an embedding w of $\mathbf{B}$ into $\mathbf{A}$ which fixes each element of $\mathbf{C}$. In particular, w embeds $\mathbf{F}$ into $\mathbf{A}$. ∎

3.3 Theorem *Let $\mathcal{C}$ be a locally finite universal class satisfying* (JEP) *and* (AP) *that has an infinite member.*

(i) *$\mathcal{C}^{ec}$ is $\aleph_0$-categorical.*

(ii) *The countably infinite member of $\mathcal{C}^{ec}$ is the unique countably infinite homogeneous $\mathcal{C}$-universal member of $\mathcal{C}$.*

Proof (i) Since $\mathcal{C}$ has an infinite member and is locally finite, this infinite member contains a countably infinite substructure which, since $\mathcal{C}$ is a universal class, is in $\mathcal{C}$. Thus $\mathcal{C}$ has at least one countably infinite member.

Let $\mathbf{A}, \mathbf{B} \in \mathcal{C}^{ec}$ be countably infinite, let $A = \{a_1, a_2, a_3, \dots\}$ and let $B = \{b_1, b_2, b_3, \dots\}$. We recursively construct, for each positive integer n, an isomorphism $u_n \colon \mathbf{A}_n \to \mathbf{B}_n$ from a finite substructure of $\mathbf{A}$ onto a finite substructure of $\mathbf{B}$ such that

(a) $\{a_1, a_2, \dots, a_n\} \subseteq A_n$,

(b) $\{b_1, b_2, \dots, b_n\} \subseteq B_n$,

(c) u_{n+1} is an extension of u_n.

Clearly, the union of the maps u_n will then be an isomorphism from $\mathbf{A}$ onto $\mathbf{B}$.

By Lemma 5.3.2, there is an isomorphism v from the substructure of $\mathbf{B}$ generated by b_1 onto a substructure $\mathbf{C}$ of $\mathbf{A}$. Let $\mathbf{A}_1$ be the substructure of $\mathbf{A}$ generated by $C \cup \{a_1\}$. By (FEP) there is an extension u_1 of v^{-1} which is an isomorphism from $\mathbf{A}_1$ onto a substructure $\mathbf{B}_1$ of $\mathbf{B}$.

Now suppose $u_n \colon \mathbf{A}_n \to \mathbf{B}_n$ satisfies (a) and (b). Taking $u = (u_n)^{-1} \colon \mathbf{B}_n \to \mathbf{A}$ in (FEP), we can extend $(u_n)^{-1}$ to an isomorphism w from the substructure of $\mathbf{B}$ generated by $B_n \cup \{b_{n+1}\}$ onto a substructure $\mathbf{C}$ of $\mathbf{A}$. Let $\mathbf{A}_{n+1}$ be the substructure of $\mathbf{A}$ generated by $C \cup \{a_{n+1}\}$. By (FEP) there is an extension u_{n+1} of w^{-1} which is an isomorphism from $\mathbf{A}_{n+1}$ onto a substructure $\mathbf{B}_{n+1}$ of $\mathbf{B}$ containing $B_n \cup \{b_{n+1}\}$. Thus u_{n+1} satisfies (a), (b) and (c).

(ii) Let $\mathbf{A} \in \mathcal{C}^{ec}$ be countably infinite. To see that $\mathbf{A}$ is $\mathcal{C}$-universal, let $\mathbf{B} \in \mathcal{C}$ be a non-trivial, countable structure where $B = \{b_1, b_2, b_3, \dots\}$. Let $\mathbf{B}_n$ denote the (finite) substructure of $\mathbf{B}$ generated by $\{b_1, b_2, \dots, b_n\}$.

By Lemma 5.3.2 there is an embedding $u_1: \mathbf{B}_1 \to \mathbf{A}$ and, by (FEP), there is an embedding $u_{n+1}: \mathbf{B}_{n+1} \to \mathbf{A}$ which is an extension of u_n. Thus the union of the embeddings $u_1, u_2, u_3, \ldots$ is an embedding of $\mathbf{B}$ into $\mathbf{A}$.

We ask the reader to check that a small modification of the proof of (i) will demonstrate that $\mathbf{A}$ is homogeneous. Let v be an isomorphism between two finite substructures of $\mathbf{A}$. Let $\mathbf{B} = \mathbf{A}$ in the proof of (i) and use (FEP) to ensure that u_1 is an extension of v.

To see that $\mathbf{A}$ is unique, suppose that $\mathbf{B}$ is a countably infinite, homogeneous $\mathfrak{C}$-universal member of $\mathfrak{C}$. Then $\mathbf{B}$ satisfies (FEP) and is therefore in $\mathfrak{C}^{\mathrm{ec}}$. Since $\mathfrak{C}^{\mathrm{ec}}$ is $\aleph_0$-categorical, $\mathbf{B}$ is isomorphic to $\mathbf{A}$. ■

Homogeneity is a very strong form of $\aleph_0$-categoricity. We refer the reader to Clark and Krauss [CK77b] for an exposition of some weaker forms of homogeneity that also imply $\aleph_0$-categoricity. A simple and very accessible illustration of all of the notions discussed so far in this section is given for the category of linear orderings in Exercise 5.3.

We now return to the setting in which $\mathfrak{C}$ is the quasi-variety $\mathcal{A} = \mathbb{ISP}\underline{\mathbf{M}}$ generated by a finite algebra $\underline{\mathbf{M}}$.

3.4 Lemma *Let* $\underline{\mathbf{M}}$ *be a non-trivial finite algebra and let* $\mathcal{A} = \mathbb{ISP}\underline{\mathbf{M}}$.

(i) $\mathcal{A}$ *is a locally finite universal class satisfying* (JEP).

(ii) *If* $\mathcal{A}$ *admits a strong duality by total structures, then* $\mathcal{A}$ *satisfies* (AP) *as well.*

Proof (i) The category $\mathcal{A}$ is locally finite by Lemma 1.3.2, and it is a universal class by the $\mathbb{ISP}$-Theorem 1.3.4. To see that $\mathcal{A}$ satisfies (JEP), we must show that for any disjoint non-empty sets I and J, we can embed $\underline{\mathbf{M}}^I$ (and similarly $\underline{\mathbf{M}}^J$) in $\underline{\mathbf{M}}^{I\cup J}$. Fix $k \in I$ and define $u: \underline{\mathbf{M}}^I \to \underline{\mathbf{M}}^{I\cup J}$ by $u(a)(i) = a(i)$ if $i \in I$ and $u(a)(j) = a(k)$ if $j \in J$.

(ii) By the Injectivity Lemma 3.2.10 we know that $\underline{\mathbf{M}}$ is injective in $\mathcal{A}$ since $\underset{\sim}{\mathbf{M}}$ is a total structure and the duality is strong. Since products of injectives are injective, each member of $\mathcal{A}$ can be embedded in an injective, that is, $\mathcal{A}$ has enough injectives. From this and the fact that $\mathcal{A}$ is closed under products, (AP) follows. (See Exercise 5.4 or Balbes and Dwinger [BD:DL], page 38.) ■

Exploiting our category-theoretic setting, we find that, under the right conditions, properties (FHP) and (FEP) of an algebra $\mathbf{A} \in \mathcal{A}$ correspond to the following dual properties of its dual $\mathbf{X} \in \mathfrak{X}$.

(FHP)* **Dual Finite Homomorphism Property** If $\mathbf{Y}, \mathbf{Z} \in \mathfrak{X}$ are finite and $\varphi: \mathbf{X} \to \mathbf{Z}$ and $\psi: \mathbf{Y} \to \mathbf{Z}$ are surjections, then there is a morphism $\lambda: \mathbf{X} \to \mathbf{Y}$ such that $\varphi = \psi\lambda$.

(FEP)* **Dual Finite Embedding Property** If $\mathbf{Y}, \mathbf{Z} \in \mathcal{X}$ are finite and $\varphi\colon \mathbf{X} \to \mathbf{Z}$ and $\psi\colon \mathbf{Y} \to \mathbf{Z}$ are surjections, then there is a surjection $\lambda\colon \mathbf{X} \to \mathbf{Y}$ such that $\varphi = \psi\lambda$.

Gathering what we have established, we can now give the central result of this section. Notice that total structures are necessary to establish (AP) in Theorem 5.3.4, which is independently required in both Theorem 5.3.1 and Theorem 5.3.3. In the next chapter we will prove the Total Structure Theorem 6.1.2 which will tell exactly when partial operations can be eliminated from a strong duality.

3.5 AC–EC Theorem *Let $\mathbf{A} \in \mathcal{A} = \mathbb{ISP}\underline{\mathbf{M}}$, and assume that a total structure $\underset{\sim}{\mathbf{M}}$ yields a strong duality on $\mathcal{A}$.*

(i) $\mathbf{A}$ *is algebraically closed if and only if $D(\mathbf{A})$ satisfies* (FHP)*.

(ii) $\mathbf{A}$ *is existentially closed if and only if $D(\mathbf{A})$ satisfies* (FEP)*.

Moreover, both $\mathcal{A}^{\mathrm{ac}}$ and $\mathcal{A}^{\mathrm{ec}}$ are $\forall\exists$-classes and $\mathcal{A}^{\mathrm{ac}}$ is $\aleph_0$-categorical with its countably infinite member being the unique countable, homogeneous $\mathcal{A}$-universal member of $\mathcal{A}^{\mathrm{ec}}$.

Proof Let $\mathbf{X} = D(\mathbf{A})$. We first claim that $\mathbf{X}$ satisfies (FHP)* or (FEP)* if and only if $\mathbf{A}$ satisfies (FHP) or (FEP), respectively. Since E and D take finite structures to finite structures, it remains to check that surjections in $\mathcal{X}$ correspond to embeddings in $\mathcal{A}$. According to the Dual Adjunction Theorem 1.5.3, surjections always go to and come from embeddings. The converse is true, according to Lemma 3.2.8, provided that $\underline{\mathbf{M}}$ is injective in $\mathcal{A}$. The Injectivity Lemma 3.2.10 assures us that $\underline{\mathbf{M}}$ is injective in $\mathcal{A}$ since $\underset{\sim}{\mathbf{M}}$ is a total structure and the duality is strong. Parts (i) and (ii) now follow from Theorem 5.3.1 and Lemma 5.3.4, and the final statement follows from Theorems 5.3.3 and Lemma 5.3.4. ■

5.4 Examples of algebraic and existential closure

Applying the AC–EC Theorem 5.3.5, we can now characterise algebraic and existential closure in many quasi-varieties where a strong

duality by total structures is available. These examples will show how a tractable dual category can greatly simplify a problem and lead to a deeper insight into its solution. They also illustrate the advantage of transferring into a category where our geometric intuitions can be used to solve what was initially given as a purely algebraic problem. And this, surely, is the whole point of constructing topological dualities!

We will use the **Cantor space** $\mathbf{X}_{\mathrm{C}} = \langle\{0, 2\}, \mathcal{T}\rangle^{\omega}$ in the construction of a number of these examples. (See Lemma B.6 for a description of direct power spaces.) If we view each point of $\mathbf{X}_{\mathrm{C}}$ as the decimal expansion of a number written in base 3, then we get an alternative description of $\mathbf{X}_C$ as a subspace of the unit interval of real numbers. A point x of a Boolean space $\mathbf{X}$ is **isolated** if $\{x\}$ is open (and therefore clopen) in $\mathbf{X}$. The space $\mathbf{X}_C$ can be characterised up to isomorphism as the Boolean space with no isolated points and countably many clopen subsets. An element a of an ordered set $\mathbf{L}$ **covers** an element b if $a > b$ and no element of $\mathbf{L}$ is between a and b. If $\mathbf{L}$ has a least element 0 and greatest element 1, then the covers of 0 are called **atoms** of $\mathbf{L}$ and the elements covered by 1 are called **coatoms** of $\mathbf{L}$.

Our first example applies Theorem 4.1.1 to the variety $\mathcal{B}$ of Boolean algebras $\mathbf{B} = \langle B; \vee, \wedge, ', 0, 1\rangle$ and provides a basis for subsequent examples. The proof is a very easy application of Theorem 5.3.5 which we illustrate in Figure 5.1.

4.1 AC–EC Boolean Algebra Theorem *Let $\mathbf{B} = E(\mathbf{X})$ be a Boolean algebra where $\mathbf{X}$ is a Boolean space.*

(i) *Every Boolean algebra is algebraically closed.*

(ii) *The following are equivalent:*

- (a) $\mathbf{B}$ *is existentially closed;*
- (b) $\mathbf{B}$ *is atomless;*
- (c) $\mathbf{X}$ *has no isolated points.*

In particular, the countable homogeneous $\mathcal{B}$-universal Boolean algebra is $E(\mathbf{X}_C)$ where $\mathbf{X}_C$ is the Cantor space.

Proof (i) To see that $\mathbf{B}$ is algebraically closed, we verify (FHP)*. Let $\lambda\colon \mathbf{X} \to \mathbf{Y}$ where, for each $z \in Z$, we have λ map $\varphi^{-1}\{z\}$ to any single point of $\psi^{-1}\{z\}$. Then $\varphi = \psi\lambda$.

(ii) Conditions (b) and (c) are equivalent since the atoms of $\mathbf{B}$ are exactly the sets $\{x\}$ where x is an isolated point in $\mathbf{X}$. To prove that (c) and (a) are equivalent, we apply the AC–EC Theorem(ii) and observe that the map λ can always be made surjective exactly when every clopen subset of $\mathbf{X}$ can be partitioned into an arbitrary finite number of non-empty clopen subsets, that is, when $\mathbf{X}$ has no isolated points. ∎

Schmid first characterised the algebraically and existentially closed distributive lattices $\mathbf{L} = \langle L; \vee, \wedge\rangle$, and we use Theorem 4.3.2 to present his result. A bounded lattice $\mathbf{L}$ is **complemented** if for every $a \in L$ there is a $b \in L$ such that $a \vee b = 1$ and $a \wedge b = 0$. An arbitrary lattice $\mathbf{L}$ is **relatively complemented** if each interval is complemented, that is, $c < a < d$ in $\mathbf{L}$ implies that there exists $b \in L$ such that $a \vee b = d$ and $a \wedge b = c$. The equivalence of the first two statements of (i) and (ii) below appears in Schmid [Sch79].

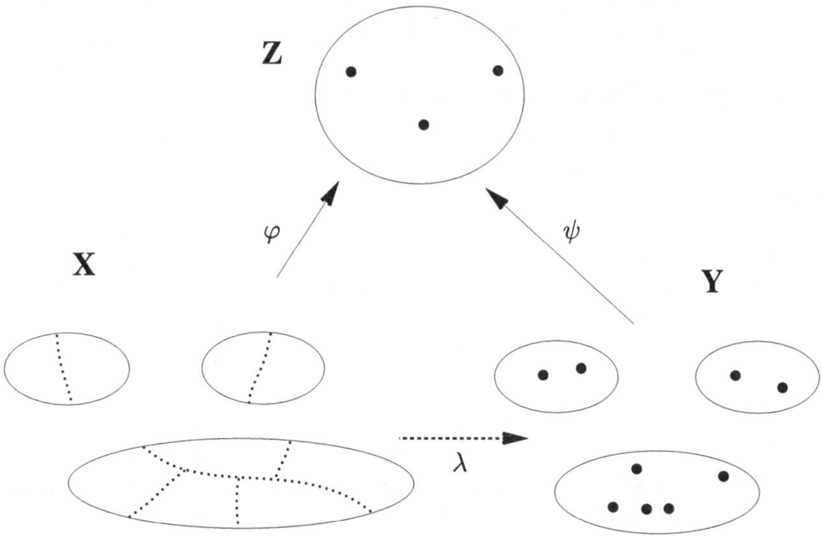

Figure 5.1 (FEP)* for Boolean algebras

4.2 AC–EC Distributive Lattice Theorem *Let* $\mathbf{L} = E(\mathbf{X})$ *be a distributive lattice where* $\mathbf{X} = \langle X; 0, 1, \leqslant, \mathcal{T}\rangle$ *is a bounded Priestley space.*

(i) *The following are equivalent:*

(a) $\mathbf{L}$ *is algebraically closed;*

(b) $\mathbf{L}$ *is relatively complemented;*

(c) *if* $x < y$ *in* $\mathbf{X}$, *then* $x = 0$ *or* $y = 1$.

(ii) *If* $\mathbf{L}$ *is algebraically closed, then the following are equivalent:*

(a) $\mathbf{L}$ *is existentially closed;*

(b) $\mathbf{L}$ *has no covers and no least or greatest element;*

(c) $\mathbf{X}$ *has no isolated points.*

In particular, the countable homogeneous $\mathfrak{D}$*-universal distributive lattice is* $E(\mathbf{X}_C^{01})$ *where* $\mathbf{X}_C^{01}$ *is the Cantor space, viewed as a subspace of the interval* $[0,1]$, *with the order* $x < y$ *if and only if* $x = 0$ *or* $y = 1$.

Proof (i) Part (a) implies (b) since $\mathbf{L}$ can be embedded into a Boolean lattice by making $X\backslash\{0,1\}$ into an antichain. We ask the reader to check that (b) implies (c), and to use (c) to establish (FHP)* which implies (a).

(ii) Assume (a) to prove (b). The projection $\pi_1 \colon \mathbf{X} \times \mathbf{X}_C^{01} \to \mathbf{X}$ is a surjection, and consequently $E(\pi_1) \colon E(\mathbf{X}) \to E(\mathbf{X}\times\mathbf{X}_C^{01})$ is an embedding of $\mathbf{L} = E(\mathbf{X})$ into a distributive lattice with no covers and no greatest element.

Now assume (b) to prove (c). If 0 were isolated, then $0 \mapsto 0$ and $X\backslash\{0\} \mapsto 1$ would be a greatest element of $\mathbf{L}$. If 1 were isolated, then, similarly, $\mathbf{L}$ would have a least element. Suppose $x \neq 0, 1$ were isolated. If $Z \subseteq X$ were any clopen set containing 0 and not 1, then, by (i)(c), the map taking $Z \cup \{x\}$ to 0 and its complement to 1 would cover the map taking $Z \backslash \{x\}$ to 0 and its complement to 1.

We leave the reader to prove (c) implies (a) by making a slight modification of the corresponding proof of the AC–EC Boolean Algebra Theorem 5.4.1. ∎

A more substantial application of the AC–EC Theorem is provided by the characterisation of algebraically and existentially closed members $\mathbf{A} = \langle A; \vee, \wedge, *, 0, 1\rangle$ of the variety $\mathcal{S}$ of Stone algebras. (See example 4.3.6 and the strong duality of Theorem 4.3.7.) Notice the simplicity of the characterisations of their dual spaces in parts (iii) of Theorems 5.4.4 and 5.4.5, which give us concrete representations of these structures. Our verification of (FHP)* and (FEP)* in this case will be greatly facilitated by the fact that coproducts in $\mathcal{X}$ have a very simple and useful representation. In the second and third sections of Chapter 6 we will determine exactly which choices of $\mathcal{A}$ admit a strong duality in which coproducts in the dual category have this kind of representation. (See Theorem 6.3.4.) We leave the proof of the following lemma for the reader.

4.3 Lemma *Let $\underset{\sim}{\mathbf{S}}$ be the structure discussed in 4.3.6. Suppose $\mathbf{X} \in \mathcal{X} = \mathbb{IS}_c\mathbb{P}^+ \underset{\sim}{\mathbf{S}}$ where $\mathbf{X}$ is a disjoint union of finitely many clopen substructures $\{\mathbf{X}_i \mid i = 1, 2, \ldots, n\}$ and, for $i \neq j$, no member of $\mathbf{X}_i$ is less than a member of $\mathbf{X}_j$. Then $\mathbf{X}$ is the coproduct of the substructures $\{\mathbf{X}_i \mid i = 1, 2, \ldots, n\}$ with respect to their inclusion maps.*

Stated in full, this says that if $\lambda_i \colon \mathbf{X}_i \to \mathbf{Y}$ is a continuous morphism for each $i = 1, 2, \ldots, n$, then the union $\bigcup\{\lambda_i \mid i = 1, 2, \ldots, n\}$ is a continuous morphism from $\mathbf{X}$ to $\mathbf{Y}$.

We consider a Stone algebra $\mathbf{A} = E(\mathbf{X})$ whose dual space is given by $\mathbf{X} = \langle X; d, \preccurlyeq, \mathcal{T}\rangle$ as described in Theorem 4.3.7. Associated with $\mathbf{A}$ is the filter of **dense** elements

$$\mathrm{Ds}(\mathbf{A}) = \{\, \delta \in A \mid \delta^* = 0 \,\} = \{\, \delta \in A \mid \delta(X) \subseteq \{a, 1\}\}$$

and the **skeleton**

$$\mathrm{Sk}(\mathbf{A}) = \{\, \sigma \in A \mid \sigma^{**} = \sigma \,\} = \{\, \sigma \in A \mid \sigma(X) \subseteq \{0, 1\}\}.$$

The equivalence (i) ⇔ (ii) in the next theorem is due to Schmitt [Sch76]. To retard the proliferation of brackets we shall write $d(x)$ as dx and $d(X)$ as dX.

4.4 AC Stone Algebra Theorem *If* $\mathbf{A} = E(\mathbf{X})$ *is a Stone algebra with dual space* $\mathbf{X} = \langle X; d, \preccurlyeq, \mathcal{T}\rangle$, *then the following are equivalent:*

(i) $\mathbf{A}$ *is algebraically closed;*

(ii) $\mathbf{A}$ *satisfies the following* $\forall\exists$*-axioms–*

(AC1) $\mathrm{Ds}(\mathbf{A})$ *is relatively complemented,*

(AC2) *if* $\gamma, \delta \in \mathrm{Ds}(\mathbf{A})$ *and* $\gamma \vee \delta = 1$, *then there is a* $\sigma \in \mathrm{Sk}(\mathbf{A})$ *such that* $\sigma \leqslant \gamma$ *and* $\sigma^* \leqslant \delta$;

(iii) d *is one-to-one on* $X \backslash dX$.

Proof Condition (i) implies condition (ii) since (ii) clearly holds if $\mathbf{A}$ is replaced by $\underline{\mathbf{S}}^X$.

Assume (ii). We first show that that $X\backslash dX$ is an antichain. Suppose that it contains a pair $x \prec y$. Let U be a clopen increasing subset of $X\backslash dX$ containing x, and let V be a clopen increasing subset of U containing y but not x. Let $\gamma \in A$ take U to a and its complement to 1, and let $\delta \in A$ take V to a and its complement to 1. Then $\gamma \prec \delta \prec 1$ in $\mathrm{Ds}(\mathbf{A})$, but δ does not have a relative complement in the interval $[\gamma, 1]$, contradicting (AC1).

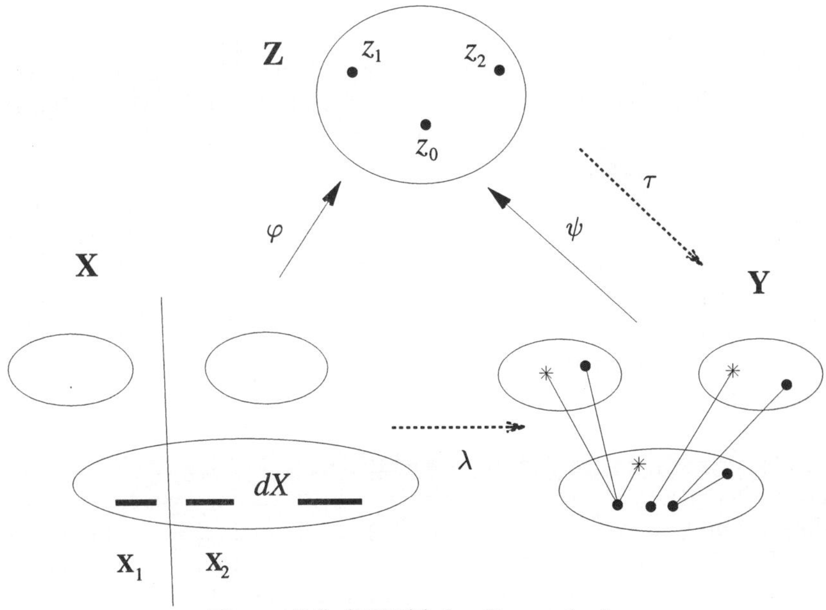

Figure 5.2 (FHP)* for Stone algebras

Now suppose that $dx = dy$ where $x \neq y$ are in $X\backslash dX$. Let U and V be disjoint clopen subsets of $X\backslash dX$ containing x and y, respectively. Let γ take U to a and its complement to 1, and let δ take V to a and its complement to 1. Then γ and δ are in $\mathbf{A}$ since $X\backslash dX$ is an antichain, and

$\gamma \vee \delta = 1$. Choose $\sigma \in \text{Sk}(\mathbf{A})$ according to (AC2). Then $\sigma(x) \preccurlyeq \gamma(x) = a$ so that $\sigma(x)$ must be 0. Similarly, $\sigma^*(y) \preccurlyeq \delta(y) = a$ so that $\sigma(y)$ must be 1. But then $0 = d0 = d\sigma(x) = \sigma(dx) = \sigma(dy) = d\sigma(y) = d1 = 1$, a contradiction. This proves (iii).

Finally, assume (iii). We use the AC–EC Theorem and prove (FHP)* in order to establish (i). Applying Lemma 5.4.3 we can partition $\mathbf{X}$, $\mathbf{Y}$ and $\mathbf{Z}$ using the order components of $\mathbf{Z}$ and their inverse images under φ and ψ, and therefore consider only the case in which $\mathbf{Z}$ has a single order component. In this case $\mathbf{Z}$ has a unique minimal element z_0 such that $dz = z_0$ for all $z \in Z$. Let $Z = \{z_0, z_1, z_2, \ldots, z_m\}$ and let $\tau\colon Z \to Y$ be any right inverse for ψ. By (iii) the closed sets $\{d\varphi^{-1}(z_j) \mid j = 1, 2, \ldots, m\}$ are disjoint subsets of $\varphi^{-1}(z_0)$. Let $\{U_1, U_2, \ldots, U_m\}$ be a clopen partition of $\varphi^{-1}(z_0)$ where $d\varphi^{-1}z_j \subseteq U_j$, and let $X_j = d^{-1}(U_j)$. According to Lemma 5.4.3, $\mathbf{X}$ is the coproduct of the disjoint, closed substructures $\{\mathbf{X}_1, \mathbf{X}_2, \ldots, \mathbf{X}_n\}$.

For $j = 1, 2, \ldots, m$, let $\lambda_j\colon \mathbf{X}_j \to \mathbf{Y}$ take $\varphi^{-1}(z_j)$ to $\tau(z_j)$ and take $X_j \cap \varphi^{-1}(z_0)$ to $d\tau(z_j)$. Then $\lambda := \lambda_1 \cup \lambda_2 \cup \cdots \cup \lambda_m\colon \mathbf{X} \to \mathbf{Y}$ and $\varphi = \psi\lambda$. (See Figure 5.2.) ∎

A set Σ of axioms is said to be **independent** over a class $\mathcal{C}$ of structures if, for each axiom in Σ, there is a member of $\mathcal{C}$ in which that axiom fails but every other axiom of Σ holds. In the context of $\mathcal{C}$, this means that no axiom of Σ is a consequence of the other members of Σ and therefore Σ irredundantly defines a subclass of $\mathcal{C}$. For example, the axioms (AC1) and (AC2) are independent over the variety $\mathcal{S}$ of Stone algebras. To see this, consider the 3-chain $\mathbf{T} \in \mathcal{X}$ and the three-element V–shaped structure $\mathbf{V} \in \mathcal{X}$. Part (iii) of the AC Stone Algebra Theorem is equivalent to the assertion that neither $\mathbf{T}$ nor $\mathbf{V}$ is embeddable into $\mathbf{X}$. The Stone algebra $E(\mathbf{T})$ satisfies (AC2) but not (AC1) while the Stone algebra $E(\mathbf{V})$ satisfies (AC1) but not (AC2). (See Figure 5.3 and Exercise 5.7.)

The first characterisation of existentially closed Stone algebras was also discovered by Schmitt [Sch76] and was subsequently elaborated and extended by Schmid [Sch79] and Clark [Cla89]. Making full use of our duality-theoretic technique, we will show that the axioms for $\mathcal{S}^{\text{ec}}$ are redundant in each of these cases by giving a yet smaller set of axioms for $\mathcal{S}^{\text{ec}}$. Then, by constructing appropriate dual spaces, we will give examples to show that our axioms are indeed irredundant. (See Exercise 5.8.)

In order to discover the existentially closed Stone algebras, we examine the proof that (iii) implies (FHP)* in the AC Stone Algebra Theorem to see what additional properties would be required of $\mathbf{X}$ to make the morphism λ surjective. We then translate these properties into an expansion of (ii) by additional $\forall\exists$-axioms. Our next theorem gives the result.

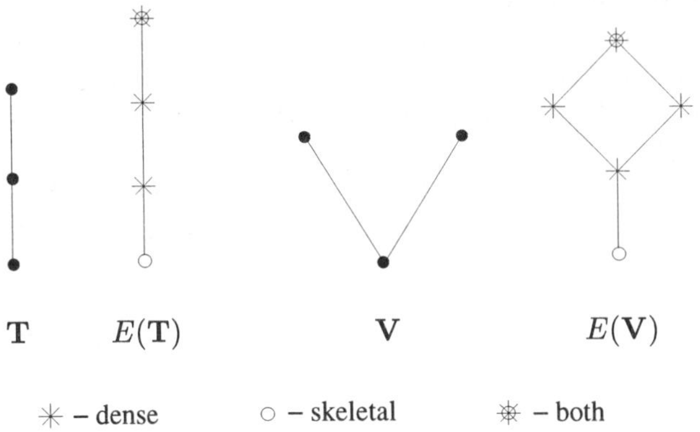

Figure 5.3 independence of (AC1) and (AC2) over $\mathcal{S}$

4.5 EC Stone Algebra Theorem *If* $\mathbf{A} = E(\mathbf{X})$ *is an algebraically closed Stone algebra with dual space* $\mathbf{X} = \langle X; d, \preccurlyeq, \mathcal{T}\rangle$, *then the following are equivalent:*

(i) $\mathbf{A}$ *is existentially closed;*

(ii) $\mathbf{A}$ *satisfies the following* $\forall\exists$*-axioms–*

(EC1) $\mathrm{Ds}(\mathbf{A})$ *has no coatoms, .*

(EC2) *if* $\sigma \in \mathrm{Sk}(\mathbf{A})$, *then there is no smallest dense element above* σ;

(iii) $X \backslash dX$ *has no isolated points and is dense in* $\mathbf{X}$.

Proof First assume (i) to prove (ii). Let $1 \neq \delta \in \mathrm{Ds}(\mathbf{A})$. Choose $x \in X$ so that $\delta(x) = a$. Now embed $\mathbf{A}$ into $\mathbf{A} \times \underline{\mathbf{S}}$ by $\beta \mapsto (\beta, \beta(x))$. Then $\delta \mapsto (\delta, a)$ and $1 \mapsto (1,1)$, but $(\delta, a) < (\delta, 1) < (1,1)$. Since $\mathbf{A}$ is existentially closed, there is also a dense element of $\mathbf{A}$ between δ and 1, proving (EC1). To prove (EC2), let $\sigma < \delta \in \mathrm{Ds}(\mathbf{A})$. Choose $x \in dX \cap \sigma^{-1}(0)$ and again embed $\mathbf{A}$ into $\mathbf{A} \times \underline{\mathbf{S}}$ by $\beta \mapsto (\beta, \beta(x))$. Then $\sigma \mapsto (\sigma, 0)$ and $\delta \mapsto (\delta, 1)$, but $(\sigma, 0) < (\delta, a) < (\delta, 1)$ where $(\delta, a) \in \mathrm{Ds}(\mathbf{A} \times \underline{\mathbf{S}})$. Since $\mathbf{A}$ is existentially closed, there is a dense element of $\mathbf{A}$ between σ and δ.

Next assume (ii) to prove (iii). If $x \in X \backslash dX$ were isolated, then $x \mapsto a$ and $X \backslash \{x\} \mapsto 1$ would be a coatom of $\mathrm{Ds}(\mathbf{A})$. Suppose that U were a non-empty clopen subset of dX. Let δ take $d^{-1}(U) \backslash U$ to a and the rest of X to 1, and let σ take $d^{-1}(U)$ to 0 and its complement to 1. Then δ and σ would be in $\mathbf{A}$, but there would be no dense element of $\mathbf{A}$ between them.

Now we assume (iii) and establish (i) by verifying (FEP)*. As in the AC Stone Algebra Theorem, we can apply Lemma 5.4.3 to reduce the proof to the case in which $\mathbf{Z}$ has a single order component and therefore a unique minimal element z_0. Let $Z = \{z_0, z_1, z_2, \ldots, z_m\}$. Since d is one-to-one on $X \backslash dX$, the closed sets $\{d\varphi^{-1}(z_j) \mid j = 1, 2, \ldots, m\}$ are disjoint subsets of $\varphi^{-1}(z_0)$. Since $X \backslash dX$ is dense in X and $dX \subseteq \varphi^{-1}(z_0)$, there is a point $x_0 \in \varphi^{-1}(z_0)$ that is not in dX. It follows that $x_0 \notin \varphi^{-1}(z_j)$ for $j = 1, 2, \ldots, m$. Let $\{U_0, U_1, U_2, \ldots, U_m\}$ be a clopen partition of $\varphi^{-1}(z_0)$ where $d\varphi^{-1}z_j \subseteq U_j$ and $x_0 \in U_0$. Again define $X_j = d^{-1}(U_j)$ for $j = 1, 2, \ldots, m$ and observe that $X_0 = d^{-1}(U_0)$ is contained in $\varphi^{-1}(z_0)$. According to Lemma 5.4.3, $\mathbf{X}$ is the coproduct of the closed substructures $\{\mathbf{X}_0, \mathbf{X}_1, \mathbf{X}_2, \ldots, \mathbf{X}_m\}$. (See Figure 5.4.)

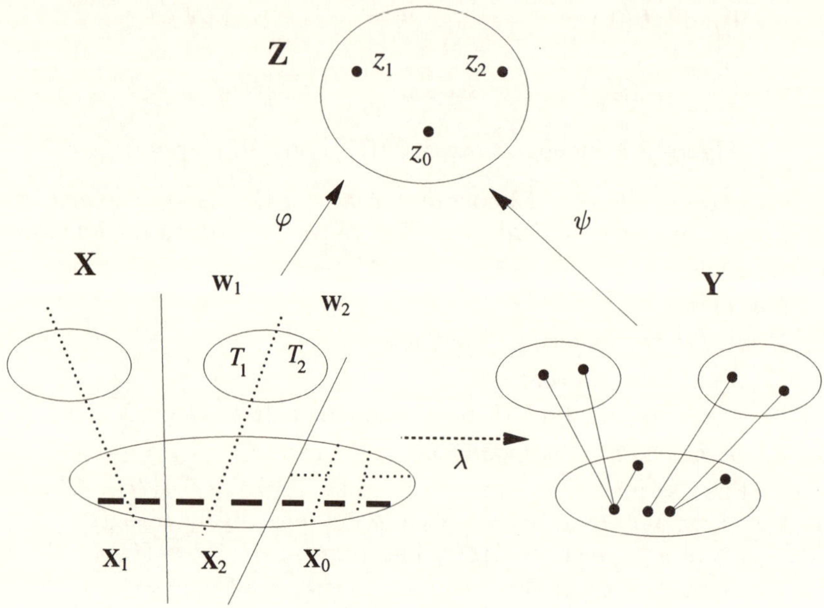

Figure 5.4 (FEP)* for Stone algebras

For $j = 1, 2, \ldots, m$, let φ_j be the restriction of φ to $\mathbf{X}_j$. We will construct the morphism $\lambda \colon \mathbf{X} \to \mathbf{Y}$ as the union of morphisms $\lambda_j \colon \mathbf{X}_j \to \mathbf{Y}$ where $\varphi_j = \psi\lambda_j$ and $\psi^{-1}(z_j) \subseteq \lambda_j(X_j)$, thus making λ surjective. For $j > 0$, let $\psi^{-1}(z_j) = \{y_1, y_2, \ldots, y_k\}$. Since $X \backslash dX$ contains no isolated points, we can partition $\varphi^{-1}(z_j)$ into k non-empty clopen sets $T_1, T_2, \ldots, T_k$. Repeating the above construction for $X_1, X_2, \ldots, X_m$, we can find a partition of $\mathbf{X}_j$ into clopen substructures $\mathbf{W}_1, \mathbf{W}_2, \ldots, \mathbf{W}_k$ where $T_i \subseteq W_i$. Then $\mathbf{X}_j$ is the coproduct of these substructures, and we can define λ_j by taking T_i to y_i and $W_i \backslash T_i$ to dy_i.

The construction of $\lambda_0 \colon \mathbf{X}_0 \to \mathbf{Y}$ is similar to the previous constructions. This time let $\psi^{-1}(z_0) \backslash d\psi^{-1}(z_0) = \{v_1, v_2, \ldots, v_k\}$ and let $\{v_1, dv_1, v_2, dv_2, \ldots, v_k, dv_k, v_{k+1}, \ldots, v_{k+q}\}$ be a list of the distinct members of $\varphi^{-1}(z_0)$. As done above, we can partition $\mathbf{X}_0$ into a coproduct of $k + q$ disjoint, non-empty substructures $\mathbf{X}_{0i}$ where $1 \leqslant i \leqslant k + q$, and then define λ_0 to map $\mathbf{X}_{0i}$ onto $\{v_i, dv_i\}$ for $1 \leqslant i \leqslant k$ and to map $\mathbf{X}_{0i}$ onto $\{v_i\}$ for $k + 1 \leqslant i \leqslant k + q$. ∎

A distinct benefit of a natural duality for a quasi-variety $\mathcal{A}$ is the ease with which it allows us to construct members of $\mathcal{A}$ with particular properties. For example, in Exercise 5.8 we describe spaces $\mathbf{X}_1$ and $\mathbf{X}_2$ such that

- $E(\mathbf{X}_1)$ and $E(\mathbf{X}_2)$ are both algebraically closed,
- $E(\mathbf{X}_1)$ satisfies (EC1) but not (EC2), and
- $E(\mathbf{X}_2)$ satisfies (EC2) but not (EC1).

Consequently (EC1) and (EC2) are independent over $\mathcal{S}^{\text{ac}}$; in particular, neither one alone will axiomatise $\mathcal{S}^{\text{ec}}$ within $\mathcal{S}^{\text{ac}}$.

We conclude this section with an explicit construction of the countable homogeneous universal Stone algebra. According to the EC Stone Algebra Theorem, this algebra will be given as $E(\mathbf{X})$ provided that $\mathbf{X}$ satisfies condition (iii) and has a countable basis of clopen sets. We construct such a space $\mathbf{X}$ as follows. Let $3 = \{0, 1, 2\}$ be the discrete space and let 3^ω denote its countably infinite direct power, where, as usual, $\omega = \{0, 1, 2, \dots\}$. For each $n > 0$, consider the subset

$$X_n := \{\, x \in 3^\omega \mid x(n) = 1 \text{ and } x(k) \neq 1 \text{ for } k > n \,\}$$

of 3^ω. The sets X_n are pairwise disjoint closed subspaces of 3^ω consisting collectively of all those $x \in 3^\omega$ for which $x^{-1}(1)$ is finite. We view $Y = \{0, \dots, \frac{1}{16}, \frac{1}{8}, \frac{1}{4}, \frac{1}{2}, 1\}$ as a Boolean subspace of the reals and order the product space $Y \times 3^\omega$ by stipulating that $(x_1, x_2) \preccurlyeq (y_1, y_2)$ if $x_1 = x_2$ and either $y_1 = y_2$ or $y_1 = 0$. The set

$$X_{\text{HU}} := \left(\bigcup_{n>0} X_n \times \left\{ \frac{1}{2^n} \right\} \right) \cup (3^\omega \times \{0\})$$

now determines a structure $\mathbf{X}_{\text{HU}} \in \mathcal{X}$ which is illustrated in Figure 5.5, where each horizontal segment represents a copy of the Cantor space $\mathbf{X}_{\text{C}}$. At a glance we can see that $\mathbf{X}_{\text{HU}}$ satisfies condition (iii) of the EC Stone Algebras Theorem, and consequently $E(\mathbf{X}_{\text{HU}})$ is the countable homogeneous $\mathcal{S}$-universal Stone algebra.

For successively more complex applications of the methods presented in this section, we refer the reader to the analyses of algebraically and existentially closed members of the variety of double Stone algebras in Clark [Cla89] and the second Lee-class $\mathcal{B}_2$ in Clark and Schmid [CS96].

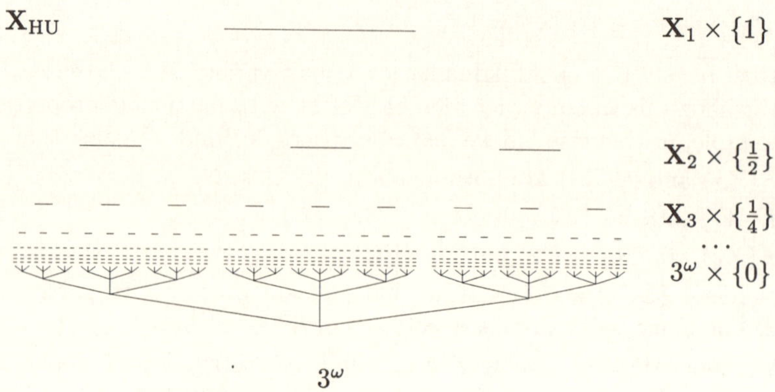

Figure 5.5 homogeneous universal Stone algebra

5.5 Injectives, weak injectives and absolute subretracts

We have already seen that injectivity plays an important role in duality theory. In this section, we shall show how a natural duality may be used to describe the injectives in the quasi-variety $\mathcal{A} = \mathbb{ISP}\underline{\mathbf{M}}$. If $\underline{\mathbf{M}}$ is injective, then, since products and retracts of injective algebras are injective, $\mathcal{A}$ will contain a profusion of injectives. Unfortunately, if $\underline{\mathbf{M}}$ is not injective, then the one-element algebras may be the only injectives in $\mathcal{A}$. For this reason we shall also study the weaker notion of absolute subretract. An algebra $\mathbf{A}$ is an **absolute subretract** in $\mathcal{A}$ if for every embedding $u\colon \mathbf{A} \to \mathbf{B}$ with $\mathbf{B} \in \mathcal{A}$ there is a homomorphism $v\colon \mathbf{B} \to \mathbf{A}$ such that $v \circ u = \mathrm{id}_{\mathbf{A}}$. Although $\underline{\mathbf{M}}$ need not be injective in $\mathcal{A}$, it is easily seen that $\underline{\mathbf{M}}$ is an absolute subretract in $\mathcal{A}$ provided it is subdirectly irreducible.

There is a large literature devoted to the study of injectives and absolute subretracts and the techniques involved vary from algebraic to topological to category-theoretic—see, for example, Banaschewski [Ban70], Davey [Dav76], Davey and Werner [DW79], Taylor [Tay72] and the references given in each. We shall employ a combination of all three. Let $\mathcal{C}$ be a category of structured sets, that is, each object in $\mathcal{C}$ is a set (with some additional structure) and each morphism in $\mathcal{C}$ is a function (which preserves the additional structure). An object $\mathbf{X} \in \mathcal{C}$ is **projective** in $\mathcal{C}$ if, for every surjective morphism $\varphi\colon \mathbf{Y} \to \mathbf{Z}$, each morphism $\alpha\colon \mathbf{X} \to \mathbf{Z}$ lifts from $\mathbf{Z}$ to $\mathbf{Y}$, that is, there exists a morphism $\gamma\colon \mathbf{X} \to \mathbf{Y}$ such that $\varphi \circ \gamma = \alpha$. Under most of the strong dualities that we have presented, the projectives in $\mathcal{X}$ are exactly the duals of the injectives in $\mathcal{A}$. The following lemma shows that this is not an accident.

5.1 Lemma *For a finite algebra* $\underline{\mathbf{M}}$ *and any choice of* $\underset{\sim}{\mathbf{M}}$,

(i) *if* $\mathbf{X}$ *is projective in* $\mathcal{X}$ *and* $\underline{\mathbf{M}}$ *is injective in* $\mathcal{A}$, *then* $E(\mathbf{X})$ *is injective in* $\mathcal{A}$, *and*

(ii) *if* $E(\mathbf{X})$ *is injective in* $\mathcal{A}$ *and* $\underset{\sim}{\mathbf{M}}$ *yields a full duality on* $\mathcal{A}$, *then* $\mathbf{X}$ *is projective in* $\mathcal{X}$.

Proof (i) Assume that $\underline{\mathbf{M}}$ is injective and that $\mathbf{X}$ is projective in $\mathcal{X}$. Let $u\colon \mathbf{A} \to \mathbf{B}$ be an embedding in $\mathcal{A}$ and let $v\colon \mathbf{A} \to E(\mathbf{X})$ be a homomorphism. The morphism $D(u)\colon D(\mathbf{B}) \to D(\mathbf{A})$ is surjective as $\underline{\mathbf{M}}$ is injective in $\mathcal{A}$. Hence, since $\mathbf{X}$ is projective in $\mathcal{X}$, the morphism $\alpha := D(v) \circ \varepsilon_{\mathbf{X}}\colon \mathbf{X} \to D(\mathbf{A})$ lifts to a morphism $\gamma\colon \mathbf{X} \to D(\mathbf{B})$ with $D(u) \circ \gamma = \alpha = D(v) \circ \varepsilon_{\mathbf{X}}$. We claim that $w := E(\gamma) \circ e_{\mathbf{B}}\colon \mathbf{B} \to E(\mathbf{X})$ is an extension of u, whence $E(\mathbf{X})$ is injective in $\mathcal{A}$. In order to prove that $v = w \circ u$ we perform the following calculation which uses Exercise 1.2 and two applications of the left-hand square of Figure 1.2.

$$\begin{aligned} w \circ u &= E(\gamma) \circ e_{\mathbf{B}} \circ u = E(\gamma) \circ ED(u) \circ e_{\mathbf{A}} = E(D(u) \circ \gamma) \circ e_{\mathbf{A}} \\ &= E(D(v) \circ \varepsilon_{\mathbf{X}}) \circ e_{\mathbf{A}} = E(\varepsilon_{\mathbf{X}}) \circ ED(v) \circ e_{\mathbf{A}} \\ &= E(\varepsilon_{\mathbf{X}}) \circ e_{E(\mathbf{X})} \circ v = \mathrm{id}_{E(\mathbf{X})} \circ v = v. \end{aligned}$$

(ii) This follows from Lemma 1.1.3(iii) which states that E is full and faithful. ∎

Since Boolean spaces play a central role in the quasi-varieties we study, it should not be surprising that the characterisation of the injectives and absolute subretracts in these quasi-varieties begins with a study of the Boolean case. According to Lemma 5.5.1 we must first describe the projectives in the dual category of Boolean spaces. Once again, denote the two-element Boolean algebra by $\underline{\mathbf{2}}$ and the two-element Boolean space by $\underset{\sim}{\mathbf{2}}$. Following the notation of Stone's Duality Theorem 4.1.2, we denote the Boolean space which is dual to a Boolean algebra $\mathbf{B}$ by $\mathbf{B}^*$ and the Boolean algebra dual to a Boolean space $\mathbf{X}$ by $\mathbf{X}^*$. The category of Boolean spaces will be denoted by $\mathcal{Z}$. Thus, D and E are reserved for the functors induced by some structure $\underset{\sim}{\mathbf{M}}$ and $\mathcal{X}$ is reserved for the topological quasi-variety $\mathbb{IS}_c\mathbb{P}^+ \underset{\sim}{\mathbf{M}}$ generated by $\underset{\sim}{\mathbf{M}}$.

We shall need the Stone–Čech compactification $\beta(S)$ of a (discretely topologised) set S. The Stone–Čech compactification is characterised by the property that it is the free Boolean space generated by the set S, that is, there is a map $\delta\colon S \to \beta(S)$ such that for every Boolean space $\mathbf{X}$ and each map $\varphi\colon S \to X$ there is a unique continuous map $\overline{\varphi}\colon \beta(S) \to \mathbf{X}$ with $\overline{\varphi} \circ \delta = \varphi$. Since the duality between $\mathcal{B}$ and $\mathcal{Z}$ is full, we can apply Exercise 3.5 to obtain a concrete realisation of this space: $\beta(S) := (\underline{\mathbf{2}}^S)^*$ with $\gamma(s)$ defined to be the natural projection $\pi_s\colon \underline{\mathbf{2}}^S \to \underline{\mathbf{2}}$ for each $s \in S$.

A topological space $\mathbf{X}$ is called **extremally disconnected** if the closure of every open set in $\mathbf{X}$ is open. Note that an extremally disconnected, compact Hausdorff space is Boolean. We now present A.M. Gleason's characterisation of the projectives in the category of compact Hausdorff spaces and in the category of Boolean spaces. (See [Gle58].)

5.2 Projective Compact Space Theorem *Let $\mathcal{X}$ be either the category of compact Hausdorff spaces or the category $\mathcal{Z}$ of Boolean spaces. Then the following are equivalent for $\mathbf{X} \in \mathcal{X}$:*

(i) $\mathbf{X}$ *is projective in* $\mathcal{X}$*;*

(ii) $\mathbf{X}$ *is a retract of* $\beta(S)$ *for some set* S*;*

(iii) $\mathbf{X}$ *is extremally disconnected.*

Proof The equivalence of (i) and (ii) is completely straightforward and is true in any category of structured sets which has free objects: use the fact that free objects are projective, retracts of projective objects are projective, and every object is the image of a free object under a surjective morphism. Assume that $\mathbf{X}$ is projective in $\mathcal{X}$. Let $U \subseteq X$ be open. We shall show that $\overline{U}$ is open in $\mathbf{X}$. Consider the following subspace of $\mathbf{X} \times \underset{\sim}{\mathbf{2}}$:

$$Y := (X \backslash U \times \{0\}) \cup (\overline{U} \times \{1\}).$$

Since Y is closed in $\mathbf{X}$, it follows that $\mathbf{Y} \in \mathcal{X}$. Since the first projection $\rho_1 \colon \mathbf{Y} \to \mathbf{X}$ is surjective and since $\mathbf{X}$ is projective in $\mathcal{X}$, there exists a continuous map $\gamma \colon \mathbf{X} \to \mathbf{Y}$ such that $\rho_1 \circ \gamma = \mathrm{id}_{\mathbf{X}}$. Clearly, $\gamma(x) = (x, 0)$ for all $x \notin \overline{U}$ and $\gamma(x) = (x, 1)$ for all $x \in U$. Since γ is continuous, we also have $\gamma(x) = (x, 1)$ for all $x \in \overline{U}$. Thus $\overline{U} = \gamma^{-1}(\overline{U} \times \{1\})$, whence $\overline{U}$ is open in $\mathbf{X}$ since $\overline{U} \times \{1\}$ is open in $\mathbf{Y}$. Thus $\mathbf{X}$ is extremally disconnected.

We have proved that (i) implies (iii). For the converse we require the following lemma.

5.3 Lemma *Let $\mathbf{X}$ be an extremally disconnected Hausdorff space, let $\mathbf{W}$ be a compact Hausdorff space and let $\psi \colon \mathbf{W} \to \mathbf{X}$ be a continuous surjection with the property that $\psi(W')$ is a proper subset of X for all proper, closed subsets W' of W. Then ψ is a homeomorphism.*

Proof By Lemma B.1(ii), it suffices to prove that, under the stated assumptions, ψ is one-to-one. Let $w_1 \neq w_2$ in W and let U_1 and U_2 be disjoint open subsets of W such that $w_i \in U_i$. By Lemma B.1(i), $X \backslash \psi(W \backslash U_i)$ is open in X. Since U_1 and U_2 are disjoint and ψ is surjective, we have

$$X \backslash \psi(W \backslash U_1) \cap X \backslash \psi(W \backslash U_2) = \varnothing.$$

As $\mathbf{X}$ is extremally disconnected, $\overline{V_1} \cap \overline{V_2} = \varnothing$ for any pair of disjoint open subsets V_1 and V_2 of X. Consequently,

$$\overline{X \backslash \psi(W \backslash U_1)} \cap \overline{X \backslash \psi(W \backslash U_2)} = \varnothing.$$

Hence, if we are able to prove that $\psi(w_i) \in \overline{X \backslash \psi(W \backslash U_i)}$ it will follow that $\psi(w_1) \neq \psi(w_2)$.

Let V be an open subset of X with $\psi(w_1) \in V$. We must show that V intersects $X \backslash \psi(W \backslash U_1)$. Since $w_1 \in U_1 \cap \psi^{-1}(V)$, the set $W \backslash (U_1 \cap \psi^{-1}(V))$ is a proper, closed subset of W and consequently $\psi(W \backslash (U_1 \cap \psi^{-1}(V)))$ is a proper subset of X. Let $x \in X \backslash \psi(W \backslash (U_1 \cap \psi^{-1}(V)))$ and let $w \in W$ with $\psi(w) = x$. Clearly, $x \in X \backslash \psi(W \backslash U_1)$. Moreover, $w \in \psi^{-1}(V)$ implies that $x = \psi(w) \in \psi(\psi^{-1}(V)) = V$, as required. Thus $\psi(w_1) \in \overline{X \backslash \psi(W \backslash U_1)}$ and, by symmetry, $\psi(w_2) \in \overline{X \backslash \psi(W \backslash U_2)}$. ■

Now let $\mathbf{X}, \mathbf{Y}$ and $\mathbf{Z}$ be in $\mathfrak{X}$ with $\mathbf{X}$ extremally disconnected, let $\varphi \colon \mathbf{Y} \to \mathbf{Z}$ be a continuous surjection and let $\alpha \colon \mathbf{X} \to \mathbf{Z}$ be continuous. Then the joint kernel

$$W := \{\, (x, y) \in X \times Y \mid \alpha(x) = \varphi(y)\}$$

is a closed subset of $X \times Y$, whence $\mathbf{W}$ is compact. Since φ is surjective, the first projection $\rho_1 \colon \mathbf{W} \to \mathbf{X}$ is surjective and hence, by Exercise B.4, there is a closed subset W_0 of W such that $\rho_1(W_0) = X$ but $\rho(W')$ is a proper subset of X for all proper, closed subsets W' of W_0. Consequently, by the lemma above, $\psi := \rho_1{\restriction}_{W_0}$ is a homeomorphism. Let $\gamma := \rho_2 \circ \psi^{-1}$. If $x \in X$, then since $\psi^{-1}(x) \in W$ we have

$$\varphi(\gamma(x)) = \varphi(\rho_2(\psi^{-1}(x))) = \alpha(\rho_1(\psi^{-1}(x))) = \alpha(x).$$

Hence $\mathbf{X}$ is projective in $\mathfrak{X}$. ■

It is now a simple matter to use Lemma 5.5.1 to convert the characterisation of the projective Boolean spaces into a characterisation of the injective Boolean algebras.

5.4 Injective Boolean Algebra Theorem *Let* $\mathbf{B}$ *be a Boolean algebra. The following are equivalent:*

(i) $\mathbf{B}$ *is injective in the class* $\mathfrak{B}$ *of all Boolean algebras;*

(ii) $\mathbf{B}$ *is an absolute subretract in* $\mathfrak{B}$*;*

(iii) $\mathbf{B}$ *is complete (as an ordered set);*

(iv) *the Boolean space* $\mathbf{B}^*$ *is extremally disconnected.*

Proof That (i) implies (ii) is a triviality. Assume that $\mathbf{B}$ is an absolute subretract in $\mathfrak{B}$. Since $\mathbf{B}$ is isomorphic to $\mathbf{B}^{**}$ which is a subalgebra of the complete Boolean algebra $\underline{\mathbf{2}}^{\mathbf{B}^*}$, it follows that $\mathbf{B}$ is a retract of a complete Boolean algebra. As a Boolean algebra retract is necessarily an order retract, Exercise 5.11(i) implies that $\mathbf{B}$ is complete. Thus (ii) implies (iii). That (iii) implies (iv) is an easy consequence of Exercise 5.14. Finally, if $\mathbf{B}^*$ is extremally disconnected, then $\mathbf{B}^*$ is projective in the category of Boolean spaces. Since $\underset{\sim}{2}$ is a total structure, the Injectivity Lemma 3.2.10 tells us that $\underline{2}$ is injective in $\mathfrak{B}$. Thus, by Lemma 5.5.1, $\mathbf{B} \cong \mathbf{B}^{**}$ is injective in $\mathfrak{B}$, whence (iv) implies (i). ■

We now develop enough general theory to enable us to transfer our knowledge of the injective Boolean algebras and projective Boolean spaces to the study of injectives and absolute subretracts in other quasi-varieties. While our main interest is in injectives and absolute subretracts, there is an intermediate concept which is particularly useful. An algebra $\mathbf{A} \in \mathcal{A}$ is **weakly injective** in $\mathcal{A}$ if, for every embedding $u \colon \mathbf{C} \to \mathbf{B}$ in $\mathcal{A}$, each surjective homomorphism $v \colon \mathbf{C} \to \mathbf{A}$ extends to a homomorphism $w \colon \mathbf{B} \to \mathbf{A}$ with $w \circ u = v$.

The basic algebraic properties of injectives, weak injectives and absolute subretracts which we require are summarised (without proof) in the following proposition. Let $u \colon \mathbf{A} \to \prod\{\mathbf{A}_i \mid i \in I\}$ be an embedding of $\mathbf{A}$ as a subdirect product; if there is a homomorphism $v \colon \prod\{\mathbf{A}_i \mid i \in I\} \to \mathbf{A}$ such that $v \circ u = \mathrm{id}_{\mathbf{A}}$, then $\mathbf{A}$ is a **subdirect retract** of the family $\{\mathbf{A}_i \mid i \in I\}$. The quasi-variety $\mathcal{A}$ has the **$\mathcal{A}$-congruence extension property** if, for all $\mathbf{B} \leqslant \mathbf{A} \in \mathcal{A}$, each $\mathcal{A}$-congruence on $\mathbf{B}$ has an extension to an $\mathcal{A}$-congruence on $\mathbf{A}$, and $\mathcal{A}$ has **factorisable $\mathcal{A}$-congruences** if, for all $\mathbf{A}, \mathbf{B} \in \mathcal{A}$, every $\mathcal{A}$-congruence on $\mathbf{A} \times \mathbf{B}$ is of the form (θ_1, θ_2) for some congruences θ_1 on $\mathbf{A}$ and θ_2 on $\mathbf{B}$ (see Exercise A.10).

5.5 Proposition *Let $\mathcal{A} = \mathbb{ISP}\underline{\mathbf{M}}$ be the quasi-variety generated by $\underline{\mathbf{M}}$.*

(i) *In $\mathcal{A}$, every injective is a weak injective and every weak injective is an absolute subretract.*

(ii) *If $\mathcal{A}$ satisfies the $\mathcal{A}$-congruence extension property, then every absolute subretract in $\mathcal{A}$ is weakly injective.*

(iii) *If $\underline{\mathbf{M}}$ is subdirectly irreducible, then it is an absolute subretract in $\mathcal{A}$.*

(iv) *Every product of (weak) injectives is a (weak) injective and every retract of an injective is injective.*

(v) *Let $\mathbf{A}_i \in \mathcal{A}$ for all $i \in I$. If $\mathcal{A}$ has factorisable $\mathcal{A}$-congruences (in particular, if $\mathrm{Con}(\mathbf{A})$ is distributive for each algebra $\mathbf{A} \in \mathcal{A}$) and $\prod\{\mathbf{A}_i \mid i \in I\}$ is an injective, a weak injective or an absolute subretract in $\mathcal{A}$, then so is $\mathbf{A}_i$ for each $i \in I$.*

(vi) *The weak injectives in $\mathcal{A}$ are closed under the formation of subdirect retracts.*

(vii) *If $\underline{\mathbf{M}}$ is injective in $\mathcal{A}$, then in $\mathcal{A}$ the injectives, the weak injectives and the absolute subretracts coincide.*

The remainder of this section extends the proof technique used in Davey [Dav76] for Heyting algebras to a general natural duality setting.

For any algebra $\mathbf{A} \in \mathcal{A}$ and Boolean space $\mathbf{X} \in \mathcal{Z}$, let $C(\mathbf{X}, \mathbf{A})$ denote the algebra of all continuous maps from $\mathbf{X}$ into $\mathbf{A}$ (endowed with the discrete topology). This defines the object half of a contravariant functor $C(-, \mathbf{A}) \colon \mathcal{Z} \to \mathcal{A}$. Indeed, when $\mathbf{A}$ is finite, $C(-, \mathbf{A})$ is the functor E induced by the purely topological structure $\underset{\sim}{\mathbf{A}}_0 := \langle A; \mathcal{T} \rangle$.

5.6 Lemma *Let $\mathbf{A}$ be a finite algebra in $\mathcal{A}$ and let $\mathbf{X} \in \mathcal{Z}$ be extremally disconnected. Then $C(\mathbf{X}, \mathbf{A})$ is a subdirect retract of copies of $\mathbf{A}$.*

Proof Let $\mathbf{X} \in \mathcal{Z}$ be extremally disconnected. By Theorem 5.5.2, $\mathbf{X}$ is a retract of $\boldsymbol{\beta}(S)$ for some set S via a retraction $\varphi \colon \boldsymbol{\beta}(S) \to \mathbf{X}$. Let $\delta \colon S \to \beta(S)$ be the inclusion of S into $\beta(S)$. For each $s \in S$, let $Y_s := \{s\}$ and let $\sigma_s \colon \mathbf{Y}_s \to \boldsymbol{\beta}(S)$ be the restriction of δ to Y_s. Then $\varphi \circ \sigma_s$ is an embedding.

Clearly, $C(\varphi, \mathbf{A})$ is an embedding and $C(\varphi \circ \sigma_s, \mathbf{A})$ is a surjection. Since functors map retracts to retracts, it follows that $\mathcal{C}(\mathbf{X}, \mathbf{A})$ is a retract of $C(\boldsymbol{\beta}(S), \mathbf{A}) \cong \prod\{\mathcal{C}(\mathbf{Y}_s, \mathbf{A}) \mid s \in S\}$. Since $C(\mathbf{Y}_s, \mathbf{A})$ is isomorphic to $\mathbf{A}$, it follows at once that $C(\mathbf{X}, \mathbf{A})$ is a subdirect retract of copies of $\mathbf{A}$. ■

For a finite algebra $\mathbf{A} \in \mathcal{A}$ and a Boolean algebra $\mathbf{B}$, the **Boolean power** $\mathbf{A}[\mathbf{B}]$ is defined to be the algebra $C(\mathbf{B}^*, \mathbf{A})$.

5.7 Lemma *Let $\mathbf{A} \in \mathcal{A}$ and let $\mathbf{B}$ be a Boolean algebra.*

(i) *If $\mathbf{A}[\mathbf{B}]$ is an injective, a weak injective or an absolute subretract in $\mathcal{A}$, then so is $\mathbf{A}$.*

(ii) *Suppose $\mathbf{B}$ is complete and $\mathbf{A}$ is finite. If $\mathbf{A}$ is an injective or a weak injective in $\mathcal{A}$, then so is the Boolean power $\mathbf{A}[\mathbf{B}]$.*

Proof (i) With $\mathbf{X} \in \mathcal{Z}$ fixed, we have a covariant functor $G := \mathcal{C}(\mathbf{X}, -)$ from $\mathcal{A}$ to $\mathcal{A}$. It is straightforward to prove that (a) G preserves embeddings and surjections, (b) the map $\eta_{\mathbf{A}} \colon \mathbf{A} \to G(\mathbf{A})$, which maps a to the constant map onto $\{a\}$, embeds $\mathbf{A}$ into $G(\mathbf{A})$ as a retract, (c) η is a natural transformation, that is, if $u \colon \mathbf{A} \to \mathbf{B}$ is a homomorphism in $\mathcal{A}$, then $\eta_{\mathbf{B}} \circ u = G(u) \circ \eta_{\mathbf{A}}$. Using these facts, the proof is a simple diagram chase for the reader.

(ii) $\mathbf{B}^*$ is extremally disconnected by Theorem 5.5.4 and therefore $\mathbf{A}[\mathbf{B}^*]$ is a subdirect retract of copies of $\mathbf{A}$ by Lemma 5.5.6. The conclusion now follows from Proposition 5.5.5, parts (iv) and (vi). ■

Our goal is to lift our knowledge of the projectives in $\mathcal{Z}$ up to a description of the projectives in $\mathcal{X}$ and then, via the functors D and E, to obtain information about the injectives, weak injectives and absolute subretracts in $\mathcal{A}$. In order to do so, we need to introduce a new concept. A (covariant) functor $\mathcal{F} \colon \mathcal{Z} \to \mathcal{X}$ is a **$\mathcal{Z}$-free functor** for $\mathcal{X}$ if $\mathcal{F}(\mathbf{X})$ is the free object in $\mathfrak{X}$ generated by the Boolean space $\mathbf{X}$, that is, there is a continuous map $\eta_{\mathbf{X}}\colon \mathbf{X} \to \mathcal{F}(\mathbf{X})$ such that, for every object $\mathbf{Y} \in \mathfrak{X}$ and each continuous map $\varphi\colon \mathbf{X} \to \mathbf{Y}$, there is a unique morphism $\overline{\varphi}\colon \mathcal{F}(\mathbf{X}) \to \mathbf{Y}$ such that $\overline{\varphi} \circ \eta_{\mathbf{X}} = \varphi$. (In category-theoretic language, $\mathcal{F}\colon \mathfrak{Z} \to \mathfrak{X}$ is a left adjoint to the forgetful functor from $\mathfrak{X}$ into $\mathfrak{Z}$. See Exercise 5.18, or Mac Lane [Mac:CW] for a detailed discussion.)

Our next result illustrates the importance of $\mathfrak{Z}$-free functors in the present context. By a category $\mathfrak{X}$ of structured Boolean spaces we mean a category in which each object is a Boolean space (with some additional structure) and each morphism is a continuous map (which preserves the additional structure). In fact, the result below requires only that there be a faithful functor $|-|\colon \mathfrak{X} \to \mathfrak{Z}$ which has a left adjoint $\mathcal{F}\colon \mathfrak{Z} \to \mathfrak{X}$. It shows that a $\mathfrak{Z}$-free functor lifts Gleason's characterisation of projectives in $\mathfrak{Z}$ to a corresponding characterisation for $\mathfrak{X}$.

5.8 Proposition *Let $\mathfrak{X}$ be a category of structured Boolean spaces and assume that $\mathcal{F}\colon \mathfrak{Z} \to \mathfrak{X}$ is a $\mathfrak{Z}$-free functor for $\mathfrak{X}$. The following are equivalent:*

(i) $\mathbf{Y}$ *is a retract of $\mathcal{F}(\mathbf{X})$ for some projective Boolean space $\mathbf{X}$;*

(ii) $\mathbf{Y}$ *is a retract of $\mathcal{F}(\beta(S))$ for some set S;*

(iii) $\mathbf{Y}$ *is a retract of $\mathcal{F}(\mathbf{X})$ for some extremally disconnected Boolean space $\mathbf{X}$;*

(iv) $\mathbf{Y}$ *is projective in $\mathfrak{X}$.*

Proof Parts (i) and (iii) are equivalent by the Projective Compact Space Theorem 5.5.2. Since $\beta(S)$ is the free Boolean space generated by the set S, it is easily seen that $\mathcal{F}(\beta(S))$ is the free object in $\mathfrak{X}$ generated by the set S. It follows easily (as in the proof of the Projective Compact Space Theorem 5.5.2) that (iv) and (ii) are equivalent.

Since $\beta(S)$ is extremally disconnected (via Theorem 5.5.4, as $\beta(S) = (\underline{\mathbf{2}}^S)^*$ and $\underline{\mathbf{2}}^S$ is complete), (ii) implies (iii). Finally, assume that $\mathbf{X} \in \mathfrak{Z}$ is extremally disconnected and let $\mathbf{Y} \in \mathfrak{X}$ be a retract of $\mathcal{F}(\mathbf{X})$. By the Projective Compact Space Theorem 5.5.2 (yet again), $\mathbf{X}$ is a retract of $\beta(S)$ for some set S, whence $\mathcal{F}(\mathbf{X})$ is a retract of $\mathcal{F}(\beta(S))$, as functors preserve retracts. As a retract of a retract is a retract, it follows that $\mathbf{Y}$ is a retract of $\mathcal{F}(\beta(S))$, whence (iii) implies (ii). ∎

If $\underset{\sim}{\mathbf{M}}$ yields a full duality on $\mathcal{A}$, then we have a convenient description of a $\mathfrak{Z}$-free functor for the category $\mathfrak{X} = \mathbb{IS}_c\mathbb{P}^+ \underset{\sim}{\mathbf{M}}$.

5.9 Proposition *Let $\underline{\mathbf{M}}$ be a finite algebra and assume that $\underset{\sim}{\mathbf{M}}$ yields a full duality on the quasi-variety $\mathcal{A}$. Then $D(C(-, \underline{\mathbf{M}}))\colon \mathfrak{Z} \to \mathfrak{X}$ is a $\mathfrak{Z}$-free functor for $\mathfrak{X}$. Conversely, if $\underset{\sim}{\mathbf{M}}$ yields a duality on $\mathcal{A}$, the algebra*

$\underline{\mathbf{M}}$ is injective in $\mathcal{A}$ and $D(\mathcal{C}(-,\underline{\mathbf{M}}))\colon \mathcal{Z} \to \mathcal{X}$ is a $\mathcal{Z}$-free functor for $\mathcal{X}$, then $\underset{\sim}{\mathbf{M}}$ yields a strong (and therefore full) duality on $\mathcal{A}$.

Proof The easiest way to see this is to quote a straightforward result from basic category theory: a composite of left adjoint functors is a left adjoint functor (see Theorem 1 on page 101 of [Mac:CW]). In order to apply this fact here, we denote the functors induced by the purely topological structure $\underset{\sim}{\mathbf{M}}_0 = \langle M; \mathcal{T}\rangle$ by $D_0\colon \mathcal{A} \to \mathcal{Z}$ and $E_0\colon \mathcal{Z} \to \mathcal{A}$. When we replace $\mathcal{A}$ by its formal opposite category $\mathcal{A}^\partial$, the bijection between $\mathcal{A}(\mathbf{A}, E_0(\mathbf{X}))$ and $\mathcal{X}(\mathbf{X}, \mathbf{D}_0(\mathbf{A}))$ becomes a bijection between $\mathcal{A}^\partial(E_0(\mathbf{X}), \mathbf{A})$ and $\mathcal{X}(\mathbf{X}, \mathbf{D}_0(\mathbf{A}))$. This says precisely that the covariant functor $E_0\colon \mathcal{X} \to \mathcal{A}^\partial$ is left adjoint to the covariant functor $D_0\colon \mathcal{A}^\partial \to \mathcal{X}$ (see Exercise 5.18). Since $\underset{\sim}{\mathbf{M}}$ yields a full duality on $\mathcal{A}$, the covariant functors $D\colon \mathcal{A}^\partial \to \mathcal{X}$ and $E\colon \mathcal{X} \to \mathcal{A}^\partial$ give a category equivalence between $\mathcal{A}^\partial$ and $\mathcal{X}$ and consequently D is both left and right adjoint to E (see Exercise 5.18 or page 91 of [Mac:CW]). Thus $D \circ E_0\colon \mathcal{Z} \to \mathcal{X}$ is left adjoint to $D_0 \circ E\colon \mathcal{X} \to \mathcal{Z}$. As $D_0E(\mathbf{X})$ is the underlying space of $DE(\mathbf{X})$ and the duality is full, there is a natural isomorphism from the functor $D_0 \circ E$ to the forgetful functor from $\mathcal{X}$ to $\mathcal{Z}$, so $D \circ E_0\colon \mathcal{Z} \to \mathcal{X}$ is left adjoint to the forgetful functor and hence is a $\mathcal{Z}$-free functor for $\mathcal{X}$. Since $DE_0(\mathbf{X}) = D(C(\mathbf{X}, \underline{\mathbf{M}}))$ we have proved the forward direction.

Now assume that $\underset{\sim}{\mathbf{M}}$ yields a duality on $\mathcal{A}$, that $\underline{\mathbf{M}}$ is injective in $\mathcal{A}$ and that $D(C(-,\underline{\mathbf{M}}))$ is a $\mathcal{Z}$-free functor for $\mathcal{X}$. Since $\beta(S)$ is the free Boolean space generated by S, it follows that $D(C(\beta(S), \underline{\mathbf{M}})) \cong D(\underline{\mathbf{M}}^S)$ is freely generated in $\mathcal{X}$ by the projections $\pi_s\colon \underline{\mathbf{M}}^S \to \underline{\mathbf{M}}$ for each $s \in S$. Hence $\underset{\sim}{\mathbf{M}}$ yields a strong duality on $\mathcal{A}$ by Exercise 3.6. ∎

We shall see later that there is a large class of algebras $\underline{\mathbf{M}}$ for which there is a choice of $\underset{\sim}{\mathbf{M}}$ which yields a duality on $\mathcal{A}$ with $D(C(-,\underline{\mathbf{M}}))$ a $\mathcal{Z}$-free functor for $\mathcal{X}$ even though $\underset{\sim}{\mathbf{M}}$ does not yield a full duality. The following result therefore applies to such algebras.

5.10 Proposition *Let $\underset{\sim}{\mathbf{M}}$ be a structure such that $D(C(-,\underline{\mathbf{M}}))\colon \mathcal{Z} \to \mathcal{X}$ is a $\mathcal{Z}$-free functor for $\mathcal{X}$. Consider the following conditions:*

(i) $\mathbf{A}$ *is injective in* $\mathcal{A}$;

(ii) $\mathbf{A}$ *is weakly injective in* $\mathcal{A}$;

(iii) $\mathbf{A}$ *is an absolute subretract in* $\mathcal{A}$;

(iv) $D(\mathbf{A})$ *is projective in* $\mathcal{X}$;

(v) $D(\mathbf{A})$ *is a retract of* $D(C(\mathbf{X}, \underline{\mathbf{M}}))$ *for some extremally disconnected Boolean space* $\mathbf{X}$.

In general, (i) $\Rightarrow$ (ii) $\Rightarrow$ (iii) $\Rightarrow$ (iv) $\Leftrightarrow$ (v), *and if $\underline{\mathbf{M}}$ is injective in $\mathcal{A}$ and $\underset{\sim}{\mathbf{M}}$ yields a duality on $\mathcal{A}$, then all of these conditions are equivalent (and $\underset{\sim}{\mathbf{M}}$ yields a strong duality on $\mathcal{A}$).*

Proof (i) $\Rightarrow$ (ii) $\Rightarrow$ (iii) by Proposition 5.5.5 and (iv) $\Leftrightarrow$ (v) by Proposition 5.5.8. Assume that $\mathcal{F} := D(C(-, \underline{\mathbf{M}}))\colon \mathcal{Z} \to \mathcal{X}$ is a $\mathcal{Z}$-free functor for $\mathcal{X}$. If $\mathbf{A}$ is an absolute subretract in $\mathcal{A}$, then, since $\mathcal{A} = \mathbb{ISP}\underline{\mathbf{M}}$, there is an embedding of $\mathbf{A}$ into $\underline{\mathbf{M}}^S$ for some set S, whence $\mathbf{A}$ is a retract of $\underline{\mathbf{M}}^S$. Hence, $D(\mathbf{A})$ is a retract of $D(\underline{\mathbf{M}}^S) \cong D(C(\beta(S), \underline{\mathbf{M}})) = \mathcal{F}(\beta(S))$. Thus $D(\mathbf{A})$ is projective in $\mathcal{X}$ by Proposition 5.5.8 and consequently (iii) $\Rightarrow$ (iv). Assume that $\underline{\mathbf{M}}$ is injective in $\mathcal{A}$ and $\underset{\sim}{\mathbf{M}}$ yields a duality on $\mathcal{A}$. If $D(\mathbf{A})$ is projective in $\mathcal{X}$, then, by Lemma 5.5.1, $\mathbf{A} \cong ED(\mathbf{A})$ is injective in $\mathcal{A}$, whence (iv) $\Rightarrow$ (i) and therefore all the conditions are equivalent. ∎

This result suggests that it is worth while characterising the projectives in $\mathcal{X}$ even when $\underset{\sim}{\mathbf{M}}$ yields a duality which is not full or $\underline{\mathbf{M}}$ is not injective in $\mathcal{A}$: even though an absolute subretract $\mathbf{A}$ in $\mathcal{A}$ need not be injective in $\mathcal{A}$, its dual $D(\mathbf{A})$ will nevertheless be projective in $\mathcal{X}$. In fact, we will see that this characterisation is more easily achieved by applying Proposition 5.5.10 to a choice of $\underset{\sim}{\mathbf{M}}$ which yields a duality that is not in general full.

In order to apply Proposition 5.5.10, we need to find a useful description of a retract of $D(C(\mathbf{X}, \underline{\mathbf{M}}))$ where $\mathbf{X}$ is an extremally disconnected Boolean space. The first step is to describe the topological structure of $D(C(\mathbf{X}, \underline{\mathbf{M}}))$, which we do by imposing some familiar conditions on $\underline{\mathbf{M}}$. Here we denote by $\mathcal{A}(C(\mathbf{X}, \underline{\mathbf{M}}), \underline{\mathbf{M}})$ the Boolean space underlying the structure $D(C(\mathbf{X}, \underline{\mathbf{M}}))$.

5.11 Proposition *Assume that $\underline{\mathbf{M}}$ generates a congruence distributive variety and that every non-trivial subalgebra of $\underline{\mathbf{M}}$ is subdirectly irreducible. Let $\{c_1, \dots, c_\ell\}$ be the set of pairwise distinct elements which form one-element subalgebras of $\underline{\mathbf{M}}$ and let $\overline{c}_1, \dots, \overline{c}_\ell$ be the corresponding constant endomorphisms of $\underline{\mathbf{M}}$. Let $\mathbf{X}$ be a Boolean space. For each $x \in X$, let $\delta_x\colon C(\mathbf{X}, \underline{\mathbf{M}}) \to \underline{\mathbf{M}}$ be the natural evaluation map, given by $\delta_x(\alpha) := \alpha(x)$ for all $\alpha \in C(\mathbf{X}, \underline{\mathbf{M}})$, and, for $i = 1, \dots, \ell$, let $\widehat{c}_i\colon C(\mathbf{X}, \underline{\mathbf{M}}) \to \underline{\mathbf{M}}$ be the constant map onto $\{c_i\}$. Let*

$$\mathcal{F}(\mathbf{X}) := ((\mathrm{End}(\underline{\mathbf{M}}) \backslash \{\overline{c}_1, \dots, \overline{c}_\ell\}) \times X) \mathbin{\dot{\cup}} \{\overline{c}_1, \dots, \overline{c}_\ell\},$$

endowed with the obvious topology, and define $\mu_{\mathbf{X}}((e, x)) = e \circ \delta_x$ and $\mu_{\mathbf{X}}(\overline{c}_i) = \widehat{c}_i$. Then $\mu_{\mathbf{X}}\colon \mathcal{F}(\mathbf{X}) \to \mathcal{A}(C(\mathbf{X}, \underline{\mathbf{M}}), \underline{\mathbf{M}})$ is a homeomorphism.

Proof We leave the proof that $\mu_{\mathbf{X}}$ is one-to-one as an easy exercise. Thus, by Lemma B.1, it remains to prove that $\mu_{\mathbf{X}}$ is surjective and continuous. Let $u\colon C(\mathbf{X}, \underline{\mathbf{M}}) \to \underline{\mathbf{M}}$ be a homomorphism and without loss of generality assume that u is not a constant map. We must find a non-constant endomorphism e and $x \in X$ such that $u = e \circ \delta_x$. Since every non-trivial subalgebra of $\underline{\mathbf{M}}$ is subdirectly irreducible, $\ker(u)$ is meet-irreducible in the congruence lattice of $C(\mathbf{X}, \underline{\mathbf{M}})$. Hence, by Jónsson's

Lemma A.15, there is an ultrafilter $\mathfrak{U}$ on X such that the congruence $\theta_{\mathfrak{U}}$ on $C(\mathbf{X}, \underline{\mathbf{M}})$, defined by $(\varphi, \psi) \in \theta_U \iff \mathrm{eq}(\varphi, \psi) \in \mathfrak{U}$, satisfies $\theta_{\mathfrak{U}} \subseteq \ker(u)$. Let $\mathfrak{U}' = \{ U \in \mathcal{U} \mid U \text{ is clopen in } \mathbf{X} \}$. Then $\mathfrak{U}'$ is an ultrafilter of the Boolean algebra $\mathbf{X}^*$ of all clopen subsets on $\mathbf{X}$. Hence, by Stone's Duality Theorem 4.1.2, there exists a unique $x \in X$ such that $\mathfrak{U}' = \varepsilon_{\mathbf{X}}(x) := \{ U \mid U \text{ is clopen in } \mathbf{X} \text{ and } x \in U \}$. Since

$$\mathrm{eq}(\varphi, \psi) = \bigcup \{ \varphi^{-1}(a) \cap \psi^{-1}(a) \mid a \in M \},$$

if $\varphi, \psi \in C(\mathbf{X}, \underline{\mathbf{M}})$, then $\mathrm{eq}(\varphi, \psi)$ is clopen in $\mathbf{X}$. Consequently,

$$(\varphi, \psi) \in \theta_{\mathfrak{U}} \iff \mathrm{eq}(\varphi, \psi) \in \mathfrak{U}' \iff \varphi(x) = \psi(x).$$

For each $a \in M$, let $k_a \in C(\mathbf{X}, \underline{\mathbf{M}})$ be the constant map onto $\{a\}$ and let e be the endomorphism of $\underline{\mathbf{M}}$ defined by $e(a) = u(k_a)$. We claim that $u = e \circ \delta_x$. If $\varphi \in C(\mathbf{X}, \underline{\mathbf{M}})$, then $e(\delta_x(\varphi)) = e(\varphi(x)) = u(k_{\varphi(x)})$. But $(\varphi, k_{\varphi(x)}) \in \theta_{\mathfrak{U}}$ since $\varphi(x) = k_{\varphi(x)}(x)$ and hence $(\varphi, k_{\varphi(x)}) \in \ker(u)$ as $\theta_{\mathfrak{U}} \subseteq \ker(u)$. Thus, $u(\varphi) = u(k_{\varphi(x)}) = e(\varphi(x)) = e(\delta_x(\varphi))$, whence $u = e \circ \delta_x$, as claimed. Thus, $\mu_{\mathbf{X}}$ is surjective.

Recall from Lemma B.6(i) that the sets of the form

$$U_{\alpha,a} := \{ u \in \mathcal{A}(C(\mathbf{X}, \underline{\mathbf{M}}), \underline{\mathbf{M}}) \mid u(\alpha) = a \}$$

for $\alpha \in C(\mathbf{X}, \underline{\mathbf{M}})$ and $a \in M$ form a subbasis for the topology on the space $\mathcal{A}(C(\mathbf{X}, \underline{\mathbf{M}}), \underline{\mathbf{M}})$. Define $V := \{ (e, x) \in \mathcal{F}(\mathbf{X}) \mid e(\alpha(x)) = a \}$. If $\{a\}$ is not a subalgebra of $\underline{\mathbf{M}}$, then $\mu_{\mathbf{X}}^{-1}(U_{\alpha,a}) = V$, and if $\{a\}$ is a subalgebra of $\underline{\mathbf{M}}$, then $\mu_{\mathbf{X}}^{-1}(U_{\alpha,a}) = V \cup \{\overline{a}\}$. Hence to prove that $\mu_{\mathbf{X}}$ is continuous, it suffices to show that V is open in $\mathcal{F}(\mathbf{X})$; but, for each $(e, x) \in V$, the set $W := \{e\} \times \alpha^{-1}(e^{-1}(a))$ is an open set in $\mathcal{F}(\mathbf{X})$ satisfying $(e, x) \in W \subseteq V$. Thus $\mu_{\mathbf{X}}$ is continuous. ■

Next we would like to choose a structure $\underset{\sim}{\mathbf{M}}$ and extend the Boolean space $\mathcal{F}(\mathbf{X})$ to a structure of the same type as $\underset{\sim}{\mathbf{M}}$ in such a way that $\mathcal{F}(\mathbf{X})$ will be isomorphic to $D(C(\mathbf{X}, \underline{\mathbf{M}}))$ and $\mathcal{F}$ will be a $\mathfrak{Z}$-free functor. Our success in doing so will depend on further restricting the choice of $\underset{\sim}{\mathbf{M}}$. Anticipating the applications we will want to make, we assume that $\underline{\mathbf{M}}$ has a $(k+1)$-ary near-unanimity term for some $k \geqslant 2$ and that every subalgebra of $\underline{\mathbf{M}}$ is subdirectly irreducible. In this context the NU Strong Duality Theorem suggests that we assume that each operation of G is either nullary or is an endomorphism of $\underline{\mathbf{M}}$, each partial operation is a partial endomorphism of $\underline{\mathbf{M}}$ and each relation in R is at most k-ary.

It is clear how to define the unary maps in $G \cup H$ on $\mathcal{F}(\mathbf{X})$: if g is an endomorphism of $\underline{\mathbf{M}}$, then $g((e, x)) := (ge, x)$ and $g(\overline{c}_i) = \overline{c}_j$, where $g(c_i) = c_j$, while if h is a partial endomorphism of $\underline{\mathbf{M}}$, then $h((e, x))$

is defined and equals (he, x) if and only if $e(M) \subseteq \mathrm{dom}(h)$, and $h(\bar{c}_i)$ is defined and equals $\bar{c}_j$ if and only if $c_i \in \mathrm{dom}(h)$ and $h(c_i) = c_j$. This definition of the (partial) operations on $\mathcal{F}(\mathbf{X})$ is independent of the set X. The structure of $\mathcal{F}(\mathbf{X})$ will be most tractable if the relations can also be defined independently of X: for $w_1, \ldots, w_n \in \mathcal{F}(\mathbf{X})$ we have $(w_1, \ldots, w_n) \in r^{\mathcal{F}(\mathbf{X})}$ if and only if at most one element $x \in X$ is involved in the elements $w_1, \ldots, w_n$ and the n-tuple of endomorphisms involved in $w_1, \ldots, w_n$ is in the relation r. This is best illustrated by a couple of examples. We have $((e, x), (f, y), (g, z)) \in r^{\mathcal{F}(\mathbf{X})}$ if and only if $x = y = z$ and $(e(a), f(a), g(a)) \in r$ for all $a \in M$, and $((e, x), \bar{c}_1, (f, y), \bar{c}_2) \in r^{\mathcal{F}(\mathbf{X})}$ if and only if $x = y$ and $(e(a), c_1, f(a), c_2) \in r$ for all $a \in M$.

We now have a structure $\mathcal{F}(\mathbf{X})$ of type $\langle G, H, R \rangle$. Since $\underline{\mathbf{M}}$ has a near-unanimity term it generates a congruence distributive variety (see Mitschke's Theorem A.14) and Proposition 5.5.11 guarantees that $\mu_{\mathbf{X}}$ is a homeomorphism. In order to ensure that $\mu_{\mathbf{X}}$ is an isomorphism from $\mathcal{F}(\mathbf{X})$ to $D(C(\mathbf{X}, \underline{\mathbf{M}}))$, and therefore that $\mathcal{F}(\mathbf{X})$ is an object in $\mathfrak{X}$, we need to impose further restrictions on $\underset{\sim}{\mathbf{M}}$. We will say that an algebraic relation $r \leqslant \underline{\mathbf{M}}^n$ (for $n \geqslant 2$) **avoids binary products** if each binary projection

$$\pi_{ij}(r) := \{\, (\pi_i(a), \pi_j(a)) \in M^2 \mid a \in r \,\},$$

for $1 \leqslant i < j \leqslant n$, contains no product $A \times B$ of non-trivial subalgebras of $\underline{\mathbf{M}}$. Many of the dualities given in Chapter 4 are given by relations which avoid binary products. This is the case, for example, for distributive lattices, median algebras, Stone algebras, Kleene algebras, double Stone algebras and de Morgan algebras. Note that the following result does not require that $\underset{\sim}{\mathbf{M}}$ yields a duality on $\mathcal{A}$.

5.12 Theorem *Assume that $\underline{\mathbf{M}}$ has a near-unanimity term and that every non-trivial subalgebra of $\underline{\mathbf{M}}$ is subdirectly irreducible. Let $\underset{\sim}{\mathbf{M}} = \langle M; G, H, R, \mathcal{T} \rangle$ and assume that each member of G is either nullary or is an endomorphism of $\underline{\mathbf{M}}$, each member of H is a partial endomorphism of $\underline{\mathbf{M}}$ and each relation $r \in R$ avoids binary products. Then the map $\mu_{\mathbf{X}} \colon \mathcal{F}(\mathbf{X}) \to D(C(\mathbf{X}, \underline{\mathbf{M}}))$ is an isomorphism in $\mathfrak{X}$. If, in addition, the endomorphisms in G, along with the constant endomorphisms corresponding to each nullary operation in G, generate the monoid $\mathrm{End}(\underline{\mathbf{M}})$, then $\mathcal{F}$ is a $\mathfrak{Z}$-free functor for $\mathfrak{X}$ with $\eta_{\mathbf{X}} \colon \mathbf{X} \to \mathcal{F}(\mathbf{X})$ defined by $\eta_{\mathbf{X}}(x) = (\mathrm{id}_{\underline{\mathbf{M}}}, x)$ for all $x \in X$.*

Proof It is almost trivial that $\mu_{\mathbf{X}}$ is an isomorphism with respect to $G \cup H$ and with respect to unary relations. We now show that, for all $n \geqslant 2$, the map $\mu_{\mathbf{X}}$ is an isomorphism with respect to an n-ary algebraic relation r provided r avoids binary products. Note that $\mu_{\mathbf{X}}$ preserves every algebraic relation $r \in R$. Assume that r avoids binary products.

Let $w_1, \ldots, w_n \in \mathcal{F}(\mathbf{X})$ and assume that $(\mu_{\mathbf{X}}(w_1), \ldots, \mu_{\mathbf{X}}(w_n)) \in r$ in $\mathcal{A}(\mathcal{C}(\mathbf{X}, \underline{\mathbf{M}}), \underline{\mathbf{M}})$. Thus

$$(*) \qquad (\mu_{\mathbf{X}}(w_1)(\alpha), \ldots, \mu_{\mathbf{X}}(w_n)(\alpha)) \in r \quad \text{for all} \quad \alpha \in C(\mathbf{X}, \underline{\mathbf{M}}).$$

We must show that $(w_1, \ldots, w_n) \in r^{\mathcal{F}(\mathbf{X})}$. Choosing α to be the constant map $k_a : X \to M$ onto $\{a\}$ as a varies over M shows that the n-tuple of endomorphisms involved in $w_1, \ldots, w_n$ is in the relation r. Assume that $w_i = (e, x)$ and $w_j = (f, y)$ for some i and j: we must show that $x = y$. Suppose that $x \neq y$. As r avoids binary products, $e(M) \times f(M) \nsubseteq \pi_{ij}(r)$, whence there exist $a, b \in M$ with $(e(a), f(b)) \notin \pi_{ij}(r)$. Let U be a clopen subset of $\mathbf{X}$ which contains x and not y and let $\alpha\colon \mathbf{X} \to \underline{\mathbf{M}}$ map U to a and $X \backslash U$ to b. By $(*)$, we have

$$(e(a), f(b)) = (e(\alpha(x)), f(\alpha(y))) = (\mu_{\mathbf{X}}(w_i)(\alpha), \mu_{\mathbf{X}}(w_j)(\alpha)) \in \pi_{ij}(r),$$

which is a contradiction. Hence $x = y$, as required.

Now assume that the endomorphisms in G, along with the constant endomorphisms corresponding to each nullary operation in G, generate the monoid $\mathrm{End}(\underline{\mathbf{M}})$. It follows that every constant c_i has a natural interpretation $c_i^{\mathbf{Y}}$ on each object $\mathbf{Y} \in \mathbb{IS}_c\mathbb{P}^+(\underset{\sim}{\mathbf{M}})$, namely, as the value of the interpretation of the constant endomorphism $\overline{c}_i$. Let $\mathbf{Y} \in \mathcal{X}$ and let $\varphi\colon \mathbf{X} \to \mathbf{Y}$ be a continuous map. Define $\overline{\varphi}\colon \mathcal{F}(\mathbf{X}) \to \mathbf{Y}$ by $\overline{\varphi}((e, x)) := e(\varphi(x))$ and $\overline{\varphi}(c_i) = c_i^{\mathbf{Y}}$. The assumptions on G guarantee that every $e \in \mathrm{End}(\underline{\mathbf{M}})$ has an interpretation on $\mathbf{Y}$ and hence $\overline{\varphi}$ is well defined. It is trivial to check that $\overline{\varphi}$ is a morphism, and is the unique morphism satisfying $\overline{\varphi} \circ \eta_{\mathbf{X}} = \varphi$. Thus, $\mathcal{F}\colon \mathcal{Z} \to \mathcal{X}$ is a $\mathcal{Z}$-free functor. ∎

We wish now to use this description of the $\mathcal{Z}$-free functor to describe the projectives in $\mathcal{X}$ and thence the injectives and absolute subretracts in $\mathcal{A}$. If we are willing to work with a duality rather than a strong duality, we can eliminate the partial operations from the type of $\underset{\sim}{\mathbf{M}}$. While this introduces additional objects into the dual category $\mathcal{X}$ it has the great advantage of simplifying the structure of $\mathcal{F}(\mathbf{X})$ and thereby making it easier to describe the projectives in $\mathcal{X}$. While the whole theory can be developed in the presence of one-element subalgebras, they force us to do some distracting bookkeeping. Since all the examples to which we wish to apply this theory are devoid of one-element subalgebras, we shall henceforth assume that $\underline{\mathbf{M}}$ has no one-element subalgebras and hence that $\mathcal{F}(\mathbf{X}) = \mathrm{End}(\underline{\mathbf{M}}) \times \mathbf{X}$.

The description of the projectives in $\mathcal{X}$ will come from the following more general result whose proof is provided, via a collection of hints, in Exercise 5.15. We shall apply this result in the case that $\mathcal{M} = \mathrm{End}(\underline{\mathbf{M}})$ in order to describe the retracts of $\mathrm{End}(\underline{\mathbf{M}}) \times \mathbf{X}$.

5.13 Proposition *Let $\mathcal{M}$ be a monoid and assume that R is a set of finitary relations on $\mathcal{M}$ which are compatible with the right translations, that is, $(h_1, \dots, h_n) \in r$ implies that $(h_1 g, \dots, h_n g) \in r$ for all $h_1, \dots, h_n, g \in \mathcal{M}$. For each Boolean space $\mathbf{X}$, endow $\mathcal{M} \times \mathbf{X}$ with the product topology, define the action of $g \in \mathcal{M}$ on $\mathcal{M} \times \mathbf{X}$ by $g((h, x)) := (gh, x)$ and define each relation $r \in R$ on $\mathcal{M} \times \mathbf{X}$ by*

$$((h_1, x_1), \dots, (h_n, x_n)) \in r \iff x_1 = \cdots = x_n \ \& \ (h_1, \dots, h_n) \in r.$$

Let S be a subset of the idempotents of $\mathcal{M}$, let $\mathcal{M}e$ be the left ideal of $\mathcal{M}$ generated by e and let $\mathbf{X}_e$, for $e \in S$, be pairwise disjoint non-empty Boolean spaces. Define

$$\mathbf{Y} := \dot{\bigcup}\{\, \mathcal{M}e \times \mathbf{X}_e \mid e \in S \,\}.$$

Then $\mathbf{Y}$ is a substructure of the structure $\mathcal{F}(\mathbf{X}) := \langle \mathcal{M} \times \mathbf{X}; \mathcal{M}, R, \mathcal{T} \rangle$, where $\mathbf{X} = \dot{\bigcup}\{\, \mathbf{X}_e \mid e \in S \,\}$, and $\tau \colon \mathcal{F}(\mathbf{X}) \to \mathbf{Y}$, defined by $\tau((g, x)) = (ge, x)$ for $x \in \mathbf{X}_e$, is a retraction. Conversely, if $\mathbf{Y}$ is a retract of $\mathcal{F}(\mathbf{X})$ for some Boolean space $\mathbf{X}$, then $\mathbf{Y}$ is of this form where each $\mathbf{X}_e$ is a retract of $\mathbf{X}$.

In order to apply the combined power of Theorem 5.5.12 and Propositions 5.5.8, 5.5.10 and 5.5.13, for the remainder of this section we shall restrict our attention to finite algebras $\underline{\mathbf{M}}$ and structures $\underset{\sim}{\mathbf{M}}$ which satisfy the following conditions:

(A) $\underline{\mathbf{M}}$ has a near-unanimity term, every non-trivial subalgebra of $\underline{\mathbf{M}}$ is subdirectly irreducible and $\underline{\mathbf{M}}$ has no one-element subalgebras,

(B) $\underset{\sim}{\mathbf{M}} = \langle M; \mathrm{End}(\underline{\mathbf{M}}), R, \mathcal{T} \rangle$, where each relation $r \in R$ avoids binary products.

Once again, the assumption that $\underline{\mathbf{M}}$ has no one-element subalgebras is one of convenience rather than necessity, and it is not necessary to assume that $\underset{\sim}{\mathbf{M}}$ yields a duality on $\mathcal{A}$.

5.14 Corollary *If $\underline{\mathbf{M}}$ and $\underset{\sim}{\mathbf{M}}$ satisfy* (A) *and* (B), *then $\mathbf{Y}$ is projective in $\mathfrak{X}$ if and only if it is isomorphic to $\dot{\bigcup}\{\, \mathrm{End}(\underline{\mathbf{M}})e \times \mathbf{X}_e \mid e \in S \,\}$, where each $e \in S$ is a retraction of $\underline{\mathbf{M}}$ onto a subalgebra and $\{\, \mathbf{X}_e \mid e \in S \,\}$ is a collection of pairwise disjoint extremally disconnected Boolean spaces.*

Proof Since $\underline{\mathbf{M}}$ has no one-element subalgebras, $\mathcal{F}(\mathbf{X}) = \mathrm{End}(\underline{\mathbf{M}}) \times X$ for each $\mathbf{X} \in \mathfrak{Z}$. By Theorem 5.5.12 we know that $\mathcal{F}$ is a $\mathfrak{Z}$-free functor. Consequently, by Proposition 5.5.8, a structure $\mathbf{Y} \in \mathfrak{X}$ is projective in $\mathfrak{X}$ if and only if it is a retract of $\mathrm{End}(\underline{\mathbf{M}}) \times \mathbf{X}$ for some extremally disconnected Boolean space $\mathbf{X}$. In order to apply Proposition 5.5.13 we observe that the idempotents of the monoid $\mathcal{M} = \mathrm{End}(\underline{\mathbf{M}})$ are exactly the retractions of $\underline{\mathbf{M}}$ onto a subalgebra.

Now if $\mathbf{Y} := \bigcup\{\, \mathrm{End}(\underline{\mathbf{M}})e \times \mathbf{X}_e \mid e \in S\,\}$ where S is a set of idempotents in $\mathrm{End}(\underline{\mathbf{M}})$ indexing a collection of pairwise disjoint extremally disconnected Boolean spaces, then the union $\mathbf{X} = \dot{\bigcup}\{\, \mathbf{X}_e \mid e \in S\,\}$ is also extremally disconnected. By Proposition 5.5.13 we conclude that $\mathbf{Y}$ is a retract of $\mathcal{F}(\mathbf{X})$ and is therefore projective in $\mathbf{X}$. Conversely, suppose $\mathbf{Y}$ is projective in $\mathcal{X}$ and is therefore a retract of $\mathcal{F}(\mathbf{X}) = \mathrm{End}(\underline{\mathbf{M}}) \times \mathbf{X}$ for some extremally disconnected Boolean space $\mathbf{X}$. By the 'Conversely' part of Proposition 5.5.13 there is a set $S \subseteq \mathrm{End}(\underline{\mathbf{M}})$ of idempotents and pairwise disjoint retracts $\mathbf{X}_e$ of $\mathbf{X}$ such that $\mathbf{Y} \cong \bigcup\{\, \mathrm{End}(\underline{\mathbf{M}})e \times \mathbf{X}_e \mid e \in S\,\}$. As each retract $\mathbf{X}_e$ of $\mathbf{X}$ is also extremally disconnected, $\mathbf{Y}$ has the required form. ■

Observe that, for each $e \in S$, the component $\mathrm{End}(\underline{\mathbf{M}})e \times \mathbf{X}_e$ is a substructure of $\mathbf{Y}$ and the relations in R never involve elements from separate components. It follows easily that $\mathbf{Y}$ is the coproduct in $\mathcal{X}$ of the components $\mathrm{End}(\underline{\mathbf{M}})e \times \mathbf{X}_e$ for $e \in S$. In Chapter 6, we return to the consideration of those $\mathcal{X}$ in which coproduct is given by disjoint union and show that this property is tightly linked to assumptions (A) and (B); see Theorem 6.3.4 and, more generally, Theorem 6.3.3. If we add the assumption that $\underset{\sim}{\mathbf{M}}$ does yield a duality on $\mathcal{A}$, then we can convert Corollary 5.5.14 into a description of the injectives, weak injectives and absolute subretracts in $\mathcal{A}$. This result is new but is based on a careful dissection of the proof for the n-element Heyting chain in [Dav76] (see Exercise 8.4).

5.15 Injectivity and Boolean Powers Theorem *Assume that* $\underline{\mathbf{M}}$ *and* $\underset{\sim}{\mathbf{M}}$ *satisfy* (A) *and* (B) *and assume that* $\underset{\sim}{\mathbf{M}}$ *yields a duality on* $\mathcal{A}$.

(i) *An algebra* $\mathbf{A} \in \mathcal{A}$ *is (weakly) injective in* $\mathcal{A}$ *if and only if it is isomorphic to* $\mathbf{M}_1[\mathbf{B}_1] \times \cdots \times \mathbf{M}_n[\mathbf{B}_n]$, *where the algebras* $\mathbf{M}_i$ *are pairwise non-isomorphic subalgebras of* $\underline{\mathbf{M}}$ *which are (weakly) injective in* $\mathcal{A}$ *and each Boolean algebra* $\mathbf{B}_i$ *is complete and non-trivial.*

(ii) *If* $\mathbf{A} \in \mathcal{A}$ *is an absolute subretract in* $\mathcal{A}$, *then it is isomorphic to* $\mathbf{M}_1[\mathbf{B}_1] \times \cdots \times \mathbf{M}_n[\mathbf{B}_n]$, *where the algebras* $\mathbf{M}_i$ *are pairwise non-isomorphic subalgebras of* $\underline{\mathbf{M}}$ *which are absolute subretracts in* $\mathcal{A}$ *and each Boolean algebra* $\mathbf{B}_i$ *is complete and non-trivial.*

Proof Let e be an idempotent in $\mathrm{End}(\underline{\mathbf{M}})$ and let $\mathbf{M}_e = e(\underline{\mathbf{M}})$. We begin by proving that, for any Boolean space $\mathbf{X}$, we have

$$E(\mathrm{End}(\underline{\mathbf{M}})e \times \mathbf{X}) = \mathcal{X}(\mathrm{End}(\underline{\mathbf{M}})e \times \mathbf{X}, \underset{\sim}{\mathbf{M}}) \cong C(\mathbf{X}, \mathbf{M}_e).$$

For $\alpha \in \mathcal{X}(\mathrm{End}(\underline{\mathbf{M}})e \times \mathbf{X}, \underset{\sim}{\mathbf{M}})$, define $\hat{\alpha} \in C(\mathbf{X}, \mathbf{M}_e)$ by $\hat{\alpha}(x) := \alpha((e,x))$, and, for $\beta \in C(\mathbf{X}, \mathbf{M}_e)$, define $\hat{\beta} \in \mathcal{X}(\mathrm{End}(\underline{\mathbf{M}})e \times \mathbf{X}, \underset{\sim}{\mathbf{M}})$ by $\hat{\beta}((f,x)) = f\beta(x)$. The reader should carry out the straightforward task of verifying that

(a) $\hat{\alpha}$ is well defined and continuous,

(b) $\hat{\beta}$ is an $\mathfrak{X}$-morphism,

(c) $\hat{\hat{\alpha}} = \alpha$ and $\hat{\hat{\beta}} = \beta$, and

(d) $\alpha \mapsto \hat{\alpha}$ is a homomorphism from $\mathfrak{X}(\mathrm{End}(\underline{\mathbf{M}})e \times \mathbf{X}, \underset{\sim}{\mathbf{M}})$ to $C(\mathbf{X}, \mathbf{M}_e)$.

If $\mathbf{A}$ is an injective, a weak injective or an absolute subretract in $\mathcal{A}$, then, by Proposition 5.5.10, $D(\mathbf{A})$ is a retract of $\mathbf{F}(\mathbf{X})$ for some extremally disconnected Boolean space $\mathbf{X}$. By Proposition 5.5.13, there are a subset S of idempotents of $\mathrm{End}(\underline{\mathbf{M}})$ and pairwise disjoint, non-empty subspaces $\mathbf{X}_e$ of $\mathbf{X}$, each of which is a retract, such that $D(\mathbf{A})$ is isomorphic to $\mathbf{Y} := \bigcup\{\, \mathrm{End}(\underline{\mathbf{M}})e \times \mathbf{X}_e \mid e \in S \,\}$. Let $S = \{e_1, \ldots, e_n\}$ and let $\mathbf{X}_i := \mathbf{X}_{e_i}$ and $\mathbf{M}_i := e_i(\underline{\mathbf{M}})$. As $\mathbf{Y}$ is the coproduct of its components and E maps coproducts to products, it follows that

$$\begin{aligned}\mathbf{A} &\cong ED(\mathbf{A}) \cong E(\mathbf{Y})\\ &\cong E(\mathrm{End}(\underline{\mathbf{M}})e_1 \times \mathbf{X}_1) \times \cdots \times E(\mathrm{End}(\underline{\mathbf{M}})e_n \times \mathbf{X}_n)\\ &\cong C(\mathbf{X}_1, \mathbf{M}_1) \times \cdots \times C(\mathbf{X}_n, \mathbf{M}_n) \cong \mathbf{M}_1[\mathbf{B}_1] \times \cdots \times \mathbf{M}_n[\mathbf{B}_n],\end{aligned}$$

where $\mathbf{B}_i := \mathbf{X}_i^*$. We may assume that the algebras $\mathbf{M}_i$ are pairwise non-isomorphic since it is clear that $C(\mathbf{X}, \mathbf{M}_i) \times C(\mathbf{Y}, \mathbf{M}_i)$ is isomorphic to $C(\mathbf{X} \,\dot{\cup}\, \mathbf{Y}, \mathbf{M}_i)$. Since $\mathbf{X}_i$ is a retract of an extremally disconnected space, it is also extremally disconnected. As $\mathbf{X}_i$ is non-empty, it follows that $\mathbf{B}_i$ is complete and non-trivial. As $\underline{\mathbf{M}}$ has a near-unanimity term, $\mathcal{A}$ is congruence distributive and so has factorisable congruences (by Exercise A.10). Thus, by Proposition 5.5.5(v) and Lemma 5.5.7, if $\mathbf{A}$ is an injective, weak injective or absolute subretract in $\mathcal{A}$, then $\mathbf{M}_i[\mathbf{B}_i]$, and therefore $\mathbf{M}_i$, is also. This proves (ii) and the forward direction of (i).

We now prove the converse direction of (i). If $\mathbf{M}_i$ is (weakly) injective in $\mathcal{A}$ and $\mathbf{B}_i$ is complete for all i, then, by Proposition 5.5.5(iv) and Lemma 5.5.7, $\mathbf{M}_1[\mathbf{B}_1] \times \cdots \times \mathbf{M}_n[\mathbf{B}_n]$ is also (weakly) injective in $\mathcal{A}$, as required. ∎

Given an algebra $\underline{\mathbf{M}}$ which satisfies (A) and a structure $\underset{\sim}{\mathbf{M}}$ which satisfies (B) and yields a duality on $\mathcal{A}$, the only question which remains is which subalgebras of $\underline{\mathbf{M}}$ are (weakly) injective in $\mathcal{A}$. Exercise 5.16 shows that a subalgebra $\mathbf{N}$ of $\underline{\mathbf{M}}$ is (weakly) injective in $\mathcal{A}$ if and only if it is (weakly) injective in $\mathbb{S}(\underline{\mathbf{M}})$, a property which, for small examples, is easily checked by hand.

A far-reaching theorem, which provides very general sufficient condition for the (weak) injectives in a variety to have the form described in the Injectivity and Boolean Powers Theorem 5.5.15, is proved in Davey and Werner [DW79]. It should be pointed out that several special cases of Theorem 5.5.15, proved via duality theory, provided crucial evidence for the authors of [DW79] that a more general theorem existed. Moreover, while the proof of the main result of [DW79] is proved without using a topological duality, it makes substantial use of a topological representation of certain algebras in $\mathcal{A}$ via sheaves.

5.6 Category equivalences

Categories $\mathcal{A}$ and $\mathcal{B}$ are said to be **equivalent**, written $\mathcal{A} \cong \mathcal{B}$, if there are covariant functors $G\colon \mathcal{A} \to \mathcal{B}$ and $H\colon \mathcal{B} \to \mathcal{A}$ such that the compositions $H \circ G$ and $G \circ H$ are naturally isomorphic to the identities on $\mathcal{A}$ and $\mathcal{B}$, respectively. (See [Mac:CW].) According to Theorem 4.1.1, every primal algebra generates a variety which is dually equivalent to the category of Boolean spaces and is therefore equivalent to the variety of Boolean algebras. Using other strong dualities it is possible to generate other pairs of quasi-varieties which are equivalent as categories by virtue of being dually equivalent to the same topological quasi-variety. Before investigating the role of strong duality in category equivalence, we list some important features that are preserved under category equivalence when the categories in question are both quasi-varieties. Recall that, for an algebra $\mathbf{A}$ in a quasi-variety $\mathcal{A}$, we denote by $\mathrm{Con}_{\mathcal{A}}(\mathbf{A})$ the lattice of $\mathcal{A}$-congruences on $\mathbf{A}$ (Theorem 1.3.5). By $\mathrm{Sub}(\mathbf{A})$ we mean the lattice formed by adding a (possibly additional) bottom element 0 to the otherwise lattice-ordered set $\mathbb{S}(\mathbf{A})$.

6.1 Equivalence Preservation Theorem *Assume $G\colon \mathcal{A} \to \mathcal{B}$ and $H\colon \mathcal{B} \to \mathcal{A}$ is an equivalence between quasi-varieties $\mathcal{A}$ and $\mathcal{B}$, and let $\mathbf{A} \in \mathcal{A}$ and $\mathbf{B} = G(\mathbf{A}) \in \mathcal{B}$.*

(i) *H preserves embeddings and surjections.*

(ii) $\mathrm{Con}_{\mathcal{A}}(\mathbf{A}) \cong \mathrm{Con}_{\mathcal{B}}(\mathbf{B})$.

(iii) $\mathrm{Sub}(\mathbf{A}) \cong \mathrm{Sub}(\mathbf{B})$.

(iv) *$\mathbf{A}$ is finitely generated if and only if $\mathbf{B}$ is finitely generated.*

(v) *$\mathbf{A}$ is finite if and only if $\mathbf{B}$ is finite.*

(vi) *If $\mathbf{A}$ and $\mathbf{B}$ are infinite, then they have the same cardinality.*

(vii) *$\mathcal{A}$ is locally finite if and only if $\mathcal{B}$ is locally finite.*

Proof (i) According to Exercise 3.11 and Exercise 3.14, the embeddings and surjections in $\mathcal{A}$ are precisely the monics and regular epics (coequalisers) in $\mathcal{A}$, and are therefore preserved under a category equivalence. Part (ii) follows since $\mathcal{A}$- and $\mathcal{B}$-congruences correspond to surjections, and part (iii) follows since subalgebras correspond to embeddings. An algebra $\mathbf{A}$ is finitely generated if and only if it is a compact element of $\mathrm{Sub}(\mathbf{A})$, so (iv) follows from (iii). To prove (v) and (vi), suppose that $\mathbf{A}$ is finite. Let $\mathbf{F}_1$ be a free algebra on one generator in $\mathcal{B}$ (Theorem A.2).

Since H is left adjoint to G, by Exercise 5.18 we have

$$|\mathbf{B}| = |\mathcal{B}(\mathbf{F}_1, G(\mathbf{A}))| = |\mathcal{A}(H(\mathbf{F}_1), \mathbf{A})|.$$

By (iv) $H(\mathbf{F}_1)$ is finitely generated. Thus $\mathbf{B}$ is finite if $\mathbf{A}$ is finite. If $\mathbf{A}$ and $\mathbf{B}$ are infinite, then $|\mathbf{B}| \leqslant |\mathbf{A}|$. Similarly, $\mathbf{B}$ is finite if $\mathbf{A}$ is finite and, otherwise, $|\mathbf{A}| \leqslant |\mathbf{B}|$. Finally, (vii) follows from (iv) and (v). ∎

These observations come from Davey and Werner [DW83], Theorem 3.3, where it is also shown that, if $\mathcal{A}$ and $\mathcal{B}$ happen to be varieties, then Mal'cev conditions not involving composition of terms are also preserved by category equivalence.

What categories can we find, for example, that are equivalent to the variety $\mathcal{D}_{01}$ of bounded distributive lattices? Wille discovered the property of these algebras needed to answer our question. We say that an algebra $\underline{\mathbf{N}}$ has a **definable** lattice structure if it has a **term-reduct** in the class of lattices, that is, there are binary terms $\vee$ and $\wedge$ of $\underline{\mathbf{N}}$ such that $\langle N; \vee, \wedge\rangle$ is a lattice. A non-trivial finite algebra $\underline{\mathbf{N}}$ is **order primal** if there is an order relation $\leqslant$ on N such that the clone of $\underline{\mathbf{N}}$ consists exactly of the set of $\leqslant$-preserving operations on $\underline{\mathbf{N}}$. When the order $\leqslant$ is a lattice order, we say that $\underline{\mathbf{N}}$ is **lattice order primal**. Note that if $\underline{\mathbf{N}}$ is lattice order primal then, as they are order-preserving, the lattice operations $\vee$ and $\wedge$ are also terms of $\underline{\mathbf{N}}$, whence the lattice structure is definable within the algebra $\underline{\mathbf{N}}$.

6.2 Lemma (Wille [Wil77]) *A finite algebra $\underline{\mathbf{N}}$ is lattice order primal if and only if it has a definable lattice structure with corresponding order $\leqslant$ such that the subalgebras of $\underline{\mathbf{N}}^2$ are precisely Δ (the diagonal), $\leqslant$, $\geqslant$ and $\underline{\mathbf{N}}^2$.*

Proof Assume that $\underline{\mathbf{N}}$ is lattice order primal with respect to $\leqslant$. As noted above, the lattice operations are definable. Let $\mathbf{A}$ be a subalgebra of $\underline{\mathbf{N}}^2$, and let 0 and 1 denote the bottom and top elements of $\underline{\mathbf{N}}$. We define two unary operations on N,

$$s(c) = \bigwedge\{a \mid (a, c) \in A\} \quad \text{and} \quad r(c) = \bigwedge\{b \mid (c, b) \in A\},$$

and claim that each of these either is the identity function or has constant value 0. Suppose that s is not the identity. Since $(c, c) \in A$ for each $c \in N$, we conclude that $s(c) \leqslant c$. In particular, $s(0) = 0$. Choose $d \in N$ such that $s(d) < d$ and define $f\colon N \to N$ by $f(c) = 1$ if $c \geqslant d$ and $f(c) = 0$ otherwise. Since f preserves $\leqslant$ it is a term function of $\underline{\mathbf{N}}$. For each $c \in N$ we have $(s(c), c) \in A$, since A is closed under $\wedge$, and therefore $(f(s(c)), f(c)) \in A$. Consequently $s(f(c)) \leqslant f(s(c))$. Taking $c = d$, we have $s(1) = s(f(d)) \leqslant f(s(d)) = 0$. Thus $s(1) = 0$, and $(0, 1) \in A$. Then, for any $c \in N$, we have $(0, c) = (0, 1) \wedge (c, c) \in A$ and therefore $s(c) = 0$. Similarly, r is either the identity or has constant value 0. It is now easy to check that the four possible values for r and s determine the four desired values of $\mathbf{A}$.

Conversely, assume that $\underline{\mathbf{N}}$ has a definable lattice structure and that Δ, $\leqslant$ and $\geqslant$ are the only proper subalgebras of $\underline{\mathbf{N}}^2$. By the NU Lemma 2.3.3 and Corollary 2.3.2, a map $t\colon N^m \to N$ is a term function of $\underline{\mathbf{N}}$ if and only if it preserves every binary algebraic relation on $\underline{\mathbf{N}}$. It follows from the assumptions on the subalgebras of $\underline{\mathbf{N}}^2$ that $\underline{\mathbf{N}}$ is order primal with respect to the order $\leqslant$. ■

6.3 Corollary *Every lattice order primal algebra $\underline{\mathbf{N}}$ generates a variety $\mathcal{A}$ that is equivalent as a category to the variety $\mathcal{D}_{01}$ of bounded distributive lattices.*

Proof Let $\vee$ and $\wedge$ be the join and meet operations on N induced by $\leqslant$. Since $\vee$, $\wedge$ and the constants all preserve $\leqslant$, they are all term functions of $\underline{\mathbf{N}}$. Consequently the median operation is also a term function of $\underline{\mathbf{N}}$, and $\underline{\mathbf{N}}$ has no proper subalgebras nor automorphisms. From Mitschke's Theorem A.14 and Theorem 1.3.6(iii) we see that $\mathcal{A}$ is also the quasi-variety generated by $\underline{\mathbf{N}}$. By the NU Strong Duality Theorem, the structure $\langle N; \mathcal{B}_2, \mathcal{T}\rangle$ yields a strong duality on $\mathcal{A}$. By Lemma 5.6.2 and the $\underset{\sim}{\mathbf{M}}$-Shift Strong Duality Lemma 3.2.3(iii)(a), we see that the reduced structure $\underset{\sim}{\mathbf{N}} = \langle N; \leq, \mathcal{T}\rangle$ also yields a strong duality on $\mathcal{A}$. A direct application of the Separation Theorem 1.4.4 shows that the dual category $\mathcal{X} = \mathbb{IS}_c\mathbb{P}^+ \underset{\sim}{\mathbf{N}}$ is exactly the category $\mathbb{IS}_c\mathbb{P}^+ \underset{\sim}{\mathbf{2}}$ of all Priestley spaces, where $\underset{\sim}{\mathbf{2}} = \langle \{0, 1\}; \leq, \mathcal{T}\rangle$. Since $\mathcal{A}$ and $\mathcal{D}_{01}$ are each dually equivalent to $\mathcal{X}$ (see Exercise 4.5), they are equivalent to each other. ■

We will see that Corollary 5.6.3 describes the *only* quasi-varieties that are equivalent to $\mathcal{D}_{01}$. More generally, we will see that, if there is a strong duality between $\mathcal{A}$ and $\mathcal{X}$, then the only quasi-varieties equivalent to $\mathcal{A}$ are those which are also dually equivalent to $\mathcal{X}$ by a strong duality.

Our theorem comes from a concrete version of a classical theorem of Freyd which we will now quote, referring the reader to Freyd [Fre66] or the appendix of [DW83] for a proof. If $\mathcal{X}$ is a category of structured sets, then we say that $\mathcal{X}$ has **concrete products** if finite products in $\mathcal{X}$ exist and coincide with Cartesian set products. Suppose that $\mathcal{B}$ is a subcategory of a variety $\mathcal{K}$ of algebras of type F containing the finitely generated free algebras $\mathbf{F}_n$ for $n \geqslant 1$, and that $S\colon \mathcal{B} \to \mathcal{X}$ and $T\colon \mathcal{X} \to \mathcal{B}$ form a dual adjunction $\langle S, T, e, \varepsilon\rangle$. Under these conditions the object $\underset{\sim}{\mathbf{N}} = S(\mathbf{F}_1)$ carries a natural algebraic structure of type F. To see this, let x be the free generator of $\mathbf{F}_1$ and let $x_1, x_2, \ldots, x_n$ be the free generators of $\mathbf{F}_n$. Notice that $\mathbf{F}_n$ is the coproduct in $\mathcal{B}$ of n copies of $\mathbf{F}_1$. Exploiting this observation, we define, for each n-ary $f \in F$, the n-ary **co-operation** $\hat{f}\colon \mathbf{F}_1 \to \mathbf{F}_n$ to be the unique homomorphism taking x to $f(x_1, x_2, \ldots, x_n)$. The set F_1 with these associated co-operations is called a **coalgebra** of type F. Since S takes coproducts to products (Lemma 1.1.4), under S each co-operation $\hat{f}\colon \mathbf{F}_1 \to \mathbf{F}_n$ becomes an operation $S(\hat{f})\colon S(\mathbf{F}_1)^n \to S(\mathbf{F}_1)$ on $S(\mathbf{F}_1)$. This makes $S(\mathbf{F}_1)$ into an *algebra* of type F. The following is not the most general form of Freyd's theorem but is sufficient for our purposes.

6.4 Freyd's Representability Theorem (Freyd [Fre66]) *Let $\mathcal{B}$ be a full subcategory of a variety $\mathcal{K}$ of algebras of type F containing the finitely generated free algebras. Suppose that $S\colon \mathcal{B} \to \mathcal{X}$ and $T\colon \mathcal{X} \to \mathcal{B}$ form a dual adjunction $\langle S, T, e, \varepsilon\rangle$ between $\mathcal{B}$ and a category $\mathcal{X}$ of structured sets which has concrete products. Let $\underset{\sim}{\mathbf{N}} = S(\mathbf{F}_1)$ have underlying set N, and let $\underline{\mathbf{N}}$ be the algebra over N induced by the coalgebra structure on F_1. Then*

(i) $\underline{\mathbf{N}} \in \mathcal{K}$,

(ii) $\mathcal{X}(\mathbf{X}, \underset{\sim}{\mathbf{N}})$ *is a subalgebra of* $\underline{\mathbf{N}}^X$ *for each* $\mathbf{X} \in \mathcal{X}$, *and*

(iii) *there is a natural isomorphism* $\eta\colon T \to \mathcal{X}(-, \underset{\sim}{\mathbf{N}})$.

Applying this theorem to our setting establishes the connection between strong duality and category equivalence.

6.5 General Equivalence Theorem (Davey and Werner [DW83]) *Assume that $\underset{\sim}{\mathbf{M}}$ yields a strong duality between $\mathcal{A} = \mathbb{ISP}\underline{\mathbf{M}}$ and $\mathcal{X} = \mathbb{IS}_c\mathbb{P}^+ \underset{\sim}{\mathbf{M}}$, and let $\mathcal{B}$ be a quasi-variety. Then the following are equivalent:*

(i) $\mathcal{A} \cong \mathcal{B}$;

(ii) *there are a finite algebra $\underline{\mathbf{N}} \in \mathcal{B}$ and a structure $\underset{\sim}{\mathbf{N}} \in \mathcal{X}$ with the same underlying set N such that*

  (a) $\underset{\sim}{\mathbf{N}}$ *is algebraic over* $\underline{\mathbf{N}}$,

  (b) $\mathcal{B} = \mathbb{ISP}\underline{\mathbf{N}}$ *and* $\mathcal{X} = \mathbb{IS}_c\mathbb{P}^+ \underset{\sim}{\mathbf{N}}$,

  (c) $\underset{\sim}{\mathbf{N}}$ *yields a strong duality on* $\mathcal{B}$;

(iii) *there are a finite algebra $\underline{\mathbf{N}} \in \mathcal{B}$ and a structure $\underset{\sim}{\mathbf{N}} \in \mathcal{X}$ with the same underlying set N such that*

  (a) $\underset{\sim}{\mathbf{N}}$ *is algebraic over* $\underline{\mathbf{N}}$,

  (b)′ $\mathcal{B} = \mathbb{ISP}\underline{\mathbf{N}}$,

  (c)′ *the structures $\underset{\sim}{\mathbf{M}}$ and $\underset{\sim}{\mathbf{N}}$ are each a retract of a finite power of the other and ($\widetilde{\text{C}}$LO) holds with respect to $\underline{\mathbf{N}}$ and $\underset{\sim}{\mathbf{N}}$.*

Moreover, under an equivalence between $\mathcal{A}$ and $\mathcal{B}$, the algebras $\underline{\mathbf{M}}$ and $\underline{\mathbf{N}}$ each map to an algebra isomorphic to the other.

Proof Assume that $\underset{\sim}{\mathbf{M}}$ yields a strong duality between $\mathcal{A}$ and $\mathcal{X}$ via the usual hom-functors $D\colon \mathcal{A} \to \mathcal{X}$ and $E\colon \mathcal{X} \to \mathcal{A}$. Statement (ii) implies (i) since both $\mathcal{A}$ and $\mathcal{B}$ are, in this case, dually equivalent to $\mathcal{X}$. Now assume (i). We shall prove (iii). Let $G\colon \mathcal{B} \to \mathcal{A}$ and $H\colon \mathcal{A} \to \mathcal{B}$ be mutually inverse equivalences and let $S = DG\colon \mathcal{B} \to \mathcal{X}$ and $T = HE\colon \mathcal{X} \to \mathcal{B}$ be the induced dual equivalence. Choose $\underline{\mathbf{N}}$ and $\underset{\sim}{\mathbf{N}}$ as in Freyd's Theorem. By Theorem 5.6.1, N is finite. Theorem A.2 allows us to apply Freyd's Representability Theorem to conclude that $\underset{\sim}{\mathbf{N}}$ can be used to represent T as a hom-functor. Part (a) follows immediately from the definition of the operations on $\underline{\mathbf{N}}$.

To prove that $\mathcal{B} = \mathbb{ISP}\underline{\mathbf{N}}$, recall (Exercise 3.5) that $D(\underline{\mathbf{M}})$ is $\mathcal{X}$-freely generated by the identity $\mathrm{id}_{\underline{\mathbf{M}}}$ and consequently the projection $\pi_{\mathrm{id}_{\underline{\mathbf{M}}}}\colon \mathcal{X}(D(\underline{\mathbf{M}}), \underset{\sim}{\mathbf{N}}) \to \underline{\mathbf{N}}$ is an isomorphism. Thus

$$\underline{\mathbf{N}} \cong \mathcal{X}(D(\underline{\mathbf{M}}), \underset{\sim}{\mathbf{N}}) \cong T(D(\underline{\mathbf{M}})) = H(ED(\underline{\mathbf{M}})) \cong H(\underline{\mathbf{M}}) \in \mathcal{B},$$

and therefore $\mathbb{ISP}\underline{\mathbf{N}} \subseteq \mathcal{B}$. Conversely, if $\mathbf{B} \in \mathcal{B}$, then

$$\mathbf{B} \cong TS(\mathbf{B}) \cong \mathcal{X}(S(\mathbf{B}), \underset{\sim}{\mathbf{N}}) \leqslant \underline{\mathbf{N}}^{S(\mathbf{B})}$$

and we conclude that $\mathcal{B} \subseteq \mathbb{ISP}\underline{\mathbf{N}}$, and therefore $\mathcal{B} = \mathbb{ISP}\underline{\mathbf{N}}$, which proves (b)′.

To see that $\underset{\sim}{\mathbf{M}}$ is a retract of a finite power of $\underset{\sim}{\mathbf{N}}$, let $n = |H(\mathbf{F}_1^{\mathcal{A}})|$, where $\mathbf{F}_1^{\mathcal{A}}$ is an $\mathcal{A}$-free algebra on 1 generator, and let $\mathbf{F}_n^{\mathcal{B}}$ be the $\mathcal{B}$-free algebra on n generators. Since $\mathbf{F}_1^{\mathcal{A}}$ is projective in $\mathcal{A}$ and G and H preserves surjections (Theorem 5.6.1), the algebra $H(\mathbf{F}_1^{\mathcal{A}})$ is projective in $\mathcal{B}$. It follows that $H(\mathbf{F}_1^{\mathcal{A}})$ is a retract of $\mathbf{F}_n^{\mathcal{B}}$. Applying DG to the associated diagram, we conclude that $\underset{\sim}{\mathbf{M}} \cong D(\mathbf{F}_1^{\mathcal{A}}) \cong DGH(\mathbf{F}_1^{\mathcal{A}})$ is a retract of $DG(\mathbf{F}_n^{\mathcal{B}}) = S(\mathbf{F}_n^{\mathcal{B}}) \cong S(\mathbf{F}_1^{\mathcal{B}})^n = \underset{\sim}{\mathbf{N}}^n$. If we apply the same argument to $\mathbf{F}_1^{\mathcal{B}}$, we see that $\underset{\sim}{\mathbf{N}} = S(\mathbf{F}_1^{\mathcal{B}}) = DG(\mathbf{F}_1^{\mathcal{B}})$ is a retract of $D(\mathbf{F}_m^{\mathcal{A}}) \cong D(\mathbf{F}_1^{\mathcal{A}})^m \cong \underset{\sim}{\mathbf{M}}^m$ for some $m \in \mathbb{N}$.

We next show that (CLO) holds with respect to $\underline{\mathbf{N}}$ and $\underset{\sim}{\mathbf{N}}$. Let $n \geqslant 1$. We have $S(\mathbf{F}_n^{\mathcal{B}}) \cong S(\mathbf{F}_1)^n$ and therefore

$$\mathbf{F}_n^{\mathcal{B}} \cong TS(\mathbf{F}_n^{\mathcal{B}}) \cong T(\underset{\sim}{\mathbf{N}}^n) \cong \mathcal{X}(\underset{\sim}{\mathbf{N}}^n, \underset{\sim}{\mathbf{N}}) \leqslant \underline{\mathbf{N}}^{N^n}.$$

As $\underset{\sim}{\mathbf{N}}$ is algebraic over $\underline{\mathbf{N}}$, we have $\mathbf{F}_n^{\mathcal{B}} \subseteq \mathcal{X}(\underset{\sim}{\mathbf{N}}^n, \underset{\sim}{\mathbf{N}})$. Since these algebras are finite, we conclude that $\mathbf{F}_n^{\mathcal{B}} = \mathcal{X}(\underset{\sim}{\mathbf{N}}^n, \underset{\sim}{\mathbf{N}})$. This shows that (CLO) holds and concludes the proof of (c)′. Hence (i) implies (iii).

Finally, we show that (iii) implies (ii). Assume that (iii) holds. Since each of $\underset{\sim}{\mathbf{M}}$ and $\underset{\sim}{\mathbf{N}}$ can be embedded into a power of the other, it is clear that $\mathcal{X} := \mathbb{IS}_c\mathbb{P}^+ \underset{\sim}{\mathbf{M}} = \mathbb{IS}_c\mathbb{P}^+ \underset{\sim}{\mathbf{N}}$. As $\underset{\sim}{\mathbf{M}}$ is injective in $\mathcal{X}$ and $\underset{\sim}{\mathbf{N}}$ is a retract of a power of $\underset{\sim}{\mathbf{M}}$, the structure $\underset{\sim}{\mathbf{N}}$ is injective in $\mathcal{X}$ (see Proposition 5.5.5(iv)). Since (CLO) holds with respect to $\underline{\mathbf{N}}$ and $\underset{\sim}{\mathbf{N}}$ by assumption, $\underset{\sim}{\mathbf{N}}$ yields a duality on $\mathcal{B}$ by the First Duality Theorem 2.2.2.

We prove that the duality is strong by showing that a closed substructure $\mathbf{X} \leqslant \underset{\sim}{\mathbf{N}}^K$ is term-closed with respect to $\underset{\sim}{\mathbf{N}}$. Let $y \in N^K \backslash X$. By (c)$'$ there are a set L and an embedding $\varphi \colon \underset{\sim}{\mathbf{N}}^K \to \underset{\sim}{\mathbf{M}}^L$. Since φ is an embedding, $\varphi(\mathbf{X})$ is a closed substructure of $\underset{\sim}{\mathbf{M}}^L$ and is therefore term-closed with respect to $\underset{\sim}{\mathbf{M}}$. Thus there are $\alpha, \beta \colon \underset{\sim}{\mathbf{M}}^L \to \underset{\sim}{\mathbf{M}}$ which agree on $\varphi(X)$ but not at $\varphi(y)$. By (c)$'$ there is an embedding of $\underset{\sim}{\mathbf{M}}$ into a power of $\underset{\sim}{\mathbf{N}}$. Following this embedding with an appropriate projection into $\underset{\sim}{\mathbf{N}}$ gives two morphisms from $\underset{\sim}{\mathbf{N}}^K$ into $\underset{\sim}{\mathbf{N}}$ which agree on $\mathbf{X}$ but not at y. Thus $\mathbf{X}$ is term-closed with respect to $\underset{\sim}{\mathbf{N}}$. This completes the proof of (ii).

Finally, the fact that $\underline{\mathbf{N}} \cong H(\underline{\mathbf{M}})$ was verified in the proof that (ii)(b)$'$ follows from (i), and a similar argument shows that $\underline{\mathbf{M}} \cong G(\underline{\mathbf{N}})$. ■

6.6 Boolean Algebra Equivalence Theorem (Hu [Hu71]) *A quasi-variety is equivalent as a category to the variety of Boolean algebras if and only if it is the variety generated by a primal algebra.*

Proof This follows at once from Theorem 5.6.5. The duality on which the equivalence is built is Stone duality for Boolean algebras, so we choose $\underline{\mathbf{M}} = \langle\{0,1\}; \vee, \wedge, ', 0, 1\rangle$, $\underset{\sim}{\mathbf{M}} = \langle\{0,1\}; \mathcal{T}\rangle$ and $\underset{\sim}{\mathbf{N}} = \langle N; \mathcal{T}\rangle$. Since $G = H = R = \varnothing$, (CLO) tells us that $\underline{\mathbf{N}}$ must be primal. Note that one direction is also given by Theorem 4.1.1. ■

6.7 BDL-Equivalence Theorem (Quackenbush [Qua79] and Davey and Rival [DR82]) *A quasi-variety is equivalent as a category to the variety of bounded distributive lattices if and only if it is the variety generated by a lattice order primal algebra.*

Proof Although one direction is given by Corollary 5.6.3, both directions now follow from Theorem 5.6.5 and Exercise 5.11. The base duality is Priestley duality for bounded distributive lattices (Exercise 4.5), so choose $\underline{\mathbf{M}} = \underline{\mathbf{D}}_{01} = \langle\{0,1\}; \vee, \wedge, 0, 1\rangle$, $\underset{\sim}{\mathbf{M}} = \underset{\sim}{\mathbf{D}}_{01} = \langle\{0,1\}; \leqslant, \mathcal{T}\rangle$ and $\underset{\sim}{\mathbf{N}} = \langle N; \leqslant, \mathcal{T}\rangle$. Condition (CLO) holds if and only if $\underline{\mathbf{N}}$ is order primal. Use Exercise 5.11 to establish the fact that $\underset{\sim}{\mathbf{N}}$ is an order retract of a finite power of $\underset{\sim}{\mathbf{D}}_{01}$ if and only if $\underset{\sim}{\mathbf{N}}$ is a lattice. ■

We refer the reader to Davey and Werner [DW83] and to the exercises below for a number of additional applications. Coming from a different perspective, McKenzie [McK96] has given a general construction for generating all (quasi-)varieties equivalent to a given (quasi-)variety.

Exercises 5

Exercises from the text. Fill in the proofs of Theorem 5.3.1(ii) and Lemma 5.4.3. Complete the proof of Theorem 5.3.3 by showing that $\mathbf{A}$ is homogeneous. By sketching the appropriate spaces, complete the details of the proof of Theorem 5.4.2. Complete the proof of Lemma 5.5.1 by proving that $E(\gamma) \circ e_{\mathbf{B}} = v$. Prove each of the claims made in Proposition 5.5.5. Prove (a)–(c) as stated in the proof of Lemma 5.5.7 and then complete the proof of the lemma. Complete the missing details in the proof of the Injectivity and Boolean Powers Theorem 5.5.15. Complete the proof of the BDL-Equivalence Theorem 5.6.7.

5.1 Give an alternative proof of the Compactness Theorem 5.1.1 using the Inverse Limit Theorem 1.3.3. For each finite $\Delta \subseteq \Sigma$, let

$$A_\Delta = \{x \in X \mid x(s) = 0 \text{ but } x(t) = t \text{ for each } t \in \Delta\}.$$

If s is not a consequence of any finite subset of Σ, then A_Γ is non-empty when Γ is finite. For finite $\Gamma \subseteq \Delta \subseteq \Sigma$, let $\varphi_{\Delta\Gamma} \colon A_\Delta \to A_\Gamma$ be the inclusion map. Check that this is an inverse system and use it to show that A_Σ would have to be non-empty.

5.2 Using the method of Section 2, show that if $\mathbf{B}_1$ and $\mathbf{B}_2$ are non-trivial Boolean algebras and $\mathrm{End}(B_1) \cong \mathrm{End}(B_2)$, then $\mathbf{B}_1 \cong \mathbf{B}_2$.

5.3 Let $\mathcal{L}$ be the class of linearly ordered sets $\mathbf{L} = \langle L; \leqslant\rangle$. We say that $\mathbf{L}$ is **unbounded** if it has no least or greatest element, and it is **dense** if it has no covers. Let $\mathcal{L}^{\mathrm{ud}}$ denote the class of unbounded dense linear orders and let $\mathbf{Q} = \langle Q; \leqslant\rangle \in \mathcal{L}^{\mathrm{ud}}$ be the rational numbers.

(i) Show that $\mathcal{L}$ is a universal class and that $\mathcal{L}^{\mathrm{ud}}$ is an $\forall\exists$-class.

(ii) Show that $\mathbf{Q}$ is $\mathcal{L}$-universal. (If $\mathbf{L} \in \mathcal{L}$ is countable, where $L = \{a_1, a_2, a_3, \dots\}$, recursively define an embedding $f \colon \mathbf{L} \to \mathbf{Q}$ beginning with an arbitrary choice of $f(a_1)$.)

(iii) Show that $\mathbf{Q}$ is homogeneous by extending the argument used in (ii).

(iv) Show that $\mathcal{L}$ and $\mathcal{L}^{\mathrm{ud}}$ both satisfy (JEP).

(v) Show that every order $\leqslant$ on a set L can be extended to a linear order on L. (Use Zorn's Lemma to find an extension $\leqslant'$ of $\leqslant$ which is maximal among all orders on L which are extensions of $\leqslant$, and then show that $\leqslant'$ must be a linear order.)

(vi) Use (v) to show that $\mathcal{L}$ satisfies (AP).

(vii) Modify the argument of part (iii) to show that the class $\mathcal{L}^{\mathrm{ud}}$ is $\aleph_0$-categorical, that is, every countably infinite unbounded dense linear ordering is isomorphic to the rationals.

5.4 Apply the strong duality directly to prove Lemma 5.3.4(ii) as follows. Since the duality is strong, we consider embeddings of the form $u = E(\alpha) \colon E(\mathbf{Z}) \to E(\mathbf{X})$ and $v = E(\beta) \colon E(\mathbf{Z}) \to E(\mathbf{Y})$ in $\mathcal{A}$ where $\alpha \colon \mathbf{X} \to \mathbf{Z}$ and $\beta \colon \mathbf{Y} \to \mathbf{Z}$ in $\mathcal{X}$. By the Injectivity Lemma

3.2.10 we again know that $\underline{\mathbf{M}}$ must be injective in $\mathcal{A}$ since $\underset{\sim}{\mathbf{M}}$ is a total structure and the duality is strong. Let

$$\mathrm{eq}(\alpha, \beta) = \{(x, y) \in X \times Y \mid \alpha(x) = \beta(y)\}$$

denote the equaliser of α and β, and let $\rho_X : \mathrm{eq}(\alpha, \beta) \to X$ and $\rho_Y : \mathrm{eq}(\alpha, \beta) \to Y$ be the two projection maps. Verify the following.

(i) α and β are surjective.

(ii) ρ_X and ρ_Y are surjective.

(iii) $\mathrm{eq}(\alpha, \beta)$ is non-empty and therefore determines a closed substructure $\mathbf{eq}(\alpha, \beta)$ of $\mathbf{X} \times \mathbf{Y}$.

(iv) $\alpha\rho_X = \beta\rho_Y$ in $\mathcal{X}$.

(v) (AP) is fulfilled by taking $\mathbf{D} = E(\mathbf{eq}(\alpha, \beta))$, taking $r = E(\rho_X)$ and taking $s = E(\rho_Y)$ in $\mathcal{A}$.

By an **instance of** (FHP)*/(FEP)* in $\mathcal{X}$ we mean a pair $\mathbf{Y}, \mathbf{Z} \in \mathcal{X}$ of finite structures and a surjection $\psi : \mathbf{Y} \to \mathbf{Z}$ such that for every surjection $\varphi : \mathbf{X} \to \mathbf{Z}$ there is a morphism/surjection $\lambda : \mathbf{X} \to \mathbf{Y}$ for which $\varphi = \psi\lambda$.

5.5 Reviewing the proof of Theorem 5.4.1, find a single instance of (FEP)* which implies that $\mathbf{X}$ has no isolated points and consequently that $\mathbf{X}$ satisfies all instances of (FEP)*.

5.6 Review the proof of Theorem 5.4.2.

(i) Find a single instance of (FHP)* which implies condition (i)(c) and consequently that $\mathbf{X}$ satisfies all instances of (FHP)*.

(ii) Assume that $\mathbf{X}$ has (FHP)*. Find three instances of (FEP)* which hold, respectively, only if 1 is not isolated, only if 0 is not isolated, and only if no other point is isolated. Together these imply condition (ii)(c) and consequently imply that $\mathbf{X}$ satisfies all instances of (FEP)*.

5.7 Verify the independence of (AC1) and (AC2) over the variety of Stone algebras by referring to Figure 5.3.

5.8 Verify the independence of (EC1) and (EC2) over the class $\mathcal{S}^{\mathrm{ac}}$ of algebraically closed Stone algebras as follows.

(i) Let $X_1 = X_{\mathrm{C}} \times \{0, 1\}$ where X_{C} is the Cantor space and $(x, a) \preccurlyeq (y, b)$ only if $x = y$ and $a \leqslant b$. Use the AC Stone Algebra Theorem 5.4.4 to show that $\mathbf{A} = E(\mathbf{X}_1) \in \mathcal{S}^{\mathrm{ac}}$, and that $\mathbf{A}$ satisfies (EC1) but not (EC2). This shows that (AC1), (AC2) and (EC1) together do not imply (EC2).

(ii) Let $\mathbf{X}_2$ be the substructure of $\mathbf{X}_{\mathrm{HU}}$ obtained by removing all but one point of $X_1 \times \{1\}$. (See Figure 5.5, p.144.) Use the AC Stone Algebra Theorem 5.4.4 to show that $\mathbf{B} = E(\mathbf{X}_2) \in \mathcal{S}^{\mathrm{ac}}$, and that $\mathbf{B}$ satisfies (EC2) but not (EC1). This shows that (AC1), (AC2) and (EC2) together do not imply (EC1).

The characterisation of algebraically and existentially closed de Morgan algebras $\mathbf{A} = \langle A; \vee, \wedge, \neg, 0, 1\rangle$ follows a pattern similar to those that we have presented. We state the results in the next two exercises, which should be done using the strong duality given in Theorem 4.3.16. Let $\mathbf{A} = E(\mathbf{X})$ be a de Morgan algebra where $\mathbf{X} = \langle X; f, \preccurlyeq, \mathcal{T}\rangle \in \mathbb{IS}_c\mathbb{P}^+ \underset{\sim}{\mathbf{dM}}$.

5.9 Show that the following are equivalent:

(i) $\mathbf{A}$ is algebraically closed;

(ii) $\langle A; \vee, \wedge, 0, 1\rangle$ is a complemented lattice;

(iii) $\langle X; \preccurlyeq\rangle$ is an antichain.

5.10 Assume that $\mathbf{A}$ is algebraically closed, and show that the following are equivalent:

(i) $\mathbf{A}$ is existentially closed;

(ii) the lattice $\langle A; \vee, \wedge, 0, 1\rangle$ has no atoms;

(iii) $\mathbf{X}$ has no isolated points.

5.11 (i) Assume that the ordered set $\mathbf{Q}$ is an order retract of a complete lattice $\mathbf{P}$, that is, there are order-preserving maps $\sigma\colon \mathbf{Q} \to \mathbf{P}$ and $\tau\colon \mathbf{P} \to \mathbf{Q}$ such that $\tau \circ \sigma = \mathrm{id}_{\mathbf{Q}}$. Show that $\mathbf{Q}$ is complete.

(ii) Show that the following are equivalent:

(a) $\mathbf{P}$ is injective in the category $\mathbf{O}$ of ordered sets;

(b) $\mathbf{P}$ is an absolute subretract in $\mathbf{O}$;

(c) $\mathbf{P}$ is complete.

5.12 Let $\underline{\mathbf{D}}_{01} = \langle\{0,1\}; \vee, \wedge, 0, 1\rangle$ and let $\underset{\sim}{\mathbf{D}}_{01} = \langle\{0,1\}; \leqslant, \mathcal{T}\rangle$. Then $\mathcal{D}_{01} := \mathbb{ISP}\,\underline{\mathbf{D}}_{01}$ is the class of all bounded distributive lattices and $\mathcal{X} := \mathbb{IS}_c\mathbb{P}^+\,\underset{\sim}{\mathbf{D}}_{01}$ is the class of all Priestley spaces. By Exercise 4.5, $\underset{\sim}{\mathbf{D}}_{01}$ yields a strong duality on $\mathcal{D}_{01}$.

(i) Prove that a finite Priestley space $\langle X; \leqslant, \mathcal{T}\rangle$ is injective in $\mathcal{X}$ if and only if the underlying ordered set $\langle X; \leqslant\rangle$ is a lattice. [Hint: Show that every finite lattice is an order retract of a finite power of the two-element chain and hence, when endowed with the discrete topology, is injective in $\mathcal{X}$.]

(ii) [In this exercise we give a duality-theoretic proof of Balbes' [Bal67] characterisation of finite, projective, bounded distributive lattices.] Let $\mathbf{A} = \langle A; \vee, \wedge, 0, 1\rangle$ be a finite bounded distributive lattice. Use (i) and Exercise 4.5 to show that the following are equivalent:

(a) $\mathbf{A}$ is projective in the class $\mathcal{D}_{01}$;

(b) the underlying ordered set of the dual $D(\mathbf{A})$ is a lattice;

(c) the ordered set of join-irreducible elements of $\mathbf{A}$ is a lattice;

(d) 1 is join-irreducible in $\mathbf{A}$ and the meet in $\mathbf{A}$ of two join-irreducible elements of $\mathbf{A}$ is again join-irreducible.

5.13 Show that every finite discrete space is injective in $\mathcal{Z}$ and hence conclude that every finite Boolean algebra is projective in $\mathcal{B}$.

5.14 Prove (iii) $\Rightarrow$ (iv) of the Injective Boolean Algebra Theorem 5.5.4 as follows. Let $\mathbf{X}$ be a Boolean space, let $\mathbf{X}^*$ denote the Boolean algebra of clopen subsets of $\mathbf{X}$ and let U_i be a clopen subset of X for each $i \in I$. Show that the least upper bound of $\{ U_i \mid i \in I \}$ exists in $\mathbf{X}^*$ if and only if the closure of $\bigcup\{ U_i \mid i \in I \}$ is open in $\mathbf{X}$.

5.15 Prove Proposition 5.5.13 by establishing each of the following claims.

(0) The map $\tau\colon \mathcal{F}(\mathbf{X}) \to \mathbf{Y}$, as defined in Proposition 5.5.13, is a retraction.

Let $\mathcal{E}$ be the set of idempotents of $\mathcal{M}$. Define a binary relation $\leq$ on $\mathcal{E}$ by $e \leqslant f \Leftrightarrow ef = e$ and let θ be the equivalence relation on $\mathcal{E}$ defined by $e\,\theta\, f \Leftrightarrow e \leqslant f \,\&\, f \leqslant e$. Then we may define an order on $\mathcal{E}/\theta$ by $e/\theta \leqslant f/\theta \Leftrightarrow (\exists e', f' \in \mathcal{E})\, e\,\theta\, e' \leqslant f'\,\theta\, f$. For all $e \in \mathcal{E}$, let e^* be a fixed element of the equivalence class e/θ. Thus $e\,\theta\, f \Leftrightarrow e^* = f^*$. Define $\mathcal{E}^* := \{ e^* \mid e \in \mathcal{E} \}$. Let $\mathbf{Y}$ be a substructure of $\mathcal{F}(\mathbf{X})$ and let $\tau\colon \mathcal{F}(\mathbf{X}) \to \mathbf{Y}$ be a retraction.

(i) If $e \in \mathcal{E}$, then $g \in \mathcal{M}e \Leftrightarrow ge = g$, and if $e, f \in \mathcal{E}$, then $e \leqslant f \Leftrightarrow \mathcal{M}e \subseteq \mathcal{M}f$.

(ii) Denote the identity element of $\mathcal{M}$ by 1. If $x \in \pi_2(Y)$, then $\tau((1, x)) = (i_x, x)$ for some $i_x \in \mathcal{E}$.

(iii) Let $g \in \mathcal{M}$. If $(g, x) \in Y$, then $(i_x, x) \in Y$ and $gi_x = g$.

(iv) Let $e, f \in \mathcal{E}$. If $(e, x) \in Y$ and $f \leqslant e$, then $(f, x) \in Y$, in particular, $(e^*, x) \in Y$

(v) Let $g \in \mathcal{M}$. If $(g, x) \in Y$, then $(i_x^*, x) \in Y$ and $gi_x^* = g$.

(vi) Define $\mathcal{E}_x^* := \{ e \in \mathcal{E}^* \mid (e, x) \in Y \}$. Thus $\mathcal{E}_x^* \neq \varnothing \Leftrightarrow x \in \pi_2(Y)$. If $x \in \pi_2(Y)$, then $i_x^* \in \mathcal{E}_x^*$ and $e \leqslant i_x^*$ for all $e \in \mathcal{E}_x^*$.

(vii) For $e \in \mathcal{E}^*$, let $Y_e := \{ x \in X \mid (e, x) \in Y \}$ and define

$$X_e := \{ x \in X \mid x \in Y_e \;\&\; (\forall f \in \mathcal{E}^*)\, f/\theta > e/\theta \Rightarrow x \notin Y_f \}.$$

Then $X_e = \{ x \in \pi_2(Y) \mid i_x^* = e \}$. Hence, if $x \in \pi_2(Y)$, then $x \in X_e$ where $e = i_x^*$

(viii) If $e, f \in \mathcal{E}^*$ and $e \neq f$, then $X_e \cap X_f = \varnothing$.

(ix) $Y = \bigcup\{ \mathcal{M}e \times X_e \mid e \in \mathcal{E}^* \}$.

(x) $\pi_2(Y) = \bigcup\{ X_e \mid e \in \mathcal{E}^* \}$ and the map $\pi_2 \circ \tau \circ \eta_{\mathbf{X}}\colon \mathbf{X} \to \mathbf{X}$ is a retraction onto $\pi_s(Y)$, where $\eta_{\mathbf{X}}\colon \mathbf{X} \to \mathcal{F}(\mathbf{X})$ is defined by $\eta_{\mathbf{X}}(x) := (1, x)$ for all $x \in X$.

(xi) The map $\kappa\colon \pi_2(\mathbf{Y}) \to \mathcal{E}$, given by $\kappa(x) := i_x$, is continuous and therefore X_e is clopen in $\pi_2(Y)$. [Express κ in terms of $\eta_{\mathbf{X}}$, τ and π_1.]

To complete the proof, simply choose $S := \{\, e \in \mathcal{E}^* \mid X_e \neq \varnothing \,\}$.

5.16 Assume $\underline{\mathbf{M}}$ generates a congruence distributive variety.

(i) Let $\mathbf{N}$ be a subdirectly irreducible subalgebra of $\underline{\mathbf{M}}$. Show that $\mathbf{N}$ is weakly injective in $\mathcal{A}$ if and only if it is weakly injective in $\mathbb{S}(\underline{\mathbf{M}})$.

(ii) Show that if $\underline{\mathbf{M}}$ is subdirectly irreducible, then it is weakly injective in $\mathcal{A}$.

(iii) Assume that every non-trivial subalgebra of $\underline{\mathbf{M}}$ is subdirectly irreducible. Show that a subalgebra $\mathbf{N}$ of $\underline{\mathbf{M}}$ is injective in $\mathcal{A}$ if and only if it is injective in $\mathbb{S}(\underline{\mathbf{M}})$.

(See Davey [Dav77] for useful generalisations of this exercise.)

5.17 Check that the algebras $\underline{\mathbf{M}}$ and the structures $\underset{\sim}{\mathbf{M}}$ described in Chapter 4 for the dualities for distributive lattices, median algebras, Stone algebras, Kleene algebras, double Stone algebras and de Morgan algebras satisfy conditions (A) and (B) given in Section 5. Use the Injectivity and Boolean Powers Theorem 5.5.15 and the previous exercise to describe the injectives in each of these classes.

5.18 Let $G\colon \mathcal{A} \to \mathcal{B}$ and $H\colon \mathcal{B} \to \mathcal{A}$ be (covariant) functors. Then G is a **left adjoint** to H and H is a **right adjoint** to G if, for all $\mathbf{A} \in \mathcal{A}$, there is a morphism $\eta_{\mathbf{A}}\colon \mathbf{A} \to HG(\mathbf{A})$ such that, for every object $\mathbf{B} \in \mathcal{B}$ and each morphism $\varphi\colon \mathbf{A} \to H(\mathbf{B})$, there is a unique morphism $\overline{\varphi}\colon G(\mathbf{A}) \to \mathbf{B}$ such that $H(\overline{\varphi}) \circ \eta_{\mathbf{A}} = \varphi$.

(i) Show that G is left adjoint to H if and only if there is a bijection $\gamma_{\mathbf{A},\mathbf{B}}\colon \mathcal{B}(G(\mathbf{A}), \mathbf{B}) \to \mathcal{A}(\mathbf{A}, H(\mathbf{B}))$, for all $\mathbf{A} \in \mathcal{A}$ and all $\mathbf{B} \in \mathcal{B}$, which is natural in both $\mathbf{A}$ and $\mathbf{B}$.

(ii) Show that if G_1 and G_2 are both left adjoints to H, then there is a natural isomorphism $\lambda\colon G_1 \to G_2$.

(iii) Show that if categories $\mathcal{A}$ and $\mathcal{B}$ are equivalent via functors $G\colon \mathcal{A} \to \mathcal{B}$ and $H\colon \mathcal{B} \to \mathcal{A}$, then each of the functors G and H is left adjoint to the other.

5.19 Show that each of the following algebras generates a variety which is equivalent as a category to the variety of bounded distributive lattices.

(i) $\underline{\mathbf{M}}_3^+ = \langle \{0, a, b, c, 1\}; \vee, \wedge, 0, a, b, c, 1 \rangle$, that is, the five-element lattice of length 2 with every element added as a nullary operation.

(ii) $\underline{\mathbf{N}} = \langle\{0, a, 1\}; \vee, \wedge, ^\sharp, ^\flat, a\rangle$, where $\langle\{0, a, 1\}; \vee, \wedge\rangle$ is a three-element chain with $0 < a < 1$ and $^\sharp$ and $^\flat$ are unary maps given by $0^\sharp = 0^\flat = 0$, $1^\sharp = 1^\flat = 1$ and $a^\sharp = 1$, $a^\flat = 0$.

5.20 Let $\underline{\mathbf{SL}} = \langle\{0, 1\}; \wedge, 1\rangle$ be the two-element meet semilattice with 1, then $\mathcal{SL} = \mathbb{ISP}\underline{\mathbf{SL}}$ is the variety of meet semilattices with 1.

(i) Show that a variety $\mathcal{B}$ is equivalent as a category to $\mathcal{SL}$ if and only if there is a non-trivial finite algebra $\underline{\mathbf{S}} \in \mathcal{B}$ such that

(a) $\mathcal{B} = \mathbb{ISP}\underline{\mathbf{S}}$,

(b) $\underline{\mathbf{S}}$ has a definable meet semilattice-with-1 structure, $\underset{\sim}{\mathbf{S}} = \langle S; \wedge, 1\rangle$,

(c) the semilattice $\underset{\sim}{\mathbf{S}}$ is a distributive lattice, and

(d) every $(\wedge, 1)$-homomorphism $t\colon \underset{\sim}{\mathbf{S}} \to \underset{\sim}{\mathbf{S}}$ is a term function of $\underline{\mathbf{S}}$.

(ii) Let $\underline{\mathbf{N}}' = \langle\{0, a, 1\}; \wedge, ^\sharp, ^\flat, ^*\rangle$ be obtained from the algebra $\underline{\mathbf{N}}$ of Exercise 5.19 by deleting both the binary operation $\vee$ and the constant a and then adding the extra unary operation * given by $0^* = a$, $a^* = 1^* = 1$. Show that the variety generated by $\underline{\mathbf{N}}'$ is equivalent as a category to $\mathcal{SL}$.

5.21 Let $\underline{\mathbf{M}}$ be a semi-primal algebra. With the help of the Semi-primal Strong Duality Theorem 3.3.14, prove that a quasi-variety $\mathcal{B}$ is equivalent as a category to $\mathcal{A} := \mathbb{ISP}\underline{\mathbf{M}}$ if and only if there is a finite algebra $\underline{\mathbf{N}} \in \mathcal{B}$ such that

(i) $\mathcal{B} = \mathbb{ISP}\underline{\mathbf{N}}$,

(ii) $\underline{\mathbf{N}}$ is semi-primal, and

(iii) there is an order isomorphism $\varphi\colon \mathrm{Sub}(\underline{\mathbf{M}}) \to \mathrm{Sub}(\underline{\mathbf{N}})$ with respect to which trivial subalgebras of $\underline{\mathbf{M}}$ correspond to trivial subalgebras of $\underline{\mathbf{N}}$.

6

What Makes a Duality Useful?

In this chapter, which is based on Clark and Davey [CD96], we again pick up the theoretical development of duality theory initiated in the first three chapters. Having discovered what makes a *natural duality*, a *full duality* and a *strong duality*, we are ready to address a more subtle question: what makes a *useful duality*? Our strong duality theorems are quite powerful, for we have seen that they can be used to produce a wealth of interesting examples. Using our applications as a guide, we can now begin to sort through these examples of strong duality in order to pick out those which are likely to be most useful to the algebraist. Our goal is to identify specific features which make a duality useful and then to determine, based on properties of $\underline{\mathbf{M}}$, exactly when $\mathcal{A}$ admits such a duality.

The general strategy of the applications presented in the last chapter was to formulate a problem within the category $\mathcal{A}$, transfer the problem to an equivalent (dual) problem in the category $\mathcal{X}$, solve the problem in $\mathcal{X}$, and finally transfer the answer back to obtain a solution to the original problem in $\mathcal{A}$. The dual equivalence provided by a strong duality gives us the mechanism that we need to make these transfers. But the success of this strategy depends on the problem in question being *easier* to solve in the category $\mathcal{X}$ than it was in the category $\mathcal{A}$. At first this might sound strange: if two categories are equivalent, how can a problem be easier to solve in one than in the other? The answer we find by surveying the applications in Chapter 5. Although the *categories* $\mathcal{A}$ and $\mathcal{X}$ are equivalent, the *objects* of $\mathcal{X}$ are sometimes much simpler than their duals in $\mathcal{A}$. When this is the case, we expect the duality to be useful in the study of properties of algebras in $\mathcal{A}$.

Consider the following example, which is typical of many strong dualities that arise as immediate consequences of our theory. Let $\underline{\mathbf{M}}$ be the n-element chain regarded as a lattice, for any $n \geqslant 2$. Then $\operatorname{Con}\underline{\mathbf{M}}$ is a Boolean lattice of order 2^{n-1}. Using the NU Strong Duality Theorem 3.3.8, we obtain a strong duality for the variety $\mathcal{A}$ of distributive lattices with $\underline{\mathbf{M}}$ as generating algebra by taking $\underset{\sim}{\mathrm{M}} = \langle M; H, R, \mathcal{T}\rangle$ where H, R and $\mathcal{T}$ are chosen as follows: H is the set of all homomorphisms from sublattices of powers $\underline{\mathbf{M}}^k$ into $\underline{\mathbf{M}}$, for some $k < n$, each homomorphism being viewed as a partial operation on M, while R is (an entailing set for) the sublattices of $\underline{\mathbf{M}}^2$ viewed as binary relations on M, and $\mathcal{T}$ is the discrete topology on M. The dual category $\mathcal{X}$ is a category of compact structures having the same type as $\underset{\sim}{\mathrm{M}}$.

From the point of view of duality theory *per se*, this is a perfectly fine example of the many strong dualities that we are now able to generate. But it is hard to imagine that this example could ever be of any use to an algebraist, for it gives a dual equivalence between the familiar category of distributive lattices and a category of topological spaces carrying very complicated structures. For it to profit the algebraist to shift from the familiar category $\mathcal{A}$ to the new category $\mathcal{X}$ of structured topological spaces, it is necessary that those spaces have a particularly simple and tractable type. This was the case in all of the examples used in Chapter 5, and in most of those presented in Chapter 4.

In this chapter we will look for choices of $\underline{\mathbf{M}}$ that admit a strong duality which is likely to prove useful because the dual category $\mathcal{X}$ has a type which is restricted in one of several ways to ensure its members will be relatively simple. As we will see, these dual categories are often the ones that have other nice properties as well. For example coproducts, which always exist in $\mathcal{X}$ under a full duality, are often much simpler and more easily recognised than the corresponding direct products in $\mathcal{A}$ if the type of $\mathcal{X}$ is suitably restricted. This can lead to illuminating information about congruences on $\mathbf{A} = E(\mathbf{X})$ since they, as subalgebras of $\mathbf{A} \times \mathbf{A}$, are dual to special images of the coproduct $\mathbf{X} * \mathbf{X}$.

We will prove ten theorems corresponding to ten different properties which are likely to make a strong duality useful. In each case, we would like to know exactly which finitely generated quasi-varieties $\mathcal{A}$ admit a strong duality which has the property in question. In each case, our results of Chapters 2 and 3 will quickly give us sufficient conditions on $\underline{\mathbf{M}}$ for the desired strong duality to exist. Our goal will be to show that these conditions are also necessary. Eight of these theorems are named after their specific restriction on the type of the structure $\underset{\sim}{\mathbf{M}}$. Each of our ten theorems gives several equivalent conditions, beginning with the desired restriction on the type of $\underset{\sim}{\mathbf{M}}$ and ending with a corresponding condition on $\underline{\mathbf{M}}$. The results are summarised in Figure 6.1, where we have ordered them by increasing strength of their hypotheses. Here we denote by $\#S$ the maximum arity of the functions or relations in S. Recall from Chapter 3 that K denotes the set consisting of those elements of M which form trivial subalgebras of $\underline{\mathbf{M}}$. For a full statement of each result, the reader is referred to the section indicated.

We are by no means prepared to claim that these results give a complete answer to the question, 'What makes a duality useful?' (Indeed, even the ultimate brute force strong duality can be useful—see the proof of Proposition 10.3.2.) Other reasonable restrictions are available for investigation, and additional equivalences to the ones that we have found might still surface. We therefore hope that our efforts will serve to stim-

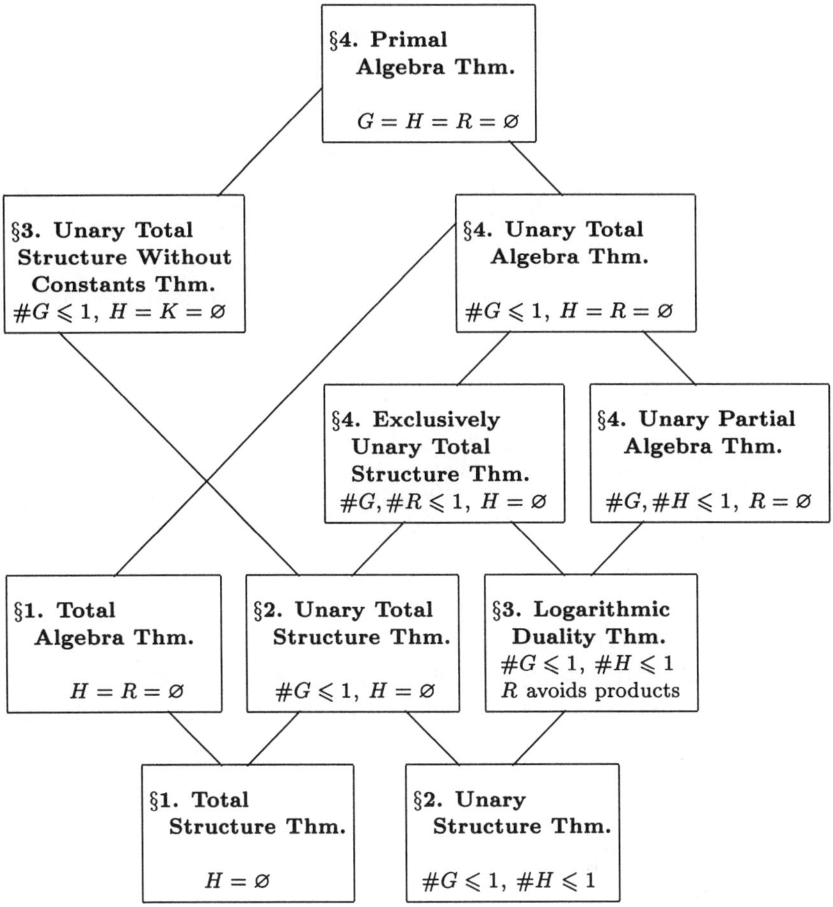

Figure 6.1 ten useful types of strong duality

ulate further investigation into the next question, 'What *else* makes a duality useful?'

6.1 Total structures and injectivity

The theory of strong dualities, as developed in Chapter 3, makes it clear that partial operations play an inherent role in the description of the dual category and can therefore generally be expected to appear in its type. Moreover there are many very tractable dualities which utilise proper partial operations. But we also saw a number of central results for producing strong dualities, culminating in the Two-for-One Strong

Duality Theorem 3.3.2, which apply only when $\underset{\sim}{\mathbf{M}}$ is a **total structure**, that is, when the set H of partial operations is empty. Recall that total structures were essential in Lemma 5.3.4, the key result needed to establish the central Theorem 5.3.5 in Section 3 of Chapter 5. As partial operations are not generally included in the study of algebraic systems, it is natural to ask when we can get by without them. More specifically, suppose that some structure $\underset{\sim}{\mathbf{M}} = \langle M; G, H, R, \mathcal{T}\rangle$ yields a strong duality on $\mathcal{A}$. Our first goal in this section is to determine when it will be the case that

(i) *some total structure $\underset{\sim}{\mathbf{M}}'$ yields a strong duality on $\mathcal{A}$.*

We begin with a sufficient condition for this to happen. Suppose that

(ii) *for each natural number n, every n-ary partial operation $h \in H$ extends to a homomorphism $g\colon \underline{\mathbf{M}}^n \to \underline{\mathbf{M}}$.*

Let $\underset{\sim}{\mathbf{M}}' = \langle M; G', R', \mathcal{T}\rangle$ be the total structure obtained from $\underset{\sim}{\mathbf{M}}$ by deleting H and, for each $h \in H$, adding to G some homomorphism g extending h and adding the relation $\operatorname{dom}(h)$ to R. According to the $\underset{\sim}{\mathbf{M}}$-Shift Strong Duality Lemma 3.2.3, the structure $\underset{\sim}{\mathbf{M}}'$ will also yield a strong duality on $\mathcal{A}$.

Notice that condition (ii) is a special case of injectivity of $\underline{\mathbf{M}}$ in $\mathcal{A}$. Thus we can eliminate partial operations whenever we find that

(iii) *$\underline{\mathbf{M}}$ is injective in $\mathcal{A}$.*

Since $\underset{\sim}{\mathbf{M}}$ yields a strong duality on $\mathcal{A}$, it is injective in $\mathcal{X}$ by the First Strong Duality Theorem. By Lemma 3.2.10 this will guarantee (iii) provided that

(iv) *the image of every morphism in $\mathcal{X}$ is a closed substructure of its codomain.*

Recall from Warning 1.4.1 that (iv) is the peculiar condition which always holds in the absence of partial operations but can easily fail when they are present. In order to find an alternative condition which will ensure (iv), we introduce a new notion. We say that arbitrary structures $\underset{\sim}{\mathbf{M}}$ and $\underset{\sim}{\mathbf{M}}'$ over the same set M are **structurally equivalent** if the same subsets of powers of M determine substructures of powers of $\underset{\sim}{\mathbf{M}}$ as determine substructures of powers of $\underset{\sim}{\mathbf{M}}'$, and the same continuous maps between those subsets are morphisms with respect to $\underset{\sim}{\mathbf{M}}$ as are with respect to $\underset{\sim}{\mathbf{M}}'$.

Returning to the context of our discussion, structural equivalence gives us a sufficient condition to ensure that (iv) will hold:

(v) *$\underset{\sim}{\mathbf{M}}$ is structurally equivalent to a total structure.*

It is easy to check that structurally equivalent structures strongly entail each other, so that if one yields a strong duality on $\mathcal{A}$ then the other does as well. Rather surprisingly, the converse is also true: if any total structure yields a strong duality on $\mathcal{A}$, then $\underset{\sim}{\mathbf{M}}$ itself is structurally equivalent to that total structure!

1.1 Lemma *Assume that* $\underset{\sim}{\mathbf{M}}$ *yields a strong duality on* $\mathcal{A}$ *and that* $\underset{\sim}{\mathbf{M}}'$ *is an arbitrary structure over the set* M. *Then* $\underset{\sim}{\mathbf{M}}'$ *yields a strong duality on* $\mathcal{A}$ *if and only if it is structurally equivalent to* $\underset{\sim}{\mathbf{M}}$.

Proof Suppose that $\underset{\sim}{\mathbf{M}}'$ also yields a strong duality on $\mathcal{A}$. Then the same sets determine substructures of powers of $\underset{\sim}{\mathbf{M}}$ and of $\underset{\sim}{\mathbf{M}}'$, namely the hom-closed sets. Assume that $X \subseteq M^S$ and $Y \subseteq M^T$ are hom-closed and that $\varphi\colon X \to Y$ preserves the structure of $\underset{\sim}{\mathbf{M}}$. Since products in $\mathfrak{X}$ are Cartesian, to see that φ preserves the structure of $\underset{\sim}{\mathbf{M}}'$, we must check that $\pi_t \circ \varphi\colon X \to M$ preserves the structure of $\underset{\sim}{\mathbf{M}}'$ for each $t \in T$. Since $\underset{\sim}{\mathbf{M}}$ is injective in $\mathfrak{X}$, we conclude that $\pi_t \circ \varphi$ extends to a morphism ψ from $\underset{\sim}{\mathbf{M}}^S$ to $\underset{\sim}{\mathbf{M}}$. Using $(\mathrm{CLO})_S$ we see that ψ must be an S-ary term function on $\underline{\mathbf{M}}$ since $\underset{\sim}{\mathbf{M}}$ yields a duality on $\mathcal{A}$. But term functions preserve all algebraic structure on M and therefore, in particular, they preserve the structure of $\underset{\sim}{\mathbf{M}}'$. ■

The facts we have established tell us exactly when we can eliminate partial operations.

1.2 Total Structure Theorem *If* $\underset{\sim}{\mathbf{M}}$ *yields a strong duality on* $\mathcal{A}$, *then the following are equivalent:*

(i) *some total structure* $\underset{\sim}{\mathbf{M}}'$ *yields a strong duality on* $\mathcal{A}$;
(ii) *for each natural number* n, *every* n*-ary partial operation* $h \in H$ *in the structure on* $\underset{\sim}{\mathbf{M}}$ *extends to a homomorphism* $g\colon \underline{\mathbf{M}}^n \to \underline{\mathbf{M}}$;
(iii) $\underline{\mathbf{M}}$ *is injective in* $\mathcal{A}$;
(iv) *the image of every morphism in* $\mathfrak{X}$ *is a closed substructure of its codomain;*
(v) $\underset{\sim}{\mathbf{M}}$ *is structurally equivalent to a total structure.*

In particular, $\underset{\sim}{\mathbf{M}}'$ *in* (i) *may be obtained from* $\underset{\sim}{\mathbf{M}}$ *by deleting* H *and using* (ii) *to replace each partial operation* $h \in H$ *with an extension* g *of* h *to a total operation in* G *and with its domain,* $\mathrm{dom}(h)$, *as a relation in* R.

Even better than obtaining a strong duality from a total structure ($H = \varnothing$) would be to obtain a strong duality from a total algebra ($R = H = \varnothing$). The Two-for-One Strong Duality Theorem gives us a list of equivalent conditions, any one of which can be checked to determine if *a particular* total algebra $\underset{\sim}{\mathbf{M}} = \langle M; G, \mathcal{T}\rangle$ yields a strong duality on $\mathcal{A}$. We now give our second description of a useful duality which tells when *some* choice of a total algebra yields a strong duality on $\mathcal{A}$. We say that $\underline{\mathbf{M}}$ satisfies the **finite equaliser property** (FEQ) if, for all $n \in \mathbb{N}$, every subalgebra of $\underline{\mathbf{M}}^n$ is an intersection of equalisers of homomorphisms from $\underline{\mathbf{M}}^n$ to $\underline{\mathbf{M}}$.

1.3 Total Algebra Theorem *For a finite algebra $\underline{\mathbf{M}}$, the following are equivalent:*

(i) *there is a total algebra $\underset{\sim}{\mathrm{M}}$ which yields a strong duality on $\mathcal{A}$;*

(ii) *there is a structure $\underset{\sim}{\mathrm{M}}$ which yields a strong duality on $\mathcal{A}$ such that*

  (a) *the image of every morphism in $\mathfrak{X}$ is a substructure,*

  (b) *every one-to-one morphism in $\mathfrak{X}$ is an embedding;*

(iii) *$\underline{\mathbf{M}}$ is injective in $\mathcal{A}$ and epics are surjective in $\mathcal{A}$;*

(iv) *$\underline{\mathbf{M}}$ is injective in $\mathcal{A}$ and, for all $\mathbf{A} \in \mathcal{A}$, every subalgebra of $\mathbf{A}$ is an intersection of equalisers (of homomorphisms from $\mathbf{A}$ into $\underline{\mathbf{M}}$);*

(iv)′ *$\underline{\mathbf{M}}$ is injective in $\mathcal{A}$ and, for all $\mathbf{A} \in \mathcal{A}$, every subalgebra of $\mathbf{A}$ is an equaliser of a pair of homomorphisms from $\mathbf{A}$ into $\mathbf{B}$ for some $\mathbf{B} \in \mathcal{A}$;*

(v) *$\underline{\mathbf{M}}$ is injective in $\mathcal{A}$ and satisfies* (FEQ);

(v)′ *$\underline{\mathbf{M}}$ is injective in $\mathcal{A}$ and, for all $n \geqslant 1$, every subalgebra of $\underline{\mathbf{M}}^n$ is an equaliser of a pair of homomorphisms from $\underline{\mathbf{M}}^n$ into $\mathbf{B}$ for some finite $\mathbf{B} \in \mathcal{A}$.*

Moreover, if $\underset{\sim}{\mathrm{M}}$ satisfies (ii), *then $\underset{\sim}{\mathrm{M}}$ is structurally equivalent to a total algebra $\underset{\sim}{\mathrm{M}}'$ which may be obtained from $\underset{\sim}{\mathrm{M}}$ by first replacing each partial map h by its domain,* $\mathrm{dom}(h)$, *and any total extension, g, of the map h and then replacing each n-ary relation r by a finite family, G_r, of total functions, $g\colon M^n \to M$, such that r is an intersection of equalisers of pairs of functions from G_r.*

Proof That (i) implies (ii) is trivial.

Assume that (ii) holds. By the Total Structure Theorem, (ii)(a) implies that $\underline{\mathbf{M}}$ is injective in $\mathcal{A}$. If $f\colon \mathbf{A} \to \mathbf{B}$ is an epic in $\mathcal{A}$, then $D(f)\colon D(\mathbf{B}) \to D(\mathbf{A})$ is one-to-one and hence, by (ii)(b), is an embedding. Since $\underset{\sim}{\mathrm{M}}$ is injective in $\mathfrak{X}$, the map $ED(f)$ is surjective, whence $f = e_{\mathbf{B}}^{-1} \circ ED(f) \circ e_{\mathbf{A}}$ is surjective. Thus (ii) implies (iii).

Let $\mathbf{B} \leqslant \mathbf{A} \in \mathcal{A}$ and define

$$C := \bigcap\{\mathrm{eq}(x_1, x_2) \mid x_1, x_2 \in \mathcal{A}(\mathbf{A}, \underline{\mathbf{M}}) \text{ and } x_1{\restriction}_{\mathbf{B}} = x_2{\restriction}_{\mathbf{B}}\}.$$

We claim that, provided $\underline{\mathbf{M}}$ is injective in $\mathcal{A}$, the inclusion of $\mathbf{B}$ into $\mathbf{C}$ is an epic. Let $y_1, y_2 \in \mathcal{A}(\mathbf{C}, \underline{\mathbf{M}})$ with $y_1{\restriction}_{\mathbf{B}} = y_2{\restriction}_{\mathbf{B}}$ and suppose that $y_1(c) \neq y_2(c)$ for some $c \in C \backslash B$. Since $\underline{\mathbf{M}}$ is injective, there exist $x_1, x_2 \in \mathcal{A}(\mathbf{A}, \underline{\mathbf{M}})$ with $x_i{\restriction}_C = y_i$. Thus $x_1{\restriction}_B = x_2{\restriction}_B$ and $c \notin \mathrm{eq}(x_1, x_2)$, whence $c \notin C$, a contradiction. Thus the inclusion of $\mathbf{B}$ into $\mathbf{C}$ is an epic. Consequently, (iii) implies (iv).

Again, (iv) implies (v) is trivial.

Assume that $\underline{\mathbf{M}}$ is injective in $\mathcal{A}$ and satisfies (FEQ). Let the structure on $\underset{\sim}{\mathrm{M}}$ be the enriched hom-clone of $\underline{\mathbf{M}}$, that is,

$$\underset{\sim}{\mathrm{M}} := \langle M; K \cup \bigcup\{\mathcal{A}(\underline{\mathbf{M}}^n, \underline{\mathbf{M}}) \mid n \in \mathrm{N}\}, \mathcal{T}\rangle.$$

We shall prove that $\underset{\sim}{\mathbf{M}}$ yields a strong duality on $\mathcal{A}$. Assume, without loss of generality, that $\underset{\sim}{\mathbf{M}}$ has named constants. If $\underline{\mathbf{M}} = \langle M; F\rangle$, then we define

$$\underline{\mathbf{M}}' := \langle M; K \cup \bigcup\{\, \mathcal{A}(\underline{\mathbf{M}}^n, \underline{\mathbf{M}}) \mid n \in \mathbb{N}\}\rangle \text{ and } \underset{\sim}{\mathbf{M}}' := \langle M; F, \mathcal{T}\rangle.$$

Since $\underline{\mathbf{M}}$ is injective it follows at once that (IC) holds with respect to $\underset{\sim}{\mathbf{M}}'$ and $\underline{\mathbf{M}}'$. The statement that every subalgebra of each finite power of $\underline{\mathbf{M}}$ is an intersection of equalisers is precisely (FTC) with respect to $\underset{\sim}{\mathbf{M}}'$. Thus, by the Two-for-One Strong Duality Theorem 3.3.2, $\underset{\sim}{\mathbf{M}}$ yields a strong duality on $\mathcal{A}$. Hence, (v) implies (i).

The equivalence of (iv) and (iv)′ (respectively, (v) and (v)′) follows from the fact that $\mathcal{A} = \mathbb{ISP}\underline{\mathbf{M}}$.

To prove the 'Moreover', assume that $\underline{\mathbf{M}}$ satisfies (ii). By (v) we can construct the total algebra $\underset{\sim}{\mathbf{M}}'$, and by the $\underset{\sim}{\mathbf{M}}$-Shift Strong Duality Lemma we know that $\underset{\sim}{\mathbf{M}}'$ also yields a strong duality on $\mathcal{A}$. Lemma 6.1.1 now tells us that $\underset{\sim}{\mathbf{M}}'$ is structurally equivalent to $\underset{\sim}{\mathbf{M}}$. ■

The condition (FEQ) alone, without the injectivity of $\underline{\mathbf{M}}$, implies that $\underline{\mathbf{M}}$ is dualisable by a total algebra (see Exercise 6.2). Dualisability by a total algebra is very restrictive. It implies, via (CLO), that the clone of $\underline{\mathbf{M}}$ is determined by a set of algebraic total operations (and therefore by the hom-clone of $\underline{\mathbf{M}}$). By Burris and Willard [BW87], up to term equivalence, on a fixed finite set there is only a finite number of algebras with this property!

6.2 Unary structures, coproducts and near-unanimity

Most of the examples of strong dualities that we have presented have dual categories that are tractable in a quite different way: all operations and partial operations are at most unary. This restriction can result in an especially nice dual category. At most unary operations means that closed substructures of any structure in $\mathcal{X}$ are closed under unions and consequently, since we allow the empty structure in $\mathcal{X}$, form a *distributive* lattice. The NU Strong Duality Theorem tells us that a strong duality meeting this restriction exists provided that $\underline{\mathbf{M}}$ has a near-unanimity term and every non-trivial subalgebra of $\underline{\mathbf{M}}$ is subdirectly irreducible. In this section we will first prove the Unary Structure Theorem which asserts that these conditions are all equivalent. Then we will add the hypothesis that $\underset{\sim}{\mathbf{M}}$ is a total structure as well. The Unary Total Structure Theorem will tell us that this is also equivalent to asserting that coproducts in $\mathcal{X}$ are always given as unions of the cofactors.

2.1 Lemma *Let* $\mathbf{A}$ *be a subalgebra of* $\underline{\mathbf{M}}^S$, *let* $\varnothing \neq T \subseteq S$ *and let*

$$X_T := \{\, u\colon A \to M \mid \text{for all } a, b \in A, \text{if } a{\restriction}_T = b{\restriction}_T, \text{then } u(a) = u(b)\}$$

be the set of maps from A *to* M *that depend only on* T. *Then* X_T *is a closed substructure of* $\underset{\sim}{\mathbf{M}}^A$.

Proof Let $h \in G \cup H$ and let $(x_1, \dots, x_n) \in \mathrm{dom}_{X_T}(h)$. Consequently, the n-tuple $(x_1, \dots, x_n)$ is in $\mathrm{dom}_{\underset{\sim}{\mathbf{M}}^A}(h)$ and $x_1, \dots, x_n \in X_T$. Let $a, b \in A$ with $a{\restriction}_T = b{\restriction}_T$. Thus $x_i(a) = x_i(b)$ for all i, and hence

$$\begin{aligned} h(x_1, \dots, x_n)(a) &= h(x_1(a), \dots, x_n(a)) \\ &= h(x_1(b), \dots, x_n(b)) = h(x_1, \dots, x_n)(b). \end{aligned}$$

Thus $h(x_1, \dots, x_n)$ depends only on T and is therefore in X_T. Hence X_T is a substructure of $\underset{\sim}{\mathbf{M}}^A$.

Let $u \in M^A \backslash X_T$. Thus there exist $a, b \in A$ with $a{\restriction}_T = b{\restriction}_T$ and $u(a) \neq u(b)$. Now the set of points in M^A which agree with u at a and b is an open set which contains u and is disjoint from X_T. Thus X_T is closed in M^A. ∎

An early observation of the general theory of natural dualities was that the presence of a near-unanimity term for $\underline{\mathbf{M}}$ was sufficient to ensure that it admitted a duality. In Chapter 10 we will present a partial converse of this result due to Davey, Heindorf and McKenzie [DHM95] (the Big NU Obstacle Theorem 10.2.2): *if the lattice of* $\mathcal{A}$*-congruences on each member of* $\mathcal{A}$ *is distributive and* $\mathcal{A}$ *admits a duality, then* $\underline{\mathbf{M}}$ *must have a near-unanimity term.*

The proof of our next theorem depends on this fact. Notice that our list of equivalences begins with the desired condition on the type of $\underset{\sim}{\mathbf{M}}$ and ends with a finitely verifiable condition on $\underline{\mathbf{M}}$.

2.2 Unary Structure Theorem *The following are equivalent:*

(i) *there is a structure* $\underset{\sim}{\mathbf{M}}$ *with each operation in* $G \cup H$ *at most unary which yields a strong duality on* $\mathcal{A}$;

(ii) *there is a structure* $\underset{\sim}{\mathbf{M}}$ *such that the substructures of powers of* $\underset{\sim}{\mathbf{M}}$ *are closed under unions and* $\underset{\sim}{\mathbf{M}}$ *yields a strong duality on* $\mathcal{A}$;

(iii) *the variety generated by* $\underline{\mathbf{M}}$ *is congruence distributive, every non-trivial subalgebra of* $\underline{\mathbf{M}}$ *is subdirectly irreducible and* $\underline{\mathbf{M}}$ *is dualisable;*

(iv) $\underline{\mathbf{M}}$ *has a near-unanimity term and every non-trivial subalgebra of* $\underline{\mathbf{M}}$ *is subdirectly irreducible.*

Proof The assertion (iii)⇒(iv) follows from the Big NU Duality Theorem 10.2.2, and the assertion (iv)⇒(iii) follows from the NU Strong Duality Theorem 3.3.8 and Mitschke's Theorem A.14. Again, (iv)⇒(i) follows from the NU Strong Duality Theorem and (i) trivially implies (ii).

Now assume (ii) to prove (iv). For each $\mathbf{A} \in \mathcal{A}$ the lattice of $\mathcal{A}$-congruences on $\mathbf{A}$ is, by Theorem 3.2.1, dually isomorphic to the lattice of closed substructures of $D(\mathbf{A})$ under union and intersection, and this is a distributive lattice. From Theorem 10.2.2 we conclude that $\underline{\mathbf{M}}$ has a near-unanimity term.

In order to complete the proof of (ii)⇒(iv), it remains to show that every non-trivial subalgebra $\mathbf{Q}$ of $\underline{\mathbf{M}}$ is subdirectly irreducible. Let $\theta_1, \theta_2 \in \operatorname{Con} \mathbf{Q}$ such that $\theta_1 \cap \theta_2 = \mathbf{0}^Q$. We will show that either $\theta_1 = \mathbf{0}^Q$ or $\theta_2 = \mathbf{0}^Q$. Let $\mathbf{A} := \theta_1 \cdot \theta_2$, a subalgebra of $\underline{\mathbf{M}}^2$. Since $\theta_1 \cap \theta_2 = \mathbf{0}^Q$, for all $(a, b) \in A$ there is a unique $c \in M$ such that $a\theta_1 c\theta_2 b$. Hence we may define a map $h\colon A \to M$ by $h(a, b) = c$. Since θ_1 and θ_2 are congruences, h is a homomorphism. Define

$$X_1 := \{\, u\colon A \to M \mid u \text{ depends only on the first coordinate} \,\},$$

$$X_2 := \{\, u\colon A \to M \mid u \text{ depends only on the second coordinate} \,\}.$$

By Lemma 6.2.1, the sets $\mathbf{X}_1$ and $\mathbf{X}_2$ are closed substructures of $\underset{\sim}{\mathbf{M}}^A$. By (ii), we have that $\mathbf{X} = \mathbf{X}_1 \cup \mathbf{X}_2$ is also a closed substructure of $\underset{\sim}{\mathbf{M}}^A$.

From the definition of strong duality we know that $\mathbf{X}$ is hom-closed in $\underset{\sim}{\mathbf{M}}^A$. Let $\rho_i\colon A \to M$ be the projection for $i = 1, 2$. Then $\rho_1, \rho_2 \in \mathbf{X}$ and clearly $(\rho_1, \rho_2) \in \operatorname{dom}_X(h)$ since

$$(\rho_1, \rho_2)(a, b) = (\rho_1(a, b), \rho_2(a, b)) = (a, b) \in \mathbf{A} = \operatorname{dom}(h)$$

for all $(a, b) \in \mathbf{A}$. Since $\mathbf{X}$ is hom-closed, $h = h(\rho_1, \rho_2) \in \mathbf{X}$. Without loss of generality, assume that $h \in X_1$, that is, h depends only on its first coordinate. We claim that $\theta_1 = \mathbf{0}^Q$. Indeed,

$$\begin{aligned}(a, b) \in \theta_1 &\implies a\theta_1 a\theta_2 a \text{ and } a\theta_1 b\theta_2 b \\ &\implies h(a, a) = a \text{ and } h(a, b) = b \\ &\implies a = b\end{aligned}$$

since $h(a, a) = h(a, b)$. Hence $\mathbf{Q}$ is subdirectly irreducible, as required. ∎

The best natural dualities seem to be the ones that combine the restrictions of the two preceding theorems, as we saw, for example, in the cases of the strong dualities for distributive lattices, Kleene algebras, de Morgan algebras, Stone and double Stone algebras and the variety generated by a finite field (in Chapter 4). These dualities share yet another special property which contributes further to their usefulness. This property concerns products.

In the category $\mathcal{A}$ direct products (which are concrete, that is, formed pointwise) always exist and are denoted by

$$\mathbf{A} \underset{\pi_1}{\leftarrow} \mathbf{A} \times \mathbf{B} \underset{\pi_2}{\rightarrow} \mathbf{B}$$

where π_1 and π_2 are the associated surjective projections. Under a full duality (and therefore also under a strong duality) direct products in $\mathcal{A}$ are always dual to coproducts in $\mathfrak{X}$ (Lemma 1.1.4). In particular, *coproducts always exist in* $\mathfrak{X}$ and will be denoted by

$$\mathbf{X} \underset{\sigma_1}{\rightarrow} \mathbf{X} * \mathbf{Y} \underset{\sigma_2}{\leftarrow} \mathbf{Y}$$

where σ_1 and σ_2 are the associated maps. Ready access to the coproduct $D(\mathbf{A}) * D(\mathbf{B})$ means a convenient representation for the direct product $\mathbf{A} \times \mathbf{B}$ in $\mathcal{A}$. Under the conditions of the Unary Structure Theorem coproducts in $\mathfrak{X}$ have a very nice property which by itself has nothing to do with duality.

2.3 Lemma *Suppose that the coproduct*

$$\mathbf{X} \underset{\sigma_1}{\rightarrow} \mathbf{X} * \mathbf{Y} \underset{\sigma_2}{\leftarrow} \mathbf{Y}$$

of members $\mathbf{X}$ *and* $\mathbf{Y}$ *of* $\mathfrak{X}$ *exists in* $\mathfrak{X}$. *Then the morphisms* σ_1 *and* σ_2 *are both embeddings. If in addition every operation in* $G \cup H$ *is at most unary, then* $\mathbf{X} * \mathbf{Y} = \sigma_1(\mathbf{X}) \cup \sigma_2(\mathbf{Y})$ *is the union of the images.*

Proof To see that σ_1 is an embedding, we must show that distinct elements, unrelated elements and elements outside of the domain of a partial operation are kept so under a morphism from $\mathbf{X}$ into $\underset{\sim}{\mathbf{M}}$. We use the fact that $\mathbf{X}$ can be embedded into a power of $\underset{\sim}{\mathbf{M}}$. As a result we can find, for example, a projection of $\mathbf{X}$ into $\underset{\sim}{\mathbf{M}}$ that separates two arbitrary distinct elements. Since this projection factors through σ_1, we conclude that σ_1 must also separate those two elements. The other parts of the argument are similar.

Since σ_1 and σ_2 are embeddings, the images $\sigma_1(\mathbf{X})$ and $\sigma_2(\mathbf{Y})$ are both closed substructures of $\mathbf{X} * \mathbf{Y}$. Since the operations of $G \cup H$ are at most unary, their union $\mathbf{Z} = \sigma_1(\mathbf{X}) \cup \sigma_2(\mathbf{Y})$ is a closed substructure of $\mathbf{X} * \mathbf{Y}$. It follows directly from the definition of coproduct that $\mathbf{Z}$ is also a coproduct of $\mathbf{X}$ and $\mathbf{Y}$.

To see that $\mathbf{Z}$ must be all of $\mathbf{X} * \mathbf{Y}$, consider the commuting diagram of Figure 6.2. Let $\eta\colon \mathbf{Z} \to \mathbf{X} * \mathbf{Y}$ be the inclusion and let $\alpha\colon \mathbf{X} * \mathbf{Y} \to \mathbf{Z}$ complete the lower part of the diagram. Since $\eta \circ \alpha$ is the unique map that completes the outer diagram, it must be the identity on $\mathbf{X} * \mathbf{Y}$. Thus η is surjective. ∎

If coproducts in $\mathfrak{X}$ satisfy $\mathbf{X} * \mathbf{Y} = \sigma_1(\mathbf{X}) \cup \sigma_2(\mathbf{Y})$, we shall say that **coproducts in $\mathfrak{X}$ are given by union**. Lemma 6.2.3 tells us that the conditions of the Unary Structure Theorem imply that coproducts in $\mathfrak{X}$ are given by union. If $\underset{\sim}{\mathbf{M}}$ is a total structure as well, we will show that the

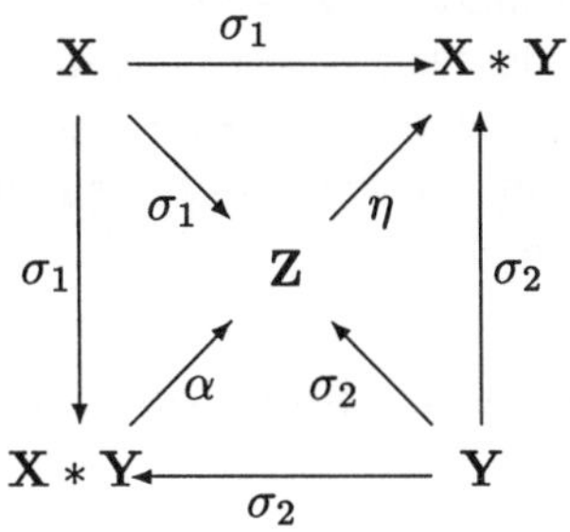

Figure 6.2 coproducts as unions

converse also holds. Recall that $\underline{\mathbf{M}}$ is called self injective if every homomorphism from a subalgebra of $\underline{\mathbf{M}}$ into $\underline{\mathbf{M}}$ extends to an endomorphism of $\underline{\mathbf{M}}$. It turns out, in the context of the Unary Structure Theorem, that this simple finitary condition actually implies full injectivity of $\underline{\mathbf{M}}$ in $\mathcal{A}$! (See Exercise 5.16(iii) for the algebraic import of this result.)

2.4 Unary Total Structure Theorem *The following are equivalent:*

(i) *there is a total structure* $\underset{\sim}{\mathbf{M}}$ *with* $G \subseteq K \cup \mathrm{End}(\underline{\mathbf{M}})$ *(and* R *finite) which yields a strong duality on* $\mathcal{A}$*;*

(ii) *there is a total structure* $\underset{\sim}{\mathbf{M}}$ *(of finite type) such that substructures of powers of* $\underset{\sim}{\mathbf{M}}$ *are closed under unions and* $\underset{\sim}{\mathbf{M}}$ *yields a strong duality on* $\mathcal{A}$*;*

(iii) *there is a total structure* $\underset{\sim}{\mathbf{M}}$ *(of finite type) such that coproducts in* $\mathcal{X}$ *are given by union and* $\underset{\sim}{\mathbf{M}}$ *yields a strong duality on* $\mathcal{A}$*;*

(iv) $\underline{\mathbf{M}}$ *has a near-unanimity term, every non-trivial subalgebra of* $\underline{\mathbf{M}}$ *is subdirectly irreducible and* $\underline{\mathbf{M}}$ *is injective in* $\mathcal{A}$*;*

(v) $\underline{\mathbf{M}}$ *has a near-unanimity term, every non-trivial subalgebra of* $\underline{\mathbf{M}}$ *is subdirectly irreducible and* $\underline{\mathbf{M}}$ *is self injective.*

Proof The equivalence of (i), (ii) and (iv) follows from the corresponding equivalence in the Unary Structure Theorem together with the Total Structure Theorem, (i) implies (iii) by Lemma 6.2.3, and (iv) trivially implies (v). We shall complete the proof by showing that (v) implies (iv) and that (iii) implies (ii).

Assume (v). By the NU Strong Duality Theorem we have a structure $\underset{\sim}{\mathbf{M}}$ yielding a strong duality on $\mathcal{A}$ in which each operation and partial operation is at most unary. Applying the fact that $\underline{\mathbf{M}}$ is self injective together with the Total Structure Theorem ('In particular,... '), we can replace $\underset{\sim}{\mathbf{M}}$ with a total structure $\underset{\sim}{\mathbf{M}}'$ which yields a strong duality on $\mathcal{A}$. Now (iv) follows from the implication (i)$\Rightarrow$(iii) of the Total Structure Theorem.

Finally assume (iii) and let $\mathbf{X}$ and $\mathbf{Y}$ be substructures of $\underset{\sim}{\mathbf{M}}^S$. By the definition of coproduct, there is a morphism $\varphi\colon \mathbf{X} * \mathbf{Y} \to \underset{\sim}{\mathbf{M}}^S$ such that $\varphi \circ \sigma_1$ and $\varphi \circ \sigma_2$ are the inclusion maps of $\mathbf{X}$ and $\mathbf{Y}$ into $\underset{\sim}{\mathbf{M}}^S$, respectively. By part (iv) of the Total Structure Theorem, the image of φ, namely $\mathbf{X} \cup \mathbf{Y}$, is a substructure of $\underset{\sim}{\mathbf{M}}^S$. ■

6.3 Better coproducts and logarithmicity

As we observed in the last section, coproducts always exist in $\mathcal{X}$ under a strong duality and are dual to direct products in $\mathcal{A}$. We could therefore ask that a useful duality should provide a simple means of constructing the coproduct of any two structures in $\mathcal{X}$, thereby giving us a nice representation of a finite product of algebras from representations of the factors. For example, the coproduct of two Boolean spaces is just their disjoint union. Under the conditions of the Unary Total Structure Theorem coproducts are always obtainable as the union of copies of the cofactors. Unfortunately, this in itself is not enough information to allow us to construct a coproduct from its cofactors for two reasons:

(i) we need to know what the intersection of the cofactors is to be;

(ii) we need to know how to define non-unary relations on the union.

The case of Boolean algebras is easy because the intersection is empty and there are no non-unary relations in the dual category. In this section we will determine exactly when there is a simple resolution to (ii). But first we show that there is always a simple answer to (i).

Recall from Chapter 3 that, in the presence of a full duality, the (possibly empty) set K of elements which form trivial subalgebras of $\underline{\mathbf{M}}$ constitutes a special substructure $\mathbf{K}$ of $\underset{\sim}{\mathbf{M}}$. Lemma 3.1.2(iii) told us that, for $\mathbf{X} \in \mathcal{X}$, there is a unique embedding of $\mathbf{K}$ into $\mathbf{X}$ whose image is denoted by $\mathbf{K}_{\mathbf{X}}$. If $\mathbf{A} \in \mathcal{A}$ and $a \in K$, we take $a_{D(\mathbf{A})}$ to be the constant map from A to a. Then

$$\mathbf{K}_{D(\mathbf{A})} = \{a_{D(\mathbf{A})} \mid a \in K\}.$$

We can now describe the intersection of cofactors in a coproduct under a full duality.

3.1 Lemma *If $\underset{\sim}{\mathbf{M}}$ yields a full duality on $\mathcal{A}$, then for any coproduct*

$$\mathbf{X} \underset{\sigma_1}{\to} \mathbf{X} * \mathbf{Y} \underset{\sigma_2}{\leftarrow} \mathbf{Y}$$

in $\mathcal{X}$, we have $\sigma_1(\mathbf{X}) \cap \sigma_2(\mathbf{Y}) = \mathbf{K}_{\mathbf{X}*\mathbf{Y}}$.

Proof We represent $\mathbf{X}$ and $\mathbf{Y}$ as $D(\mathbf{A})$ and $D(\mathbf{B})$, represent $\mathbf{X} * \mathbf{Y}$ as $D(\mathbf{A} \times \mathbf{B})$, and σ_1 and σ_2 as $D(\pi_1)$ and $D(\pi_2)$. Then for $u \in D(\mathbf{A} \times \mathbf{B})$

we observe that

$$\begin{aligned} &u \in \sigma_1(\mathbf{X}) \cap \sigma_2(\mathbf{Y}) = D(\pi_1)(D(\mathbf{A})) \cap D(\pi_2)(D(\mathbf{B})) \\ \iff &u = f \circ \pi_1 = g \circ \pi_2 \text{ for some } f \in D(\mathbf{A}) \text{ and } g \in D(\mathbf{B}) \\ \iff &u(c,d) = f(c) = g(d) \text{ for all } c \in A, d \in B \end{aligned}$$

which says precisely that $u = a_{D(\mathbf{A}\times\mathbf{B})}$ for some $a \in K$. ∎

Assume the conditions of the Unary Structure Theorem: $\underset{\sim}{\mathbf{M}}$ yields a strong duality on $\mathcal{A}$ and each member of $G \cup H$ is at most unary. How can we obtain a constructive description of a coproduct of two structures $\mathbf{X}$ and $\mathbf{Y}$ in $\mathfrak{X}$? In order to do this we shall define the **direct union** of $\mathbf{X}$ and $\mathbf{Y}$ to be the structure whose base set is the union of their base sets with $\mathbf{K_X}$ and $\mathbf{K_Y}$ amalgamated, whose topology is the unique topology which makes the inclusion maps embeddings, and whose operations, partial operations and relations are each the union of the corresponding operations, partial operations and relations of $\mathbf{X}$ and $\mathbf{Y}$. If each member of $G \cup H$ is at most unary and the coproduct of every pair of structures in $\mathfrak{X}$ is their direct union, we shall say that **coproducts in $\mathfrak{X}$ are given by direct union**. In general, of course, it is not necessary that $\mathfrak{X}$ is even closed under direct unions. When it is, we can use the following observation which is actually independent of duality.

3.2 Lemma *If each member of $G \cup H$ is at most unary and $\mathfrak{X}$ is closed under direct unions, then coproducts in $\mathfrak{X}$ are given by direct union.*

Proof Assume that $\mathfrak{X}$ is closed under direct unions. Let $\mathbf{X}, \mathbf{Y} \in \mathfrak{X}$ and let

$$\mathbf{X} \underset{\gamma_1}{\rightarrow} \mathbf{Z} \underset{\gamma_2}{\leftarrow} \mathbf{Y}$$

be embeddings of $\mathbf{X}$ and $\mathbf{Y}$ into their direct union $\mathbf{Z}$. To show that $\mathbf{Z}$ is their coproduct, consider any $\mathbf{W} \in \mathfrak{X}$ and morphisms $\rho_1 \colon \mathbf{X} \to \mathbf{W}$ and $\rho_2 \colon \mathbf{Y} \to \mathbf{W}$. Let $\tau \colon Z \to W$ be the unique map such that $\tau \circ \gamma_i = \rho_i$ for $i = 1, 2$. From the definition of direct union it is now immediately verifiable that τ is a morphism in $\mathfrak{X}$. ∎

If $\underset{\sim}{\mathbf{M}}$ yields a strong duality and coproducts in $\mathfrak{X}$ are given by direct union, then we may think of the duality as relating products in $\mathcal{A}$ to 'sums' in $\mathfrak{X}$. Under these conditions we will say that $\underset{\sim}{\mathbf{M}}$ yields a **logarithmic duality** on $\mathcal{A}$. If $\underset{\sim}{\mathbf{M}}$ yields a strong duality with every member of $G \cup H$ at most unary, then Lemmas 6.2.3 and 6.3.1 combine to give us a fairly complete description of coproducts: the coproduct of $\mathbf{X}$ and $\mathbf{Y}$ is a structure whose base set is the union of their base sets with $\mathbf{K_X}$ and $\mathbf{K_Y}$ amalgamated, whose topology is the unique topology which makes the inclusion maps embeddings, and whose operations and partial operations are each the union of the corresponding operations and partial operations of $\mathbf{X}$ and $\mathbf{Y}$. It remains only to describe relations on the coproduct. For example, if in addition every member of R is unary, then $\mathbf{X} * \mathbf{Y}$ is exactly the direct union of $\mathbf{X}$ and $\mathbf{Y}$ and we have a logarithmic duality. On the other hand, if some non-unary relation of R relates elements of $\mathbf{X} * \mathbf{Y}$ that are neither all in $\sigma_1(X)$ nor all in $\sigma_2(Y)$, then $\mathbf{X} * \mathbf{Y}$ will not be the direct union.

In the next section we will examine conditions under which relations are also unary. Unfortunately, many interesting instances of the Unary Structure Theorem that we have seen require binary relations, often in the form of order relations. Consider, for example, the duality between distributive lattices $\langle L; \vee, \wedge \rangle$ and bounded Priestley spaces $\langle X; 0, 1, \leq, \mathcal{T} \rangle$ of Theorem 1.2.5. In this case $K = \{0, 1\}$ and it is immediately verifiable that $\mathfrak{X}$ is closed under direct union. Applying Lemma 6.3.2 we conclude that, in spite of the presence of a binary relation, Priestley duality is indeed logarithmic.

Rather surprisingly we find that Priestley duality is not unique in this respect. Examining them case by case, we find that all of the strong dualities mentioned in the previous section as illustrations of the Unary Total Structure Theorem turn out to be logarithmic as well! This phenomenon is explained by the following theorem, which gives a simple finitary test for logarithmicity. The key turns out to be a special property of the relations in R. Recall from Chapter 5 that an n-ary algebraic relation r (for $n \geq 2$) **avoids binary products** if each binary projection

$$\pi_{ij}(r) := \{ (\pi_i(a), \pi_j(a)) \in M^2 \mid a \in r \},$$

for $1 \leqslant i < j \leqslant n$, contains no product $A \times B$ of non-trivial subalgebras of $\underline{\mathbf{M}}$. The same notion is precisely what is needed here.

3.3 Logarithmic Duality Theorem *Assume the conditions of the Unary Structure Theorem: $\underline{\mathbf{M}}$ has a $(k+1)$-ary near-unanimity term and every non-trivial subalgebra of $\underline{\mathbf{M}}$ is subdirectly irreducible. Then the following are equivalent:*

(i) *there is an $\underset{\sim}{\mathbf{M}}$ which yields a logarithmic duality on $\mathcal{A}$;*

(ii) *there is an $\underset{\sim}{\mathbf{M}}$ which yields a strong duality on $\mathcal{A}$ where the coproduct in $\mathfrak{X}$ of every pair of finite structures is given by direct union;*

(iii) *there is an $\underset{\sim}{\mathbf{M}}$ which yields a strong duality on $\mathcal{A}$ where for all $\mathbf{A}, \mathbf{B} \leqslant \underline{\mathbf{M}}$, the coproduct of $D(\mathbf{A})$ and $D(\mathbf{B})$ is given by direct union;*

(iv) *there is an $\underset{\sim}{\mathbf{M}}$ which yields a strong duality on $\mathcal{A}$ where for all $n \geqslant 2$, every n-ary relation $r \in R$ avoids binary products;*

(v) *the set of k-ary algebraic relations r which avoid binary products entails every k-ary algebraic relations on $\underline{\mathbf{M}}$.*

Moreover, any choice of $\underset{\sim}{\mathbf{M}}$ *satisfying one of* (i),(ii),(iii) *or* (iv) *also satisfies the others.*

Proof The implications (i)$\Rightarrow$(ii)$\Rightarrow$(iii) are trivial. Assume that (iii) holds. Taking the same choice of $\underset{\sim}{\mathbf{M}}$, we prove (iv). Let $r \in R$ be n-ary for some $n \geqslant 2$ and assume that there are subalgebras $\mathbf{A}$ and $\mathbf{B}$ of $\underline{\mathbf{M}}$ such that $A \times B \subseteq \pi_{ij}(r)$ for some i and j. We will show that either $\mathbf{A}$ or $\mathbf{B}$ must be trivial. We represent the coproduct of $D(\mathbf{A})$ and $D(\mathbf{B})$, with its associated morphisms, as

$$D(\mathbf{A}) \underset{\sigma_1}{\to} D(\mathbf{A} \times \mathbf{B}) \underset{\sigma_2}{\leftarrow} D(\mathbf{B})$$

where $\sigma_1(x) = x \circ \pi_1$ and $\sigma_2(y) = y \circ \pi_2$ for $x \in D(\mathbf{A})$ and $y \in D(\mathbf{B})$. Now choose $x \colon \mathbf{A} \to \underline{\mathbf{M}}$ and $y \colon \mathbf{B} \to \underline{\mathbf{M}}$ to be the two inclusions maps. Then we have

$$\begin{aligned} & A \times B \subseteq \pi_{ij}(r) \\ \Longrightarrow\ & (x(a), y(b)) \in \pi_{ij}(r) \text{ for all } (a,b) \in A \times B \\ \Longrightarrow\ & (x \circ \pi_1(a,b), y \circ \pi_2(a,b)) \in \pi_{ij}(r) \text{ for all } (a,b) \in A \times B \\ \Longrightarrow\ & (x \circ \pi_1, y \circ \pi_2) \in \pi_{ij}(r) \text{ on } D(\mathbf{A} \times \mathbf{B}) \\ \Longrightarrow\ & (\sigma_1(x), \sigma_2(y)) \in \pi_{ij}(r) \text{ on } D(\mathbf{A} \times \mathbf{B}). \end{aligned}$$

Since $D(\mathbf{A} \times \mathbf{B})$ is the direct union of $\sigma_1(D(\mathbf{A}))$ and $\sigma_2(D(\mathbf{B}))$ it follows that either $x \circ \pi_1 = \sigma_1(x) \in K_{D(\mathbf{A}\times\mathbf{B})}$ or $y \circ \pi_2 = \sigma_2(y) \in K_{D(\mathbf{A}\times\mathbf{B})}$. Thus one of the inclusions x or y maps to a single element of $\mathbf{K}$, and therefore either $\mathbf{A}$ or $\mathbf{B}$ is trivial. This proves that (iv) holds.

Now assume that (iv) holds in order to prove (i). Let $\mathbf{A}, \mathbf{B} \in \mathcal{A}$, let $r \in R$ be n-ary for some $n \geqslant 2$ and let $z_1, \dots, z_n \in D(\mathbf{A} \times \mathbf{B})$ with $(z_1, \dots, z_n) \in r$. Assume that there exist i and j such that $z_i \in \sigma_1(D(\mathbf{A}))$ and $z_j \in \sigma_2(D(\mathbf{B}))$. We must show that either $z_i \in K_{D(\mathbf{A}\times\mathbf{B})}$ or $z_j \in K_{D(\mathbf{A}\times\mathbf{B})}$.

Let $z_i = x \circ \pi_i$ and $z_j = y \circ \pi_j$ for some $x \in D(\mathbf{A})$ and $y \in D(\mathbf{B})$. Then

$$\begin{aligned} & (z_1, \dots, z_n) \in r \\ \Longrightarrow\ & (z_i(a,b), z_j(a,b)) \in \pi_{ij}(r) \text{ for all } (a,b) \in A \times B \\ \Longrightarrow\ & (x \circ \pi_i(a,b), y \circ \pi_j(a,b)) \in \pi_{ij}(r) \text{ for all } (a,b) \in A \times B \\ \Longrightarrow\ & (x(a), y(b)) \in \pi_{ij}(r) \text{ for all } (a,b) \in A \times B \\ \Longrightarrow\ & x(A) \times y(B) \subseteq \pi_{ij}(r). \end{aligned}$$

Thus, by (iv), either $x(A)$ is trivial, and hence $z_i = x \circ \pi_i \in K_{D(\mathbf{A}\times\mathbf{B})}$, or $y(B)$ is trivial, in which case $z_j = y \circ \pi_j \in K_{D(\mathbf{A}\times\mathbf{B})}$. Finally, (iv) $\Leftrightarrow$ (v) by the NU Strong Duality Theorem. ■

According to Lemma 6.3.1, images of cofactors in a coproduct in $\mathcal{X}$ will be disjoint exactly when $\mathbf{K} = \underset{\sim}{\varnothing}$. If coproducts in $\mathcal{X}$ are given by unions and $\mathbf{K} = \underset{\sim}{\varnothing}$, we say that **coproducts in $\mathcal{X}$ are given by disjoint union**. Combining the representation of the congruence lattices given in Theorem 3.2.1 with the Unary Total Structure Theorem we immediately obtain a full description of the conditions under which this happens.

3.4 Unary Total Structure Without Constants Theorem *The following are equivalent:*

(i) *there is a total structure $\underset{\sim}{\mathbf{M}}$ with $G \subseteq \operatorname{End}(\underline{\mathbf{M}})$ (and R finite), such that the monoid of maps generated by G contains no constant maps, which yields a strong duality on $\mathcal{A}$;*

(ii) *there is a total structure $\underset{\sim}{\mathbf{M}}$ (of finite type) such that substructures of powers of $\underset{\sim}{\mathbf{M}}$ are closed under unions, the enriched partial clone of $\underset{\sim}{\mathbf{M}}$ contains no constant maps, and $\underset{\sim}{\mathbf{M}}$ yields a strong duality on $\mathcal{A}$;*

(iii) *there is a total structure $\underset{\sim}{\mathbf{M}}$ (of finite type) such that coproducts in $\mathcal{X}$ are given by disjoint union and $\underset{\sim}{\mathbf{M}}$ yields a strong duality on $\mathcal{A}$;*

(iv) *$\underline{\mathbf{M}}$ has a near-unanimity term, every subalgebra of $\underline{\mathbf{M}}$ is subdirectly irreducible, $\underline{\mathbf{M}}$ has no one-element subalgebras and is injective in $\mathcal{A}$;*

(v) *$\underline{\mathbf{M}}$ has a near-unanimity term, every subalgebra of $\underline{\mathbf{M}}$ is subdirectly irreducible, $\underline{\mathbf{M}}$ has no one-element subalgebras and is self injective.*

6.4 Exclusively unary structures and arithmeticity

In this last section we will look for strong dualities in which the operations are nullary or unary, and the partial operations *and* relations of $\underset{\sim}{\mathbf{M}}$ are unary. We then say that the members of $\mathcal{X}$ are **exclusively unary structures**. Such dualities will be logarithmic. The Arithmetic Strong Duality Theorem 3.3.11 gives us sufficient conditions for $\mathcal{A}$ to admit a strong duality by exclusively unary structures if $\mathcal{A}$ is a variety: non-trivial subalgebras of $\underline{\mathbf{M}}$ must be subdirectly irreducible and $\mathcal{A}$ must be arithmetical. We will show that these conditions are also necessary.

4.1 Lemma *Suppose that each member of $G \cup H$ is at most unary and each member of R is either unary or is a congruence on a subalgebra of $\underline{\mathbf{M}}$. If $\underset{\sim}{\mathbf{M}}$ yields a duality on $\mathcal{A}$ and $\underset{\sim}{\mathbf{M}}$ is injective in $\mathcal{X}_{\text{fin}}$, then $\underline{\mathbf{M}}$ has a Pixley arithmeticity term.*

Proof Let $X \subseteq M^3$ consist of all triples (a,b,b), (a,b,a) and (b,b,a) where $a,b \in M$. Since each member of $G \cup H$ is at most unary, X determines a substructure $\mathbf{X}$ of $\underset{\sim}{\mathbf{M}}^3$. Define $\sigma\colon X \to M$ by $\sigma(a,b,b) = \sigma(a,b,a) = \sigma(b,b,a) = a$. Clearly σ preserves every distinguished element, every unary (partial) operation and every unary relation. It is also straightforward to check that σ preserves every congruence relation on a subalgebra of $\underline{\mathbf{M}}$. Consequently σ is a morphism which, by injectivity, extends to a morphism from $\underset{\sim}{\mathbf{M}}^3$ into $\underset{\sim}{\mathbf{M}}$. By (CLO), this morphism is a term function of $\underline{\mathbf{M}}$. ∎

We will now see that if a strong duality satisfying the conditions of this lemma is available at all, then relations can be eliminated altogether.

4.2 Unary Partial Algebra Theorem *Assume that every homomorphic image of a subalgebra of $\underline{\mathbf{M}}$ is in $\mathcal{A}$. Then the following are equivalent:*

(i) *there is an $\underset{\sim}{\mathbf{M}}$ which yields a strong (necessarily logarithmic) duality on $\mathcal{A}$ where each member of $G \cup H$ is at most unary and R is empty;*

(ii) *there is an $\underset{\sim}{\mathbf{M}}$ which yields a strong duality on $\mathcal{A}$ where each member of $G \cup H$ is at most unary and each member of R is either unary or is a congruence;*

(iii) *$\underline{\mathbf{M}}$ generates an arithmetical variety and every non-trivial subalgebra of $\underline{\mathbf{M}}$ is subdirectly irreducible.*

Proof Trivially (i) implies (ii), (ii) implies (iii) by Lemma 6.4.1 and the Unary Structure Theorem and (iii) implies (i) by the Arithmetic Strong Duality Theorem 3.3.11 and Theorem 6.3.3. ∎

Many strong dualities via unary partial algebras arise from the Quasi-primal Strong Duality Theorem 3.3.13 and from the Heyting Strong Duality Theorem 4.2.2.

Of course the hypothesis of the Unary Partial Algebra Theorem is not in general available, for often the quasi-variety $\mathcal{A}$ is not closed under homomorphic images and is therefore much smaller than the variety generated by $\underline{\mathbf{M}}$. That so many of the quasi-varieties we consider do turn out to be varieties is due to a rather surprising fact: strong duality via an exclusively unary *total* structure actually implies closure of $\mathcal{A}$ under homomorphisms! The proof of this fact is a fine example of dualities at work. Our argument depends upon viewing a congruence θ on an algebra $\mathbf{A}$ as a subalgebra of $\mathbf{A} \times \mathbf{A}$, and then considering the dual $D(\theta)$ and the coproduct $D(\mathbf{A}) * D(\mathbf{A})$.

4.3 Theorem *Assume that an exclusively unary total structure $\underset{\sim}{\mathbf{M}}$ yields a strong duality on $\mathcal{A}$. Then $\mathcal{A}$ is closed under homomorphic images and hence is the variety generated by $\underline{\mathbf{M}}$.*

Proof Represent $\mathbf{A} \in \mathcal{A}$ via the bijection $e_{\mathbf{A}} : \mathbf{A} \to ED(\mathbf{A}) \leqslant \underline{\mathbf{M}}^{D(A)}$. Under a strong duality Theorem 3.2.1 asserts that every $\mathcal{A}$-congruence on $\mathbf{A}$ is induced by the projection onto a closed substructure $\mathbf{Z} \leqslant D(\mathbf{A})$.

Under a strong duality via an exclusively unary total structure, we will show that *every* congruence θ on $\mathbf{A}$ is induced by such a projection.

By the Unary Total Structure Theorem the coproduct

$$D(\mathbf{A}) \underset{D(\pi_1)}{\rightarrow} D(\mathbf{A}) * D(\mathbf{A}) \underset{D(\pi_2)}{\leftarrow} D(\mathbf{A})$$

is the union $(D(\mathbf{A}) \circ \pi_1) \cup (D(\mathbf{A}) \circ \pi_2)$ of two copies of $D(\mathbf{A})$. Let ρ_i be the restriction of π_i to θ. By the Total Structure Theorem, $\underline{\mathbf{M}}$ is injective in $\mathcal{A}$ and consequently

$$D(\theta) = (D(\mathbf{A}) \circ \rho_1) \cup (D(\mathbf{A}) \circ \rho_2).$$

Let $Z = \{x \in D(\mathbf{A}) \mid \theta \subseteq \ker x\}$ and let $\theta_Z = \bigcap\{\ker x \mid x \in Z\}$. For $a, b \in A$, notice that $(a, b) \in \theta_Z$ if and only if $x(a) = x(b)$ for all $x \in Z$ if and only if $e_{\mathbf{A}}(a)(x) = e_{\mathbf{A}}(b)(x)$ for all $x \in Z$, that is, $e_{\mathbf{A}}(a)$ and $e_{\mathbf{A}}(b)$ agree on Z. Thus θ_Z is just the congruence induced by the projection onto Z.

Clearly $\theta \subseteq \theta_Z$. We will show that $\theta = \theta_Z$. Let $(a, b) \in \theta_Z$ and let $\gamma\colon D(\theta) \to \underset{\sim}{\mathbf{M}}$ where $\gamma(u \circ \rho_1) = u(a)$ and $\gamma(v \circ \rho_2) = v(b)$. To see that γ is well defined, assume that $u \circ \rho_1 = v \circ \rho_2$. Then, for any $c \in A$, we have $(c, c) \in \theta$ and therefore

$$u(c) = (u \circ \rho_1)(c, c) = (v \circ \rho_2)(c, c) = v(c).$$

Thus $u = v$. Now for every $(c, d) \in \theta$, we observe that

$$u(c) = (u \circ \rho_1)(c, d) = (v \circ \rho_2)(c, d) = v(d) = u(d).$$

This means that $\theta \subseteq \ker u$ and, since $(a, b) \in \theta_Z$, that $u(a) = u(b) = v(b)$.

To see that γ is a continuous homomorphism it is sufficient, since $D(\theta)$ is an exclusively unary structure, to check that its restriction to each $D(\mathbf{A}) \circ \rho_i$ is a continuous homomorphism. For $u \in D(\mathbf{A})$ we have

$$(\gamma \circ D(\pi_1))(u) = \gamma(D(\pi_1)(u)) = \gamma(u \circ \rho_1) = u(a) = e_{\mathbf{A}}(a)(u).$$

Consequently on $D(\mathbf{A}) \circ \rho_1$ we see that γ agrees with the continuous homomorphism $e_{\mathbf{A}}(a) \circ D(\pi_1)^{-1}$. Similarly, γ agrees with $e_{\mathbf{A}}(b) \circ D(\pi_2)^{-1}$ on $D(\mathbf{A}) \circ \rho_2$.

Since $\gamma\colon D(\theta) \to \underset{\sim}{\mathbf{M}}$ is in $ED(\theta)$, we conclude from duality that there is a pair $(c, d) \in \theta$ such that $\gamma = e_\theta(c, d)$. Thus $u(c) = (u \circ \rho_1)(c, d) = \gamma(u \circ \rho_1) = u(a)$ for all $u \in D(\mathbf{A})$. Since $\mathbf{A} \in \mathcal{A}$, we conclude that $c = a$. Similarly, $d = b$ and $(a, b) = (c, d) \in \theta$, showing that $\theta = \theta_Z$ as claimed. Thus $\mathbf{A}/\theta$ is embedded in $\underline{\mathbf{M}}^Z$ and is therefore in $\mathcal{A}$. ∎

4.4 Exclusively Unary Total Structure Theorem *The following are equivalent:*

(i) *there is an exclusively unary total structure* $\underset{\sim}{\mathrm{M}}$ *that yields a (necessarily logarithmic) strong duality on* $\mathcal{A}$*;*

(ii) $\mathcal{A}$ *is an arithmetical variety in which* $\underline{\mathbf{M}}$ *is injective and every non-trivial subalgebra of* $\underline{\mathbf{M}}$ *is subdirectly irreducible;*

(iii) $\underline{\mathbf{M}}$ *has a Pixley arithmeticity term,* $\underline{\mathbf{M}}$ *is self injective, and if* $\mathbf{Q}$ *is a non-trivial subalgebra of* $\underline{\mathbf{M}}$ *then* $\mathbf{Q}$ *is subdirectly irreducible and every homomorphic image of* $\mathbf{Q}$ *is in* $\mathcal{A}$.

Proof Self injectivity in (iii) implies injectivity in (ii) by the Unary Total Structure Theorem. Otherwise the equivalence of (ii) and (iii) follows from Pixley's Theorem A.17 and Jónsson's Second Theorem A.16. Now assume (i). Then Theorem 6.4.3 and the Unary Partial Algebra Theorem 6.4.2 assure us that $\mathcal{A}$ is an arithmetical variety. By the Total Structure Theorem, $\underline{\mathbf{M}}$ is injective in $\mathcal{A}$, and every non-trivial subalgebra of $\underline{\mathbf{M}}$ is subdirectly irreducible by the Unary Structure Theorem. This proves (ii). Finally, (ii) implies (i) by the Arithmetic Strong Duality Theorem 3.3.11, the Total Structure Theorem 6.1.2 and Theorem 6.3.3. ∎

Many examples of the Exclusively Unary Total Structure Theorem can be found as applications of the Semi-primal Strong Duality Theorem 3.3.14. In Theorem 4.2.3 we saw examples of Heyting algebras which are strongly dualisable in an even better sense: each member of G is at most unary, that is, $G \subseteq K \cup \mathrm{End}(\underline{\mathbf{M}})$, and H and R are both empty. When this occurs we said that $\underline{\mathbf{M}}$ was **strongly endodualisable**.

4.5 Unary Total Algebra Theorem *The following are equivalent:*

(i) $\underline{\mathbf{M}}$ *is strongly endodualisable;*

(ii) $\underline{\mathbf{M}}$ *has a Pixley arithmeticity term, is self injective and each non-trivial subalgebra* $\mathbf{Q}$ *of* $\underline{\mathbf{M}}$ *satisfies these conditions–*

   (a) $\mathbf{Q}$ *is subdirectly irreducible,*

   (b) *each homomorphic image of* $\mathbf{Q}$ *is in* $\mathcal{A}$,

   (c) $\mathbf{Q}$ *is an intersection of equalisers of endomorphisms of* $\underline{\mathbf{M}}$.

Proof Assuming (i), parts (a) and (b) of (ii) follow from the Exclusively Unary Total Structure Theorem and part (c) follows from the Total Algebra Theorem 6.1.3. Now assume (ii). From the Exclusively Unary Total Structure Theorem we see that $\underset{\sim}{\mathrm{M}}' = \langle M; K \cup \mathrm{End}(\underline{\mathbf{M}}), \mathcal{B}_1, \mathcal{T}\rangle$ yields a strong duality on $\mathcal{A}$. To see that $\underset{\sim}{\mathrm{M}} = \langle M; K \cup \mathrm{End}(\underline{\mathbf{M}}), \mathcal{T}\rangle$ also yields a strong duality on $\mathcal{A}$ we must, by the $\underset{\sim}{\mathrm{M}}$-Shift Strong Duality Theorem, check that any endomorphism-preserving operation on M also preserves subalgebras of $\underline{\mathbf{M}}$. For non-trivial subalgebras of $\underline{\mathbf{M}}$, this is a consequence of (c). For trivial subalgebras this follows from the fact that a trivial subalgebra is a fixpointset of a constant endomorphism. ■

Finally, we prove the converse of our first example from Chapter 4 which characterises strong dualities that satisfy the ultimate constraint on the type of the dual structures.

4.6 Primal Algebra Theorem *The following are equivalent:*

(i) *the structure* $\underset{\sim}{\mathbf{M}}$ *with no operations, partial operations or relations yields a strong duality on* $\mathcal{A}$;

(ii) $\underline{\mathbf{M}}$ *is a primal algebra.*

Proof By (CLO) (see the First Duality Theorem 2.2.2), $\underline{\mathbf{M}}$ is primal if $\underset{\sim}{\mathbf{M}}$ yields a duality on $\mathcal{A}$, and (ii) implies (i) by Theorem 4.1.1(i). ■

Exercises 6

6.1 Prove the following sharpened local version of the equivalence between (ii) and (iv) of the Total Structure Theorem 6.1.2 by exploiting the schizophrenic nature of partial maps as both morphisms in $\mathcal{A}$ and partial operations on the objects in $\mathfrak{X}$.

Proposition *Let* $n \in \mathbb{N}$, *let* $\mathbf{A}$ *be a subalgebra of* $\underline{\mathbf{M}}^n$, *denote the inclusion map by* $u\colon \mathbf{A} \to \underline{\mathbf{M}}^n$ *and let* $h\colon \mathbf{A} \to \underline{\mathbf{M}}$ *be a homomorphism. Then* h *extends to a homomorphism* $g\colon \underline{\mathbf{M}}^n \to \underline{\mathbf{M}}$ *if and only if the image of* $D(\underline{\mathbf{M}}^n)$ *under the morphism* $D(u)$ *is closed under* h.

The remaining exercises provide refinements of the Total Structure Theorem 6.1.2 and the Total Algebra Theorem 6.1.3 and thereby address the problem of whether every full duality is strong. We shall see that the injectivity of $\underline{\mathbf{M}}$ and $\underset{\sim}{\mathbf{M}}$ is intimately connected with the surjectivity of epics. See Exercise 3.11 to Exercise 3.14 for basic results concerning epics morphisms.

6.2 (i) [Dan97] Prove that every finite algebra $\underline{\mathbf{M}}$ which satisfies (FEQ) is dualisable. [Hint: Show that the (enriched) hom-clone of $\underline{\mathbf{M}}$ entails $\mathcal{B}$, then apply the Second Duality Theorem.]

(ii) [Dan97] Show that if every algebra in $\mathcal{A}_{\text{fin}}$ is injective in $\mathcal{A}_{\text{fin}}$, then $\underline{\mathbf{M}}$ satisfies (FEQ) and therefore is strongly dualised by a total algebra.

6.3 Assume that $\underset{\sim}{\mathbf{M}}$ yields a duality on $\mathcal{A}$ and that $\underline{\mathbf{M}}$ is injective in $\mathcal{A}$ with respect to epic embeddings. Show that $u\colon \mathbf{A} \to \mathbf{B}$ is a surjection in $\mathcal{A}$ if (and only if) $D(u)\colon D(\mathbf{B}) \to D(\mathbf{A})$ is an embedding in $\mathfrak{X}$.

6.4 Show that the following are equivalent given that $\underset{\sim}{\mathbf{M}}$ yields a full duality on $\mathcal{A}$:

(i) epics are surjective in $\mathcal{A}$;

(ii) (a) $\underline{\mathbf{M}}$ is injective in $\mathcal{A}$ with respect to epic embeddings, and
(b) every one-to-one morphism in $\mathcal{X}$ is an embedding.

6.5 Show that the following are equivalent given that $\underset{\sim}{\mathbf{M}}$ yields a full duality on $\mathcal{A}$:

(i) epics are surjective in $\mathcal{X}$;

(ii) (a) $\underset{\sim}{\mathbf{M}}$ is injective in $\mathcal{X}$ with respect to epic embeddings, and
(b) the image of every morphism in $\mathcal{X}$ is a (closed) substructure;

(iii) $\underline{\mathbf{M}}$ is injective in $\mathcal{A}$;

(iv) (a) $\underset{\sim}{\mathbf{M}}$ is injective in $\mathcal{X}$, and
(b) the image of every morphism in $\mathcal{X}$ is a (closed) substructure;

(v) (a) if $\mathbf{Y} \in \mathcal{X}$ and $\mathbf{X} \lneqq \mathbf{Y}$, then there exist morphisms $\alpha, \beta \colon \mathbf{Y} \to \underset{\sim}{\mathbf{M}}$ such that $X \subseteq \mathrm{eq}(\alpha, \beta) \subsetneqq Y$, and
(b) the image of every morphism in $\mathcal{X}$ is a (closed) substructure;

(vi) $\underline{\mathbf{M}}$ is injective in $\mathcal{A}$ and $\underset{\sim}{\mathbf{M}}$ is injective in $\mathcal{X}$;

(vii) $\underset{\sim}{\mathbf{M}}$ is structurally equivalent to a total structure and $\underset{\sim}{\mathbf{M}}$ is injective in $\mathcal{X}$;

(viii) some total structure yields a strong duality on $\mathcal{X}$.

6.6 Assume that $\underset{\sim}{\mathbf{M}}$ yields a full duality on $\mathcal{A}$ and consider the following:

(0) $\underline{\mathbf{M}}$ is injective in $\mathcal{A}$;

(1) epics are surjective in $\mathcal{X}$;

(1)* epics are surjective in $\mathcal{A}$;

(2) $\underline{\mathbf{M}}$ is injective in $\mathcal{A}$ with respect to epic embeddings;

(3) $u \colon \mathbf{A} \to \mathbf{B}$ is a surjection in $\mathcal{A}$ if (and only if) its dual $D(u) \colon D(\mathbf{B}) \to D(\mathbf{A})$ is an embedding in $\mathcal{X}$;

(4) $\underset{\sim}{\mathbf{M}}$ is injective in $\mathcal{X}$;

(5) for each proper closed substructure $\mathbf{X}$ of $\mathbf{Y} \in \mathcal{X}$, there exist morphisms $\alpha, \beta \colon \mathbf{Y} \to \underset{\sim}{\mathbf{M}}$ with $X \subseteq \mathrm{eq}(\alpha, \beta) \subsetneqq Y$;

(6) epic embeddings are surjective in $\mathcal{X}$.

(i) Show that in general we have (1)* $\Rightarrow$ (2) and

$$(0) \iff (1) \Longrightarrow (2) \Longrightarrow (3) \iff (4) \Longrightarrow (5) \Longrightarrow (6).$$

(ii) Show that if the image of every morphism in $\mathcal{X}$ is a substructure, then (0) $\iff$ (1) $\iff$ (2) $\iff$ (3) $\iff$ (4) $\iff$ (5) $\iff$ (6).

(iii) Show that if every one-to-one morphism in $\mathcal{X}$ is an embedding, then (1)* $\iff$ (2) $\iff$ (3) $\iff$ (4) $\iff$ (5) $\iff$ (6).

7

Piggyback Dualities

Piggybacking is a technique which, loosely speaking, says that if $\underline{\mathbf{M}}$ is dualisable, then an efficient way to find a structure $\underset{\sim}{\mathbf{M}}$ which yields a duality on $\mathcal{A} = \mathbb{ISP}\underline{\mathbf{M}}$ is to ride piggyback on a known duality for an algebra $\underline{\mathbf{D}}$ which is closely related to $\underline{\mathbf{M}}$ by using the relationship between $\underline{\mathbf{M}}$ and $\underline{\mathbf{D}}$ to extract $\underset{\sim}{\mathbf{M}}$ from the structure $\underset{\sim}{\mathbf{D}}$ which dualises $\underline{\mathbf{D}}$. For example, if $\underline{\mathbf{M}} \in \mathbb{ISP}\underline{\mathbf{D}}$, then we could try to convert a dualising structure $\underset{\sim}{\mathbf{D}}$ for $\underline{\mathbf{D}}$ into a dualising structure $\underset{\sim}{\mathbf{M}}$ for $\underline{\mathbf{M}}$ by using the fact that $\underline{\mathbf{M}}$ can be embedded into a power of $\underline{\mathbf{D}}$. Alternatively, $\underline{\mathbf{M}}$ might have an underlying bounded distributive lattice structure in which case we would choose $\underline{\mathbf{D}}$ to be the two-element bounded distributive lattice and we would try to obtain $\underset{\sim}{\mathbf{M}}$ from our knowledge of Priestley duality on the bounded distributive lattice underlying $\underline{\mathbf{M}}$.

7.1 Multisorted dualities

In the next section we shall develop a piggyback theorem for algebras with an underlying bounded distributive lattice. Since many of the applications of this result in the literature are to varieties rather than to quasi-varieties, we now give a brief overview of the modifications to our general theory which are required to extend it from the quasi-variety generated by $\underline{\mathbf{M}}$ to the variety generated by $\underline{\mathbf{M}}$.

Consider a finite algebra $\underline{\mathbf{M}}$ which possesses a near-unanimity term. The NU Strong Duality Theorem 3.3.8 shows that there is a strong duality for the quasi-variety $\mathcal{A} = \mathbb{ISP}\underline{\mathbf{M}}$ generated by $\underline{\mathbf{M}} = \langle M; F^M \rangle$. More often than not, the variety $\mathcal{V} = \mathbb{HSP}\underline{\mathbf{M}}$ generated by $\underline{\mathbf{M}}$ is larger than the quasi-variety and it is reasonable to ask how we might modify our approach so as to obtain a duality for $\mathcal{V}$. Since, by Mitschke's Theorem A.14, the existence of a near-unanimity term guarantees that $\mathcal{V}$ is congruence distributive, Jónsson's Second Theorem A.16 tells us that the subdirectly irreducible algebras in $\mathcal{V}$ lie in $\mathbb{HS}(\underline{\mathbf{M}})$, whence they are all finite and (up to isomorphism) finite in number. Hence, if we choose $\underline{\mathfrak{M}}$ to consist of one representative from each of the isomorphism classes of the maximal subdirectly irreducible algebras in $\mathcal{V}$, then $\underline{\mathfrak{M}}$ is a finite set of finite algebras and, by Birkhoff's Theorem A.10, is minimal with respect to the property that $\mathcal{V} = \mathbb{ISP}\underline{\mathfrak{M}}$. Thus it is natural to develop a duality for $\mathcal{V}$ in which the role of the finite algebra $\underline{\mathbf{M}}$ is played by the finite set $\underline{\mathfrak{M}}$ of finite algebras. This leads us immediately to multisorted dual structures.

Let $\mathfrak{M}$ be a finite set of finite algebras of type F and let $\mathcal{A} := \mathbb{ISP}\mathfrak{M}$ be the quasi-variety generated by $\mathfrak{M}$. For $\underline{\mathbf{M}}_1, \underline{\mathbf{M}}_2 \in \mathfrak{M}$, we allow the possibility that $\underline{\mathbf{M}}_1$ and $\underline{\mathbf{M}}_2$ are isomorphic but, for notational convenience, we assume that the underlying sets M_1 and M_2 are disjoint. Although the finiteness assumptions on $\mathfrak{M}$ can be relaxed (see Davey and Priestley [DP87] on which this and the next section are based), we shall not do so here as the stipulation that $\mathfrak{M}$ is a finite set of finite algebras ensures that the results of the previous chapters can be lifted up to this more general setting.

The structure $\underset{\sim}{\mathbf{M}}$ must now be replaced with a **multisorted structure**

$$\underset{\sim}{\mathfrak{M}} = \langle \dot{\bigcup}\{\, M \mid M \in \mathfrak{M} \,\}; G^{\underset{\sim}{\mathfrak{M}}}, H^{\underset{\sim}{\mathfrak{M}}}, R^{\underset{\sim}{\mathfrak{M}}}, \mathcal{T}^{\underset{\sim}{\mathfrak{M}}} \rangle,$$

where

(i) $G^{\underset{\sim}{\mathfrak{M}}}$ consists of a homomorphism $g^{\underset{\sim}{\mathfrak{M}}} : \underline{\mathbf{M}}_1 \times \cdots \times \underline{\mathbf{M}}_n \to \underline{\mathbf{M}}_{n+1}$ (where $\underline{\mathbf{M}}_i \in \mathfrak{M}$ for all i), for each n-ary operation symbol $g \in G$,

(ii) $H^{\underset{\sim}{\mathfrak{M}}}$ consists of a homomorphism $h^{\underset{\sim}{\mathfrak{M}}} : \mathbf{D} \to \underline{\mathbf{M}}_{n+1}$ (where $\mathbf{D}$ is a proper subalgebra of $\underline{\mathbf{M}}_1 \times \cdots \times \underline{\mathbf{M}}_n$ and $\underline{\mathbf{M}}_i \in \mathfrak{M}$ for all i), for each n-ary partial operation symbol $h \in H$,

(iii) $R^{\underset{\sim}{\mathfrak{M}}}$ consists of an n-ary relation $r^{\underset{\sim}{\mathfrak{M}}}$ which forms a subalgebra of $\underline{\mathbf{M}}_1 \times \cdots \times \underline{\mathbf{M}}_n$ (where $\underline{\mathbf{M}}_i \in \mathfrak{M}$ for all i), for each n-ary relation symbol $r \in R$, and

(iv) $\mathcal{T}^{\underset{\sim}{\mathfrak{M}}}$ is the discrete topology on $\dot{\bigcup}\{\, M \mid M \in \mathfrak{M} \,\}$.

By analogy with the case where $\mathfrak{M}$ consists of a single algebra, we refer to the elements of $G^{\underset{\sim}{\mathfrak{M}}}$ as (**multisorted**) **operations**, to the elements of $H^{\underset{\sim}{\mathfrak{M}}}$ as (**multisorted**) **partial operations** and to the elements of $R^{\underset{\sim}{\mathfrak{M}}}$ as (**multisorted**) **relations** on $\mathfrak{M}$ and we summarise (i)–(iii) by saying that the structure on $\underset{\sim}{\mathfrak{M}}$ is **algebraic over** $\mathfrak{M}$.

Assume that $\langle X; \mathcal{T} \rangle$ is a topological space written as a disjoint union indexed by $\mathfrak{M}$:

$$X = \dot{\bigcup}\{\, X_M \mid M \in \mathfrak{M} \,\}.$$

If, for every operation $g^{\underset{\sim}{\mathfrak{M}}} : \underline{\mathbf{M}}_1 \times \cdots \times \underline{\mathbf{M}}_n \to \underline{\mathbf{M}}_{n+1}$ in $G^{\underset{\sim}{\mathfrak{M}}}$, there is a corresponding map $g^{\mathbf{X}} : X_{M_1} \times \cdots \times X_{M_n} \to X_{M_{n+1}}$, and similarly for the partial operations in $H^{\underset{\sim}{\mathfrak{M}}}$ and relations in $R^{\underset{\sim}{\mathfrak{M}}}$, then we refer to $\mathbf{X} = \langle X; G^{\mathbf{X}}, H^{\mathbf{X}}, R^{\mathbf{X}}, \mathcal{T} \rangle$ as an $\mathfrak{M}$-**sorted structure** of the same type as $\underset{\sim}{\mathfrak{M}}$ and X_M is called the M-**sort** of $\mathbf{X}$. If $\mathbf{X}$ and $\mathbf{Y}$ are $\mathfrak{M}$-sorted structures of the same type as $\underset{\sim}{\mathfrak{M}}$, then a map $\varphi : X \to Y$ is a **morphism** provided it is continuous and preserves the sorts (that is, φ maps X_M into Y_M) and the (partial) operations and relations in the obvious sense.

The concepts of **substructure**, **isomorphism**, **embedding** and **product** also have natural multisorted definitions. In particular, if S is a non-empty set, then the **power** $\underset{\sim}{\mathfrak{M}}^S$ is the $\mathfrak{M}$-sorted structure whose underlying set is

$$\mathfrak{M}^S := \dot{\bigcup}\{\, M^S \mid M \in \mathfrak{M}\,\}$$

with the obvious topology, and (partial) operations and relations extended pointwise in the natural manner.

As expected, we define $\mathfrak{X}$ to be the category of all $\mathfrak{M}$-sorted structures of the same type as $\underset{\sim}{\mathfrak{M}}$ which are isomorphic to a closed substructure of some power $\underset{\sim}{\mathfrak{M}}^S$ of $\underset{\sim}{\mathfrak{M}}$ with $S \neq \varnothing$, in symbols, $\mathfrak{X} = \mathbb{IS}_c\mathbb{P}^+ \underset{\sim}{\mathfrak{M}}$. It is straightforward to check that, for every $\mathbf{A} \in \mathcal{A}$,

$$D(\mathbf{A}) := \dot{\bigcup}\{\, \mathcal{A}(\mathbf{A}, \underline{\mathbf{M}}) \mid M \in \mathfrak{M}\,\}$$

is a closed substructure of $\underset{\sim}{\mathfrak{M}}^A$, and for every $\mathbf{X} \in \mathfrak{X}$, the homset $\mathfrak{X}(\mathbf{X}, \underset{\sim}{\mathfrak{M}})$ forms a subalgebra of $\prod\{\, \underline{\mathbf{M}}^{X_M} \mid M \in \mathfrak{M}\,\}$. Hence, we obtain contravariant functors $D\colon \mathcal{A} \to \mathfrak{X}$ and $E\colon \mathfrak{X} \to \mathcal{A}$ which form a dual adjunction $\langle D, E, e, \varepsilon\rangle$ between $\mathcal{A}$ and $\mathfrak{X}$. For all $\mathbf{A} \in \mathcal{A}$ and all $\mathbf{X} \in \mathfrak{X}$, the morphisms

$$e_{\mathbf{A}}\colon \mathbf{A} \to ED(\mathbf{A}) \quad \text{and} \quad \varepsilon_{\mathbf{X}}\colon \mathbf{X} \to DE(\mathbf{X}),$$

which are given by evaluation, are embeddings. As in the single-sorted case, the dual of the algebra in $\mathcal{A}$ freely generated by a non-empty set S is isomorphic in $\mathfrak{X}$ to $\underset{\sim}{\mathfrak{M}}^S$.

Almost all of the results proved elsewhere in this text extend naturally and easily to the multisorted setting. An exception is the Two-For-One Strong Duality Theorem and results which depend upon it. Fortunately, the NU Duality Theorem and the NU Strong Duality Theorem do extend.

For any non-empty set S, each S-ary term of type F induces a morphism $t\colon \underset{\sim}{\mathfrak{M}}^S \to \underset{\sim}{\mathfrak{M}}$ which is the disjoint union over $M \in \mathfrak{M}$ of the induced S-ary term functions $t^{\underline{\mathbf{M}}}\colon M^S \to M$. We refer to such a morphism as an S-ary **term function on** $\underline{\mathfrak{M}}$. The definitions of **entailment**, **hom-entailment** and **strong entailment** extend in the obvious way to multisorted versions.

1.1 Multisorted NU Duality Theorem *Let $k \geqslant 2$ and assume that $\underline{\mathfrak{M}}$ has a $(k{+}1)$-ary near-unanimity term. If*

$$\underset{\sim}{\mathfrak{M}} = \langle \dot{\bigcup}\{M \mid M \in \mathfrak{M}\}; \mathcal{B}_k, \mathcal{T}\rangle,$$

where

$$\mathcal{B}_k = \bigcup\{\, \mathbb{S}(\underline{\mathbf{M}}_1 \times \cdots \times \underline{\mathbf{M}}_k) \mid M_1, \ldots, M_k \in \mathfrak{M}\,\},$$

then any structure which entails $\underset{\sim}{\mathfrak{M}}$ yields a duality on $\mathcal{A}$.

The **irreducibility index**, $\mathrm{Irr}(\underline{\mathfrak{M}})$, of $\underline{\mathfrak{M}}$ is defined to be the maximum value of $\mathrm{Irr}(\underline{\mathbf{M}})$ as $\underline{\mathbf{M}}$ varies over $\underline{\mathfrak{M}}$. Define $\mathcal{P}_n$ to be the set of all n-ary algebraic multisorted partial operations on $\underline{\mathfrak{M}}$ and define K to be the set of all elements which form a one-element subalgebra of some $\underline{\mathbf{M}} \in \underline{\mathfrak{M}}$.

1.2 Multisorted NU Strong Duality Theorem *Let $k \geqslant 2$ and assume that $\underline{\mathfrak{M}}$ has a $(k{+}1)$-ary near-unanimity term. If*

$$\underset{\sim}{\mathfrak{M}} = \langle \dot{\bigcup}\{M \mid M \in \mathfrak{M}\}; \mathcal{B}_k, H, \mathcal{T} \rangle,$$

where

$$H = K \cup \bigcup\{\, \mathcal{P}_n \mid 1 \leqslant n \leqslant \mathrm{Irr}(\underline{\mathfrak{M}})\},$$

then any structure which strongly entails $\underset{\sim}{\mathfrak{M}}$ yields a strong duality on $\mathcal{A}$.

It follows at once that any finitely generated variety of algebras which have an underlying lattice structure has a strong multisorted duality. In the next section we shall show that if the algebras have an underlying distributive lattice then the NU Duality Theorems may be bypassed and we can obtain a duality by riding piggyback on Priestley's duality for the underlying distributive lattices.

7.2 Piggyback dualities for distributive-lattice-based algebras

Most of the algebras which arise in algebraic logic are bounded distributive lattices with additional operations corresponding to implication or various types of negation. In this section we shall present a theorem which applies to a finitely generated variety of algebras of this sort. In fact, the original theorem as proved in Davey and Priestley [DP87] applies to algebras based in any quasi-variety which itself has a multisorted duality. Since almost all applications in the literature have been to produce (multisorted) dualities for algebras based in bounded distributive lattices, it is this version which we present here.

To avoid excessive use of subscripts and superscripts, throughout this chapter we shall denote the two-element bounded distributive lattice by $\underline{\mathbf{D}} := \langle \{0,1\}; \vee, \wedge, 0, 1\rangle$ and the two-element Priestley space by $\underset{\sim}{\mathbf{D}} := \langle \{0,1\}; \leqslant, \mathcal{T}\rangle$. Priestley's duality tells us that $\underset{\sim}{\mathbf{D}}$ yields a strong duality between $\mathcal{D} := \mathbb{ISP}\underline{\mathbf{D}}$, which now denotes the variety of *bounded* distributive lattices, and $\mathcal{P} := \mathbb{IS}_c\mathbb{P}^+ \underset{\sim}{\mathbf{D}}$, the category of Priestley spaces (see Exercise 4.5). Denote the hom-functors which yield this duality by $H \colon \mathcal{D} \to \mathcal{P}$ and $K \colon \mathcal{P} \to \mathcal{D}$ and, for each $\mathbf{A} \in \mathcal{D}$ and $\mathbf{Y} \in \mathcal{P}$, denote the corresponding evaluation maps by $k_{\mathbf{A}} \colon \mathbf{A} \to KH(\mathbf{A})$ and $\kappa_{\mathbf{Y}} \colon \mathbf{Y} \to HK(\mathbf{Y})$. We shall say that the class $\underline{\mathfrak{M}}$ has a **term-reduct in $\mathcal{D}$** if there are binary terms $\vee$ and $\wedge$ and constant unary terms z and u

of type F such that $\underline{\mathbf{M}}^\flat := \langle M; \vee^{\mathbf{M}}, \wedge^{\mathbf{M}}, 0^{\mathbf{M}}, 1^{\mathbf{M}}\rangle$ is a bounded distributive lattice for each algebra $\underline{\mathbf{M}} \in \underline{\mathfrak{M}}$, where $0^{\mathbf{M}}$ and $1^{\mathbf{M}}$ are the values in M of the constant unary term functions $z^{\mathbf{M}}$ and $u^{\mathbf{M}}$, respectively.

The single-sorted case of the following theorem in which $\underline{\mathfrak{M}} = \{\underline{\mathbf{M}}\}$ and $|\Omega_M| = 1$ is due to Davey and Werner ([DW85] and [DW86]). The theorem is a special case of the full multisorted version given in Davey and Priestley [DP87]. The formulation here comes from Davey [Dav93] in the single-sorted case and Priestley [Pri95a] in the multisorted case.

2.1 Piggyback Duality Theorem *Assume that $\mathfrak{M}$ is a finite set of finite algebras which has a term-reduct in $\mathcal{D}$ and let $\mathcal{A} := \mathbb{ISP}\underline{\mathfrak{M}}$. For each $M \in \mathfrak{M}$, let Ω_M be a subset of $\mathcal{D}(\underline{\mathbf{M}}^\flat, \underline{\mathbf{D}})$. Let*

$$\underset{\sim}{\mathfrak{M}} = \langle \bigcup\{ M \mid M \in \mathfrak{M}\}; G, R, \mathcal{T}\rangle,$$

where

(i) *R is the set of all $\mathcal{A}$-subalgebras of $\underline{\mathbf{M}}_1 \times \underline{\mathbf{M}}_2$ which are maximal in*

$$(\omega_1, \omega_2)^{-1}(\leqslant) := \{ (a, b) \in M_1 \times M_2 \mid \omega_1(a) \leqslant \omega_2(b)\},$$

where $\omega_1 \in \Omega_{M_1}$, $\omega_2 \in \Omega_{M_2}$ and $M_1, M_2 \in \mathfrak{M}$,

(ii) *$G \subseteq \bigcup\{ \mathcal{A}(\underline{\mathbf{M}}_1, \underline{\mathbf{M}}_2), \mid M_1, M_2 \in \mathfrak{M}\}$ satisfies the separation condition*

(S) *for all $M_1 \in \mathfrak{M}$ and all $a \neq b$ in M_1, we have $\omega(a) \neq \omega(b)$, for some $\omega \in \Omega_{M_1}$, or $\omega(g(a)) \neq \omega(g(b))$, where $\omega \in \Omega_{M_2}$ for some $M_2 \in \mathfrak{M}$ and $g \in \mathcal{A}(\underline{\mathbf{M}}_1, \underline{\mathbf{M}}_2)$ is a composite of a finite number of maps from G,*

(iii) *$\mathcal{T}$ is the discrete topology.*

Then $\underset{\sim}{\mathfrak{M}}$ yields a duality on $\mathcal{A}$.

Proof The idea behind the proof is extremely simple. Priestley duality tells us that, for each $\mathbf{A} \in \mathcal{A}$, the evaluation map $k_{\mathbf{A}^\flat} \colon \mathbf{A}^\flat \to KH(\mathbf{A}^\flat)$ is surjective. We shall show that, given the surjectivity of $k_{\mathbf{A}^\flat}$, our choice of $\underset{\sim}{\mathfrak{M}}$ guarantees the surjectivity of the map $e_{\mathbf{A}} \colon \mathbf{A} \to ED(\mathbf{A})$. We shall construct a one-to-one map $\Delta \colon ED(\mathbf{A}) \to KH(\mathbf{A}^\flat)$ satisfying $\Delta \circ e_{\mathbf{A}} = k_{\mathbf{A}^\flat}$, for each $\mathbf{A} \in \mathcal{A}$. The simplest of diagram chases then shows that $e_{\mathbf{A}}$ is surjective. The lattice homomorphisms in Ω_M will enable us to define the map Δ as follows. Let $\Omega := \bigcup\{\Omega_M \mid M \in \mathfrak{M}\}$, and note that Ω is finite. For $\omega \in \Omega_M$, let $\Phi_\omega \colon \mathcal{A}(\mathbf{A}, \underline{\mathbf{M}}) \to \mathcal{D}(\mathbf{A}^\flat, \underline{\mathbf{D}})$ be given by $\Phi_\omega(x) := \omega \circ x$ for all $x \in \mathcal{A}(\mathbf{A}, \underline{\mathbf{M}})$. Since $\underline{\mathfrak{M}}$ has a term-reduct in $\mathcal{D}$, the map Φ_ω is well defined. If the maps Φ_ω for $\omega \in \Omega$ are jointly surjective, then for each $y \in \mathcal{D}(\mathbf{A}^\flat, \underline{\mathbf{D}})$ there exist $\omega \colon \underline{\mathbf{M}}^\flat \to \underline{\mathbf{D}}$ in Ω and $x \in \mathcal{A}(\mathbf{A}, \underline{\mathbf{M}})$ such that $\Phi_\omega(x) := \omega \circ x = y$. Now let $\alpha \in ED(\mathbf{A})$ be a multisorted morphism and let $\alpha_M \colon \mathcal{A}(\mathbf{A}, \underline{\mathbf{M}}) \to M$ be its M-component. We may *attempt* to define $\Delta(\alpha) \colon \mathfrak{D}(\mathbf{A}^\flat, \underline{\mathbf{D}}) \to \underset{\sim}{\mathbf{D}}$ by $\Delta(\alpha)(y) := \omega(\alpha_M(x))$: see Figure 7.1.

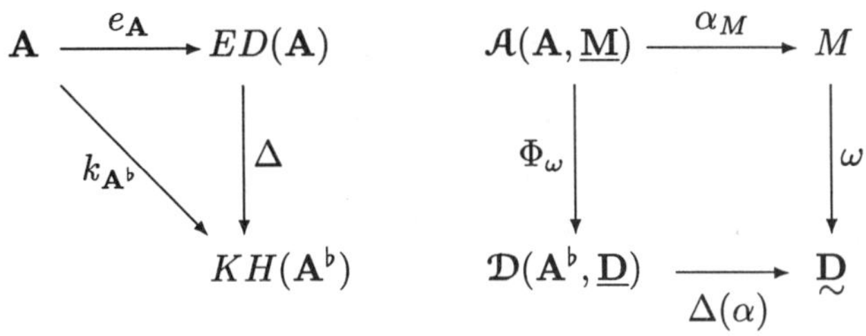

Figure 7.1 the definition of Δ

To complete the proof we must now verify that

(1) the maps Φ_ω for $\omega \in \Omega$ are jointly surjective,

(2) $\Delta(\alpha)$ is well defined (that is, independent of the choice of M, of $\omega \in \Omega_M$ and of $x \in \mathcal{A}(\mathbf{A}, \underline{\mathbf{M}})$ such that $\Phi_\omega(x) = y$),

(3) $\Delta(\alpha)$ is $\leqslant$-preserving,

(4) $\Delta(\alpha)$ is continuous,

(5) Δ is one-to-one, and

(6) $(\Delta \circ e_{\mathbf{A}})(a) = k_{\mathbf{A}^\flat}(a)$ for all $a \in A$.

The following lemma, from which (1) will follow, is of independent interest.

2.2 Lemma *Assume that $\underline{\mathfrak{M}}$ is a finite set of finite algebras which has a term-reduct in $\mathfrak{D}$ and let $\mathcal{A} := \mathbb{ISP}\underline{\mathfrak{M}}$. For each $M \in \mathfrak{M}$, let Ω_M be a subset of $\mathfrak{D}(\underline{\mathbf{M}}^\flat, \underline{\mathbf{D}})$ and let $\Omega := \bigcup\{\, \Omega_M \mid M \in \mathfrak{M} \,\}$. Let $\mathbf{A} \in \mathcal{A}$. The maps*

$$\Phi_\omega \colon \mathcal{A}(\mathbf{A}, \underline{\mathbf{M}}) \to \mathfrak{D}(\mathbf{A}^\flat, \underline{\mathbf{D}}) \text{ with } \omega \in \Omega,$$

where $\Phi_\omega(x) := \omega \circ x$, for all $x \in \mathcal{A}(\mathbf{A}, \underline{\mathbf{M}})$, are jointly surjective if and only if

(J) *for all $a \neq b$ in A there exist $\underline{\mathbf{M}} \in \underline{\mathfrak{M}}$, $\omega \in \Omega_M$ and $x \in \mathcal{A}(\mathbf{A}, \underline{\mathbf{M}})$ such that $\omega(x(a)) \neq \omega(x(b))$.*

Proof Let $Z := \bigcup\{\, \mathrm{im}(\Phi_\omega) \mid \omega \in \Omega \,\} \subseteq H(\mathbf{A}^\flat)$. Since Ω is finite, Z is closed in $H(A^\flat)$ and hence $\mathbf{Z} \in \mathfrak{P}$. Let $\mu \colon \mathbf{Z} \to H(\mathbf{A}^\flat)$ be the inclusion map. Since $\underset{\sim}{\mathbf{D}}$ yields a strong duality on $\mathfrak{D}$, by Lemmas 3.2.8 and 3.2.10

we have

$$\begin{aligned}
&\text{the maps } \Phi_\omega \text{ for } \omega \in \Omega \text{ are jointly surjective}\\
\iff &\mu \text{ is surjective}\\
\iff &K(\mu)\colon KH(\mathbf{A}^\flat) \to K(\mathbf{Z}) \text{ is one-to-one}\\
\iff &K(\mu) \circ k_{\mathbf{A}^\flat}\colon A \to K(\mathbf{Z}) \text{ is one-to-one}\\
\iff &a \neq b \text{ in } A \Rightarrow (\exists M \in \mathfrak{M})(\exists \omega \in \Omega_M)(\exists x \in \mathcal{A}(\mathbf{A}, \underline{\mathbf{M}}))\\
&\Phi_\omega(x)(a) \neq \Phi_\omega(x)(b)\\
\iff &a \neq b \text{ in } A \Rightarrow (\exists M \in \mathfrak{M})(\exists \omega \in \Omega_M)(\exists x \in \mathcal{A}(\mathbf{A}, \underline{\mathbf{M}}))\\
&\omega(x(a)) \neq \omega(x(b)),
\end{aligned}$$

as required. ∎

Since $\mathbf{A} \in \mathbb{ISP}\underline{\mathfrak{M}}$, the homomorphisms from $\mathbf{A}$ into the algebras in $\underline{\mathfrak{M}}$ separate the points of A. Hence condition (S) and the lemma above combine to prove (1).

Since $\leqslant \cap \geqslant$ equals $=$, in order to prove both (2) and (3) it suffices to show that for all $M_1, M_2 \in \mathfrak{M}$, for all $\omega_1 \in \Omega_{M_1}$, for all $\omega_2 \in \Omega_{M_2}$, for all $x_1 \in \mathcal{A}(\mathbf{A}, \underline{\mathbf{M}}_1)$, and for all $x_2 \in \mathcal{A}(\mathbf{A}, \underline{\mathbf{M}}_2)$, we have

$$\begin{aligned}
&\Phi_{\omega_1}(x_1) = \omega_1 \circ x_1 = y_1 \leqslant y_2 = \omega_2 \circ x_2 = \Phi_{\omega_2}(x_2)\\
&\qquad\qquad \Longrightarrow \Delta(\alpha)(y_1) = \omega_1(\alpha_{M_1}(x_1)) \leqslant \omega_2(\alpha_{M_2}(x_2)) = \Delta(\alpha)(y_2).
\end{aligned}$$

To this end, for $i = 1, 2$, let $M_i \in \mathfrak{M}$, let $\omega_i \in \Omega_{M_i}$ and let $x_i \in \mathcal{A}(\mathbf{A}, \underline{\mathbf{M}}_i)$ with $\Phi_{\omega_1}(x_1) = \omega_1 \circ x_1 = y_1 \leqslant y_2 = \omega_2 \circ x_2 = \Phi_{\omega_2}(x_2)$. Then, since $\leqslant$ is defined pointwise in $H(\mathbf{A}^\flat)$, we have $\omega_1(x_1(a)) \leqslant \omega_2(x_2(a))$ for all $a \in A$. Thus the set $s := \{\, (x_1(a), x_2(a)) \in M_1 \times M_2 \mid a \in A \,\}$ forms a subalgebra of $\underline{\mathbf{M}}_1 \times \underline{\mathbf{M}}_2$ which is contained in $(\omega_1, \omega_2)^{-1}(\leqslant)$. Let $\mathbf{r}$ be a subalgebra of $\underline{\mathbf{M}}_1 \times \underline{\mathbf{M}}_2$ which is maximal in $(\omega_1, \omega_2)^{-1}(\leqslant)$ and contains s. Thus $r \in R$. As $r^{D(\mathbf{A})}$ is defined pointwise, we have $(x_1, x_2) \in r^{D(\mathbf{A})}$ whence $(\alpha_{M_1}(x_1), \alpha_{M_2}(x_2)) \in r$ as α preserves r, by assumption. Since $r \subseteq (\omega_1, \omega_2)^{-1}(\leqslant)$, we have $\omega_1(\alpha_{M_1}(x_1)) \leqslant \omega_2(\alpha_{M_2}(x_2))$, as required.

To show that $\Delta(\alpha)$ is continuous, let V be closed in $D = \{0, 1\}$. Then $\Delta(\alpha)^{-1}(V) = \bigcup_{\omega \in \Omega} \Phi_\omega(\alpha_M^{-1}(\omega^{-1}(V)))$ which is closed in $H(\mathbf{A}^\flat)$ since Ω is finite, α_M and ω are continuous and, by Lemma B.1, as Φ_ω is continuous it maps closed sets to closed sets. Thus (4) holds.

We now establish that Δ is one-to-one. Let $\alpha, \beta\colon D(\mathbf{A}) \to \underset{\sim}{\mathfrak{M}}$ be distinct morphisms. Then there exist $\underline{\mathbf{M}}_1 \in \underline{\mathfrak{M}}$ and $x \in \mathcal{A}(\mathbf{A}, \underline{\mathbf{M}}_1)$ such that $\alpha_{M_1}(x) \neq \beta_{M_1}(x)$. By (S), there exist $\underline{\mathbf{M}}_2 \in \underline{\mathfrak{M}}$, $g \in \mathcal{A}(\underline{\mathbf{M}}_1, \underline{\mathbf{M}}_2)$, which is a composite of maps from G, and $\omega \in \Omega_{M_2}$ such that $\omega(g(\alpha_{M_1}(x))) \neq \omega(g(\beta_{M_1}(x)))$. Since α and β preserve g, this yields

$$(\omega \circ \alpha_{M_2})(g(x)) = \omega(\alpha_{M_2}(g(x))) \neq \omega(\beta_{M_2}(g(x))) = (\omega \circ \beta_{M_2})(g(x)),$$

whence $\Delta(\alpha)\circ\Phi_\omega = \omega\circ\alpha_{M_2} \neq \omega\circ\beta_{M_2} = \Delta(\beta)\circ\Phi_\omega$, yielding $\Delta(\alpha) \neq \Delta(\beta)$ and thereby proving (5).

Finally, we show that $(\Delta \circ e_{\mathbf{A}})(a) = k_{\mathbf{A}^\flat}(a)$ for all $a \in A$, which proves (6) and completes the proof of the theorem. Let $y \in H(\mathbf{A}^\flat) = \mathfrak{D}(\mathbf{A}^\flat, \underline{\mathbf{D}})$. As the maps Φ_ω for $\omega \in \Omega$ are jointly surjective, there exist $M \in \mathfrak{M}$, $\omega \in \Omega_M$ and $x \in \mathcal{A}(\mathbf{A}, \underline{\mathbf{M}})$ such that $y = \omega \circ x$. Hence,

$$\begin{aligned}(\Delta \circ e_{\mathbf{A}})(a)(y) = \Delta(e_{\mathbf{A}}(a))(y) &= \omega((e_{\mathbf{A}}(a))(x)) \\ &= \omega(x(a)) = y(a) = k_{\mathbf{A}^\flat}(a)(y),\end{aligned}$$

completing the proof of the Piggyback Duality Theorem. ■

The maps $\omega \in \Omega_M$ are referred to as the **carriers** of the piggyback duality, as they allow us to transport the structure from $\underset{\sim}{\mathbf{D}}$ up to the multisorted structure $\underset{\sim}{\mathfrak{M}}$, while the relations in R are referred to as the **piggyback relations** on $\underset{\sim}{\mathfrak{M}}$.

Even in the single-sorted case, there is often more than one choice for the sets Ω_M and G which guarantee that the separation condition (S) holds. In practice, we often try to minimise the size of the sets Ω_M at the expense of increasing the size of the set G as this reduces the size of R. At the other end of the spectrum, there is a canonical choice which guarantees that (S) holds: $\Omega_M = \mathfrak{D}(\underline{\mathbf{M}}^\flat, \underline{\mathbf{D}})$ for each $M \in \mathfrak{M}$ and $G = \varnothing$.

2.3 Notation To simplify our notation, from now on we shall discontinue the use of the $\mathbf{A}^\flat$ notation and will not distinguish notationally between the algebra $\mathbf{A}$ and its term-reduct in $\mathfrak{D}$. In any given situation it will be clear from the context which persona, $\mathbf{A}$ or $\mathbf{A}^\flat$, is involved.

7.3 Piggyback dualities for varieties of Heyting algebras

We shall delay an example of a piggyback-induced multisorted natural duality until the next section. In this section we shall investigate when the Piggyback Duality Theorem in its simplest form (that is, with a single algebra $\underline{\mathbf{M}}$ and a single carrier ω) may be applied to yield a (single-sorted) natural duality for the variety generated by a finite subdirectly irreducible Heyting algebra $\underline{\mathbf{M}}$. En route, we shall discover a new proof of the Heyting Strong Duality Theorem 4.2.2. The results in this section come from Davey and Werner ([DW85] and [DW86]).

Since Con($\underline{\mathbf{M}}$) is isomorphic to the lattice of filters of $\underline{\mathbf{M}}$, if $\underline{\mathbf{M}}$ is a finite subdirectly irreducible Heyting algebra, then $\underline{\mathbf{M}} = \mathbf{L} \oplus \mathbf{1}$ is obtained by adjoining a new top, 1, to a finite distributive lattice $\mathbf{L}$. It is important to note that, while $\mathbf{L}$ is a Heyting algebra in its own right, it is not a subalgebra of $\underline{\mathbf{M}}$. Take $\Omega_M := \{\omega\}$ where $\omega = \chi_{\{1\}}\colon M \to D$ is the characteristic function of $\{1\}$. The Heyting subalgebras of $\underline{\mathbf{M}}^2$ which are contained in $(\omega,\omega)^{-1}(\leqslant)$ (which we abbreviate to $\omega^{-1}(\leqslant)$) have a very simple structure.

3.1 Lemma *Let* $\underline{\mathbf{M}} = \mathbf{L} \oplus \mathbf{1}$ *be a finite subdirectly irreducible Heyting algebra. A subalgebra* $\mathbf{r}$ *of* $\underline{\mathbf{M}}^2$ *is contained in* $\omega^{-1}(\leqslant)$ *if and only if* r *is the graph of a (partial) endomorphism of* $\underline{\mathbf{M}}$.

Proof Note that $\omega^{-1}(\leqslant) = (L\times L)\cup(L\times\{1\})\cup\{(1,1)\} = M^2\backslash(\{1\}\times L)$. Since a (partial) endomorphism e of $\underline{\mathbf{M}}$ maps 1 to 1, we certainly have $r := \operatorname{graph}(e) \subseteq \omega^{-1}(\leqslant)$. Conversely, assume that $\mathbf{r}$ is a subalgebra of $\underline{\mathbf{M}}^2$ with $r \subseteq \omega^{-1}(\leqslant)$. Then

$$\begin{aligned}(b,c),(b,d) \in r \text{ implies } (1, c \Rightarrow d) &= (b \Rightarrow b, c \Rightarrow d) = (b,c) \Rightarrow (b,d) \in r\\ \text{implies } c \Rightarrow d &= 1\\ \text{implies } c &\leqslant d.\end{aligned}$$

By symmetry, we conclude that $(b,c),(b,d) \in r$ implies $c = d$, whence r is the graph of a (partial) endomorphism of $\underline{\mathbf{M}}$. ■

Hence a piggyback relation, that is, a subalgebra of $\underline{\mathbf{M}}^2$ which is maximal in $\omega^{-1}(\leqslant)$, is either the graph of an endomorphism or the graph of a non-extendable proper partial endomorphism of $\underline{\mathbf{M}}$. It now remains to discover conditions on $\underline{\mathbf{M}}$ which will guarantee that the endomorphisms of $\underline{\mathbf{M}}$ when composed with ω separate the points of M.

Recall that the **linear sum** $\mathbf{P} \oplus \mathbf{Q}$ of two ordered sets $\mathbf{P}$ and $\mathbf{Q}$ is obtained by taking the disjoint union of the two ordered sets and then declaring every element of P to be less than every element of Q. If $\mathbf{P}$ has a top and $\mathbf{Q}$ has a bottom, then the **vertical sum** $\mathbf{P} \overline{\oplus} \mathbf{Q}$ of $\mathbf{P}$ and $\mathbf{Q}$ is obtained from $\mathbf{P} \oplus \mathbf{Q}$ by identifying the top of $\mathbf{P}$ with the bottom of $\mathbf{Q}$. In this notation, $\underline{\mathbf{M}} = \mathbf{L} \oplus \mathbf{1} = \mathbf{L} \overline{\oplus} \mathbf{2}$.

3.2 Lemma *Let* $\mathbf{L}$ *be a finite Heyting algebra.*

(i) *The variety and the quasi-variety of Heyting algebras generated by* $\underline{\mathbf{M}} := \mathbf{L} \oplus \mathbf{1}$ *coincide, that is,* $\mathcal{V}(\underline{\mathbf{M}}) = \mathbb{ISP}\underline{\mathbf{M}}$, *if and only if*

  ($\mathcal{V}$) *every subdirectly irreducible subalgebra of a homomorphic image of* $\mathbf{L}$ *can be embedded into* $\mathbf{L} \oplus \mathbf{1}$.

(ii) *The class of finite Heyting algebras* $\mathbf{L}$ *satisfying* ($\mathcal{V}$) *is closed under linear sums, vertical sums and finite direct products.*

(iii) *If* $\mathbf{L}$ *satisfies* ($\mathcal{V}$), *then the endomorphisms of the Heyting algebra* $\underline{\mathbf{M}} := \mathbf{L} \oplus \mathbf{1}$ *when composed with* ω *separate the points of* M.

Proof We leave (i) and (ii) as exercises for the reader. To prove (iii), assume that $\mathbf{L}$ satisfies ($\mathcal{V}$) and let $a \not\geqslant b$ in M. Since $\omega \circ \mathrm{id}_{\underline{\mathbf{M}}}$ separates

a and 1, we may assume that $a, b \in L$. Since every filter of a Heyting algebra is the class of some congruence, there is a surjective homomorphism $u\colon \mathbf{L} \to \mathbf{L}_1$ with $u^{-1}(1) = {\uparrow}b$. Since $u(a) \neq u(b) = 1$, by Birkhoff's theorem A.10, there is a further surjective homomorphism $v\colon \mathbf{L}_1 \to \mathbf{L}_2$, with $\mathbf{L}_2$ subdirectly irreducible, such that $v(u(a)) \neq 1$. By $(\mathcal{V})$ there is an embedding $w\colon \mathbf{L}_2 \to \underline{\mathbf{M}}$. Let $k\colon \underline{\mathbf{M}} \to \mathbf{L}$ be the homomorphism which collapses $1 \in M$ to the top of $\mathbf{L}$ and satisfies $k(c) = c$ for all $c \in L$. Then $e := w \circ v \circ u \circ k\colon \underline{\mathbf{M}} \to \underline{\mathbf{M}}$ is an endomorphism of $\underline{\mathbf{M}}$ such that $\omega \circ e$ separates a and b. ■

Part (ii) of the lemma above yields a plethora of Heyting algebras satisfying $(\mathcal{V})$: for example, finite chains and finite Boolean algebras satisfy $(\mathcal{V})$ as they can be obtained from the two-element chain via vertical sum and direct product. Nevertheless, it is easy to find finite Heyting algebras which satisfy $(\mathcal{V})$ but cannot be obtained via any combination of linear sum, vertical sum and direct product from the two-element chain: if $\mathbf{3}$ denotes the three-element chain with $0 < a < 1$, then the Heyting algebra whose underlying ordered set is $\mathbf{3}^2 \backslash \{(0,1)\}$ is the simplest example.

The previous two lemmas combine with the Piggyback Duality Theorem and the NU Strong Duality Corollary 3.3.9 (or, more directly, Theorem 3.3.7) to provide the promised alternative proof of the Heyting Strong Duality Theorem 4.2.2.

3.3 Heyting Strong Duality Theorem Again *Assume that* $\underline{\mathbf{M}} = \mathbf{L} \oplus \mathbf{1}$ *is a finite subdirectly irreducible Heyting algebra and that* $\mathbf{L}$ *satisfies* $(\mathcal{V})$. *Let* G *be a set of endomorphisms of* $\underline{\mathbf{M}}$ *and* H *be a set of partial endomorphisms of* $\underline{\mathbf{M}}$ *such that* $G \cup H$ *generates the monoid of all partial endomorphisms of* $\underline{\mathbf{M}}$. *Then* $\underset{\sim}{\mathbf{M}} = \langle M; G, H, \mathcal{T}\rangle$ *yields a strong duality on* $\mathcal{A} = \mathbb{ISP}\underline{\mathbf{M}}$.

Proof By Lemma 7.3.1, $G \cup H$ entails all of the piggyback relations on $\underline{\mathbf{M}}$, and by Lemma 7.3.2, the endomorphisms of $\underline{\mathbf{M}}$ composed with ω separate the points of M. Thus the Piggyback Duality Theorem guarantees that $\underset{\sim}{\mathbf{M}}$ yields a duality on $\mathcal{A}$. Since every non-trivial subalgebra of $\underline{\mathbf{M}}$ is subdirectly irreducible, the NU Strong Duality Corollary 3.3.9 or Theorem 3.3.7 tells us that the duality is strong. ■

This theorem applies to varieties of relative Stone algebras, that is, when $\underline{\mathbf{M}} = \underline{\mathbf{C}}_n$ is an n-element Heyting chain. In Chapter 8, we prove a result which allows us to establish very easily that, in fact, $\underline{\mathbf{C}}_n$ is endodualisable (see 8.1.5). That is, while the endomorphisms and partial endomorphisms yield a strong duality on the (quasi-)variety generated by $\underline{\mathbf{C}}_n$, the endomorphisms alone still yield a duality (which for $n \geqslant 4$ will not be strong).

7.4 Restricted Priestley dualities

Whenever we have a class $\mathcal{A}$ of algebras, like Ockham algebras, which have a term-reduct in the class $\mathcal{D}$ of bounded distributive lattices, we can regard $\mathcal{A}$ as a subcategory of the category $\mathcal{D}$ of bounded distributive lattices. Provided we can describe (up to isomorphism) the objects in the category $\mathcal{P}$ of Priestley spaces which are of the form $H(\mathbf{A})$ for $\mathbf{A} \in \mathcal{A}$ and can describe the morphisms in $\mathcal{P}$ which are of the form $H(u)$ for some $u \in \mathcal{A}(\mathbf{A}, \mathbf{B})$, then we may form the category $\mathcal{Y}$ consisting of all (isomorphic copies of) such objects and morphisms. Restricting the Priestley functors H and K to $\mathcal{A}$ and $\mathcal{Y}$ yields a dual equivalence between $\mathcal{A}$ and $\mathcal{Y}$. This is known as the **restricted Priestley duality** for $\mathcal{A}$.

In practice, it is often convenient to identify $H(\mathbf{A})$ with the set of prime filters of $\mathbf{A}$ and to identify $K(\mathbf{Y})$ with the set of clopen increasing subsets of $\mathbf{Y}$. The following result gathers together some particularly useful properties of the Priestley duality for $\mathcal{D}$. Denote the order-theoretic duals of an ordered set $\mathbf{P} = \langle P; \leqslant\rangle$ and a bounded distributive lattice $\mathbf{L} = \langle L; \vee, \wedge, 0, 1\rangle$ by $\mathbf{P}^\partial$ and $\mathbf{L}^\partial$ respectively, that is, $\mathbf{P}^\partial = \langle P; \geqslant\rangle$ and $\mathbf{L}^\partial = \langle L; \wedge, \vee, 1, 0\rangle$.

4.1 Properties of Priestley Duality *Let* $\mathbf{A}$ *and* $\mathbf{B}$ *be bounded distributive lattices, let* $\mathbf{X}$, $\mathbf{X}_i$ *and* $\mathbf{Y}$ *be Priestley spaces, let* u *be a bounded-distributive-lattice homomorphism and let* φ *and* φ_i *be Priestley space morphisms.*

(i) $H(\mathbf{A} \oplus \mathbf{1}) \cong \mathbf{1} \oplus H(\mathbf{A})$ *and* $K(\mathbf{1} \oplus \mathbf{X}) \cong K(\mathbf{X}) \oplus \mathbf{1}$.

(ii) $H(\mathbf{A}^\partial) \cong H(\mathbf{A})^\partial$ *and* $K(\mathbf{X}^\partial) \cong K(\mathbf{X})^\partial$.

(iii) $\operatorname{Con}(\mathbf{A}) \cong \mathcal{C}\ell(H(\mathbf{A}))^\partial$ *and* $\operatorname{Con}(K(\mathbf{X})) \cong \mathcal{C}\ell(\mathbf{X})^\partial$, *where* $\mathcal{C}\ell(\mathbf{X})$ *denotes the lattice of closed subsets of* $\mathbf{X}$.

(iv) (a) $H(\mathbf{A} \times \mathbf{B}) \cong H(\mathbf{A}) \dot\cup H(\mathbf{B})$, *and* (b) $K(\mathbf{X} \dot\cup \mathbf{Y}) \cong K(\mathbf{X}) \times K(\mathbf{Y})$.

(v) (a) *A homomorphism* $u\colon \mathbf{A} \to \mathbf{B}$ *is surjective* (*an embedding*) *if and only if* $H(u)\colon H(\mathbf{B}) \to H(\mathbf{A})$ *is an embedding* (*surjective*).

(b) *A morphism* $\varphi\colon \mathbf{X} \to \mathbf{Y}$ *is surjective* (*an embedding*) *if and only if* $K(\varphi)\colon H(\mathbf{Y}) \to H(\mathbf{X})$ *is an embedding* (*surjective*).

(vi) *If the images of the morphisms* $\varphi_i\colon \mathbf{X}_i \to \mathbf{Y}$, *for* $i \in I$, *are jointly dense in* $\mathbf{Y}$, *that is, if* $\bigcup_{i\in I} \varphi_i(X_i)$ *is dense in* $\mathbf{Y}$, *then the natural map* $\sqcap_{i\in I} K(\varphi_i)\colon K(\mathbf{Y}) \to \prod_{i\in I} K(\mathbf{X}_i)$ *is an embedding. In particular, if the morphisms* $\varphi_i\colon \mathbf{X}_i \to, \mathbf{Y}$, *for* $i \in I$, *are jointly surjective, then* $\sqcap_{i\in I} K(\varphi_i)$ *is an embedding.*

Proof For (i) and (ii), it is easiest to work with prime filters and clopen increasing sets. Then (i) is obvious and (ii) follows from the fact that F is a prime filter of $\mathbf{A}$ if and only if $A \backslash F$ is a prime ideal of $\mathbf{A}$ and therefore a prime filter of $\mathbf{A}^\partial$, and, similarly, U is a clopen increasing subset of $\mathbf{X}$

if and only if $X \backslash U$ is a clopen decreasing subset of $\mathbf{X}$ and thus a clopen increasing subset of $\mathbf{X}^{\partial}$. Part (iii) is a consequence of 3.2.1, (iv) is the statement that coproducts in $\mathcal{P}$ are given by disjoint union (see 6.3.4), (v) follows from Lemmas 3.2.6 and 3.2.8, while (vi) is an exercise. ■

4.2 Ockham Algebras An algebra $\mathbf{A} = \langle A; \vee, \wedge, f, 0, 1\rangle$ is an **Ockham algebra** if $\langle A; \vee, \wedge, 0, 1\rangle$ is a bounded distributive lattice and f corresponds to a negation operation which satisfies the de Morgan laws, that is,

$$f(0) = 1, \quad f(1) = 0, \quad f(a \vee b) = f(a) \wedge f(b) \quad \text{and} \quad f(a \wedge b) = f(a) \vee f(b),$$

for all $a, b \in A$. The variety of Ockham algebras is denoted by $\mathcal{O}$. The varieties of Stone algebras (see 4.3.6), of de Morgan algebras (see 4.3.15) and of Kleene algebras (see 4.3.9) are subvarieties of $\mathcal{O}$. There is an extensive literature on the variety of Ockham algebras and its subvarieties: see Blyth and Varlet [BV:OA] and the references given there. The restricted Priestley duality for Ockham algebras is particularly easy to describe.

4.3 Restricted Priestley Duality for $\mathcal{O}$ The operation f on an Ockham algebra $\mathbf{A}$ is nothing more than a dual endomorphism of the distributive lattice term-reduct of $\mathbf{A}$, that is, $f \in \mathcal{D}(\mathbf{A}, \mathbf{A}^{\partial})$. Hence, by (ii) of the proposition above, the operation f corresponds to a morphism $g \in \mathcal{P}(H(\mathbf{A}), H(\mathbf{A})^{\partial})$. Thus the objects of the restricted category $\mathcal{Y}_{\mathcal{O}}$, known as **Ockham spaces**, are of the form $\langle Y; g, \leqslant, \mathcal{T}\rangle$, where $\langle Y; \leqslant, \mathcal{T}\rangle$ is a Priestley space and $g\colon Y \to Y$ is a continuous, order-reversing map. Let $c(d) = d'$ be the usual Boolean complement on $\{0, 1\}$, then, for $\mathbf{A} \in \mathcal{O}$, the map g is defined on the set $H(\mathbf{A}) = \mathcal{D}(\mathbf{A}, \underline{\mathbf{D}})$ by $g(y) := c \circ y \circ f$ and, for $\mathbf{Y} \in \mathcal{Y}_{\mathcal{O}}$, the map f is defined on $K(\mathbf{Y}) = \mathcal{P}(\mathbf{Y}, \underset{\sim}{\mathbf{D}})$ by $f(\alpha) := c \circ \alpha \circ g$. The morphisms of $\mathcal{Y}_{\mathcal{O}}$ are continuous order-preserving maps which preserve g.

Extensive use has been made of the restricted Priestley duality for $\mathcal{O}$, for example, in Urquhart [Urq79], where it was first written down, in Blyth and Varlet [BV:OA] and in Davey and Priestley [DP87]. To illustrate its use we shall present Urquhart's characterisation of congruences and of subdirectly irreducible algebras. The details are assigned to the exercises.

4.4 An Application A subset Z of an Ockham space $\mathbf{Y}$ is called a g-**subset** if it is closed under g. Since, by 7.4.1(iii), the lattice of lattice congruences on $K(\mathbf{Y})$ is isomorphic to the order-theoretic dual of the lattice of closed subsets of $\mathbf{Y}$, it is easily seen that the lattice of Ockham algebra congruences on $K(\mathbf{Y})$ is isomorphic to the order-theoretic dual of the lattice of all closed g-subsets of $\mathbf{Y}$: if Z is a closed subset of $\mathbf{Y}$, then the projection congruence $\theta_Z := \{\, (\alpha, \beta) \mid \alpha{\restriction}_Z = \beta{\restriction}_Z \,\}$ is an Ockham congruence if and only if Z is a closed g-subset and, moreover, if Z is a closed g-subset of $\mathbf{Y}$ then $H(K(\mathbf{Y})/\theta_Z)$ is isomorphic as an Ockham space to $\mathbf{Z}$. For $y \in Y$, define $g^\omega\{y\} := \{\, g^n(y) \mid n \geqslant 0 \,\}$. Since the (topological) closure of a g-closed subset of Y is g-closed, the closure $\overline{g^\omega\{y\}}$ of $g^\omega\{y\}$ is the smallest closed g-subset of Y containing y. Define

$$\overline{Z := \{\, y \in Y \mid \overline{g^\omega\{y\}} \neq Y \,\}}.$$

Then $K(\mathbf{Y})$ is a subdirectly irreducible Ockham algebra if and only if $Z \neq Y$. In fact, Z is a closed g-subset of $\mathbf{Y}$ and if $K(\mathbf{Y})$ is subdirectly irreducible, then θ_Z is the smallest non-trivial congruence on $K(\mathbf{Y})$. If $\mathbf{Y}$ is finite, then $K(\mathbf{Y})$ is subdirectly irreducible if and only if $\mathbf{Y}$ is **one-generated**, that is, $g^\omega\{y\} = Y$ for some $y \in Y$. Consider the map γ_n defined on $\mathbb{Z}_m = \{0, \dots, m-1\}$ by $\gamma_n(k) = k + 1$ for $0 \leqslant k < m - 1$ and $\gamma_n(m-1) = n$: see Figure 7.2. If $\leqslant$ is an order on $\mathbb{Z}_m$ with respect to which γ_n is order-reversing, then $\mathbf{W} = \langle \mathbb{Z}_m; \gamma_n, \leqslant, \mathcal{T} \rangle$, where $\mathcal{T}$ is the discrete topology, is a one-generated Ockham space, and, up to isomorphism, every finite one-generated Ockham space arises in this way.

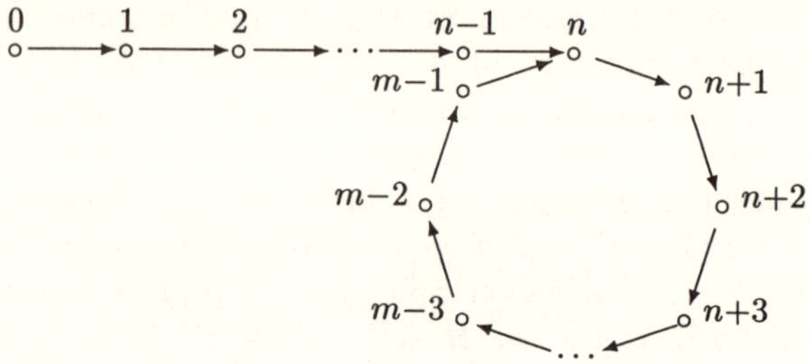

Figure 7.2 the action of γ_n on $\mathbb{Z}_m$

In Figure 7.3, three examples of one-generated Ockham spaces are given, along with the corresponding subdirectly irreducible Ockham algebras. Note that $K(\mathbf{W}^1) \cong \underline{\mathbf{S}}$ is the three-element Stone algebra while $K(\mathbf{Z}) \cong \underline{\mathbf{K}}$ is the three-element Kleene algebra. Since $\mathbf{W}^1$ is a g-subset of $\mathbf{W}^0$, the algebra $K(\mathbf{W}^1)$ is a homomorphic image of $K(\mathbf{W}^0)$. The maps g and f are indicated via arrows except that their fixpoints are simply indicated by solid circles. A map α in $K(\mathbf{W}^0)$, for example, is indicated by the string $\alpha(0)\alpha(1)\alpha(2)$.

In order to highlight the differences between natural dualities and restricted Priestley dualities, we shall compare them on the variety $\mathcal{K}$ of Kleene algebras. Recall from 4.3.9 that $\mathcal{K} = \mathbb{ISP}\underline{\mathbf{K}}$ is the subvariety of $\mathcal{O}$ generated by the three-element Kleene algebra $\underline{\mathbf{K}}$. To find the restricted

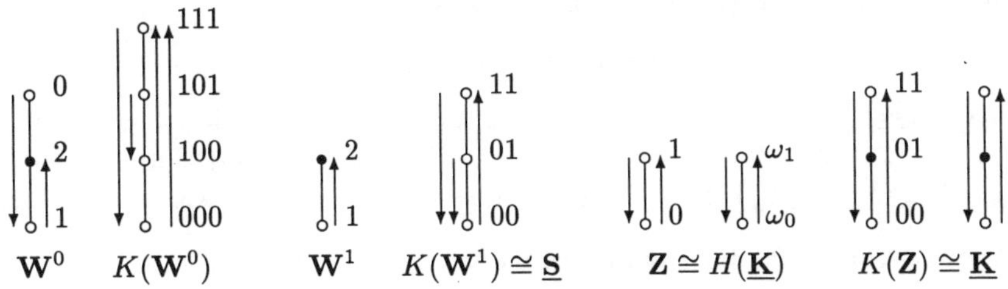

Figure 7.3 three one-generated Ockham spaces

Priestley duality for $\mathcal{K}$, it remains to characterise Ockham spaces $\mathbf{Y}$ such that $K(\mathbf{Y})$ is a Kleene algebra. An alternative proof of the following result is given in Exercise 7.1.

4.5 Restricted Priestley Duality for $\mathcal{K}$ *Let $\mathbf{Y} = \langle Y; g, \leqslant, \mathcal{T}\rangle$ be an Ockham space. Then $\mathbf{Y}$ satisfies*

(K) $\quad (\forall y \in Y)\ g(g(y)) = y$ *and* $(\forall y \in Y)\ y \leqslant g(y)$ *or* $y \geqslant g(y)$.

if and only if $\mathbf{Y}$ is the Priestley dual of a Kleene algebra, that is, $K(\mathbf{Y})$ is a Kleene algebra.

Proof Let $\mathbf{Y}$ be an Ockham space satisfying (K) and let $\mathbf{Z}$ be the Ockham space dual to $\underline{\mathbf{K}}$ as shown in Figure 7.3. We must prove that $K(\mathbf{Y})$ can be embedded into a power of $\underline{\mathbf{K}} \cong K(\mathbf{Z})$. By part (vi) of 7.4.1, it suffices to find a jointly surjective family of Ockham space morphisms $\{\, \varphi_i \colon \mathbf{Z} \to \mathbf{Y} \mid i \in I \,\}$. (If the maps φ_i are Ockham space morphisms, then the product map $\sqcap_{i\in I}\varphi_i \colon K(\mathbf{Y}) \to K(\mathbf{Z})^I \cong \underline{\mathbf{K}}^I$ is an Ockham homomorphism.) Let $y \in Y$. Since $\mathbf{Y}$ satisfies (K), we have either $y \leqslant g(y)$ or $y \geqslant g(y)$ and we may define $\varphi_y \colon Z \to Y$ by $\varphi_y(0) = y$ and $\varphi_y(1) = g(y)$, if $y \leqslant g(y)$, and $\varphi_y(0) = g(y)$ and $\varphi_y(1) = y$, if $y \geqslant g(y)$. Since $g^2(y) = y$, the map φ_y is an Ockham space morphism. Clearly the family $\{\, \varphi_y \colon \mathbf{Z} \to \mathbf{Y} \mid y \in Y \,\}$ is jointly surjective, as required.

Conversely, let $\mathbf{A}$ be a Kleene algebra. We must prove that $\mathbf{Y} = H(\mathbf{A})$ satisfies (K). Since $\mathbf{A}$ is isomorphic to a subalgebra of a power of $\underline{\mathbf{K}}$, the set $\mathcal{K}(\mathbf{A}, \underline{\mathbf{K}})$ separates the points of $\mathbf{A}$. Note that $H(\underline{\mathbf{K}}) = \mathcal{D}(\underline{\mathbf{K}}, \underline{\mathbf{D}}) = \{\omega_0, \omega_1\}$, where $\omega_0 = \chi_{\{1\}}$ and $\omega_1 = \chi_{\{a,1\}}$: see Figure 7.3. Since $\mathcal{D}(\underline{\mathbf{K}}, \underline{\mathbf{D}})$ separates the points of $\underline{\mathbf{K}}$, it follows that condition (J) of Lemma 7.2.2 holds with $\mathfrak{M} = \{\underline{\mathbf{K}}\}$ and $\Omega_K = \{\omega_0, \omega_1\}$. By Lemma 7.2.2, the maps

$$\Phi_{\omega_0} \colon \mathcal{K}(\mathbf{A}, \underline{\mathbf{K}}) \to \mathcal{D}(\mathbf{A}, \underline{\mathbf{D}}) \text{ and } \Phi_{\omega_1} \colon \mathcal{K}(\mathbf{A}, \underline{\mathbf{K}}) \to \mathcal{D}(\mathbf{A}, \underline{\mathbf{D}})$$

are jointly surjective. Let $y \in H(\mathbf{A}) = \mathcal{D}(\mathbf{A}, \underline{\mathbf{D}})$. Thus, for some $i \in \{0, 1\}$, there exists $x \in \mathcal{K}(\mathbf{A}, \underline{\mathbf{K}})$ such that $\omega_i \circ x = \Phi_{\omega_i}(x) = y$.

Now $\varphi_y := H(x)\colon H(\underline{\mathbf{K}}) \to H(\mathbf{A})$ satisfies

$$\varphi_y(\omega_i) = H(x)(\omega_i) = \omega_i \circ x = y.$$

Since $H(\underline{\mathbf{K}}) \cong \mathbf{Z}$ (see Figure 7.3), $g^2(\omega_i) = \omega_i$ which implies that $g^2(y) = g^2(\varphi_y(\omega_i)) = \varphi_y(g^2(\omega_i)) = \varphi_y(\omega_i) = y$. Similarly, we find that $y \leqslant g(y)$ if $i = 0$ and $y \geqslant g(y)$ if $i = 1$. Thus $\mathbf{Y} = H(\mathbf{A})$ satisfies (K). ■

The restricted Priestley dual category for $\mathcal{K}$ is therefore the full subcategory $\mathcal{Y}_{\mathcal{K}}$ of $\mathcal{Y}_{\mathcal{O}}$ consisting of the Ockham spaces which satisfy (K). It is an easy matter to construct (finite or infinite) Ockham spaces in $\mathcal{Y}_{\mathcal{K}}$.

4.6 Pros and Cons We now have available to us both the natural duality (via 4.3.10) and the restricted Priestley duality (via 7.4.5). Which should we use and when? The restricted Priestley duality is a powerful tool in the study of individual Kleene algebras, particularly if we are concerned with the underlying distributive lattice. But for the study of more global properties of the class $\mathcal{K}$, the natural duality has many strengths over the restricted Priestley duality. Let $\mathcal{X}$ be the natural dual category as described in 4.3.10.

(1) Products in the natural dual $\mathcal{X}$ are simply the usual Cartesian products with structure defined pointwise from the factors. Although products exist in $\mathcal{Y}_{\mathcal{K}}$ (since coproducts exist in $\mathcal{K}$), they need not be Cartesian: the Cartesian product of two spaces satisfying (K) need not satisfy (K)!

(2) Free objects exist in $\mathcal{X}$ (see Exercise 3.5) while this fails in $\mathcal{Y}_{\mathcal{K}}$ (see Exercise 7.2).

(3) The dual in $\mathcal{X}$ of the free Kleene algebra $\mathbf{F}_{\underline{\mathbf{K}}}(S)$ is simply $\underset{\sim}{\mathbf{K}}^S$. It is not at all clear how to find the dual in $\mathcal{Y}_{\mathcal{K}}$ of $\mathbf{F}_{\underline{\mathbf{K}}}(S)$!

4.7 Working Together Of course, we could ask for the best of both worlds. Since the categories $\mathcal{X}$ and $\mathcal{Y}_{\mathcal{K}}$ are both dually equivalent to $\mathcal{K}$, they are equivalent to one another. If we could give an explicit description of the functors between $\mathcal{X}$ and $\mathcal{Y}_{\mathcal{K}}$ which yield this equivalence, we would have a **translation process** between $\mathcal{X}$ and $\mathcal{Y}_{\mathcal{K}}$ permitting us to use the two dualities in tandem. We could then, for example, find (the Priestley dual of) the underlying distributive lattice of $\mathbf{F}_{\underline{\mathbf{K}}}(S)$ by applying the translation process to the object $\underset{\sim}{\mathbf{K}}^S$ in $\mathcal{X}$. This translation process has been worked out in a number of cases: for example, in the next section we shall give the description, from Davey and Priestley [DP87], of the translation between the restricted Priestley duality and a multisorted natural duality for the variety of Ockham algebras generated by a finite subdirectly irreducible algebra $\underline{\mathbf{M}}$, and the translation process in the case where $\underline{\mathbf{M}}$ is an n-valued Łukasiewicz algebra is described in Priestley [Pri95b].

It is often the case that the restricted Priestley duality for the quasi-variety generated by $\underline{\mathbf{M}}$ or, more generally, by $\underline{\mathfrak{M}}$ is known and we want to apply the Piggyback Duality Theorem 7.2.1 to obtain a natural duality. Once we have chosen sets G and Ω_M, for $\underline{\mathbf{M}} \in \underline{\mathfrak{M}}$, which satisfy the separation condition (S) of the Piggyback Duality Theorem, two questions need to be addressed.

(i) How can we conveniently describe the piggyback relations on $\underline{\mathfrak{M}}$?

(ii) How do we find a 'small' subset R' of R such that $G \cup R'$ entails R (and therefore $\underset{\sim}{\mathfrak{M}}' = \langle \dot{\bigcup}\{ M \mid M \in \mathfrak{M} \}; G, R', \mathcal{T}\rangle$ yields a duality on the quasi-variety generated by $\underline{\mathfrak{M}}$)?

In some cases we can use algebraic techniques as we did in the case of Heyting algebras in Lemma 7.3.1. Alternatively, we can attempt to answer these questions by riding piggyback on the *known* restricted Priestley duality. This technique was first used to answer question (i) in Davey and Priestley ([DP87] and [DP93a]), and to answer question (ii) in Davey and Priestley [DP93b]. We illustrate the technique in Sections 5 and 6 below.

A very useful $\mathcal{D}$–$\mathcal{P}$ 'dictionary', which describes the restricted Priestley dual for a number of classes, is given at the end of Priestley [Pri84]. A list of 239 references which 'are related to, develop, and/or apply Priestley or natural dualities' is given by Adams and Dziobiak at the end of the *Studia Logica* Special Issue on Priestley Duality [AD96].

7.5 Piggyback dualities for varieties of Ockham algebras

Suppose we wish to find a natural duality for the variety $\mathcal{V}$ generated by a finite subdirectly irreducible Ockham algebra $\underline{\mathbf{M}}^0$. By the discussion in Section 1, $\mathcal{V} = \mathbb{ISP}\underline{\mathfrak{M}}$ where $\underline{\mathfrak{M}}$ is a transversal of the isomorphism classes of maximal subdirectly irreducible algebras in $\mathbb{HS}(\underline{\mathbf{M}}^0)$. As is often the case, the easiest way to find the maximal subdirectly irreducible algebras in $\mathbb{HS}(\underline{\mathbf{M}}^0)$ is via the restricted Priestley duality. By 7.4.4, up to isomorphism $\underline{\mathbf{M}}^0 = K(\mathbf{W}^0)$, where $\mathbf{W}^0 = \langle \mathbb{Z}_m; \gamma_n, \leqslant, \mathcal{T}\rangle$ is a one-generated Ockham space. Let $W^1 := \{\gamma_n^k(1) \mid k \geqslant 0\}$ and note that $W^1 = \mathbb{Z}_m$ if $n = 0$ and $W^1 = \mathbb{Z}_m \backslash \{0\}$ if $1 \leqslant n \leqslant m-1$. Thus $\mathbf{W}^1 = \langle W^1; \gamma_n, \leqslant, \mathcal{T}\rangle$ is the substructure of $\mathbf{W}^0$ generated by 1. Let $\underline{\mathbf{M}}^1 = K(\mathbf{W}^1)$. We find (see Exercise 7.3) that

(a) $\underline{\mathfrak{M}} = \{\underline{\mathbf{M}}^0\}$, whence $\mathcal{V} = \mathbb{ISP}\underline{\mathbf{M}}^0$, if $n \in \{0, 1\}$ or if γ_n is order-preserving, for example, when $\underline{\mathbf{M}}^0$ is the three-element Kleene algebra (see **Z** in Figure 7.3), and

(b) $\underline{\mathfrak{M}} = \{\underline{\mathbf{M}}^0, \underline{\mathbf{M}}^1\}$, whence $\mathcal{V} = \mathbb{ISP}\{\underline{\mathbf{M}}^0, \underline{\mathbf{M}}^1\}$, if $n \notin \{0, 1\}$ and γ_n is not order-preserving, for example, when $\mathbf{W}^0$ and $\mathbf{W}^1$ are as shown in Figure 7.3.

As we wish to illustrate the way in which the restricted Priestley duality can be used to describe the piggyback relations, we first need a description of the restricted Priestley dual $\mathcal{Y}_{\mathcal{V}}$ of $\mathcal{V}$ similar to the description of the restricted Priestley dual $\mathcal{Y}_{\mathcal{K}}$ of $\mathcal{K}$. Note that the formulæ

$$\sigma_0(y) \equiv \mathop{\&}_{0\leqslant k,\ell\leqslant m} \{ g^k(y) \leqslant g^\ell(y) \mid \gamma_n^k(0) \leqslant \gamma_n^\ell(0) \text{ in } \mathbf{W}^0 \} \quad \text{and}$$

$$\sigma_1(y) \equiv \mathop{\&}_{0\leqslant k,\ell\leqslant m} \{ g^k(y) \leqslant g^\ell(y) \mid \gamma_n^k(1) \leqslant \gamma_n^\ell(1) \text{ in } \mathbf{W}^1 \}$$

when evaluated in the dual space $\mathbf{Z}$ of $\underline{\mathbf{K}}$ reduce to

$$\sigma_0(y) \equiv \big(g^2(y) = y \;\&\; y \leqslant g(y)\big) \text{ and } \sigma_1(y) \equiv \big(g^2(y) = y \;\&\; y \geqslant g(y)\big),$$

whence the formula (K), which by 7.4.5 describes the objects of $\mathcal{Y}_{\mathcal{K}}$, is equivalent to the disjunction $\sigma_0(y) \vee \sigma_1(y)$. Since $\underline{\mathbf{M}}^i = K(\mathbf{W}^i) = \mathcal{P}(\mathbf{W}^i, \underset{\sim}{\mathbf{D}})$, we may define $\omega_i \colon \underline{\mathbf{M}}^i \to \underline{\mathbf{D}}$ by $\omega_i(\alpha) := \alpha(i)$. Under the isomorphism $\kappa_{\mathbf{W}^i} \colon \mathbf{W}^i \to HK(\mathbf{W}^i)$, the element $i \in W^i$ corresponds to the morphism $\omega_i \in H(\underline{\mathbf{M}}^i) = HK(\mathbf{W}^i)$. Motivated by these observations, Goldberg [Gol81] and Davey and Priestley [DP86] proved the following beautiful description of $\mathcal{Y}_{\mathcal{V}}$.

5.1 Restricted Priestley Duality for $\mathcal{V}$ *For $i = 0, 1$, let $\underline{\mathbf{M}}^i$, $\mathbf{W}^i$, σ_i and ω_i be as defined above. Let $\mathcal{V} = \mathbb{HSP}\underline{\mathbf{M}}^0 = \mathbb{ISP}\{\underline{\mathbf{M}}^0, \underline{\mathbf{M}}^1\}$ be the variety generated by $\underline{\mathbf{M}}^0$ and let $\mathcal{Y}_{\mathcal{V}} = \{ \mathbf{Y} \in \mathcal{Y}_{\mathcal{O}} \mid K(\mathbf{Y}) \in \mathcal{V} \}$ be the restricted Priestley dual of $\mathcal{V}$. The following are equivalent for an Ockham space $\mathbf{Y} = \langle Y; g, \leqslant, \mathcal{T} \rangle$:*

(i) $\mathbf{Y} \in \mathcal{Y}_{\mathcal{V}}$;

(ii) $\mathbf{Y}$ *satisfies* $(\forall y)\,(\sigma_0(y) \vee \sigma_1(y))$;

(iii) Y *is a (not necessarily disjoint) union of subsets Y^0 and Y^1, where*

$$\begin{aligned} Y^i &:= \{ y \in Y \mid \mathbf{Y} \vDash \sigma_i(y) \} \\ &= \{ y \in Y \mid (\exists \varphi_y \in \mathcal{Y}_{\mathcal{O}}(\mathbf{W}^i, \mathbf{Y})\; \varphi_y(i) = y \}. \end{aligned}$$

Moreover, we have $H(\mathbf{A})^i = \{ y \in H(\mathbf{A}) \mid \big(\exists x \in \mathcal{V}(\mathbf{A}, \underline{\mathbf{M}}^i)\big)\; \omega_i \circ x = y \}$, for all $\mathbf{A} \in \mathcal{V}$.

Proof This is a relatively straightforward generalisation of the proof of 7.4.5 and the details are left to the reader. ∎

We shall take $\underline{\mathfrak{M}} = \{\underline{\mathbf{M}}^0, \underline{\mathbf{M}}^1\}$ and produce a two-sorted natural duality for the variety $\mathcal{V}$. In case (a), when $\underline{\mathbf{M}}^0$ is the only maximal subdirectly irreducible algebra, there is a natural embedding of $\underline{\mathbf{M}}^1$ into $\underline{\mathbf{M}}^0$ and the two-sorted duality can be converted into a single-sorted duality. As we shall see, even in this case, the two-sorted duality has the advantage of making the translation process more transparent. We take $\Omega_{M_0} = \{\omega_0\}$ and $\Omega_{M_1} = \{\omega_1\}$ and let $G = \{g_0, g_1\}$, where $g_0 \colon \underline{\mathbf{M}}^0 \to \underline{\mathbf{M}}^1$ and $g_1 \colon \underline{\mathbf{M}}^1 \to \underline{\mathbf{M}}^0$ are defined by

$$g_0(\alpha) = \alpha{\restriction}_{W^1}, \text{ for all } \alpha \in \mathcal{P}(\mathbf{W}^0, \underset{\sim}{\mathbf{D}}), \quad \text{and}$$
$$g_1(\alpha) = \alpha \circ \gamma_n^2, \text{ for all } \alpha \in \mathcal{P}(\mathbf{W}^1, \underset{\sim}{\mathbf{D}}).$$

Observe that $g_0 = K(\iota)$, where $\iota \colon \mathbf{W}^1 \to \mathbf{W}^0$ is the inclusion map, and $g_1 = K(\gamma_n^2)$. Since both ι and γ_n^2 are Ockham space morphisms, g_0 and g_1 are Ockham algebra homomorphisms. Our first task is to establish the separation condition (S).

5.2 Lemma *The sets* $\Omega_{M_0} = \{\omega_0\}$, $\Omega_{M_1} = \{\omega_1\}$ *and* $G = \{g_0, g_1\}$ *satisfy condition* (S) *of the Piggyback Duality Theorem.*

Proof We must prove that if $\alpha \neq \beta$ in $\underline{\mathbf{M}}^i$ then there exist j and a homomorphism $g \colon \underline{\mathbf{M}}^i \to \underline{\mathbf{M}}^j$ which is a k-fold composite of g_0 and g_1, for some $k \geqslant 0$, such that $(\omega_j \circ g)(\alpha) \neq (\omega_j \circ g)(\beta)$. Assume that $\alpha \neq \beta$ in $\underline{\mathbf{M}}^0$ and let k be the least element of $\mathbb{Z}_m$ such that $\alpha(k) \neq \beta(k)$. We ask the reader to confirm that if $k = 2j$, then $(\omega_0 \circ (g_1 \circ g_0)^j)(\alpha) \neq (\omega_0 \circ (g_1 \circ g_0)^j)(\beta)$, and if $k = 2j+1$, then $(\omega_1 \circ g_0 \circ (g_1 \circ g_0)^j)(\alpha) \neq (\omega_1 \circ g_0 \circ (g_1 \circ g_0)^j)(\beta)$. Similarly, if $\alpha \neq \beta$ in $\underline{\mathbf{M}}^1$, we let k be the least element of W^1 such that $\alpha(k) \neq \beta(k)$. If $k = 0$, then $W^1 = \mathbb{Z}_m$ and consequently $\underline{\mathbf{M}}^1 = \underline{\mathbf{M}}^0$, whence we may use the previous argument. Assume that $k \in \mathbb{Z}_m \backslash \{0\}$. Again, we ask the reader to confirm that if $k = 2j$, then $(\omega_0 \circ (g_1 \circ g_0)^{j-1} \circ g_1)(\alpha) \neq (\omega_0 \circ (g_1 \circ g_0)^{j-1} \circ g_1)(\beta)$, and if $k = 2j+1$, then $(\omega_1 \circ (g_0 \circ g_1)^j)(\alpha) \neq (\omega_1 \circ (g_0 \circ g_1)^j)(\beta)$. ■

5.3 The Alternating Structure Let $\mathbf{W} = \langle W; \gamma, \leqslant, \mathcal{T} \rangle$ be a finite one-generated Ockham space generated by $e \in W$. (The element e is usually referred to as an **end** of $\mathbf{W}$ and is unique unless the directed graph of γ, see Figure 7.2, is a loop in which case every element of W is an end.) A significant role is played by the **alternating structure** induced by the map γ on $\{0,1\}^W$. Define $\underset{\sim}{\mathbf{2}}^W = \langle \{0,1\}^W; g, \ll, \mathcal{T} \rangle$, where $\mathcal{T}$ is the discrete topology, $g(\alpha) := \alpha \circ \gamma$ for all $\alpha \in \{0,1\}^W$, and $\ll$ is the **alternating order** induced by γ on $\{0,1\}^W$: for all $\alpha, \beta \in \{0,1\}^W$,

$$\alpha \ll \beta \iff (\forall k \geqslant 0) \begin{cases} \alpha(\gamma^k(e)) \leqslant \beta(\gamma^k(e)), & \text{if } k \text{ is even,} \\ \alpha(\gamma^k(e)) \geqslant \beta(\gamma^k(e)), & \text{if } k \text{ is odd.} \end{cases}$$

Several points must be noted. First, the structure $\underset{\sim}{\mathbf{2}}^W$ depends only upon the number $m = |W|$, the map γ and, when γ is a loop, the choice of the end e: the order on the Ockham space $\mathbf{W}$ plays no role in the definition of $\underset{\sim}{\mathbf{2}}^W$. In the definition of $\ll$, the variable k is quantified over all non-negative integers, not just over $\mathbb{Z}_m$. Consequently, the nature of the alternating order changes radically according to whether $m-n$ is even or odd. If $m-n$ is even, then

$$\alpha \ll \beta \iff (\forall k \in \mathbb{Z}_m) \begin{cases} \alpha(\gamma^k(e)) \leqslant \beta(\gamma^k(e)), & \text{if } k \text{ is even,} \\ \alpha(\gamma^k(e)) \geqslant \beta(\gamma^k(e)), & \text{if } k \text{ is odd,} \end{cases}$$

and if $m-n$ is odd, then

$$\alpha \ll \beta \iff (\forall k \in \mathbb{Z}_m) \begin{cases} \alpha(\gamma^k(e)) \leqslant \beta(\gamma^k(e)), & \text{if } k < n \text{ is even,} \\ \alpha(\gamma^k(e)) \geqslant \beta(\gamma^k(e)), & \text{if } k < n \text{ is odd,} \\ \alpha(\gamma^k(e)) = \beta(\gamma^k(e)), & \text{if } n \leqslant k \leqslant m-1. \end{cases}$$

By Exercise 7.4, $\underset{\sim}{\mathbf{2}}^W$ is an Ockham space. Our next lemma is a lovely example of schizophrenia at work.

5.4 Lemma $M^0 = \{\alpha \in \{0,1\}^{W^0} \mid \underset{\sim}{\mathbf{2}}^{W^0} \models \sigma_0(\alpha)\}$.

Proof Let $\alpha \in M^0 := \mathcal{P}(\mathbf{W}^0, \underset{\sim}{\mathbf{D}})$. In order to prove that $\underset{\sim}{\mathbf{2}}^{W^0} \models \sigma_0(\alpha)$, we must verify that $\gamma_n^k(0) \leqslant \gamma_n^\ell(0)$ in $\mathbf{W}^0$ implies that $g^k(\alpha) \ll g^\ell(\alpha)$ in $\underset{\sim}{\mathbf{2}}^{W^0}$. But

$$\begin{aligned} \gamma_n^k(0) \leqslant \gamma_n^\ell(0) &\implies (\forall j \geqslant 0) \begin{cases} \gamma_n^k(j) \leqslant \gamma_n^\ell(j), & \text{if } j \text{ is even,} \\ \gamma_n^k(j) \geqslant \gamma_n^\ell(j), & \text{if } j \text{ is odd,} \end{cases} \\ &\implies (\forall j \geqslant 0) \begin{cases} \alpha(\gamma_n^k(j)) \leqslant \alpha(\gamma_n^\ell(j)), & \text{if } j \text{ is even,} \\ \alpha(\gamma_n^k(j)) \geqslant \alpha(\gamma_n^\ell(j)), & \text{if } j \text{ is odd,} \end{cases} \\ &\implies (\forall j \geqslant 0) \begin{cases} g^k(\alpha)(j) \leqslant g^\ell(\alpha)(j), & \text{if } j \text{ is even,} \\ g^k(\alpha)(j) \geqslant g^\ell(\alpha)(j), & \text{if } j \text{ is odd,} \end{cases} \end{aligned}$$

by the definition of g in $\underset{\sim}{\mathbf{2}}^{W^0}$. That is, $g^k(\alpha) \ll g^\ell(\alpha)$.

Conversely, assume that $\alpha \colon W^0 \to \{0,1\}$ is such that $\underset{\sim}{\mathbf{2}}^{W^0} \models \sigma_0(\alpha)$. If $k \leqslant \ell$ in $\mathbf{W}^0$, then $\gamma_n^k(0) \leqslant \gamma_n^\ell(0)$ and therefore $g^k(\alpha) \ll g^\ell(\alpha)$. Hence $\alpha(k) = \alpha(\gamma_n^k(0)) = g^k(\alpha)(0) \leqslant g^\ell(\alpha)(0) = \alpha(\gamma_n^\ell(0)) = \alpha(\ell)$, whence α is order-preserving. ■

The piggyback relations are the Ockham subalgebras of $\underline{\mathbf{M}}^i \times \underline{\mathbf{M}}^j$ which are maximal in $(\omega_i, \omega_j)^{-1}(\leqslant)$. Fortunately, such maximal subalgebras are unique: if $\mathbf{B}$ is a $\{0,1\}$-sublattice of an Ockham algebra $\mathbf{A}$, then

$$B^\circ := \{\, b \in B \mid (\forall k \geqslant 0)\, f^k(b) \in B \,\}$$

is an Ockham subalgebra of $\mathbf{A}$ and is the largest subalgebra of $\mathbf{A}$ contained in B (see Exercise 7.5). Since (S) holds, the Piggyback Duality Theorem guarantees that there is a two-sorted duality for $\boldsymbol{\mathcal{V}}$ given by the two maps g_0 and g_1 along with four multisorted binary relations.

5.5 Two-Sorted Piggyback Duality for $\mathcal{V}$ [DP87] *For $i = 0, 1$, let $\underline{\mathbf{M}}^i$, $\mathbf{W}^i$, ω_i and g_i be as defined above. Let $\mathcal{V} = \mathbb{HSP}\underline{\mathbf{M}}^0 = \mathbb{ISP}\underline{\mathfrak{M}}$, where $\underline{\mathfrak{M}} = \{\underline{\mathbf{M}}^0, \underline{\mathbf{M}}^1\}$, be the variety generated by $\underline{\mathbf{M}}^0$ and let*

$$\underset{\sim}{\mathfrak{M}} := \langle M^0 \dot{\cup} M^1; g_0, g_1, \ll_0, \ll_1, \ll_{01}, \ll_{10}, \mathcal{T}\rangle,$$

where

(i) $\ll_0 \subseteq M^0 \times M^0$ *is the restriction of the alternating order on* $\underset{\sim}{\mathbf{2}}^{W^0}$,

(ii) $\ll_1 \subseteq M^1 \times M^1$ *is the restriction of the alternating order on* $\underset{\sim}{\mathbf{2}}^{W^1}$,

(iii) $\ll_{01} = \{ (\alpha, \beta) \in M^0 \times M^1 \mid \alpha \ll_0 \beta \circ \gamma_n \text{ in } \underset{\sim}{\mathbf{2}}^{W^0} \}$, *and*

(iv) $\ll_{10} = \{ (\beta, \alpha) \in M^1 \times M^0 \mid \beta \circ \gamma_n \ll_0 \alpha \text{ in } \underset{\sim}{\mathbf{2}}^{W^0} \}$.

Then $\underset{\sim}{\mathfrak{M}}$ yields a duality on $\mathcal{V}$.

Proof By the remarks above, it remains only to prove that the relations $\ll_0$, $\ll_1$, $\ll_{01}$, $\ll_{10}$ are the unique maximal subalgebras of $(\omega_i, \omega_j)^{-1}(\leqslant)$ for $i, j \in \{0, 1\}$. In fact,

$$\ll_0 = \omega_0^{-1}(\leqslant)^\circ, \ \ll_1 = \omega_1^{-1}(\leqslant)^\circ,$$
$$\ll_{01} = (\omega_0, \omega_1)^{-1}(\leqslant)^\circ, \ \ll_{10} = (\omega_1, \omega_0)^{-1}(\leqslant)^\circ.$$

We shall verify only the first assertion. The others are proved analogously. Let $\alpha, \beta \in \underline{\mathbf{M}}^0$. Then $(\alpha, \beta) \in \omega_0^{-1}(\leqslant)$ if and only if $\alpha(0) = \omega_0(\alpha) \leqslant \omega_0(\beta) = \beta(0)$. Thus, since $f^k(\alpha) = \alpha \circ \gamma_n^k$, if k is even, and $f^k(\alpha) = c \circ \alpha \circ \gamma_n^k$, if k is odd,

$$\begin{aligned}
(\alpha, \beta) \in \omega_0^{-1}(\leqslant)^\circ &\iff (\forall k \geqslant 0)\, f^k((\alpha, \beta)) \in \omega_0^{-1}(\leqslant) \\
&\iff (\forall k \geqslant 0)\, f^k(\alpha)(0) \leqslant f^k(\beta)(0) \\
&\iff (\forall k \geqslant 0) \begin{cases} \alpha(\gamma_n^k(0)) \leqslant \beta(\gamma_n^k(0)) & \text{if } k \text{ is even,} \\ c(\alpha(\gamma_n^k(0))) \leqslant c(\beta(\gamma_n^k(0))) & \text{if } k \text{ is odd,} \end{cases} \\
&\iff (\forall k \geqslant 0) \begin{cases} \alpha(\gamma_n^k(0)) \leqslant \beta(\gamma_n^k(0)) & \text{if } k \text{ is even,} \\ \alpha(\gamma_n^k(0))) \geqslant \beta(\gamma_n^k(0)) & \text{if } k \text{ is odd,} \end{cases} \\
&\iff \alpha \ll \beta \text{ in } \underset{\sim}{\mathbf{2}}^{W^0}.
\end{aligned}$$

Thus $\ll_0 = \omega_0^{-1}(\leqslant)^\circ$, as claimed. ∎

Note that if $n = 0$, as in the case of Kleene algebras, then $\underline{\mathbf{M}}^0 = \underline{\mathbf{M}}^1$ and $\ll_1$ is the converse of $\ll_0$.

We now have available to us the restricted Priestley dual $H(\mathbf{A})$ and the two-sorted natural dual $D(\mathbf{A})$. Part (iii) of 7.5.1 hints at the translation process between these two dual structures. We split $\mathbf{Y} = \langle Y; g, \leqslant, \mathcal{T}\rangle \in \mathcal{Y}_{\mathcal{V}}$ into the closed subsets Y^0 and Y^1 (see Exercise 7.6) and form a two-sorted structure based on $Y^0 \dot{\cup} Y^1$: the maps g_0 and g_1

are the restrictions of g to Y^0 and Y^1 and the relations $\ll_0$, $\ll_1$, $\ll_{01}$ and $\ll_{10}$ show how the original Ockham space order $\leqslant$ relates within and between the sorts Y^0 and Y^1. The translation in the opposite direction begins with the observation that on $\underset{\sim}{\mathfrak{M}}$ the intersection of $\ll_{10}$ with the converse, $\gg_{10}$, of $\ll_{01}$ is the graph of a one-to-one partial map I from M^1 to M^0 given by '$I(\beta)$ is defined and $I(\beta) = \beta \circ \gamma_n$ if and only if $\beta \circ \gamma_n \in M^0$.' It follows that $\ll_{10}^{\mathbf{X}} \cap \gg_{10}^{\mathbf{X}}$ is the graph of a one-to-one partial map from X^1 to X^0, for all $\mathbf{X} \in \mathcal{X} = \mathbb{IS}_c\mathbb{P}^+ \underset{\sim}{\mathfrak{M}}$. Given $\mathbf{X} \in \mathcal{X}$, we form an Ockham space based on $X^0 \dot\cup X^1$ modulo the identification given by the partial map I: modulo this identification, g is the union of g_0 and g_1 and $\leqslant$ is the union of $\ll_0$, $\ll_1$, $\ll_{01}$ and $\ll_{10}$. All of this is formalised in the following result which we state without proof, referring the reader to Davey and Priestley [DP87] for the details. Since $\mathrm{Irr}(\underset{\sim}{\mathfrak{M}}) = 1$, we could apply 7.1.2 and upgrade the two-sorted duality for $\mathcal{V}$ to a strong duality by adding further partial maps between $\underline{\mathbf{M}}^i$ and $\underline{\mathbf{M}}^j$, for $i, j \in \{0, 1\}$, to the structure on $\underset{\sim}{\mathfrak{M}}$. Rather than do this we shall restrict our attention to the general translation functors on $\mathcal{Y}_\mathcal{V}$ and $D(\mathcal{V})$.

5.6 Translation Between $\mathcal{Y}_\mathcal{V}$ and $\mathcal{X}$ *For* $\mathbf{Y} = \langle Y; g, \leqslant, \mathcal{T}\rangle \in \mathcal{Y}_\mathcal{V}$, *define* $U(\mathbf{Y}) = \langle Y^0 \dot\cup Y^1; g_0, g_1, \ll_0, \ll_1, \ll_{01}, \ll_{10}, \mathcal{T}\rangle$, *where*

(i) $g_0 = g{\restriction}_{Y^0}$ *and* $g_1 = g{\restriction}_{Y^1}$,

(ii) $\ll_0 = \leqslant \cap (Y^0 \times Y^0)$, $\ll_1 = \leqslant \cap (Y^1 \times Y^1)$, $\ll_{01} = \leqslant \cap (Y^0 \times Y^1)$, $\ll_{10} = \leqslant \cap (Y^1 \times Y^0)$, *and*

(iii) $\mathcal{T}$ *is the disjoint union topology.*

For $\mathbf{X} = \langle X^0 \dot\cup X^1; g_0, g_1, \ll_0, \ll_1, \ll_{01}, \ll_{10}, \mathcal{T}\rangle \in \mathcal{X} = \mathbb{IS}_c\mathbb{P}^+ \underset{\sim}{\mathfrak{M}}$, *define* $V(\mathbf{X}) = \langle (X^0 \dot\cup X^1)/\Theta; g, \leqslant, \mathcal{T}\rangle$, *where*

(a) Θ *is the equivalence relation given by* $(x, y) \in \Theta$ *if and only if*

$$x = y \text{ or } (x, y) \in \ll_{10}^{\mathbf{X}} \cap \gg_{10}^{\mathbf{X}} \text{ or } (y, x) \in \ll_{10}^{\mathbf{X}} \cap \gg_{10}^{\mathbf{X}},$$

(b) $g(x/\Theta) = g_i(x)/\Theta$, *if* $x \in X^i$,

(c) $x/\Theta \leqslant y/\Theta$ *if and only if* $(x, y) \in \ll_0 \cup \ll_1 \cup \ll_{01} \cup \ll_{10}$, *and*

(d) $\mathcal{T}$ *is the finest topology on* $(X^0 \dot\cup X^1)/\Theta$ *such that the induced maps* $X^i \to (X^0 \dot\cup X^1)/\Theta$ *are continuous.*

Then $U(\mathbf{Y})$ *and* $V(\mathbf{X})$ *are well defined,* $U(\mathbf{Y}) \in \mathcal{X}$, *and for all* $\mathbf{A} \in \mathcal{V}$, $V(D(\mathbf{A})) \in \mathcal{Y}_\mathcal{V}$, $U(H(\mathbf{A})) \cong D(\mathbf{A})$ *and* $V(D(\mathbf{A})) \cong H(\mathbf{A})$.

Since $D(\mathbf{F}_{\underline{\mathbf{M}}^0}(1)) \cong \underset{\sim}{\mathfrak{M}}$, the Ockham space dual to $\mathbf{F}_{\underline{\mathbf{M}}^0}(1)$ is isomorphic to $V(\underset{\sim}{\mathfrak{M}}) = \langle (M^0 \dot\cup M^1)/\Theta; g, \leqslant, \mathcal{T}\rangle$, as defined above. Let $\underline{\mathbf{N}}^0 = K(\overline{\mathbf{W}}^0)$, where $\overline{\mathbf{W}}^0 = \langle \mathbb{Z}_m; \gamma_n, =, \mathcal{T}\rangle$ is the *flat* Ockham space corresponding to $\mathbf{W}^0$, and let $\mathcal{V}'$ be the variety generated by $\underline{\mathbf{N}}^0$. Since the identity map is an Ockham space morphism from $\overline{\mathbf{W}}^0$ onto $\mathbf{W}^0$, the algebra $\underline{\mathbf{M}}^0$ embeds into $\underline{\mathbf{N}}^0$. Thus $\mathcal{V} \subseteq \mathcal{V}'$ and consequently $\mathbf{F}_{\underline{\mathbf{M}}^0}(1) \cong \mathbf{F}_{\underline{\mathbf{N}}^0}(1)/\theta$, for some congruence θ. The following result, which uses the translation process to give a very convenient description of $H(\mathbf{F}_{\underline{\mathbf{M}}^0}(1))$ in terms of $\underset{\sim}{\mathbf{2}}^{\mathbf{W}^0}$, says, in particular, that $H(\mathbf{F}_{\underline{\mathbf{N}}^0}(1)) \cong \underset{\sim}{\mathbf{2}}^{\mathbf{W}^0}$ and that the congruence θ corresponds to the g-subset $\mathbf{Y}$ of $\underset{\sim}{\mathbf{2}}^{\mathbf{W}^0}$. For every non-empty set S, we endow $(\underset{\sim}{\mathbf{2}}^{W^0})^S$ with the pointwise structure.

5.7 Free Algebras via the Translation *Let S be a non-empty set and define $Y := \{y \in (\{0,1\}^{W^0})^S \mid (\underset{\sim}{\mathbf{2}}^{W^0})^S \models \sigma_0(y) \vee \sigma_1(y)\}$. Then Y is a closed g-subset of $(\underset{\sim}{\mathbf{2}}^{W^0})^S$ and the Ockham space $\mathbf{Y} = \langle Y; g, \ll, \mathcal{T}\rangle$ is isomorphic to the restricted Priestley dual, $H(\mathbf{F}_{\underline{\mathbf{M}}^0}(S))$, of $\mathbf{F}_{\underline{\mathbf{M}}^0}(S)$.*

Proof We shall sketch the proof in the case that $|S| = 1$ and leave it to the reader to check that the general case then follows from the fact that the functor D maps coproducts in $\mathcal{V}$ to products in $\mathcal{X}$ (see Exercise 7.7). Define $\varphi\colon M^0 \dot\cup M^1 \to \{0,1\}^{W^0}$ by $\alpha \mapsto \alpha$ for $\alpha \in M^0$ and $\beta \mapsto \beta \circ \gamma_n$ for $\beta \in M^1$. As observed in the discussion preceding 7.5.6, on $M^0 \dot\cup M^1$, the equivalence relation Θ identifies $\beta \in M^1$ with $\alpha \in M^0$ whenever $\alpha = \beta \circ \gamma_n$. It follows easily that $\Theta = \ker(\varphi)$. To see that Y is the image of φ, we first need to prove

(i) $\underset{\sim}{\mathbf{2}}^{W^1} \models \sigma_1(\beta) \implies \underset{\sim}{\mathbf{2}}^{W^0} \models \sigma_1(\beta \circ \gamma_n)$, and

(ii) $\underset{\sim}{\mathbf{2}}^{W^0} \models \sigma_1(\alpha) \implies (\exists \beta \in M^1)\, \alpha = \beta \circ \gamma_n$.

To prove (i), let $\underset{\sim}{\mathbf{2}}^{W^1} \models \sigma_1(\beta)$. We must prove that if $\gamma_n^k(1) \leqslant \gamma_n^\ell(1)$ in $\mathbf{W}^1$, then $g^k(\beta \circ \gamma_n) \ll g^\ell(\beta \circ \gamma_n)$ in $\underset{\sim}{\mathbf{2}}^{W^0}$:

$$\begin{aligned}
\gamma_n^k(1) \leqslant \gamma_n^\ell(1) &\implies g^k(\beta) \ll g^\ell(\beta) \text{ in } \underset{\sim}{\mathbf{2}}^{W^1}, \text{ since } \underset{\sim}{\mathbf{2}}^{W^1} \models \sigma_1(\beta),\\
&\implies (\forall j \geqslant 1)\begin{cases} \beta(\gamma_n^k(j)) \leqslant \beta(\gamma_n^\ell(j)), & \text{if } j \text{ is odd},\\ \beta(\gamma_n^k(j)) \geqslant \beta(\gamma_n^\ell(j)), & \text{if } j \text{ is even},\end{cases}\\
&\implies (\forall j \geqslant 0)\begin{cases} \beta(\gamma_n^{k+1}(j)) \leqslant \beta(\gamma_n^{\ell+1}(j)), & \text{if } j \text{ is even},\\ \beta(\gamma_n^{k+1}(j)) \geqslant \beta(\gamma_n^{\ell+1}(j)), & \text{if } j \text{ is odd},\end{cases}\\
&\implies g^k(\beta \circ \gamma_n) \ll g^\ell(\beta \circ \gamma_n) \text{ in } \underset{\sim}{\mathbf{2}}^{W^0}.
\end{aligned}$$

Hence $\underset{\sim}{\mathbf{2}}^{W^0} \models \sigma_1(\beta \circ \gamma_n)$.

To prove (ii), let $\underset{\sim}{\mathbf{2}}^{W^0} \models \sigma_1(\alpha)$ and define $\beta\colon W^1 \to \{0,1\}$ by $(\forall k \in W^1)\, \beta(k) := \alpha(k \ominus 1)$, where $k \ominus 1 := k - 1$, if $k \neq 0$, and $k \ominus 1 := m - 1$, if $k = 0$. (The latter case applies only if $n = 0$, in which case the directed graph of γ_n is a loop and $\underline{\mathbf{M}}^0 = \underline{\mathbf{M}}^1$.) We must prove that $\beta \in \mathcal{P}(\mathbf{W}^1, \underset{\sim}{\mathbf{D}})$, that is, β preserves the order $\leqslant$ on $\mathbf{W}^1$. Let $k, \ell \in W^1$; then

$$k \leqslant \ell \text{ in } W^1 \implies \gamma_n^{k\ominus 1}(1) \leqslant \gamma_n^{\ell\ominus 1}(1) \text{ in } W^1$$

$$
\begin{aligned}
&\Longrightarrow g^{k\ominus 1}(\alpha) \ll g^{\ell\ominus 1}(\alpha) \text{ in } \underset{\sim}{2}^{W^0}, \text{ since } \underset{\sim}{2}^{W^0} \models \sigma_1(\alpha),\\
&\Longrightarrow g^{k\ominus 1}(\alpha)(0) \leqslant g^{\ell\ominus 1}(\alpha)(0)\\
&\Longrightarrow \alpha(\gamma_n^{k\ominus 1}(0)) \leqslant \alpha(\gamma_n^{\ell\ominus 1}(0))\\
&\Longrightarrow \alpha(k\ominus 1) \leqslant \alpha(\ell\ominus 1)\\
&\Longrightarrow \beta(k) \leqslant \beta(\ell).
\end{aligned}
$$

Thus $\beta \in M^1$ and, since $\alpha = \beta \circ \gamma_n$, (ii) holds. It follows easily from (i), (ii) and Lemma 7.5.4, applied to both M^0 and M^1, that $\varphi(M^0 \dot{\cup} M^1) = Y$.

Let $\psi\colon (M^0 \dot{\cup} M^1)/\Theta \to Y$ be the natural bijection: $\alpha/\Theta \mapsto \varphi(\alpha)$. If $\alpha \in M^0$, then

$$\psi(g(\alpha/\Theta)) = \varphi(g_0(\alpha)) = \alpha{\restriction}_{W^1} \circ \gamma_n = \alpha \circ \gamma_n = g(\varphi(\alpha)) = g(\psi(\alpha/\Theta)),$$

and if $\beta \in M^1$, then

$$\psi(g(\beta/\Theta)) = \varphi(g_1(\beta)) = \beta \circ \gamma_n^2 = g(\varphi(\beta)) = g(\psi(\beta/\Theta)),$$

whence ψ preserves g. Thus ψ is an Ockham space isomorphism from $V(\underset{\sim}{\mathfrak{M}})$ onto $\mathbf{Y}$, since it is immediate that it is an order isomorphism. ■

We close this section by applying 7.5.6 and 7.5.7 to describe the restricted Priestley duals of $\mathbf{F}_{\underline{\mathbf{K}}}(1)$ and $\mathbf{F}_{\underline{\mathbf{M}}^0}(1)$, where $\underline{\mathbf{K}} = K(\mathbf{Z})$ and $\underline{\mathbf{M}}^0 = K(\mathbf{W}^0)$ for the Ockham spaces $\mathbf{Z}$ and $\mathbf{W}^0$ of Figure 7.3.

5.8 Two Examples First consider the three-element Kleene algebra. Since $\mathbf{Z} = \mathbf{Z}^0 = \mathbf{Z}^1$, we have $K^1 = K^0 = K(\mathbf{Z}) = \{00, 01, 11\}$. Thus the two-sorted natural duality given by 7.5.5 with $\underset{\sim}{\mathfrak{M}} = \{\underline{\mathbf{K}}^0, \underline{\mathbf{K}}^1\}$ is equivalent to a single-sorted natural duality which could have been obtained by applying the Piggyback Duality Theorem 7.2.1 to the single algebra $\underline{\mathbf{K}}$ with $\Omega_K = \{\omega_0, \omega_1\}$. As we shall see, the induced single-sorted natural duality reduces to the dual structure

$$\underset{\sim}{\mathbf{K}} = \langle \{0, a, 1\}; \preccurlyeq, \sim, K_0, \mathcal{T} \rangle$$

obtained by hand in Theorem 4.3.10. Identify $\underline{\mathbf{K}}$ with $K(\mathbf{Z})$ by labelling the elements of $K(\mathbf{Z})$ as $0 = 01$, $a = 01$ and $1 = 11$, as usual. The two-sorted structure $\underset{\sim}{\mathfrak{M}} = D(\mathbf{F}_{\underline{\mathbf{K}}}(1))$ is shown diagrammatically on the left of Figure 7.4. The points below indicate how to interpret the figure.

(i) Since $K^0 = K^1 = K$, the map g_0 is the identity map on K, indicated by the upward arrows in the figure, and since $\gamma_0^2 = \mathrm{id}_Z$, the map g_1 is also the identity map on K, indicated by the downward arrows on the figure. It is noteworthy that, while the maps g_0 and g_1 are essential to the description of the translation form $\mathfrak{X}$ to $\mathfrak{Y}_V$, once we

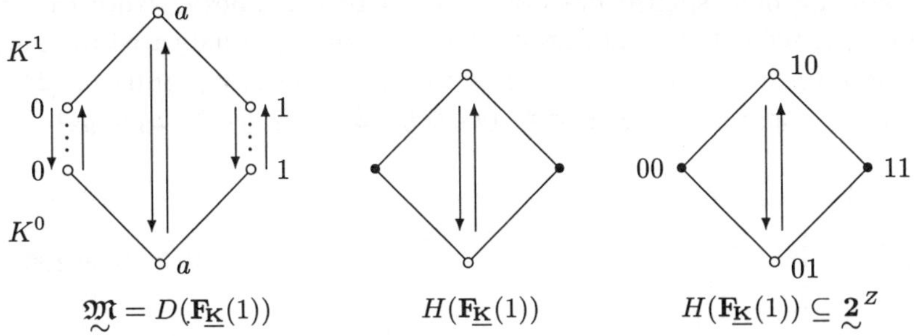

Figure 7.4 two-sorted duality for $\underline{\mathbf{K}}$

reduce from the two-sorted duality to a single-sorted one they can be deleted with no effect upon the duality.

(ii) The orders $\ll_0$ on K^0, in the lower half of the figure, and $\ll_1$ on K^1, in the upper half of the figure, are obtained via 7.5.3 with $e = 0$ and $e = 1$ as the end respectively. Note that $\ll_1$ is the converse of $\ll_0$ and is the order $\preccurlyeq$ on $\underset{\sim}{\mathbf{K}}$ defined in 4.3.10.

(iii) Calculating $\ll_{01}$ and $\ll_{10}$ from the descriptions in 7.5.5, we find that $\ll_{01} = K^2 \backslash \{(0,1),(1,0)\}$, which is precisely the relation $\sim$ on $\underset{\sim}{\mathbf{K}}$ defined in 4.3.10, while $\ll_{10} = \{(0,0),(1,1)\}$, which is precisely the relation Δ_{K_0} where $K_0 = \{0,1\}$.

(iv) Since $\ll_{10} \cap \gg_{10} = \{(0,0),(1,1)\}$, the one-to-one partial map I from K^1 to K^0 is simply the identity map on $\{0,1\}$. The identification given by the map I is indicated on the figure by the vertical dotted lines. Thus $\ll_{01}$ and $\ll_{10}$ give the comparabilities between elements of K^0, in the upper half, and K^1, in the lower half of the figure, modulo this identification.

(v) By 7.5.6, $H(\mathbf{F}_{\underline{\mathbf{K}}}(1)) = V(D(\mathbf{F}_{\underline{\mathbf{K}}}(1)))$ is the Ockham space shown in the middle of Figure 7.4. Consequently, the underlying distributive lattice of $\mathbf{F}_{\underline{\mathbf{K}}}(1)$ is isomorphic to $K(\mathbf{2}^2) \cong \mathbf{1} \oplus \mathbf{2}^2 \oplus \mathbf{1}$.

(vi) Applying 7.5.7 we obtain $H(\mathbf{F}_{\underline{\mathbf{K}}}(1))$ as the subset of $\underset{\sim}{\mathbf{2}}^Z$ shown on the right of the figure.

Now consider the algebra $\underline{\mathbf{M}}^0 = K(\mathbf{W}^0)$ of Figure 7.3. The formulæ $\sigma_0(y)$ and $\sigma_1(y)$ are

$$\sigma_0(y) \equiv \left(g(y) \leqslant g^2(y) \leqslant y \ \ \& \ \ g^3(y) = g^2(y)\right),$$
$$\sigma_1(y) \equiv y \leqslant g(y) = g^2(y).$$

With the elements of M^0 labelled $0 = 000$, $a = 100$, $b = 101$ and $1 = 111$ and the elements of M^1 labelled $0 = 00$, $c = 01$ and $1 = 11$, the structure $\underset{\sim}{\mathfrak{M}} = D(\mathbf{F}_{\underline{\mathbf{M}}^0}(1))$ is as shown on the left of Figure 7.5. The interpretation follows the lines spelled out for $\underline{\mathbf{K}}$ in (i) to (vi) above. Note that this time g_0 is indicated by downward arrows while g_1 is indicated by upward arrows. Since $m - n = 3 - 2 = 1$ is odd, the alternating order on $\underset{\sim}{\mathbf{2}}^{W^0}$ is given by $x_1x_2x_3 \ll y_1y_2y_3 \iff (x_1 \leqslant y_1 \;\&\; x_2 \geqslant y_2 \;\&\; x_3 = y_3)$.

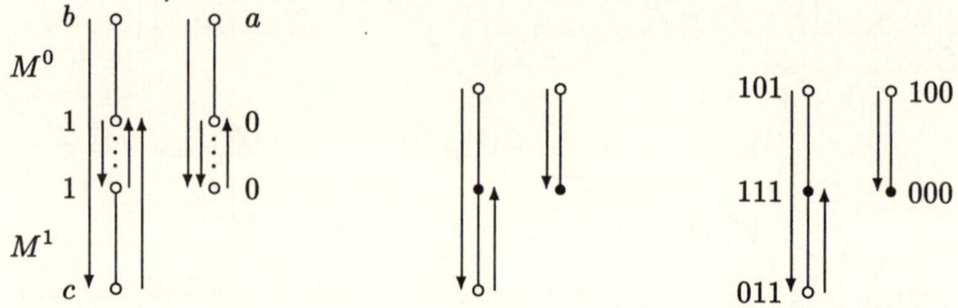

$\underset{\sim}{\mathfrak{M}} = D(\mathbf{F}_{\underline{\mathbf{M}}^0}(1))$ $H(\mathbf{F}_{\underline{\mathbf{M}}^0}(1))$ $H(\mathbf{F}_{\underline{\mathbf{M}}^0}(1)) \subseteq \underset{\sim}{\mathbf{2}}^{W^0}$

Figure 7.5 two-sorted duality for $\underline{\mathbf{M}}^0$

Bounded distributive lattices with extra operations which form a monoid of endomorphisms or dual endomorphisms, as occurs for Ockham algebras, come under the very general umbrella established in Cornish's monograph [Cor:AA]. There, amongst many other things, he gives a unified approach to the restricted Priestley duality for such classes. This part of Cornish's work is extended in Priestley [Pri00]: piggyback natural dualities are described and the translation process is discussed.

7.6 Piggyback dualities for varieties of distributive p-algebras

In Section 3, we obtained a description of the piggyback relations by purely algebraic means. Then, in Section 4, we employed a combination of algebra and an application of the restricted Priestley duality to find the piggyback relations. In this section, which is based on Davey and Priestley [DP93a], we shall illustrate a general technique for finding the piggyback relations via the restricted Priestley duality.

Recall, from 4.3.17, that the proper non-trivial subvarieties of the variety of distributive p-algebras are of the form $\mathcal{B}_n = \mathbb{ISP}\underline{\mathbf{B}}_n$, where $\underline{\mathbf{B}}_n$ is obtained by adjoining a new top, 1, to the n-atom Boolean algebra, whose top is denoted by a. Since $\mathcal{B}_0$ is simply the class of Boolean algebras, throughout this section we shall assume that $n \geqslant 1$. The restricted Priestley dual $\mathcal{Y}_n$ of $\mathcal{B}_n$ is described in Exercise 7.8. For our purposes, it suffices to know that, for finite ordered sets $\mathbf{X}$ and $\mathbf{Y}$,

(1) $\mathbf{X} \in \mathcal{Y}_n$ if and only if $|\max(x)| \leqslant n$, for all $x \in X$, where $\max(x)$ denotes the set of all maximals above x, and

(2) $\varphi \in \mathcal{Y}_n(\mathbf{X}, \mathbf{Y})$ if and only if φ is a p-morphism, that is φ is order-preserving and satisfies $\varphi(\max(x)) = \max(\varphi(x))$, for all $x \in X$.

The Priestley dual, $H(\underline{\mathbf{B}}_n)$, of $\underline{\mathbf{B}}_n$ is the ordered set, $\mathbf{V}_n$, shown in Figure 7.6. The n maximal elements $1, \dots, n$ correspond to the maps $\chi_{\uparrow b}$, where b is an atom of $\underline{\mathbf{B}}_n$, and 0 denotes the map $\omega := \chi_{\{1\}}$. Henceforth, we identify $\underline{\mathbf{B}}_n$ with $KH(\underline{\mathbf{B}}_n) = K(\mathbf{V}_n) = \mathcal{P}(\mathbf{V}_n, \underset{\sim}{\mathbf{D}})$, in which case the map ω is given by $\alpha \mapsto \alpha(0)$.

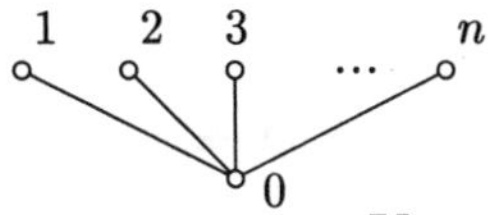

$\mathbf{V}_n = H(\underline{\mathbf{B}}_n)$

Figure 7.6 the Priestley dual of $\underline{\mathbf{B}}_n$

We shall obtain a (single-sorted) piggyback duality for $\mathcal{B}_n$ by taking $\Omega_{B_n} = \{\omega\}$. In order to establish the separation condition (S) of the Piggyback Duality Theorem 7.2.1, we need to describe the endomorphisms of $\underline{\mathbf{B}}_n$. This is (of course!) most easily achieved via the duality.

The endomorphisms of $\underline{\mathbf{B}}_n$ correspond to the endomorphisms of $\mathbf{V}_n$. It is easily seen that $\varphi\colon V_n \to V_n$ is a p-morphism if and only if either $\varphi(0) \neq 0$, in which case φ is constant, or $\varphi(0) = 0$, in which case the restriction of φ to $\{1, \dots, n\}$ is a permutation of $\{1, \dots, n\}$. For $n = 1$, there is just one non-identity endomorphism. For $n = 2$, there is one non-identity automorphism τ, which maps 0 to 0 and whose restriction to $\{1, 2\}$ is the transposition $(1\,2)$, and there are the two constant endomorphisms onto $\{1\}$ and onto $\{2\}$. Recall that, for $n \geqslant 3$, the symmetric group $\mathbf{S}_n$ is generated by the cycle $(1\,2 \dots n)$ and the transposition $(1\,2)$, and note that any constant endomorphism can be obtained by composing the constant endomorphism e onto $\{1\}$ with an appropriate automorphism. Hence, we deduce that the endomorphism monoid of $\mathbf{V}_n$ is generated by three maps: the constant endomorphism e and the two automorphisms σ, whose restriction to $\{1, \dots, n\}$ is the cycle $(1\,2 \dots n)$, and τ, whose restriction to $\{1, \dots, n\}$ is the transposition $(1\,2)$.

6.1 Lemma *The set $\Omega_{B_n} = \{\omega\}$ and the set $G = \{e\}$, when $n = 1$, $G = \{\tau, e\}$, when $n = 2$, and $G = \{\sigma, \tau, e\}$, when $n \geqslant 3$, satisfy condition* (S) *of the Piggyback Duality Theorem.*

Proof Let $\alpha, \beta \in \underline{\mathbf{B}}_n = K(\mathbf{V}_n)$ with $\alpha \neq \beta$. If $\alpha(0) \neq \beta(0)$, then, by definition, $\omega(\alpha) \neq \omega(\beta)$. If $\alpha(0) = \beta(0)$, then there exists $k \geqslant 1$ such that $\alpha(k) \neq \beta(k)$. Let e_k be the constant endomorphism of $\mathbf{V}_n$ onto $\{k\}$ (obtained by composing e with an automorphism of $\mathbf{V}_n$) and let $g := K(e_k)$ be the corresponding endomorphism of $\underline{\mathbf{B}}_n$. Then $g(\alpha) = K(e_k)(\alpha) = \alpha \circ e_k \neq \beta \circ e_k = K(e_k)(\beta) = g(\beta)$, as required. ∎

It follows immediately from the Piggyback Duality Theorem that if we take T to be the set of all piggyback relations on $\underline{\mathbf{B}}_n$ induced by the map ω, then the structure $\underset{\sim}{\mathbf{B}}_n := \langle B_n; G, T, \mathcal{T}\rangle$ yields a duality of $\mathcal{B}_n$. We need to describe the relations in T and shall achieve this via another application of the restricted Priestley duality. Since T is enormous, we shall also use entailment to delete relations from T without destroying the duality.

By Properites 7.4.1, every subalgebra $\mathbf{r}$ of $\underline{\mathbf{B}}_n^2$ is determined by a pair of jointly surjective p-morphisms $\varphi_1, \varphi_2 \colon \mathbf{V}_n \to \mathbf{Y_r}$, where $\mathbf{Y_r} = H(\mathbf{r})$; the binary relation r on B_n is then given by

$$r = r(\varphi_1, \varphi_2) := \{(\alpha \circ \varphi_1, \alpha \circ \varphi_2) \mid \alpha \in H(\mathbf{Y_r})\} \leqslant K(\mathbf{V}_n)^2 = \underline{\mathbf{B}}_n^2.$$

(The maps φ_1 and φ_2 are dual to the projections $\rho_i \colon \mathbf{r} \to \underline{\mathbf{B}}_n$.) The following two lemmas tell us when $r(\varphi_1, \varphi_2)$ is a piggyback relation. The first holds more generally (see Exercise 7.9).

6.2 Lemma *The relation $r(\varphi_1, \varphi_2)$ is contained in $\omega^{-1}(\leqslant)$ if and only if $\varphi_1(0) \leqslant \varphi_2(0)$.*

Proof For all $\mathbf{X} \in \mathcal{P}$, in particular for $\mathbf{V}_n$, we have $x \leqslant y$ in $\mathbf{X}$ if and only if $\alpha(x) \leqslant \alpha(y)$ in $\underset{\sim}{\mathbf{D}}$, for all $\alpha \in K(\mathbf{X})$. This proves the first equivalence below, and the remainder follow by definition.

$$\begin{aligned} \varphi_1(0) \leqslant \varphi_2(0) &\iff (\forall \alpha \in K(\mathbf{V}_n))\ \alpha(\varphi_1(0)) \leqslant \alpha(\varphi_2(0)) \\ &\iff (\forall \alpha \in K(\mathbf{V}_n))\ \omega(\alpha \circ \varphi_1) \leqslant \omega(\alpha \circ \varphi_2) \\ &\iff (\forall \alpha \in K(\mathbf{V}_n))\ (\alpha \circ \varphi_1, \alpha \circ \varphi_2) \in \omega^{-1}(\leqslant) \\ &\iff r(\varphi_1, \varphi_2) \subseteq \omega^{-1}(\leqslant) \end{aligned}$$

as required. ∎

The (Priestley duals of the) piggyback relations on $\underline{\mathbf{B}}_n$ are now very easy to describe.

6.3 Lemma *The relation $r(\varphi_1, \varphi_2)$ is maximal in $\omega^{-1}(\leqslant)$ if and only if*

(i) $\varphi_1(0) < \varphi_2(0)$,

(ii) $\varphi_2(0)$ *is non-maximal, and*

(iii) $|\varphi_1(\{1, 2, \ldots, n\})| = n$.

Proof If $r(\varphi_1, \varphi_2) \subseteq \omega^{-1}(\leqslant)$, then r will be non-maximal in $\omega^{-1}(\leqslant)$ if and only if there exist jointly surjective p-morphisms $\psi_1, \psi_2 \colon \mathbf{V}_n \to \mathbf{Z}$, with $|Z| > |Y_\mathbf{r}|$, and a p-morphism $\psi \colon \mathbf{Z} \to \mathbf{Y_r}$ such that $\varphi_i = \psi \circ \psi_i$

for $i = 1, 2$. We leave it to the reader to verify that the conditions in the statement of the lemma are exactly those needed to ensure that $\mathbf{Y_r}$ cannot be 'expanded' in this way. ■

From this lemma it is clear that if $r = r(\varphi_1, \varphi_2)$ is a piggyback relation, then $|Y_\mathbf{r}| = n + 2$ and $\mathbf{Y_r}$ has the shape shown in Figure 7.7 for some ℓ with $1 \leqslant \ell \leqslant n$. Note that, for fixed n, the ordered set in Figure 7.7 depends only upon the integer ℓ and hence is denoted by $\mathbf{Y}_\ell$. It is helpful to think of φ_1 and φ_2 as providing new labellings of the maximals of Y_ℓ: if $\varphi_1(i) = j$, then label j with the symbol i, and if $\varphi_2(i) = j$ then label j with the symbol $\widetilde{i}$. This view of $\mathbf{Y_r}$ enables us to give an algebraic description of the piggyback relation $r(\varphi_1, \varphi_2)$: see Exercise 7.10.

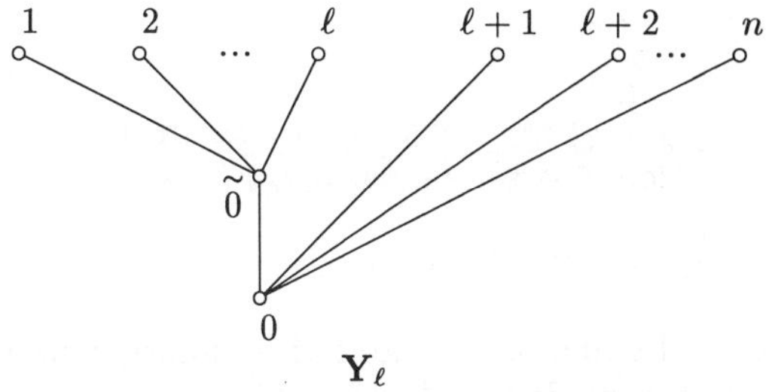

Figure 7.7 the ordered set $\mathbf{Y}_\ell$

A **partition** $\pi = (k_1, k_2, \dots, k_\ell)$ of the natural number n is an ℓ-tuple of natural numbers whose sum is n. If $k_1 \geqslant k_2 \geqslant \dots \geqslant k_\ell$, then the partition π gives rise to a **left-packed** piggyback relation $r(k_1, k_2, \dots, k_\ell) := r(\lambda_1, \lambda_2)$ corresponding to the pair of jointly surjective p-morphisms $\lambda_1, \lambda_2 \colon \mathbf{V}_n \to \mathbf{Y}_\ell$ defined as follows:

$$\lambda_1(i) = i \text{ for all } i \in V_n = \{0, 1, \dots, n\},$$

$$\lambda_2(0) = \widetilde{0} \text{ and } \lambda_2(i) = j \text{ for all } i \in X_j, \text{ where}$$

$$\begin{aligned} X_1 &= \{1, \dots, k_1\}, \\ X_2 &= \{k_1 + 1, \dots, k_1 + k_2\}, \\ X_3 &= \{k_1 + k_2 + 1, \dots, k_1 + k_2 + k_3\}, \\ &\vdots \\ X_\ell &= \{n - k_\ell + 1, \dots, n\}. \end{aligned}$$

The ordered set $\mathbf{Y}_\ell$ with the left-packed labelling coming from the partition $\pi = (k_1, k_2, \dots, k_\ell)$ of the integer n is denoted by $\mathbf{Y}_\pi$. To illustrate the notation, consider the relations r and r' associated with the partitions

(2, 2) and (3, 1) of $n = 4$. The corresponding labellings of $\mathbf{Y}_2$ are shown in Figure 7.8. The jointly surjective maps λ_1, λ_2 are given by

$$\begin{aligned} &\lambda_1(i) = i \text{ for all } i \in \{0,1,2,3,4\}, \text{ for both } r \text{ and } r', \\ &\lambda_2(1) = \lambda_2(2) = 1,\ \lambda_2(3) = \lambda_2(4) = 2, \text{ for } r, \\ &\lambda_2(1) = \lambda_2(2) = \lambda_2(3) = 1,\ \lambda_2(4) = 2, \text{ for } r'. \end{aligned}$$

The left-packed relation $r(k_1, k_2, \dots, k_\ell)$ on B_n is described in terms of digraphs in Davey and Priestley [DP93a].

Figure 7.8 the left-packed relations $\mathbf{Y}_{(2,2)}$ and $\mathbf{Y}_{(3,1)}$

Every labelling of $\mathbf{Y}_\ell$ induced by jointly surjective p-morphisms can be converted into a left-packed labelling by composing with a permutation of the labels $\{1, 2, \dots, n\}$ and a permutation of the labels $\{\tilde{1}, \tilde{2}, \dots, \tilde{n}\}$. Put more formally, if $r = r(\varphi_1, \varphi_2)$ for some pair $\varphi_1, \varphi_2 \colon \mathbf{V}_n \to \mathbf{Y}_\ell$ of jointly surjective p-morphisms, then there exist a left-packed relation $s = r(k_1, k_2, \dots, k_\ell) = r(\lambda_1, \lambda_2)$ and automorphisms σ_1 and σ_2 of $\mathbf{V}_n$ such that $\lambda_1 = \varphi_1 \circ \sigma_1$ and $\lambda_2 = \varphi_2 \circ \sigma_2$. For $i = 1, 2$, let g_i be the automorphism $K(\sigma_i)$ of $\underline{\mathbf{B}}_n = K(\mathbf{V}_n)$. Thus, $g_i(\alpha) := \alpha \circ \sigma_i$ for all $\alpha \in K(\mathbf{V}_n)$. Hence,

$$\begin{aligned} r &= r(\varphi_1, \varphi_2) \\ &= \{(\beta_1, \beta_2) \mid (\exists \alpha \in K(\mathbf{Y}_\ell))\ \beta_1 = \alpha \circ \varphi_1 \ \&\ \beta_2 = \alpha \circ \varphi_2\} \\ &= \{(\beta_1, \beta_2) \mid (\exists \alpha \in K(\mathbf{Y}_\ell))\ \beta_1 = \alpha \circ \lambda_1 \circ \sigma_1^{-1} \ \&\ \beta_2 = \alpha \circ \lambda_2 \circ \sigma_2^{-1}\} \\ &= \{(\beta_1, \beta_2) \mid (\exists \alpha \in K(\mathbf{Y}_\ell))\ \beta_1 \circ \sigma_1 = \alpha \circ \lambda_1 \ \&\ \beta_2 \circ \sigma_2 = \alpha \circ \lambda_2\} \\ &= \{(\beta_1, \beta_2) \mid (\beta_1 \circ \sigma_1, \beta_2 \circ \sigma_2) \in r(\lambda_1, \lambda_2)\} \\ &= \{(\beta_1, \beta_2) \mid (g_1(\beta_1), g_2(\beta_2)) \in r(\lambda_1, \lambda_2)\} \\ &= (g_1 \times g_2)^{-1}(r(\lambda_1, \lambda_2)) \\ &= (g_1 \times g_2)^{-1}(s). \end{aligned}$$

It follows that $\{s, g_1, g_2\}$ entails r. Indeed, it is easy to check that $(g_1 \times g_2)^{-1}(s) = (g_2 \cdot (g_1 \cdot s)^{\smile})^{\smile}$ (see Exercise 8.15), and hence $\{s, g_1, g_2\}$

entails $(g_1 \times g_2)^{-1}(s)$ as converse and action by an endomorphism are admissible constructs (see Constructs for Entailment 2.4.5). We have proved the following theorem.

6.4 Piggyback Duality for $\mathcal{B}_n$ *Let $\underset{\sim}{\mathbf{B}}_n = \langle B_n; G, R, \mathcal{T}\rangle$ where $G = \{e\}$, when $n = 1$, $G = \{\tau, e\}$, when $n = 2$, and $G = \{\sigma, \tau, e\}$, when $n \geqslant 3$, and T is the set of all left-packed binary relations on $\mathbf{B}_n$. Then $\underset{\sim}{\mathbf{B}}_n$ yields a duality on $\mathcal{B}_n$.*

The number $p(n)$ of partitions of n is inherently exponential. Since the relations in T are in a one-to-one correspondence with the partitions of n, for large n this duality is of little practical use and we are led to ask whether further relations can be safely deleted from R. In Davey and Priestley [DP93b] the restricted Priestley duality is applied to show that, provided R contains a (left-packed) relation based on $\mathbf{Y}_\ell$ for $1 \leqslant \ell \leqslant n$, then the duality will not be destroyed. (This result is discussed from a purely algebraic viewpoint in Section 6 of the next chapter.) Thus there is a duality with $|R| = n$, a substantial improvement on $p(n)$. For example, the partitions of 4 are $(1,1,1,1)$, $(2,1,1)$, $(2,2)$, $(3,1)$ and (4), whence $p(4) = 5$. Since $(2,2)$ and $(3,1)$ are both two-part partitions they are both based on $\mathbf{Y}_2$ (see Figure 7.8). Consequently, either of the corresponding relations $r(2,2)$ and $r(3,1)$ can be deleted without destroying the duality.

7.7 Completely dualisable quasi-varieties

We shall say that a finite algebra $\underline{\mathbf{D}}$ generates a **completely dualisable quasi-variety** if every finite algebra $\underline{\mathbf{M}}$ in $\mathcal{A} = \mathbb{ISP}\underline{\mathbf{D}}$ is dualisable. Almost every algebra known to be dualisable is actually known to generate a completely dualisable quasi-variety. While this is trivial if $\underline{\mathbf{D}}$ possesses a near-unanimity term, there is no apparent reason why this should be true for non-NU dualities. We therefore pose the following problem.

7.1 Completely Dualisable Quasi-variety Problem *Give an example of a dualisable algebra $\underline{\mathbf{D}}$ such that the quasi-variety generated by $\underline{\mathbf{D}}$ is not completely dualisable. Find general sufficient conditions for $\underline{\mathbf{D}}$ to generate a completely dualisable quasi-variety.*

This section is based on Davey [Dav96] and Davey and Pitkethly [DP00] and addresses the second part of this problem. Following the philosophy of this chapter, we want to find conditions on $\underline{\mathbf{D}}$ and $\underline{\mathbf{M}} \in \mathcal{A} := \mathbb{ISP}\underline{\mathbf{D}}$ which allow us to obtain a dualising structure $\underset{\sim}{\mathbf{M}}$ for $\underline{\mathbf{M}}$ by riding piggyback on a given dualising structure $\underset{\sim}{\mathbf{D}}$ for $\underline{\mathbf{D}}$. One possibility would be to apply the general Piggyback Duality Theorem, of which our Piggyback Duality Theorem 7.2.1 is a consequence, from Davey and Priestley [DP87].

The version given in [DP87] applies to any dualisable algebra $\underline{\mathbf{D}}$, not just to algebras with a term-reduct in $\mathfrak{D}$, and would require us to find a subset Ω of $\mathcal{A}(\underline{\mathbf{M}}, \underline{\mathbf{D}})$ such that, for all $\mathbf{A} \in \mathcal{A}$, the induced maps $\Phi_\omega : \mathcal{A}(\mathbf{A}, \underline{\mathbf{M}}) \to \mathcal{A}(\mathbf{A}, \underline{\mathbf{D}})$, for $\omega \in \Omega$, are jointly surjective. It turns out (see Exercise 7.11) that this can be done if and only if $\underline{\mathbf{D}}$ is a retract of $\underline{\mathbf{M}}$. Rather than follow that route, and call upon a result not proved here, we shall take the assumption that $\underline{\mathbf{D}}$ is a retract of $\underline{\mathbf{M}}$ as our starting point and show how to convert $\underset{\sim}{\mathbf{D}}$ into $\underset{\sim}{\mathbf{M}}$. Observe that if $\underline{\mathbf{D}}$ can be embedded into $\underline{\mathbf{M}} \in \mathbb{ISP}\underline{\mathbf{D}}$, then $\underline{\mathbf{D}}$ and $\underline{\mathbf{M}}$ generate the same quasi-variety.

An algebra $\mathbf{B} \in \mathcal{A}$ is a **subretract** of $\mathbf{A} \in \mathcal{A}$ if $\mathbf{B}$ is a subalgebra of $\mathbf{A}$ and there is a **subretraction** $\omega : \mathbf{A} \to \mathbf{B}$, that is a homomorphism such that $\omega{\restriction}_B = \mathrm{id}_{\mathbf{B}}$; thus, $\mathbf{B}$ is a retract of $\mathbf{A}$ with the map $\omega : \mathbf{A} \to \mathbf{B}$ as the retraction and the inclusion map $\iota : \mathbf{B} \to \mathbf{A}$ as the coretraction.

Let $\underline{\mathbf{D}}$ be a subalgebra of $\underline{\mathbf{M}}$ and let r be an n-ary algebraic operation on $\underline{\mathbf{D}}$. Since $\mathbf{r} \leqslant \underline{\mathbf{D}}^n$ and $\underline{\mathbf{D}}^n \leqslant \underline{\mathbf{M}}^n$ we may view r as an n-ary algebraic relation on $\underline{\mathbf{M}}$. For emphasis, we shall denote this relation *on* $\underline{\mathbf{M}}$ by r_D. Similarly, if g is a total (or partial) operation on D then g_D denotes g regarded as a *partial* operation on M. Define $R_D := \{ r_D \mid r \in R \}$, $G_D := \{ g_D \mid g \in G \}$, and $H_D := \{ h_D \mid h \in H \}$. If $\underline{\mathbf{D}}$ is a subretract of $\underline{\mathbf{M}}$ via a subretraction $\omega : \underline{\mathbf{M}} \to \underline{\mathbf{D}}$, then we may extend the (n-ary) partial map g_D to a total map $\hat{g}_D : M^n \to M$ by defining $\hat{g}_D := g_D \circ \omega^n$. Similarly, we may extend the (n-ary) partial map h_D to $\hat{h}_D : \mathrm{dom}(\hat{h}_D) \to M$, where $\mathrm{dom}(\hat{h}_D) = (\omega^n)^{-1}(\mathrm{dom}(h))$ and $\hat{h}_D := h_D \circ \omega^n{\restriction}_{\mathrm{dom}(\hat{h}_D)}$. Define $\widehat{G}_D := \{ \hat{g}_D \mid g \in G \}$ and $\widehat{H}_D := \{ \hat{h}_D \mid h \in H \}$.

7.2 First $\underset{\sim}{\mathbf{D}}$-to-$\underset{\sim}{\mathbf{M}}$ Theorem *Let $\underline{\mathbf{M}}$ be a finite algebra in $\mathcal{A} := \mathbb{ISP}\underline{\mathbf{D}}$ and assume that $\underline{\mathbf{D}}$ is a subretract of $\underline{\mathbf{M}}$. If $\underline{\mathbf{D}}$ is dualisable, then $\underline{\mathbf{M}}$ is dualisable. More specifically, assume that $\underset{\sim}{\mathbf{D}} = \langle D; G, H, R, \mathcal{T} \rangle$ yields a duality on $\mathcal{A}$ and that $\omega_0 : \underline{\mathbf{M}} \to \underline{\mathbf{D}}$ is a subretraction. If Ω is a subset of* $\mathrm{End}(\underline{\mathbf{M}})$ *such that $\omega_0 \in \Omega$, each $\omega \in \Omega$ maps into D and Ω separates the points of M, then each of the structures $\underset{\sim}{\mathbf{M}} := \langle M; \Omega \cup \widehat{G}_D, \widehat{H}_D, R_D, \mathcal{T} \rangle$ and $\underset{\sim}{\mathbf{M}}' := \langle M; \Omega, G_D \cup H_D, R_D, \mathcal{T} \rangle$ yields a duality on $\mathbb{ISP}\underline{\mathbf{M}} = \mathcal{A}$.*

Proof For each n-ary map $g \in G$, the set $\{\omega_0, \hat{g}_D\}$ entails g_D, since $\mathrm{fix}(\omega_0) = D$ and $\hat{g}_D{\restriction}_{D^n} = g_D$, and similarly for the partial operations in H. Thus $\underset{\sim}{\mathbf{M}}$ entails $\underset{\sim}{\mathbf{M}}'$, whence it suffices to prove that $\underset{\sim}{\mathbf{M}}'$ yields a duality on $\mathcal{A}$. Let $\mathbf{A} \in \mathcal{A}$ and let $\alpha : D(\mathbf{A}) \to \underset{\sim}{\mathbf{M}}$ be an $\underset{\sim}{\mathbf{M}}'$-morphism, where $D(\mathbf{A}) := \mathcal{A}(\mathbf{A}, \underline{\mathbf{M}}) \leqslant (\underset{\sim}{\mathbf{M}}')^A$. Since $\underline{\mathbf{D}} \leqslant \underline{\mathbf{M}}$, the set $H(\mathbf{A}) := \mathcal{A}(\mathbf{A}, \underline{\mathbf{D}})$ is a subset of $D(\mathbf{A})$. If $x \in H(\mathbf{A})$, then, since ω_0 is a subretraction of $\underline{\mathbf{M}}$ onto $\underline{\mathbf{D}}$, we have $x = \omega_0 \circ x$ and hence, since α preserves ω_0,

$$\alpha(x) = \alpha(\omega_0 \circ x) = \alpha(\omega_0(x)) = \omega_0(\alpha(x)) \in D.$$

Thus $\alpha{\restriction}_{H(\mathbf{A})}: H(\mathbf{A}) \to \underset{\sim}{\mathrm{D}}$ is a $\underset{\sim}{\mathrm{D}}$-morphism. Since, by assumption, $\underset{\sim}{\mathrm{D}}$ yields a duality on $\mathcal{A}$, it follows that there exists $a \in A$ such that $\alpha(x) = x(a)$ for all $x \in H(\mathbf{A})$. Since Ω separates the points of M, in order to show that $\alpha(x) = x(a)$ for all $x \in D(\mathbf{A})$ it suffices to show that $\omega(\alpha(x)) = \omega(x(a))$ for all $\omega \in \Omega$. Let $x \in D(\mathbf{A})$ and let $\omega \in \Omega$. Then

$$\begin{aligned}
\omega(\alpha(x)) &= \alpha(\omega(x)) && \text{as } \alpha \text{ preserves } \omega \\
&= \alpha(\omega \circ x) && \text{by the definition of } \omega(x) \\
&= (\omega \circ x)(a) && \text{as } \omega \circ x \in H(A) \\
&= \omega(x(a)),
\end{aligned}$$

as required. ■

Note that $\underline{\mathbf{M}} \in \mathbb{ISP}\underline{\mathbf{D}}$ if and only if the homomorphisms from $\underline{\mathbf{M}}$ into $\underline{\mathbf{D}}$ separate the points of M. Thus, provided $\underline{\mathbf{D}}$ is a subretract of $\underline{\mathbf{M}}$, we can always choose Ω in the theorem above to be $\mathcal{A}(\underline{\mathbf{M}}, \underline{\mathbf{D}}) \subseteq \mathrm{End}(\underline{\mathbf{M}})$. Since we can add extra endomorphisms without destroying the duality, we obtain the following corollary.

7.3 Corollary *Let $\underline{\mathbf{M}}$ be a finite algebra in $\mathcal{A} := \mathbb{ISP}\underline{\mathbf{D}}$ which has $\underline{\mathbf{D}}$ as a subretract. If $\underset{\sim}{\mathrm{D}} = \langle D; G, H, R, \mathcal{T}\rangle$ yields a duality on $\mathcal{A}$, then $\underset{\sim}{\mathrm{M}} := \langle M; \mathrm{End}(\underline{\mathbf{M}}) \cup \widehat{G}_D, \widehat{H}_D, R_D, \mathcal{T}\rangle$ yields a duality on $\mathbb{ISP}\underline{\mathbf{M}} = \mathcal{A}$.*

7.4 Corollary *The quasi-variety $\mathcal{A} = \mathbb{ISP}\underline{\mathbf{D}}$ is completely dualisable if it has a dualisable generator $\underline{\mathbf{D}}$ which is a retract of every finite member of $\mathcal{A}$.*

The type of the structure $\underset{\sim}{\mathrm{M}} := \langle M; \Omega \cup \widehat{G}_D, \widehat{H}_D, R_D, \mathcal{T}\rangle$, in the First $\underset{\sim}{\mathrm{D}}$-to-$\underset{\sim}{\mathrm{M}}$ Theorem and its corollaries, is obtained from the type of $\underset{\sim}{\mathrm{D}}$ by adding some unary total operation symbols. Unfortunately, the term-reduct $\underset{\sim}{\mathrm{M}}^{\circ} := \langle M; \widehat{G}_D, \widehat{H}_D, R_D, \mathcal{T}\rangle$ of $\underset{\sim}{\mathrm{M}}$, although of the same type as $\underset{\sim}{\mathrm{D}}$, need not belong to $\mathbb{IS}_c\mathbb{P}^+\underset{\sim}{\mathrm{D}}$. At the cost of the loss of a little generality, we can guarantee that $\underset{\sim}{\mathrm{M}}$ is a structure from $\mathbb{IS}_c\mathbb{P}^+\underset{\sim}{\mathrm{D}}$ endowed with some endomorphisms of $\underline{\mathbf{M}}$ as additional total operations.

Choose $\underline{\mathbf{M}}$ to be a subalgebra of a power $\underline{\mathbf{D}}^k$. In place of the relation r_D we can consider the usual pointwise extension, r^M, of the relation r on D to an n-ary relation on M. For example, let $\leqslant$ be the order on the two-element lattice $\underline{\mathbf{D}}$ and identify D with the subset $\{(0, \dots, 0), (1, \dots, 1)\}$ of D^k. Then $\leqslant_D$ is the order on the two-element sublattice $\underline{\mathbf{D}}$ of $\underline{\mathbf{M}}$ while $\leqslant^M$ is the order on the lattice $\underline{\mathbf{M}}$ itself. We leave the reader to derive the following result from the First $\underset{\sim}{\mathrm{D}}$-to-$\underset{\sim}{\mathrm{M}}$ Theorem.

7.5 Second $\underset{\sim}{\mathbf{D}}$-to-$\underset{\sim}{\mathbf{M}}$ Theorem *Let $\underline{\mathbf{D}}$ be a finite algebra, let $\underline{\mathbf{M}}$ be a subalgebra of $\underline{\mathbf{D}}^k$ and assume that $\underset{\sim}{\mathbf{D}} = \langle D; G, H, R, \mathcal{T}\rangle$ yields a duality on $\mathcal{A} := \mathbb{ISP}\underline{\mathbf{D}}$. Assume that $\underline{\mathbf{D}}$ is a retract of $\underline{\mathbf{M}}$ via $\omega_0 \colon \underline{\mathbf{M}} \to \underline{\mathbf{D}}$ and $v \colon \underline{\mathbf{D}} \to \underline{\mathbf{M}}$ and that W is a subset of $\mathcal{A}(\underline{\mathbf{M}}, \underline{\mathbf{D}})$ which contains ω_0 and separates the points of M. Define $\Omega := \{v \circ w \mid w \in W\} \subseteq \mathrm{End}(\underline{\mathbf{M}})$. Assume further that M is closed under the total and partial operations of $\underset{\sim}{\mathbf{D}}^k$ and that $v \colon D \to M$ is an embedding with respect to the structure on $\underset{\sim}{\mathbf{D}}$. Then $\underset{\sim}{\mathbf{M}} := \langle M; \Omega \cup G, H, R, \mathcal{T}\rangle$ (and therefore $\underset{\sim}{\mathbf{M}}' := \langle M; \mathrm{End}(\underline{\mathbf{M}}) \cup G, H, R, \mathcal{T}\rangle$) yields a duality on $\mathbb{ISP}\underline{\mathbf{M}} = \mathcal{A}$.*

7.6 Examples Using these results we can give a number of non-NU examples of completely dualisable quasi-varieties. One way to ensure that the assumptions of the Second $\underset{\sim}{\mathbf{D}}$-to-$\underset{\sim}{\mathbf{M}}$ Theorem are met is to let $\underline{\mathbf{M}}$ be a subalgebra of $\underline{\mathbf{D}}^k$ which contains the diagonal Δ^k_D, take v to be the natural map, $d \mapsto (d, \dots, d)$, and then take, for the operations in $G \cup H$, maps which are simultaneously homomorphisms and term functions (or restrictions of term functions in the case of partial maps). Any projection will serve for the retraction ω_0. Provided $\underset{\sim}{\mathbf{D}}$ yields a duality on $\mathcal{A}$, the assumptions of the theorem will be satisfied by taking W to consist of the k projections. This applies, for example, in the case of abelian groups, vector spaces over a finite field and semilattices. Every finite **abelian group** $\underline{\mathbf{A}}$ is dualisable. By the representation theorem for finite abelian groups, there is a cyclic group $\underline{\mathbf{A}}_m$ such that $\underline{\mathbf{A}} \in \mathcal{A}_m := \mathbb{ISP}\underline{\mathbf{A}}_m$ and such that $\underline{\mathbf{A}}_m$ is a direct factor and therefore a retract of $\underline{\mathbf{A}}$. Since $\underset{\sim}{\mathbf{A}}_m = \langle A_m; +, 0, \mathcal{T}\rangle$ yields a duality on $\mathcal{A}_m$ (see Theorem 4.4.2), it follows immediately from the Second $\underset{\sim}{\mathbf{D}}$-to-$\underset{\sim}{\mathbf{M}}$ Theorem that $\underset{\sim}{\mathbf{A}} := \langle A; \mathrm{End}(\mathbf{A}), +, 0, \mathcal{T}\rangle$ dualises $\underline{\mathbf{A}}$. Since the one-dimensional **vector space** is a retract of every finite non-trivial vector space, Theorem 4.4.4 and the Second $\underset{\sim}{\mathbf{D}}$-to-$\underset{\sim}{\mathbf{M}}$ Theorem combine to show that if $\underline{\mathbf{V}} = \langle V; +, 0, \{\lambda_a\}_{a\in F}\rangle$ is a finite vector space over a finite field $\mathbb{F}$, then $\underset{\sim}{\mathbf{V}} := \langle V; \mathrm{End}(\underline{\mathbf{V}}), +, 0, \{\lambda_a\}_{a\in F}, \mathcal{T}\rangle$ dualises $\underline{\mathbf{V}}$. Similarly, using Theorem 4.4.7, every finite **meet semilattice** $\underline{\mathbf{S}} = \langle S; \wedge, 1\rangle$ is dualised by $\underset{\sim}{\mathbf{S}} := \langle S; \mathrm{End}(\underline{\mathbf{S}}), \wedge, 1, \mathcal{T}\rangle$.

Both Corollary 7.7.3 and the Second $\underset{\sim}{\mathbf{D}}$-to-$\underset{\sim}{\mathbf{M}}$ Theorem apply to **Boolean algebras** and (bounded) **distributive lattices**. For Boolean algebras, they combine with Stone Duality (4.1.2) to show that $\underset{\sim}{\mathbf{B}} := \langle B; \mathrm{End}(\underline{\mathbf{B}}), \mathcal{T}\rangle$ dualises a finite Boolean algebra $\underline{\mathbf{B}}$. Since a nullary in G is entailed by the corresponding constant endomorphism, for (bounded) distributive lattices they combine with Priestley duality (4.3.1 and Exercise 4.5) to show that a finite (bounded) distributive lattice $\underline{\mathbf{L}}$ is dualised by both $\underset{\sim}{\mathbf{L}} := \langle L; \mathrm{End}(\underline{\mathbf{L}}), \leqslant_{\{0,1\}}, \mathcal{T}\rangle$ and $\underset{\sim}{\mathbf{L}}' := \langle L; \mathrm{End}(\underline{\mathbf{L}}), \leqslant^L, \mathcal{T}\rangle$.

7.8 Endodualisable algebras

For each $k \in \mathbb{N}$, an algebra $\underline{\mathbf{M}}$ is k-**endoprimal** if the only functions from M^k to M which preserve the endomorphisms of $\underline{\mathbf{M}}$ are the k-ary term functions of $\underline{\mathbf{M}}$. If $\underline{\mathbf{M}}$ is k-endoprimal for all $k \in \mathbb{N}$, then we say that $\underline{\mathbf{M}}$ is **endoprimal**. A finite algebra $\underline{\mathbf{M}}$ is called **endodualisable** if $\underset{\sim}{\mathbf{M}} := \langle M; \mathrm{End}(\underline{\mathbf{M}}), \mathcal{T}\rangle$ dualises $\underline{\mathbf{M}}$. Of course, by (CLO), every endodualisable algebra is endoprimal. In Examples 7.7.6 above it is shown that every finite Boolean algebra is endodualisable and we shall see in the next chapter (see 8.1.5) that every finite Heyting chain is endodualisable. The results of the previous section show that the endomorphisms play an important role in lifting the dualisability of an algebra $\underline{\mathbf{D}}$ up to the dualisability of an algebra $\underline{\mathbf{M}} \in \mathbb{ISP}\underline{\mathbf{D}}$. Since our primary concern here is dualisability rather than full or strong dualisability, we lose no generality if we replace partial operations and non-unary total operations by their graphs (via Lemma 2.1.2) and assume that $\underset{\sim}{\mathbf{D}} = \langle D; \mathrm{End}(\underline{\mathbf{D}}), R, \mathcal{T}\rangle$. The strategy now will be to prove that the algebra $\underline{\mathbf{M}}$ is endodualisable by first applying Corollary 7.7.3 to show that $\langle M; \mathrm{End}(\underline{\mathbf{M}}), R_D, \mathcal{T}\rangle$ dualises $\underline{\mathbf{M}}$ and then showing that, under certain circumstances, the relations in R_D may be deleted as they will be entailed by the endomorphisms of $\underline{\mathbf{M}}$. The results presented here come from Davey [Dav96] and Davey and Pitkethly [DP00].

If $G \subseteq \mathrm{End}(\underline{\mathbf{D}})$, then $\widehat{G}_D \subseteq \mathrm{End}(\underline{\mathbf{M}})$, whence we obtain the following special case of Corollary 7.7.3.

8.1 Proposition *Let $\underline{\mathbf{M}}$ be a finite algebra in $\mathcal{A} := \mathbb{ISP}\underline{\mathbf{D}}$ which has $\underline{\mathbf{D}}$ as a subretract. If $\underset{\sim}{\mathbf{D}} = \langle D; \mathrm{End}(\underline{\mathbf{D}}), R, \mathcal{T}\rangle$ yields a duality on $\mathcal{A}$, then $\underset{\sim}{\mathbf{M}} := \langle M; \mathrm{End}(\underline{\mathbf{M}}), R_D, \mathcal{T}\rangle$ yields a duality on $\mathbb{ISP}\underline{\mathbf{M}} = \mathcal{A}$.*

The case where $|R| \leqslant 1$ is particularly important here. We say that $\underline{\mathbf{M}}$ is **almost endodualisable** if there is some algebraic relation s on $\underline{\mathbf{M}}$ such that $\underset{\sim}{\mathbf{M}} = \langle M; \mathrm{End}(\underline{\mathbf{M}}), s, \mathcal{T}\rangle$ yields a duality on $\mathbb{ISP}\underline{\mathbf{M}}$.

8.2 First Endodualisability Theorem *Let $\underline{\mathbf{M}}$ be a finite algebra in $\mathcal{A} := \mathbb{ISP}\underline{\mathbf{D}}$ and assume that $\underline{\mathbf{D}}$ is a retract of $\underline{\mathbf{M}}$.*

(i) *If $\underline{\mathbf{D}}$ is endodualisable, then so is $\underline{\mathbf{M}}$.*

(ii) *If $\mathbf{s} \leqslant \underline{\mathbf{D}}^n$ and $\underline{\mathbf{D}}$ is almost endodualisable via $\underset{\sim}{\mathbf{D}} = \langle D; \mathrm{End}(\underline{\mathbf{D}}), s, \mathcal{T}\rangle$, then there exists $\mathbf{t} \leqslant \underline{\mathbf{M}}^n$, with $\mathbf{t}$ isomorphic to $\mathbf{s}$, such that $\underline{\mathbf{M}}$ is almost endodualisable via $\underset{\sim}{\mathbf{M}} = \langle M; \mathrm{End}(\underline{\mathbf{M}}), t, \mathcal{T}\rangle$.*

Proof Assume that $\underline{\mathbf{D}}$ is a retract of $\underline{\mathbf{M}}$. There is some $\underline{\mathbf{M}}' \in \mathcal{A}$ such that $\underline{\mathbf{M}}'$ is isomorphic to $\underline{\mathbf{M}}$ and $\underline{\mathbf{D}}$ is a subretract of $\underline{\mathbf{M}}'$. If $\underline{\mathbf{D}}$ is endodualisable, then Proposition 7.8.1, with $R = \varnothing$, implies that $\underline{\mathbf{M}}'$ is endodualisable and consequently $\underline{\mathbf{M}}$ is endodualisable. If $\underline{\mathbf{D}}$ is almost endodualisable via $\underset{\sim}{\mathbf{D}} = \langle D; \mathrm{End}(\underline{\mathbf{D}}), s, \mathcal{T}\rangle$, Proposition 7.8.1, with $R = \{s\}$, implies that $\underline{\mathbf{M}}'$ is almost endodualisable via $\underset{\sim}{\mathbf{M}}' = \langle M'; \mathrm{End}(\underline{\mathbf{M}})', s, \mathcal{T}\rangle$, and consequently $\underline{\mathbf{M}}$ is almost endodualisable via $\underset{\sim}{\mathbf{M}} = \langle M; \mathrm{End}(\underline{\mathbf{M}}), t, \mathcal{T}\rangle$ where $\mathbf{t}$ is the subalgebra of $\underline{\mathbf{M}}^n$ corresponding to the subalgebra $\mathbf{s}$ of $(\underline{\mathbf{M}}')^n$ under the isomorphism between $\underline{\mathbf{M}}'$ and $\underline{\mathbf{M}}$. ∎

Our main endodualisability result is delightfully schizophrenic and yields a plethora of endodualisable (and therefore endoprimal) algebras.

8.3 Second Endodualisability Theorem *Assume that $\underline{\mathbf{D}}$ is almost endodualisable via the structure $\underset{\sim}{\mathbf{D}} = \langle D; \mathrm{End}(\underline{\mathbf{D}}), s, \mathcal{T}\rangle$ and let $\underline{\mathbf{M}}$ be a finite algebra in $\mathcal{A} := \mathbb{ISP}\underline{\mathbf{D}}$. If both $\underline{\mathbf{D}}$ and $\mathbf{s}$ are retracts of $\underline{\mathbf{M}}$, then $\underline{\mathbf{M}}$ is endodualisable. In particular, if $\underline{\mathbf{D}}$ is a retract of $\underline{\mathbf{M}}$ and $\underline{\mathbf{M}}$ is isomorphic to $\mathbf{s}$, then $\underline{\mathbf{M}}$ is endodualisable.*

Proof If $\underline{\mathbf{D}}$ is almost endodualisable via $\underset{\sim}{\mathbf{D}} = \langle D; \mathrm{End}(\underline{\mathbf{D}}), s, \mathcal{T}\rangle$ and $\underline{\mathbf{D}}$ is a retract of $\underline{\mathbf{M}}$, then, by the First Endodualisability Theorem, $\underline{\mathbf{M}}$ is almost endodualisable via $\underset{\sim}{\mathbf{M}} = \langle M; \mathrm{End}(\underline{\mathbf{M}}), t, \mathcal{T}\rangle$ where $\mathbf{t}$ is isomorphic to $\mathbf{s}$. Since $\mathbf{s}$ is a retract of $\underline{\mathbf{M}}$ so is $\mathbf{t}$. We claim that $\mathrm{End}(\underline{\mathbf{M}})$ entails t and hence t can be deleted from $\underset{\sim}{\mathbf{M}}$ without destroying the duality. While this will follow from the theory of entailment developed in Chapter 9 (see Exercise 9.1), we shall present a simple direct proof.

Assume that $\mathbf{t} \leqslant \underline{\mathbf{M}}^n$ and that $\mathbf{t}$ is a retract of $\underline{\mathbf{M}}$ via homomorphisms $w\colon \underline{\mathbf{M}} \to \mathbf{t}$ and $v\colon \mathbf{t} \to \underline{\mathbf{M}}$. Let $\mathbf{A} \in \mathcal{A}$ and assume that $\alpha\colon \mathcal{A}(\mathbf{A}, \underline{\mathbf{M}}) \to M$ preserves $\mathrm{End}(\underline{\mathbf{M}})$. We must show that α preserves the relation t. Let $x_1, \dots, x_n \in \mathcal{A}(\mathbf{A}, \underline{\mathbf{M}})$ with $(x_1, \dots, x_n) \in t^{\mathcal{A}(\mathbf{A}, \underline{\mathbf{M}})}$. By Lemma 2.2.8, $x_1 \sqcap \dots \sqcap x_n\colon \mathbf{A} \to \mathbf{t}$ is well defined, whence $y := v \circ (x_1 \sqcap \dots \sqcap x_n) \in \mathcal{A}(\mathbf{A}, \underline{\mathbf{M}})$. Define endomorphisms $e_i := \rho_i \circ w$, where $\rho_i\colon \mathbf{t} \to \underline{\mathbf{M}}$ is the restriction of the projection $\pi_i\colon \underline{\mathbf{M}}^n \to \underline{\mathbf{M}}$. Since $w \circ v = \mathrm{id}_{\mathbf{t}}$ and since α preserves e_i, for all i, we have

$$\begin{aligned}\alpha(x_i) &= \alpha(\rho_i \circ (x_1 \sqcap \dots \sqcap x_n)) = \alpha(\rho_i \circ w \circ v \circ (x_1 \sqcap \dots \sqcap x_n)) \\ &= \alpha(e_i \circ y) = \alpha(e_i(y)) = e_i(\alpha(y)) = \rho_i(w(\alpha(y))),\end{aligned}$$

from which we conclude that

$$(\alpha(x_1), \dots, \alpha(x_n)) = (\rho_1(w(\alpha(y))), \dots, \rho_n(w(\alpha(y)))) = w(\alpha(y)) \in t;$$

whence α preserves t. ∎

In Davey [Dav96], a technique was used which permitted the almost endodualisability of $\underline{\mathbf{D}}$ to be lifted up to a characterisation of the endodualisable algebras in $\mathcal{A} = \mathbb{ISP}\underline{\mathbf{D}}$. The technique was formalised in Davey and Pitkethly [DP00] by finding the right finite algebra $\mathbf{E} \in \mathcal{A}$ for which we can conveniently identify the members of $\mathcal{A}$ having $\mathbf{E}$ as a retract.

8.4 Third Endodualisability Theorem *Assume that* $\underline{\mathbf{D}}$ *is almost endodualisable via* $\langle D; \mathrm{End}(\underline{\mathbf{D}}), s, \mathcal{T}\rangle$, *that* $\mathbf{E} \in \mathcal{A} = \mathbb{ISP}\underline{\mathbf{D}}$ *with* $\mathbf{E}$ *isomorphic to* $\mathbf{s}$, *and that* $\underline{\mathbf{D}}$ *is a retract of* $\mathbf{E}$. *Suppose there is a* $k \in \mathbb{N}$ *such that if* $\mathbf{A} \in \mathcal{A}$ *is* k*-endoprimal and has* $\underline{\mathbf{D}}$ *as a retract, then* $\mathbf{A}$ *also has* $\mathbf{E}$ *as a retract. Then, for any finite* $\underline{\mathbf{M}} \in \mathcal{A}$ *which has* $\underline{\mathbf{D}}$ *as a retract, the following are equivalent:*

(i) $\underline{\mathbf{M}}$ *is endodualisable;*

(ii) $\underline{\mathbf{M}}$ *is endoprimal;*

(iii) $\underline{\mathbf{M}}$ *is* k*-endoprimal;*

(iv) $\mathbf{E}$ *is a retract of* $\underline{\mathbf{M}}$*;*

Proof Since (i) $\Rightarrow$ (ii) $\Rightarrow$ (iii) $\Rightarrow$ (iv) is trivial, we assume (iv) to prove (i). Since $\underline{\mathbf{D}}$ is a retract of $\mathbf{E}$, (iv) implies that $\underline{\mathbf{D}}$ is a retract of $\underline{\mathbf{M}}$ and (i) follows from the Second Endodualisability Theorem. ■

In Davey and Pitkethly [DP00], it is proved that if $\mathbf{B} \in \mathbb{ISP}\mathbf{A}$ and $\mathbf{A}$ is a retract of $\mathbf{B}$, then $\mathbf{B}$ is endoprimal provided $\mathbf{A}$ is (see Exercise 7.14). As this applies even when the algebras are infinite, it allowed the authors of [DP00] to extend the Third Endodualisability Theorem from a characterisation of the (necessarily finite) endodualisable algebras in $\mathcal{A}$ up to a characterisation of endoprimality in general.

8.5 Examples We shall discuss a selection of the examples from [Dav96] and [DP00] and leave the reader to consult the exercises or, in some cases, the original papers for the details of the proofs.

Vector Spaces over a Finite Field The one-dimensional vector space $\underline{\mathbf{F}} = \langle F; +, 0, \{\lambda_a\}_{a\in F}\rangle$ over a finite field $\mathbb{F}$ is dualised by the structure $\underset{\sim}{\mathbf{F}} := \langle F; +, 0, \{\lambda_a\}_{a\in F}, \mathcal{T}\rangle$. Since each λ_a is an endomorphism of $\underline{\mathbf{F}}$ and since 0 is the value of a constant endomorphism, $\underline{\mathbf{F}}$ is almost endodualisable via $\langle F; \mathrm{End}(\underline{\mathbf{F}}), \mathrm{graph}(+), \mathcal{T}\rangle$. Thus, the Third Endodualisability Theorem, with $\mathbf{E} = \underline{\mathbf{F}}^2 \cong \mathrm{graph}(+)$, shows that a finite vector space $\underline{\mathbf{V}}$ is endodualisable if and only if it is not one-dimensional (Exercise 7.15).

Distributive Lattices In [MP93], [DHP95b] and [Dav96], endoprimality and endodualisability were considered for distributive lattices with or without bounds. Define $\underline{\mathbf{D}}_C = \langle \{0,1\}; \vee, \wedge, C\rangle$ to be the two-element distributive lattice with $C \subseteq \{0,1\}$. Denote the variety of C-bounded distributive lattices by $\mathcal{D}_C$. Then $\mathcal{D}_C = \mathbb{ISP}\underline{\mathbf{D}}_C$, and Priestley duality for $\mathcal{D}_C$ (see 4.3.1, Exercise 4.4 and Exercise 4.5) shows that $\underline{\mathbf{D}}_C$ is almost endodualisable via $\underset{\sim}{\mathbf{D}}_C := \langle \{0,1\}; \mathrm{End}(\underline{\mathbf{D}}_C), \leqslant_{\{0,1\}}, \mathcal{T}\rangle$. Since $\underline{\mathbf{D}}_C$ is injective in $\mathcal{D}_C$, it is a retract of every non-trivial algebra in $\mathcal{D}_C$. As a sublattice of $\underline{\mathbf{D}}_C^2$, the relation $\leqslant$ is isomorphic to the three-element chain $\mathbf{3}_C$. Thus we are set up to apply the Third Endodualisability Theorem with $\mathbf{E} = \mathbf{3}_C$ (see

Exercise 7.16): a finite lattice in $\mathcal{D}_C$ is endodualisable if and only if it is endoprimal if and only if it is not relatively complemented.

Semilattices Define $\underline{\mathbf{D}}_C := \langle\{0,1\}; \wedge, C\rangle$ to be the two-element meet semilattice with bounds $C \subseteq \{0,1\}$, and denote the variety of C-bounded meet semilattices by $\mathcal{S}_C$. Define the topological semilattices

$$\underset{\sim}{\mathbf{D}} := \langle\{0,1\}; \wedge, 0, 1, \mathcal{T}\rangle, \qquad \underset{\sim}{\mathbf{D}}_0 := \langle\{0,1\}; \wedge, 0, \mathcal{T}\rangle,$$
$$\underset{\sim}{\mathbf{D}}_1 := \langle\{0,1\}; \wedge, 1, \mathcal{T}\rangle, \qquad \underset{\sim}{\mathbf{D}}_{01} := \langle\{0,1\}; \wedge, \mathcal{T}\rangle.$$

Simple variants of Exercise 4.10 show that $\underset{\sim}{\mathbf{D}}_C$ yields a duality on $\mathcal{S}_C = \mathbb{ISP}\underline{\mathbf{D}}_C$. Thus, for all choices of $C \subseteq \{0,1\}$, the semilattice $\underline{\mathbf{D}}_C$ is almost endodualisable via $\underset{\sim}{\mathbf{D}}'_C := \langle\{0,1\}; \mathrm{End}(\underline{\mathbf{D}}_C), s, \mathcal{T}\rangle$, where $s = \mathrm{graph}(\wedge)$. In the case of $\mathcal{S}_1$ and $\mathcal{S}$ we can apply the Third Endodualisability Theorem to obtain the characterisations of endodualisability in $\mathcal{S}_1$ and $\mathcal{S}$ given in [Dav96]. Since $\mathbf{s}$ is isomorphic to $\underline{\mathbf{D}}_C^2$, we choose $\mathbf{E}$ to be $\underline{\mathbf{D}}_C^2$. Note that $\underline{\mathbf{D}}_C$ is a retract of every non-trivial semilattice in $\mathcal{S}_C$ and that $\underline{\mathbf{D}}_C^2$ is a retract of $\mathbf{S} \in \mathcal{S}_C$ if and only if $\mathbf{S}$ has a subalgebra isomorphic to $\underline{\mathbf{D}}_C^2$. We find (see Exercise 7.17) that a finite non-trivial semilattice in $\mathcal{S}_1$ (in $\mathcal{S}$) is endodualisable if and only if it is endoprimal if and only if it is not a chain (there exists $a \in S$ such that $\downarrow a$ is not a chain).

We cannot apply the Third Endodualisability Theorem to $\mathcal{S}_0$ and $\mathcal{S}_{01}$ as it applies only when endodualisability and endoprimality coincide. Descriptions of endodualisability in $\mathcal{S}_0$ and $\mathcal{S}_{01}$ were obtained in [DP00]: see Exercise 7.18. A non-trivial semilattice $\underline{\mathbf{S}} \in \mathcal{S}_{01}$ is endodualisable if and only if 0 is meet-reducible in $\underline{\mathbf{S}}$, and a non-trivial semilattice $\underline{\mathbf{S}} \in \mathcal{S}_0$ is endodualisable if and only if it has a subsemilattice isomorphic to $\underline{\mathbf{D}}_0^2$.

It is proved in [DP00] that a non-trivial semilattice in $\mathcal{S}_{01}$ (in $\mathcal{S}_0$) is endoprimal if and only if it is not a chain (there exists $a \in S$ such that $\downarrow a$ is not a chain). Consequently, the semilattice $\mathbf{1} \oplus \mathbf{2}^2$, regarded as either a bounded meet semilattice or a lower-bounded meet semilattice, is endoprimal, almost endodualisable but not endodualisable.

Sets Since every set is free, it follows by Proposition 2.2.3 that a set $\underline{\mathbf{S}}$ is endodualisable if and only if it is endoprimal. A two-element set is not endoprimal since the majority function preserves $\mathrm{End}(\underline{\mathbf{S}})$ but is not a projection. If $|S| = 3$, then $\underline{\mathbf{S}}$ is endoprimal (see Exercise 7.19) and so is endodualisable. If $|S| \geqslant 3$, then $\underline{\mathbf{S}}$ has a three-element retract and hence is endodualisable by the First Endodualisability Theorem. Thus a finite set $\underline{\mathbf{S}}$ is endodualisable if and only if it is endoprimal if and only if $|S| \neq 2$.

Most natural dualities have been proved by establishing an interpolation condition (IC). Sets provide a natural example to demonstrate that (IC) is not a necessary condition for dualisability. (See also Exercise 4.11.)

8.6 Lemma *Let $\underline{\mathbf{S}}$ be a non-trivial finite set and $\underset{\sim}{\mathbf{S}} := \langle S; \mathrm{End}(\underline{\mathbf{S}}), \mathcal{T}\rangle$.*

(i) (IC) *does not hold with respect to $\underline{\mathbf{S}}$ and $\underset{\sim}{\mathbf{S}}$.*

(ii) *$\underset{\sim}{\mathbf{S}}$ does not yield a full duality on $\mathbb{ISP}\underline{\mathbf{S}}$.*

Proof Define $X = \{(a,a,b) \mid a,b \in S\} \cup \{(b,a,a) \mid a,b \in S\} \subseteq S^3$. Then X is closed under the endomorphisms of $\underline{\mathbf{S}}$ and so X determines a substructure of $\underset{\sim}{\mathbf{S}}^3$. Define $\alpha\colon X \to S$ by $\alpha(a,b,c) = c$ if $a = b$ and $\alpha(a,b,c) = a$ if $b = c$, for all $(a,b,c) \in X$. It is easy to check that α is well defined and preserves the endomorphisms of $\underline{\mathbf{S}}$, whence α is a morphism. As $\underline{\mathbf{S}}$ is non-trivial, there are distinct elements $a, b \in S$. We have $\alpha(a,a,b) = b$ and $\alpha(b,a,a) = b$, so $\alpha \notin \{\pi_1{\restriction}_X, \pi_2{\restriction}_X, \pi_3{\restriction}_X\}$. Thus α cannot be extended to a ternary term function of $\underline{\mathbf{S}}$, and consequently (IC) does not hold with respect to $\underline{\mathbf{S}}$ and $\underset{\sim}{\mathbf{S}}$.

A term-closed subset of $\underset{\sim}{\mathbf{S}}^3$ is an intersection of equalisers of projections, and is therefore isomorphic to $\underset{\sim}{\mathbf{S}}$, $\underset{\sim}{\mathbf{S}}^2$ or $\underset{\sim}{\mathbf{S}}^3$. Define $n := |S|$. Since we have $|X| = 2n^2 - n$ and $n \neq 1$, the set X is a substructure of $\underset{\sim}{\mathbf{S}}^3$ which is not isomorphic to a term-closed set. So $\underset{\sim}{\mathbf{S}}$ does not yield a full duality on $\mathbb{ISP}\underline{\mathbf{S}}$, by the Full Duality Theorem 3.1.7. ■

As explained above, $\underset{\sim}{\mathbf{S}} = \langle S; \mathrm{End}(\underline{\mathbf{S}}), \mathcal{T}\rangle$ dualises the three-element set $\underline{\mathbf{S}}$. However, by Lemma 7.8.6, $\underline{\mathbf{S}}$ and $\underset{\sim}{\mathbf{S}}$ do not satisfy (IC), whence (IC) is not a necessary condition for dualisability.

Exercises 7

Exercises from the text. Prove Lemma 7.3.2(i)(ii) (see [DW85]). Prove 7.4.1(vi). Verify the claims, in 7.4.3, concerning the restricted Priestley dual of $\mathcal{O}$. Confirm all assertions in 7.4.4 concerning congruences on Ockham algebras and subdirectly irreducible Ockham algebras. Using the proof of the Kleene case (7.4.5) as a guide, show that the objects of $\mathcal{Y}_{\mathcal{V}}$ are as described in 7.5.1. Complete the proof of 7.5.5 (the Two-Sorted Piggyback Duality for $\mathcal{V}$). Confirm all the claims made in 7.5.8(i)–(vi) concerning the two-sorted natural duality for Kleene algebras. Do all the calculations needed to show that Figure 7.5 is correct. Complete the proof of Lemma 7.6.3. Prove the Second $\underset{\sim}{\mathbf{D}}$-to-$\underset{\sim}{\mathbf{M}}$ Theorem 7.7.5. Confirm all of the claims made concerning the examples given in 7.7.6.

7.1 Let $\mathbf{Y} = \langle Y; g, \leqslant, \mathcal{T}\rangle$ be an Ockham space. Show that $K(\mathbf{Y})$ satisfies the laws $f(f(x)) \approx x$ and $x \wedge f(x) \lesssim y \vee f(y)$ which define Kleene algebras if and only if $\mathbf{Y}$ satisfies condition (K) of 7.4.5. (Compare this 'syntactic' proof of 7.4.5 with the 'semantic' proof given earlier.)

7.2 Prove that, for every non-empty set S, the free object generated by S, in the restricted Priestley dual $\mathcal{Y}_{\mathcal{K}}$ of the variety $\mathcal{K}$ of Kleene algebras, does not exist. [Hint: It is enough to consider $|S| = 1$.]

7.3 [Gol81, DP87] Let $\mathbf{W}^0 = \langle \mathbb{Z}_m; \gamma_n, \leqslant, \mathcal{T} \rangle$ be a one-generated Ockham space, let $\mathbf{W}^1 = \langle \{\gamma_n^k(1) \mid k \geqslant 0\}; \gamma_n, \leqslant, \mathcal{T} \rangle$ be the substructure of $\mathbf{W}^0$ generated by 1. Define $\underline{\mathbf{M}}^0 = K(\mathbf{W}^0)$ and $\underline{\mathbf{M}}^1 = K(\mathbf{W}^1)$ and let $\mathcal{V}$ be the variety generated by $\underline{\mathbf{M}}^0$. Use the restricted Priestley duality to prove that

(i) every non-trivial algebra in $\mathbb{HS}(\underline{\mathbf{M}}^0)$ is subdirectly irreducible,

(ii) up to isomorphism, there are at most two maximal subdirectly irreducible algebras in $\mathcal{V}$, namely $\underline{\mathbf{M}}^0$ and $\underline{\mathbf{M}}^1$,

(iii) $\underline{\mathbf{M}}^0$ is always a maximal subdirectly irreducible algebra in $\mathcal{V}$ and $\underline{\mathbf{M}}^1$ is a maximal subdirectly irreducible algebra in $\mathcal{V}$ if and only if $n \notin \{0, 1\}$ and $\gamma_n \colon \mathbf{W}^0 \to \mathbf{W}^1$ is not order-preserving.

7.4 Prove that the alternating structure $\underset{\sim}{\mathbf{2}}^W$ defined in Section 5 is an Ockham space, that is, verify that $\ll$ is an order which g reverses.

7.5 Let $\mathbf{A} = \langle A; \vee, \wedge, F \rangle$ be an algebra such that $\vee$ and $\wedge$ are binary operations and F is a family of unary operations each of which is either an endomorphism or a dual endomorphism (in the obvious sense) of $\langle A; \vee, \wedge \rangle$. Let $B \subseteq A$. Define $B_0 = B$ and for all $k \geqslant 0$ define $B_{k+1} := B_k \backslash \{ a \in B_k \mid (\exists f \in F)\, f(a) \notin B_k \}$. Show that if B is closed under $\vee$ and $\wedge$, then $B^\circ := \bigcap_{k \geqslant 0} B_k$ is either empty or is the largest subalgebra of $\mathbf{A}$ contained in B.

7.6 Show that for each $\mathbf{Y} \in \mathcal{Y}$ the subsets Y^0 and Y^1, defined as in 7.5.1, are closed.

7.7 [DP87] Let $\mathcal{V}$ be the variety generated by a finite subdirectly irreducible Ockham algebra as in Section 5. Let $\mathbf{A}_i \in \mathcal{V}$ for all $i \in I$ and let $\mathbf{Z} := \prod\{ H(\mathbf{A}_i) \mid i \in I \}$ with the pointwise structure. Show that $Y := \{ y \in Z \mid \mathbf{Z} \models \sigma_0(y) \vee \sigma_1(y)\}$ is a closed g-subset of $\mathbf{Z}$ and use the translation process between $D(\mathcal{V})$ and $\mathcal{Y}_{\mathcal{V}}$ to show that $H(\coprod\{ \mathbf{A}_i i \in I \})$ is isomorphic to $\mathbf{Y}$.

7.8 [Pri75] Let $\mathbf{X}$ and $\mathbf{Y}$ be Priestley spaces and let $\varphi \colon \mathbf{X} \to \mathbf{Y}$ be a continuous order-preserving map.

(i) Prove that $K(\mathbf{X})$ is pseudocomplemented if and only if $\mathbf{X}$ is a p-**space**, that is, for every increasing subset U of $\mathbf{X}$ the set ${\downarrow}U := \{ y \in X \mid (\exists x \in U)\, y \leqslant x \}$ is clopen, in which case $X \backslash {\downarrow}U$ is the pseudocomplement of U.

(ii) Let $\mathbf{X}$ and $\mathbf{Y}$ be p-spaces. Show that $K(\varphi) \colon K(\mathbf{Y}) \to K(\mathbf{X})$ is a p-algebra homomorphism if and only if φ is a p-morphism, that is , for all $x \in X$, we have $\varphi(\max(x)) = \max(\varphi(x))$. Here $\max(z)$ denotes the set of all maximal elements above z.

(iii) [Lee70] Let $\mathbf{X}$ be a p-space. Show that $K(\mathbf{X})$ belongs to $\mathcal{B}_n$ if and only if $|\max(x)| \leqslant n$, for all $x \in X$.

7.9 Let $\mathbf{V}_1$ and $\mathbf{V}_2$ be Priestley spaces, let $v_1 \in V_1$ and $v_2 \in V_2$ and define $\omega_i \colon K(\mathbf{V}_i) \to \underset{\sim}{\mathbf{D}}$ by $\alpha \mapsto \alpha(v_i)$. Given $\varphi_i \in \mathcal{P}(\mathbf{V}_1, \mathbf{Y})$ define

$$r(\varphi_1, \varphi_2) := \{\, (\alpha \circ \varphi_1, \alpha \circ \varphi_2) \mid \alpha \in H(\mathbf{Y}_\mathbf{r})\} \leqslant K(\mathbf{V}_1) \times K(\mathbf{V}_2).$$

Prove that $r(\varphi_1, \varphi_2) \subseteq (\omega_1, \omega_2)^{-1}(\leqslant)$ if and only if $\varphi_1(v_1) \leqslant \varphi_2(v_2)$.

7.10 Use the restricted Priestley duality for the variety $\mathcal{B}_n$ of distributive p-algebras to prove that if s is a subalgebra of $\underline{\mathbf{B}}_n^2$ which is maximal in $\omega^{-1}(\leqslant)$, then s is of the form

$$E(g) := \{(b, g(b)) \mid b \in {\downarrow}a\} \cup \{(b, 1) \mid g(b) = a\} \cup \{(1, 1)\},$$

where $g \colon {\downarrow}a \to {\downarrow}a$ is a Boolean endomorphism.

7.11 [Dav96] Assume that $\underline{\mathbf{M}}$ is a finite algebra in the quasi-variety $\mathcal{A} = \mathbb{ISP}\underline{\mathbf{D}}$. Show that the following are equivalent:

(i) the maps $\Phi_\alpha \colon \mathcal{A}(\mathbf{A}, \underline{\mathbf{M}}) \to \mathcal{A}(\mathbf{A}, \underline{\mathbf{D}})$ are jointly surjective for all $\mathbf{A} \in \mathcal{A}$;

(ii) there exists $\alpha \in \Omega$ such that $\Phi_\alpha \colon \mathcal{A}(\mathbf{A}, \underline{\mathbf{M}}) \to \mathcal{A}(\mathbf{A}, \underline{\mathbf{D}})$ is surjective for all $\mathbf{A} \in \mathcal{A}$;

(iii) some $\alpha \in \Omega$ is a retraction of $\underline{\mathbf{M}}$ onto $\underline{\mathbf{D}}$.

7.12 [Dav96] Assume that $\underset{\sim}{\mathbf{D}} = \langle D; R, \mathcal{T}\rangle$ dualises $\underline{\mathbf{D}}$, with R finite, and let s be the product of the relations in R.

(i) Show that if $\mathcal{A}(\mathbf{r}_1, \mathbf{r}_2) \neq \varnothing$ for all $r_1, r_2 \in R$, then the natural projection $\pi_r \colon \mathbf{s} \to \mathbf{r}$ is a retraction for all $r \in R$.

(ii) Show that $\underset{\sim}{\mathbf{D}}' := \langle D; s, \mathcal{T}\rangle$ dualises $\underline{\mathbf{D}}$ provided the natural projection $\pi_r \colon \mathbf{s} \to \mathbf{r}$ is a retraction for all $r \in R$. [Hint: Either prove directly that s entails r, for all $r \in R$, or look ahead to Chapter 9 and use construct (17), retractive projection.]

(iii) Show that if $\underline{\mathbf{D}}$ is a retract of $\mathbf{r}$, for every $r \in R$, then every algebra $\underline{\mathbf{M}} \in \mathcal{A}$ which has $\mathbf{s}$ as a retract is endodualisable.

7.13 [DP00] Let $\mathbf{B} \in \mathbb{ISP}\mathbf{A}$ be such that $\mathbf{A}$ is a subretract of $\mathbf{B}$. Show that if $\alpha \colon B^k \to B$ preserves the endomorphisms of $\mathbf{B}$, then α preserves the unary relation A and for all endomorphism-preserving maps $\beta \colon B^k \to B$ we have $\beta{\restriction}_{A^k} = \alpha{\restriction}_{A^k}$ implies that $\beta = \alpha$.

7.14 [DP00] Use the previous exercise to show that if $\mathbf{A}$ is k-endoprimal and is a retract of $\mathbf{B} \in \mathbb{ISP}\mathbf{A}$, then $\mathbf{B}$ is k-endoprimal.

7.15 [Dav96] Show that the one-dimensional vector space $\underline{\mathbf{F}}$ over a field $\mathbb{F}$ is not 2-endoprimal, by showing that the map $\alpha \colon F^2 \to F$, defined by $\alpha(0, y) := y$ and $\alpha(x, y) := x$ for $x \neq 0$, preserves the endomorphisms of $\underline{\mathbf{F}}$ but is not a term function. Hence, use the Third Endodualisability Theorem to prove that a finite-dimensional vector space $\underline{\mathbf{V}}$ over a finite field $\mathbb{F}$ is endodualisable if and only if it is endoprimal if and only if it does not have dimension 1.

7.16 (i) Prove that a distributive lattice $\mathbf{L} \in \mathcal{D}_C$ has $\mathbf{3}_C$ as a retract if and only if $\mathbf{L}$ has a pair of comparable prime filters if and only if $\mathbf{L}$ is not relatively complemented.

(ii) Show that if $\mathbf{L} \in \mathcal{D}_C$ is non-trivial and relatively complemented, then $\mathbf{L}$ is not k-endoprimal, where $k = 3 - |C|$. [Hint: If (a) $C = \varnothing$, (b) $C = \{0\}$, (c) $C = \{0, 1\}$, then an endomorphism-preserving map $\alpha\colon L^3 \to L$ which is not a term function is obtained by declaring $\alpha(a, b, c)$ to be the complement of a in (a) the interval $[a \wedge b \wedge c, a \vee b \vee c]$, (b) the interval $[0, a \vee b]$, (c) the lattice $\mathbf{L}$.]

7.17 (i) Show that a semilattice $\mathbf{S}$ in $\mathcal{S}_1$ (in $\mathcal{S}$) has $\underline{\mathbf{D}}_1^2$ ($\underline{\mathbf{D}}^2$) as a retract if and only if $\mathbf{S}$ is not a chain (there exists $a \in S$ such that ${\downarrow}a$ is not a chain).

(ii) Let $\mathbf{S}$ be a non-trivial semilattice in $\mathcal{S}_C$. Show that if $\mathbf{S}$ is a chain (if ${\downarrow}a$ is a chain, for all $a \in S$), then $\mathbf{S}$ is not 2-endoprimal (is not 3-endoprimal). [Hint: Consider the binary map $\alpha(a, b) := a \vee b$ and the ternary map $\beta(a, b, c) := (a \wedge b) \vee (a \wedge c)$, which is well defined since ${\downarrow}a$ is a chain.]

7.18 (i) Show that a non-trivial semilattice $\underline{\mathbf{S}} \in \mathcal{S}_{01}$ is endodualisable if and only if 0 is meet-reducible in $\underline{\mathbf{S}}$. [Hint: Assume that 0 is meet-irreducible in $\underline{\mathbf{S}}$ and consider the term-closed set $X := \operatorname{eq}(\pi_1 \wedge \pi_2, 0) \subseteq S^2$. Show that $\alpha\colon X \to S$, given by $\alpha(a, b) = a \vee b$, is well defined, preserves $\operatorname{End}(\underline{\mathbf{S}})$ and cannot be extended to a binary term function $t\colon S^2 \to S$.]

(ii) Show that a non-trivial semilattice $\underline{\mathbf{S}} \in \mathcal{S}_0$ is endodualisable if and only if $\underline{\mathbf{S}}$ has a subalgebra isomorphic to $\underline{\mathbf{D}}_0^2$. [Hint: Assume that $\underline{\mathbf{D}}_0^2$ does not embed into $\underline{\mathbf{S}}$ and consider the term-closed set $X := \operatorname{eq}(\pi_2 \wedge \pi_3, 0) \subseteq S^3$. Show that $\alpha\colon X \to S$, given by $\alpha(a, b, c) = (a \wedge b) \vee (a \wedge c)$, is well defined, preserves $\operatorname{End}(\underline{\mathbf{S}})$ and does not extend to a ternary term function $t\colon S^3 \to S$.]

7.19 [DP00] Show that the three-element set $\underline{\mathbf{S}} = \langle \{0, 1, a\}; \varnothing \rangle$ is endoprimal. [Hint: Use Exercise 7.13 to show that, for all $k \geqslant 1$, a map $\alpha\colon S^k \to S$ is a projection (that is, a term function of $\mathbf{S}$) if and only if α preserves every map $e\colon S \to \{0, 1\}$.]

7.20 [DP00] Use the Second Endodualisability Theorem, Exercise 4.9 and the fact that the two-element Boolean algebra is term equivalent to the Sheffer-stroke algebra to show that every finite set with at least four elements is endodualisable.

8

Optimal Dualities and Entailment

Many natural dualities, including the majority of those presented in Chapter 4, may be established by following four steps.

(1) Show that $\underline{\mathbf{M}}$ has a majority term, for example, by simply observing that $\underline{\mathbf{M}}$ has an underlying lattice structure.

(2) Apply the NU Duality Theorem 2.3.4 to conclude that the crude structure $\underset{\sim}{\mathbf{M}}' := \langle M; \Omega, \mathcal{T} \rangle$, where $\Omega = \mathbb{S}(\underline{\mathbf{M}}^2)$, yields a duality on $\mathcal{A}$.

(3) Calculate the irreducibility index, $\mathrm{Irr}(\underline{\mathbf{M}})$, of $\underline{\mathbf{M}}$ and apply the NU Strong Duality Theorem 3.3.8 to conclude that $\underset{\sim}{\mathbf{M}}'' := \langle M; P, \Omega, \mathcal{T} \rangle$ yields a strong duality on $\mathcal{A}$, where $P := \bigcup \{ \mathcal{P}_n \restriction n \leqslant \mathrm{Irr}(\underline{\mathbf{M}}) \}$.

(4) Find small and serviceable sets G of operations, H of partial operations and R of relations such that, firstly, $G \cup H$ hom-entails P and, secondly, $G \cup H \cup R$ entails Ω.

Hopefully, the structure $\underset{\sim}{\mathbf{M}} = \langle M; G, H, R, \mathcal{T} \rangle$ obtained after step (4) will be 'good' enough to guarantee that the strong duality it yields is of some practical use in the study of the quasi-variety $\mathcal{A}$.

The theory we developed in the first three chapters gives us a streamlined path through steps (1), (2) and (3). Unfortunately step (4) can require considerable hands-on effort in each case, employing various *ad hoc* combinations of the $\underset{\sim}{\mathbf{M}}$-Shift Lemmas 2.4.2, 3.2.3 and admissible constructs such as those listed in the Constructs for Entailment 2.4.5. The success of this program for the lattice-based examples in Chapter 4 was strongly influenced by the fact that in each case the irreducibility index of $\underline{\mathbf{M}}$ was 1. This guaranteed that a strong duality would be given by a structure $\underset{\sim}{\mathbf{M}}$ whose operations and partial operations had arity at most 1 and whose relations had arity at most 2. If the irreducibility index of $\underline{\mathbf{M}}$ is high, we are forced to end our quest for a strong duality and settle for an optimal duality which is not strong. (We saw in Chapter 5 that useful information may be obtained via a natural duality even when it is not strong.)

Recall that $\underset{\sim}{\mathbf{M}}$ yields an **optimal duality** on $\mathcal{A}$ if $\underset{\sim}{\mathbf{M}}$yields a duality but as soon as any relation or (partial) operation is eliminated from the structure on $\underset{\sim}{\mathbf{M}}$ the resulting structure $\underset{\sim}{\mathbf{M}}^*$ no longer yields a duality. In this chapter we will exhibit a systematic method for reducing any dualising structure of finite type to one which yields an optimal duality on $\mathcal{A}$.

According to the Duality and Entailment Theorem 2.4.3, a dualising set $G \cup H \cup R$ yields an optimal duality if and only if no member of

$G \cup H \cup R$ is entailed by the rest of $G \cup H \cup R$. This observation suggests that we examine the Constructs for Entailment 2.4.5 to see if they are indeed sufficient to generate all possible entailments. In fact they are not, as we will see in Chapter 9 where we extend them to a larger list that is sufficient. Applying these constructs will remain the most expedient way to find optimal dualities when, as in most of the examples of Chapter 4, we begin with a small and manageable dualising set Ω. But larger sets Ω begin to demand that we find an algorithmic method to reduce them to optimal dualities.

Our quest for optimal dualities begins with a simple question. Is there a finite algorithm to determine, for an algebraic relation s and finite sets G, H and R, whether or not $G \cup H \cup R$ entails s? Notice that the definition of entailment does not provide such an algorithm since it asks that $G \cup H \cup R$ entail s on $D(\mathbf{A})$ for *every* $\mathbf{A} \in \mathcal{A}$. We will obtain a finite algorithm to determine entailment by establishing the startling fact that it is always sufficient to consider only the single finite algebra $\mathbf{A} = \mathbf{s}$! Capitalising on this observation, we will obtain a method to compute from a given finite dualising set all of its subsets which yield an optimal duality. This method will be fully illustrated in the cases of semi-primal algebras and Kleene algebras. This chapter is based on Davey and Priestley [DP96a].

8.1 Test algebras—schizophrenia strikes again!

In this section we show that the question of deciding whether or not a finite set $G \cup H \cup R$ entails a relation s is answerable by a finite algorithm.

The secret is a simple but lovely application of schizophrenia. Since the n-ary *relation* s is algebraic we may consider its pointwise extension $s^{D(\mathbf{s})}$ to the dual $D(\mathbf{s})$ of the *algebra* $\mathbf{s} \leqslant \underline{\mathbf{M}}^n$. We shall see that $G \cup H \cup R \vdash s$ if and only if $G \cup H \cup R$ entails s on $D(\mathbf{s})$. For this reason, we refer to the algebra $\mathbf{s}$ corresponding to an algebraic relation $s \in \mathcal{B}$ as a **test algebra**. Since $D(\mathbf{s})$ and M are finite, there is only a finite number of maps from $D(\mathbf{s})$ to M and consequently whether or not $G \cup H \cup R$ entails s is finitely determined provided $G \cup H \cup R$ is finite.

Recall the following two items of notation from Chapter 1. For $i = 1, \dots, n$, we let $\rho_i := \pi_i{\restriction}_{\mathbf{s}}$ denote the restriction of the natural projection $\pi_i \colon \underline{\mathbf{M}}^n \to \underline{\mathbf{M}}$ to the subalgebra $\mathbf{s}$. Note that $\rho_1, \dots, \rho_n \in D(\mathbf{s})$. Given a set T and maps $z_1, \dots, z_n \colon T \to M$ the map $z_1 \sqcap \cdots \sqcap z_n \colon T \to M^n$ is given by $(z_1 \sqcap \cdots \sqcap z_n)(t) := (z_1(t), \dots, z_n(t))$ for all $t \in T$. If $Z \subseteq M^T$ and $z_1, \dots, z_n \in Z$, then, since s^Z is defined pointwise, we have $(z_1, \dots, z_n) \in s^Z$ if and only if $(z_1 \sqcap \cdots \sqcap z_n)(T) \subseteq s$ (Lemma 2.2.8).

Our first lemma not only shows that is it natural to consider the relation s on $D(\mathbf{s})$, but that within $D(\mathbf{s})$ there is a universal instance of the relation s!

1.1 Lemma *Let Z be a hom-closed subset of M^T for some set T and let $s \in \mathcal{B}$.*

(i) $(\rho_1, \dots, \rho_n) \in s^{D(\mathbf{s})}$.

(ii) *Let $z_1, \dots, z_n \in Z$. Then the following are equivalent (and may be interpreted as asserting that $(\rho_1, \dots, \rho_n) \in s^{D(\mathbf{s})}$ is a universal instance in the class $\{ Z \subseteq M^T \mid T \text{ is a set and } Z \text{ is hom-closed} \}$ of the relation s):*

  (a) $(z_1, \dots, z_n) \in s^Z$;

  (b) *there is a (necessarily unique) map $v\colon T \to \mathbf{s}$, namely $v = z_1 \sqcap \dots \sqcap z_n$, such that $z_i = \rho_i \circ v$ for all i;*

  (c) *there is a (necessarily unique) map $\varphi\colon D(\mathbf{s}) \to Z$ such that φ preserves every finitary algebraic relation on $\underline{\mathbf{M}}$ and $\varphi(\rho_i) = z_i$ for all i;*

  (d) *there is a (necessarily unique) map $\varphi\colon D(\mathbf{s}) \to Z$ such that φ preserves every algebraic n-ary partial operation with domain s and satisfies $\varphi(\rho_i) = z_i$ for all i;*

  (e) *there is an s-preserving map $\varphi\colon D(\mathbf{s}) \to Z$ such that $\varphi(\rho_i) = z_i$ for all i.*

Proof For (i) simply observe that $(\rho_1, \dots, \rho_n) \in s^{D(\mathbf{s})}$ if and only if $(\rho_1 \sqcap \dots \sqcap \rho_n)(a) \in s$ for all $a \in \mathbf{s}$. This holds because $(\rho_1 \sqcap \dots \sqcap \rho_n)(a) = a$ by definition.

We now prove (ii). Assume that $(z_1, \dots, z_n) \in s^Z$. It follows that $(z_1 \sqcap \dots \sqcap z_n)(T) \subseteq s$ and hence the map $v := z_1 \sqcap \dots \sqcap z_n \colon T \to s$ is well defined and satisfies $z_i = \rho_i \circ v$ for all i. The categorical definition of product asserts that v is the unique map satisfying $z_i = \rho_i \circ v$ for all i. Thus (a) implies (b).

Now assume (b) and define $\varphi\colon D(\mathbf{s}) \to M^T$ by $\varphi(y) := y \circ v$ for all $y \in D(\mathbf{s})$. For $y \in D(\mathbf{s})$, let $\widehat{y}$ be y regarded as an n-ary partial operation on $\underline{\mathbf{M}}$. Then

$$\varphi(y) = y \circ v = y \circ (z_1 \sqcap \dots \sqcap z_n) = \widehat{y}(z_1, \dots, z_n) \in Z,$$

since Z is hom-closed. Thus $\varphi\colon D(\mathbf{s}) \to Z$ is well defined and satisfies $\varphi(\rho_i) = z_i$ for all i. We now prove that φ preserves every finitary algebraic relation on $\underline{\mathbf{M}}$. Let $\mathbf{r} \leqslant \underline{\mathbf{M}}^k$ and let $(y_1, \dots, y_k) \in r^{D(\mathbf{s})}$. Thus $y_1 \sqcap \dots \sqcap y_k \colon \mathbf{s} \to \mathbf{r}$ is a well-defined homomorphism (by Lemma 2.2.8 with $\mathbf{A}$ replaced by $\mathbf{s}$). To prove that $(\varphi(y_1), \dots, \varphi(y_k)) \in r^Z$ it suffices to show that $(\varphi(y_1) \sqcap \dots \sqcap \varphi(y_k))(T) \subseteq r$. But

$$\begin{aligned}\varphi(y_1) \sqcap \dots \sqcap \varphi(y_k) &= y_1 \circ (z_1 \sqcap \dots \sqcap z_n) \sqcap \dots \sqcap y_k \circ (z_1 \sqcap \dots \sqcap z_n) \\ &= (y_1 \sqcap \dots \sqcap y_k) \circ (z_1 \sqcap \dots \sqcap z_n),\end{aligned}$$

whence $(\varphi(y_1) \sqcap \cdots \sqcap \varphi(y_k))(T) \subseteq r$ as the image of $y_1 \sqcap \cdots \sqcap y_k$ is a subset of r. Thus φ preserves r. We now prove that φ is determined by its values on $\{\rho_1, \ldots, \rho_n\}$. Let $y \in D(\mathbf{s})$ and again let $\widehat{y}$ be y regarded as an n-ary partial operation on $\underline{\mathbf{M}}$. Thus, as φ preserves all finitary algebraic relations on $\underline{\mathbf{M}}$, the map φ preserves $\widehat{y}$. Since $\widehat{y}(\rho_1, \ldots, \rho_n) = y$ we have $\varphi(y) = \varphi(\widehat{y}(\rho_1, \ldots, \rho_n)) = \widehat{y}(\varphi(\rho_1), \ldots, \varphi(\rho_n)) = \widehat{y}(z_1, \ldots, z_n)$, whence φ is indeed determined by its values on $\{\rho_1, \ldots, \rho_n\}$. Thus (b) implies (c) and (c) implies (d).

To prove that (d) implies (e) note that the map φ guaranteed by (d) preserves the map $\rho_1 \colon s \to M$ by assumption and hence preserves its domain, namely s. Finally we show that (e) implies (a). Let φ be the map given by (e). Since $(\rho_1, \ldots, \rho_n) \in s^{D(\mathbf{s})}$ by (i), it follows immediately that $(z_1, \ldots, z_n) = (\varphi(\rho_1), \ldots, \varphi(\rho_n)) \in s^Z$. ■

We frequently apply Lemma 8.1.1 when $Z = D(\mathbf{A})$. The case when both algebras are test algebras is particularly useful and consequently we state it explicitly: replace $\mathbf{s}$ by $\mathbf{r}$ and $\mathbf{A}$ by $\mathbf{s}$.

1.2 r-on-s Lemma *Let $\mathbf{r} \leqslant \underline{\mathbf{M}}^m$, $\mathbf{s} \leqslant \underline{\mathbf{M}}^n$ and let $z_1, \ldots, z_m \in D(\mathbf{s})$. Then the following are equivalent:*

(i) $(z_1, \ldots, z_m) \in r^{D(\mathbf{s})}$;

(ii) *there is a (necessarily unique) homomorphism from $\mathbf{s}$ to $\mathbf{r}$, namely $\varphi = z_1 \sqcap \cdots \sqcap z_m$, such that $z_i = \rho_i \circ \varphi$ for all i;*

(iii) *there is an r-preserving map $\varphi \colon D(\mathbf{r}) \to D(\mathbf{s})$ such that $\varphi(\rho_i) = z_i$ for all i.*

By definition, the set $G \cup H \cup R$ entails a relation s on $Z = D(\mathbf{A})$ if every continuous map $\varphi \colon Z \to M$ that preserves $G \cup H \cup R$ also preserves s. The only property of $D(\mathbf{A})$ that is required for this definition to make sense is that $D(\mathbf{A})$ is closed under all the finitary algebraic (partial) operations on $\underline{\mathbf{M}}$. Since hom-closed sets are closed under *all* algebraic (partial) operations, it therefore makes sense to consider entailment on an arbitrary hom-closed subset Z of M^T. Entailment defined with respect to the class of all hom-closed subsets of powers of M would appear to be stronger than entailment defined with respect to the smaller class $\{\, D(\mathbf{A}) \mid \mathbf{A} \in \mathcal{A} \,\}$. In fact, these two versions of entailment are equivalent.

1.3 Test Algebra Lemma *Let $G \cup H \subseteq \mathcal{P}$, let $R \subseteq \mathcal{B}$ and let $s \in \mathcal{B}$. Then the following are equivalent:*

(i) *$G \cup H \cup R$ entails s on each hom-closed subset of every power of M;*

(ii) *$G \cup H \cup R$ entails s;*

(iii) *$G \cup H \cup R$ entails s on $D(\mathbf{s})$.*

Proof Certainly, (i) $\Rightarrow$ (ii) and (ii) $\Rightarrow$ (iii) are trivial, so it remains to prove that (iii) implies (i). Assume that $G \cup H \cup R$ entails s on $D(\mathbf{s})$. Let Z be a hom-closed subset of M^T and let $\alpha\colon Z \to M$ be continuous and preserve $G \cup H \cup R$. Let $(z_1, \ldots, z_n) \in s^Z$. By Lemma 8.1.1 there exists a $(G \cup H \cup R)$-preserving map $\varphi\colon D(\mathbf{s}) \to Z$ which satisfies $\varphi(\rho_i) = z_i$ for each i. Hence $\alpha \circ \varphi\colon D(\mathbf{s}) \to M$ is $(G \cup H \cup R)$-preserving and so preserves s, by assumption. Thus

$$(\alpha(z_1), \ldots, \alpha(z_n)) = ((\alpha \circ \varphi)(\rho_1), \ldots, (\alpha \circ \varphi)(\rho_n)) \in s,$$

whence α preserves s. Consequently $G \cup H \cup R$ entails s on Z. ■

Notice that if G, H and R are all finite, then the verification of (iii) above is a finite process which provides the promised algorithm to test entailment. We close this section with a simple but important consequence of the Test Algebra Lemma along with an illustration of its usefulness.

1.4 Corollary *In order to prove that $G \cup H \cup R$ entails s it suffices to prove that $G \cup H \cup R$ yields a duality on some isomorphic copy of the test algebra* $\mathbf{s}$.

Proof Assume that $G \cup H \cup R$ yields a duality on some isomorphic copy of the test algebra $\mathbf{s}$. Then $G \cup H \cup R$ also yields a duality on the test algebra $\mathbf{s}$ itself. Consequently, every $(G \cup H \cup R)$-preserving map $\alpha\colon D(\mathbf{s}) \to M$ is an evaluation map and thus preserves every finitary algebraic relation, and in particular preserves s. Therefore $G \cup H \cup R$ entails s on $D(\mathbf{s})$, and so $G \cup H \cup R$ entails s by the Test Algebra Lemma. ■

When combined with the piggyback dualities for varieties of Heyting algebras proved in Section 3 of the previous chapter, this result provides a very short proof of the endodualisability of the n-element Heyting chain.

1.5 Endodualisable Heyting Chain Theorem *The n-element Heyting algebra chain $\underline{\mathbf{C}}_n$ is endodualisable.*

Proof By Theorem 7.3.3 the endomorphisms plus the partial endomorphisms of $\underline{\mathbf{C}}_n$ yield a (strong) duality on $\mathbb{ISP}\underline{\mathbf{C}}_n$. Hence, to prove that $\underline{\mathbf{C}}_n$ is endodualisable we must prove that $\mathrm{End}(\underline{\mathbf{C}}_n)$ entails the partial endomorphisms of $\underline{\mathbf{C}}_n$. A simple argument using two easy exercises from Chapter 2 and the corollary above shows that $\mathrm{End}(\underline{\mathbf{C}}_n)$ entails a relation s provided the test algebra $\mathbf{s}$ is a retract of $\underline{\mathbf{C}}_n$ (see Exercise 8.1). The test algebra corresponding to the graph of a map is isomorphic to the domain of the map. Thus, it suffices to show that every subalgebra of $\underline{\mathbf{C}}_n$ is a retract of $\underline{\mathbf{C}}_n$. We ask the reader to check that ${\downarrow}a \cup \{1\}$ forms a subretract of $\underline{\mathbf{C}}_n$ for all $a \in C_n$, whence every subalgebra of $\underline{\mathbf{C}}_n$ is isomorphic to a subretract and hence is a retract of $\underline{\mathbf{C}}_n$. ■

1.6 Historical Note Proofs of the fact that the Heyting algebra $\underline{\mathbf{C}}_n$ is endodualisable have been seminal in the development of the theory of natural dualities. The first, bare-hands proof was in Davey [Dav76]. This proof was generalised and simplified in Davey [Dav82]. Next, Davey and Werner [DW83] gave a proof via the NU Duality Theorem. Based on an analysis of the proofs in [Dav76] and [Dav82], Davey and Werner [DW85, DW86] began the development of the piggyback approach to obtaining natural dualities (which led to the theory expounded in Chapter 7) and used it to give a new proof of the endodualisability of $\underline{\mathbf{C}}_n$ and to prove that the Heyting algebra $\mathbf{2}^2 \oplus \mathbf{1}$ is also endodualisable (see Exercise 8.2). Davey and Priestley [DP96b] gave the proof presented above and showed that $\underline{\mathbf{C}}_n$ and $\mathbf{2}^2 \oplus \mathbf{1}$ are the only subdirectly irreducible endodualisable Heyting algebras. Finally, Davey [Dav96] and Davey and Pitkethly [DP00] proved that every finite Heyting algebra in $\mathbb{ISP}\,\underline{\mathbf{C}}_n$, for all $n \in \mathbb{N}$, and every finite Heyting algebra in $\mathbb{ISP}(\mathbf{2}^2 \oplus \mathbf{1})$ are endodualisable (see Exercise 8.3).

8.2 Entailment closure and the 'topology' of failsets

Since we are concerned solely with duality rather than strong duality, the type of $\underset{\sim}{\mathbf{M}}$ is of no importance. Therefore, we apply Lemma 2.1.2, replacing each (partial) operation by its graph, and assume that $\underset{\sim}{\mathbf{M}} = \langle M; \Omega, \mathcal{T}\rangle$ where $\Omega \subseteq \mathcal{B}$ is a fixed set of finitary algebraic relations on $\underline{\mathbf{M}}$. In practice, Ω will be a set of relations which is known to yield a duality on $\mathcal{A}$ and which we wish to refine to an optimal duality. In our lattice-based examples, Ω is typically $\mathbb{S}(\underline{\mathbf{M}}^2)$. In the next chapter, when we discuss constructs for entailment and constructs for hom-entailment, the type of $\underset{\sim}{\mathbf{M}}$ will play a vital role.

2.1 Entailment Closure The key to the characterisation of those subsets of Ω which yield an optimal duality is the recognition that the map

$$R \longmapsto \overline{R} := \{\, s \in \Omega \mid R \vdash s \,\}$$

is a closure operator on Ω which we shall refer to as **entailment closure**. (See Exercise 8.6.) For a detailed discussion of the relationship between (algebraic) closure operators and (algebraic) complete lattices we refer the reader to Appendix A.

$$\Lambda := \{\, R \subseteq \Omega \mid \overline{R} = R \,\}$$

of all entailment-closed subsets of Ω forms a complete lattice from which we can extract many riches. Especially significant are those subsets $R \subseteq \Omega$ which are **dense** in Ω in the sense that $\overline{R} = \Omega$.

2.2 Density Lemma *Assume that $\Omega \subseteq \mathcal{B}$ yields a duality on $\mathcal{A}$ and let $R \subseteq \Omega$. Then the following are equivalent:*

(i) *R yields a duality on $\mathcal{A}$;*

(ii) *$\overline{R} = \Omega$;*

(iii) *R entails s for each $s \in \Omega \backslash R$;*

(iv) *R entails s on $D(\mathbf{s})$ for each $s \in \Omega \backslash R$.*

Proof If R yields a duality on $\mathcal{A}$, then, by Corollary 8.1.4 (or by the Duality and Entailment Theorem 2.4.3), R entails s for each $s \in \mathcal{B}$. Thus (i) implies (iii). Since (ii) is trivially equivalent to (iii) and (iii) is equivalent to (iv) by the Test Algebra Lemma, it remains only to observe that (iii) $\Rightarrow$ (i) is a consequence of the $\underset{\sim}{\mathbf{M}}$-Shift Duality Lemma 2.4.2 since Ω yields a duality on $\mathcal{A}$. ■

In general, the entailment-closed subsets of Ω are not the closed sets of a topology on Ω since they are not closed under finite unions. Nevertheless, the family Λ has some important and useful 'topological' characteristics.

2.3 Failsets Let $s \in \Omega$ and let $\gamma \colon D(\mathbf{s}) \to M$ be any map. Define

$$U = \mathrm{Fail}_{\mathbf{s}}(\gamma) := \{\, r \in \Omega \mid \gamma \text{ fails to preserve } r \,\}.$$

If $U \neq \varnothing$ we call U a **weak failset of** s (within Ω), and if $s \in U$ we call U a **failset of** s (within Ω). We refer to U as a **failset** if it is a failset of some $s \in \Omega$. For a fixed $s \in \Omega$, let

$$\mathcal{F}_s := \{\, \mathrm{Fail}_{\mathbf{s}}(\gamma) \mid \gamma \colon D(\mathbf{s}) \to M \text{ fails to preserve } s \,\}$$

be the family of all failsets of s and let

$$\mathcal{F} := \bigcup \{\, \mathcal{F}_s \mid s \in \Omega \,\}$$

be the family of all failsets.

If γ is an evaluation map, say $\gamma = e_{\mathbf{s}}(c)$ for some $c \in s$, then the set $U = \mathrm{Fail}_{\mathbf{s}}(\gamma)$ is empty. Conversely, if Ω yields a duality on $\mathbf{s}$ and $U = \varnothing$, then γ is an evaluation map. In general, the size of U gives a measure of how far γ is from being an evaluation map.

The following lemma is immediate from the definitions and tells us that failsets are 'open'.

2.4 Lemma *For any map $\gamma \colon D(\mathbf{s}) \to M$, the complement of* $\mathrm{Fail}_{\mathbf{s}}(\gamma)$ *is closed in Ω. That is, $\Omega \backslash U$ is closed in Ω for all $U \in \mathcal{F}$.*

The equivalence of (ii) and (iii) in the Test Algebra Lemma can be reinterpreted in terms of entailment closure, thereby showing that failsets serve as 'basic open sets'.

2.5 Lemma *Let $R \subseteq \Omega$ and $s \in \Omega$. The following are equivalent:*

(i) *$s \in \overline{R}$, that is, R entails s;*

(ii) *every failset of s intersects R;*

(iii) *every failset which contains s intersects R.*

Proof If R entails s, then by the definition of entailment, for all $\mathbf{A} \in \mathcal{A}$, every continuous map $\gamma\colon D(\mathbf{A}) \to M$ which fails to preserve s must also fail to preserve some $r \in R$. Hence, any failset which contains s must intersect R. Thus (i) implies (iii). That (iii) implies (ii) is trivial. The (contrapositive of) the implication (ii) $\Rightarrow$ (i) is an immediate consequence of the Test Algebra Lemma. ■

8.3 Failsets and optimal dualities

While failsets have many natural 'topological' properties, the analogy cannot be carried too far. As we shall see, even when Ω is infinite, $\mathcal{F}_s$ contains an abundance of minimal elements, which is not at all typical of the family of basic open sets containing the point s in a topological space. In fact these minimal failsets provide the key to understanding optimal dualities. To see how this works, we order both $\mathcal{F}_s$ and $\mathcal{F}$ by set inclusion. If U is a minimal element of $\mathcal{F}_s$ we call U a **locally minimal failset of** s and if U is minimal in $\mathcal{F}$ then we refer to U as a **globally minimal failset**. Clearly every globally minimal failset is a locally minimal failset, but not conversely. Since the size of $\mathrm{Fail}_{\mathbf{s}}(\gamma)$ is a measure of how far γ is from being an evaluation, we should expect minimal failsets to play an important role in our theory as each one comes from a map γ which is as close as possible to an evaluation map without actually being one.

Another notion will help us make this connection. Let $\mathcal{S}$ be a family of non-empty sets and let T be a subset of their union. Then T is called a **transversal** of $\mathcal{S}$ if T intersects each $U \in \mathcal{S}$ but no proper subset of T does. If $\mathcal{S}$ is finite, or the sets in $\mathcal{S}$ are pairwise disjoint, then a transversal of $\mathcal{S}$ certainly exists. We now show how these ideas relate to optimal dualities.

3.1 Optimal Duality by Failsets Theorem *Suppose that Ω yields a duality on $\mathcal{A}$ and $R \subseteq \Omega$.*

(i) *R yields a duality on $\mathcal{A}$ if and only if it intersects every failset.*

(ii) *R yields an optimal duality on $\mathcal{A}$ if and only if it is a transversal of the family of failsets.*

Moreover, if every failset contains a globally minimal failset—as is the case if Ω is finite—then

(iii) *R yields an optimal duality on $\mathcal{A}$ if and only if it is a transversal of the globally minimal failsets.*

Proof (i) follows from the Density Lemma and Lemma 8.2.5, while (ii) and (iii) follow immediately from (i). ■

Notice that this theorem affords us an explicit algorithm to compute *all* optimal dualising subsets of a given finite dualising set Ω provided that $G \cup H \cup R$ is finite:

- compute all of the failsets of each $s \in \Omega$;
- pick out the globally minimal failsets;
- pick out the transversals of the globally minimal failsets.

We will illustrate this method by proving that a semi-primal algebra $\underline{\mathbf{M}}$ has a *unique* optimal dualising set. Recall that the Semi-primal Strong Duality Theorem 3.3.14 states that $\underset{\sim}{\mathbf{M}} = \langle M; K, \mathbb{S}(\underline{\mathbf{M}}), \mathcal{T}\rangle$ yields a strong duality on the variety generated by a semi-primal algebra $\underline{\mathbf{M}}$, where K is the set of distinguished elements corresponding to the trivial subalgebras of $\underline{\mathbf{M}}$.

3.2 Semi-primal Optimal Duality Theorem *Let $\underline{\mathbf{M}}$ be a semi-primal algebra. Define K_0 to be the set of all distinguished elements corresponding to the trivial subalgebras of $\underline{\mathbf{M}}$ which are meet-irreducible in $\mathbb{S}(\underline{\mathbf{M}})$ and define R_0 to be the set of non-trivial subalgebras of $\underline{\mathbf{M}}$ which are meet-irreducible in $\mathbb{S}(\underline{\mathbf{M}})$. Then $\underset{\sim}{\mathbf{M}}_0 := \langle M; K_0, R_0, \mathcal{T}\rangle$ is the unique reduct of $\underset{\sim}{\mathbf{M}}$ which yields an optimal duality on the variety $\mathcal{A}$ generated by $\underline{\mathbf{M}}$.*

Proof Since $\mathbb{S}(\underline{\mathbf{M}})$ yields a duality on $\mathcal{A}$, and since every subalgebra of $\underline{\mathbf{M}}$ is an intersection of meet-irreducible subalgebras (A.5), it follows that $\underset{\sim}{\mathbf{M}}_0$ yields a duality on $\mathcal{A}$. We will show that for every $N \in K_0 \cup R_0$ we can find a map $\gamma\colon D(\mathbf{N}) \to M$ such that $\mathrm{Fail}_{\mathbf{N}}(\gamma) = \{N\}$. As this is clearly a globally minimal failset, it will follow that $K_0 \cup R_0$ is the unique transversal of the globally minimal failsets.

Assume that $\mathbf{N}$ is a meet-irreducible subalgebra of $\underline{\mathbf{M}}$, and let $\mathbf{N}^*$ be its unique cover. Choose $d \in N^* \backslash N$. If $\mathbf{N}$ is non-trivial, then $D(\mathbf{N}) = \{\mathrm{inc}_N\}$, where inc_N is the inclusion map, and we define $\gamma\colon D(\mathbf{N}) \to M$ by $\gamma(\mathrm{inc}_N) = d$. If $\mathbf{N}$ is trivial, then $D(\mathbf{N}) = \{x_k \mid k \in K\}$, where $x_k\colon N \to \{k\}$ is an isomorphism, and we define $\gamma\colon D(\mathbf{N}) \to M$ by

$$\gamma(x_k) = \begin{cases} k & \text{if } N \neq \{k\}, \\ d & \text{if } N = \{k\}. \end{cases}$$

In both cases it is easily seen that a unary algebraic relation r is in $\mathrm{Fail}_{\mathbf{N}}(\gamma)$ if and only if $N \subseteq r$ and $d \notin r$. Thus $\mathrm{Fail}_{\mathbf{N}}(\gamma) = \{N\}$. ■

We will see that there are two other very different ways to identify the globally minimal failsets, and therefore the optimal dualising sets. The first comes from the observation that entailment is in fact a local process.

3.3 Lemma *Let $R \subseteq \Omega$ and $s \in \Omega$. Then R entails s (that is, $s \in \overline{R}$) if and only if S entails s for some finite subset S of R.*

Proof Assume that $R \vdash s$. By Lemma 8.2.5, R intersects every failset of s. Since $\mathcal{F}_s$ is finite, R contains a transversal S of $\mathcal{F}_s$ which is also finite. Again by Lemma 8.2.5, $S \vdash s$. ∎

This lemma provides indirect evidence for the existence of a complete set of admissible constructs for entailment. Indeed, the existence of a complete set of admissible constructs (which we will establish in Chapter 9) yields Lemma 8.3.3 as a consequence: suppose that a complete set of constructs exists; if a relation s is entailed by R then, since the constructs are complete and each involves only a finite number of relations from R, the relation s can be constructed from some finite subset S of R, and consequently, since the constructs are admissible, the finite subset S of R entails s. This fact has an important algebraic consequence.

3.4 Proposition *The map $R \mapsto \overline{R}$ is an algebraic closure operator on Ω and consequently the lattice $\mathbf{\Lambda}$ of entailment closed subsets of Ω is algebraic.*

Proof We ask the reader to complete Exercise 8.6 and then use it to show that $R \mapsto \overline{R}$ is a closure operator on Ω. If $R \vdash s$, then, by Lemma 8.3.3, $S \vdash s$ for some finite subset S of R. This says precisely that entailment closure is algebraic, and Theorem A.8 tells us that $\mathbf{\Lambda}$ is an algebraic lattice. ∎

Whenever we have an algebraic lattice, an important role is played by the completely meet-irreducible (CMI) elements since every element of the lattice is the meet of a collection of CMI elements (Theorem A.5). The CMI elements of $\mathbf{\Lambda}$ give us an alternative description of the locally minimal failsets.

3.5 Locally Minimal Failset Theorem *Let R be completely meet-irreducible in $\mathbf{\Lambda}$ with R^* its unique upper cover. Then $\Omega\backslash R$ is a locally minimal failset of s for each $s \in R^*\backslash R$. Conversely if U is a locally minimal failset of s then $R := \Omega\backslash U$ is completely meet-irreducible, with $R^* := \overline{R \cup \{s\}}$ as its unique upper cover.*

Proof Assume that R is completely meet-irreducible in the lattice $\mathbf{\Lambda}$ and let $s \in R^*\backslash R$. Then $s \in \overline{R \cup \{r\}}$ for all $r \in \Omega\backslash R$. Since R is closed and $s \notin R$, we conclude that R does not entail s, that is,

(a) $$(\exists \gamma\colon D(\mathbf{s}) \to M)\ (s \in \mathrm{Fail}_{\mathbf{s}}(\gamma) \subseteq \Omega\backslash R).$$

The fact that $s \in \overline{R \cup \{r\}}$ for all $r \in \Omega\backslash R$ says that

(b) $$(\forall\ \lambda\colon D(\mathbf{s}) \to M)\ (s \in \mathrm{Fail}_{\mathbf{s}}(\lambda) \subseteq \Omega\backslash R \iff \mathrm{Fail}_{\mathbf{s}}(\lambda) = \Omega\backslash R).$$

Together (a) and (b) say precisely that $\Omega\backslash R = \mathrm{Fail}_{\mathbf{s}}(\gamma)$ is a locally minimal failset of s.

Conversely, assume that $U = \mathrm{Fail}_{\mathbf{s}}(\gamma)$ is a locally minimal failset of s. Since by Lemma 8.2.4 the complement of a failset is always closed, $R := \Omega\backslash U$ is closed and so $R \in \Lambda$. Let $r \in \Omega\backslash R$; we must show that $s \in \overline{R \cup \{r\}}$. Suppose that $s \notin \overline{R \cup \{r\}}$. Then, by the Test Algebra Lemma, there exists $\lambda\colon D(\mathbf{s}) \to M$ which preserves $R \cup \{r\}$ but fails to preserve s. Thus

$$s \in \mathrm{Fail}_{\mathbf{s}}(\lambda) \subseteq U\backslash\{r\} \subsetneq U = \mathrm{Fail}_{\mathbf{s}}(\gamma),$$

which contradicts the minimality of U in $\mathcal{F}_s$. ■

One important consequence of this theorem is the existence of locally minimal failsets for arbitrary sets Ω.

3.6 Corollary *Let $s \in \Omega$.*

(i) *If $U = \mathrm{Fail}_{\mathbf{r}}(\lambda)$ is a weak failset containing s, then there is a locally minimal failset U_s of s with $s \in U_s \subseteq U$.*

(ii) *Every weak failset is a union of locally minimal failsets.*

(iii) *Any weak failset U which is minimal with respect to $s \in U$ is a locally minimal failset of s.*

Proof By Lemma 8.2.4, $\Omega\backslash U$ is a closed set not containing s. By a standard Zorn's Lemma argument (using the fact that entailment closure is algebraic), there is a closed set R which is maximal with respect to not containing s. Clearly R is completely meet-irreducible and $\overline{R \cup \{s\}}$ is its unique upper cover. Thus, by Theorem 8.3.5, $U := \Omega\backslash R$ is a locally minimal failset of s. This proves (i), and (ii) and (iii) are immediate consequences. ■

Let $r, s \in \Omega$. If $r \vdash s$, then any failset which contains s must also contain r. Hence every failset of s contains

$$\Psi_s := \{\, r \in \Omega \mid r \vdash s \,\}.$$

In fact, we have more.

3.7 Lemma *Ψ_s is the intersection of the (locally minimal) failsets of s.*

Proof Certainly, $\Psi_s \subseteq \bigcap\{\, U \mid U \in \mathcal{F}_s \,\}$. Let $r \notin \Psi_s$; then r does not entail s. Hence, by Lemma 8.2.5, there exists $U \in \mathcal{F}_s$ with $r \notin U$. Thus, $\Psi_s = \bigcap\{\, U \mid U \in \mathcal{F}_s \,\}$. Since, by Corollary 8.3.6, each failset of s contains a minimal failset of s, we conclude that Ψ_s is the intersection of the minimal failsets of s. ■

Of course, Ψ_s is usually not itself a failset. It turns out that Ψ_s is a failset precisely when $\Omega\backslash\Psi_s$ is completely meet-prime in the lattice $\mathbf{\Lambda}$: see Exercise 8.12.

We now turn our attention to globally minimal failsets, the transversals of which give us optimal dualities provided that Ω is a finite dualising set. They correspond to the coatoms of the lattice $\mathbf{\Lambda}$.

3.8 Globally Minimal Failset Theorem *Let $\varnothing \neq U \subseteq \Omega$. Then the following are equivalent:*

(i) *$\Omega\backslash U$ is a coatom in the lattice $\mathbf{\Lambda}$;*

(ii) *U is a globally minimal failset;*

(iii) *U is a locally minimal failset of r for all $r \in U$;*

(iv) *U is a (weak) failset and $(\Omega\backslash U) \cup \{s\} \vdash r$, for all $r, s \in U$.*

Proof Assume that $\Omega\backslash U$ is a coatom in $\mathbf{\Lambda}$. Since coatoms are completely meet-irreducible, U is a locally minimal failset of some $r \in \Omega$, by the Locally Minimal Failset Theorem. If V is a failset with $V \subseteq U$, then $\Omega\backslash V$ is closed, by Lemma 8.2.4, and $\Omega\backslash U \subseteq \Omega\backslash V$. As $\Omega\backslash U$ is a coatom, $\Omega\backslash U = \Omega\backslash V$ and so $U = V$. Thus U is a globally minimal failset. Hence (i) implies (ii).

It is an immediate consequence of Corollary 8.3.6 that if $U = \mathrm{Fail}_{\mathbf{s}}(\gamma)$ is a globally minimal failset then it is a minimal failset of r for each $r \in U$. Thus (ii) implies (iii).

Assume (iii) and let $r, s \in U$. By Lemma 8.2.5, in order to prove that $(\Omega\backslash U) \cup \{s\} \vdash r$ it suffices to prove that every $V \in \mathcal{F}_r$ intersects $(\Omega\backslash U) \cup \{s\}$. Let $V \in \mathcal{F}_r$. If $V \subseteq U$, then $s \in V = U$, as U is a locally minimal failset of s. If $V \nsubseteq U$, then $(\Omega\backslash U) \cap V \neq \varnothing$, as required. Thus (iii) implies (iv).

Finally, we show that (iv) implies (i). Assume (iv). Since U is a (weak) failset, $\Omega\backslash U$ is closed, by Lemma 8.2.4, and is not the top of $\mathbf{\Lambda}$ as $U \neq \varnothing$. For each $s \in U$, the set $(\Omega\backslash U) \cup \{s\}$ entails each $r \in U$ and hence the closure of $(\Omega\backslash U) \cup \{s\}$ is Ω. This says precisely that $\Omega\backslash U$ is a coatom of $\mathbf{\Lambda}$. ■

A third description of the globally minimal failsets contained in a dualising set Ω arises from a determination of which relations in Ω are, in an appropriate sense, forced to occur in any optimal duality and which relations never occur in an optimal duality. Assume that Ω yields a duality on $\mathcal{A}$. We say that a subset U of Ω is **unavoidable** (within Ω) if whenever a subset R of Ω yields a duality on $\mathcal{A}$, then $R \cap U \neq \varnothing$. Notice that U is unavoidable if and only if $\Omega\backslash U$ does not yield a duality on $\mathcal{A}$. The unavoidable subsets form an ordered set (under set inclusion) with

Ω as its top. A subset U of Ω will be called a **minimal unavoidable set** (within Ω) if it is a minimal element of this ordered set.

3.9 Unavoidable Set Theorem *Assume that Ω yields a duality on $\mathcal{A}$.*

(i) *Within Ω, every failset is unavoidable.*

(ii) *Within Ω, every unavoidable set contains a failset.*

(iii) *A subset U of Ω is a minimal unavoidable set if and only if it is a globally minimal failset.*

Proof Let $U \subseteq \Omega$ be a failset. By Theorem 8.3.1(i), the complement $\Omega \backslash U$ does not yield a duality on $\mathcal{A}$ and so U is unavoidable within Ω. Now let U be an unavoidable set within Ω. Consequently, the complement $\Omega \backslash U$ does not yield a duality on $\mathcal{A}$. By Theorem 8.3.1(i), there is some failset V which is disjoint from $\Omega \backslash U$, whence $V \subseteq U$. This proves (i) and (ii). A simple argument shows that (iii) follows from the Optimal Duality by Failsets Theorem 8.3.1 (i) and (ii). ■

In the finite case the Optimal Duality by Failsets Theorem 8.3.1, the Globally Minimal Failset Theorem 8.3.8 and the Unavoidable Set Theorem 8.3.9 give us a thorough characterisation of optimal dualities.

3.10 Optimal Duality Theorem *Assume that $\Omega \subseteq \mathcal{B}$ is finite and yields a duality on $\mathcal{A}$ and let $R \subseteq \Omega$. Then the following are equivalent:*

(i) *R yields an optimal duality on $\mathcal{A}$;*

(ii) *R is a transversal of the globally minimal failsets within Ω;*

(iii) *R is a transversal of the complements of coatoms in the lattice Λ of entailment closed subsets of Ω;*

(iv) *R is a transversal of the minimal unavoidable sets within Ω.*

In almost all cases described in Chapter 4 we found that the dualising relations were either unary or binary. When $\Omega \subseteq \mathbb{S}(\underline{\mathbf{M}}) \cup \mathbb{S}(\underline{\mathbf{M}}^2)$ yields a duality on $\mathcal{A}$ there is a natural concept of absolutely unavoidable relation. If $s \in \mathbb{S}(\underline{\mathbf{M}})$ we define $s^{\smile}$ to be s. We say that $s \in \Omega \subseteq \mathbb{S}(\underline{\mathbf{M}}) \cup \mathbb{S}(\underline{\mathbf{M}}^2)$ is **absolutely unavoidable** (within Ω) if $\{s, s^{\smile}\}$ is a (necessarily minimal) unavoidable set. The following observation is an immediate consequence of Theorems 8.3.8 and 8.3.9.

3.11 Absolutely Unavoidable Relation Corollary *Let Ω be a subset of $\mathbb{S}(\underline{\mathbf{M}}) \cup \mathbb{S}(\underline{\mathbf{M}}^2)$ which yields a duality and let $s \in \Omega$. The following are equivalent:*

(i) *s is absolutely unavoidable within Ω;*

(ii) *$\{s, s^{\smile}\}$ is a (necessarily globally minimal) failset (of s) within Ω.*

If Ω yields a duality on $\mathcal{A}$, then $s \in \Omega$ is **absolutely avoidable** if $R \backslash \{s\}$ yields a duality on $\mathcal{A}$ whenever R does, for each $R \subseteq \Omega$. That is, s can always be omitted without destroying a duality. For example, all trivial relations, such as M, M^2 and Δ_M, are absolutely avoidable.

3.12 Absolutely Avoidable Relation Theorem *Assume that Ω is finite and yields a duality on $\mathcal{A}$. Then the following are equivalent:*

(i) *s is absolutely avoidable;*
(ii) *s belongs to no globally minimal failset;*
(iii) *s belongs to the intersection of the coatoms of $\mathbf{\Lambda}$;*
(iv) *s belongs to no minimal unavoidable set.*

Proof Parts (iii) and (iv) are equivalent to (ii) by Theorems 8.3.8 and 8.3.9, respectively. Suppose that (i) fails, that is, for some $R \subseteq \Omega$ the set $R \cup \{s\}$ yields a duality on $\mathcal{A}$ but R does not. Let $S \subseteq R \cup \{s\}$ yield an optimal duality. Then s must be in S, and therefore in some globally minimal failset by Theorem 8.3.1. Thus (ii) implies (i).

Assume that s is absolutely avoidable. Suppose that s belongs to some minimal unavoidable set U. We claim that $U\backslash\{s\}$ is unavoidable, contradicting the minimality of U. Assume that R yields a duality on $\mathcal{A}$. Then by assumption $R^* := R\backslash\{s\}$ also yields a duality on $\mathcal{A}$. Since U is unavoidable, R^* intersects U and consequently R intersects $U\backslash\{s\}$. Hence $U\backslash\{s\}$ is unavoidable, as claimed. Thus (i) implies (iv). ■

8.4 Kleene algebras once again

In 4.3.9 and 4.3.10 in Chapter 4 we introduced the class, $\mathcal{K}$, of Kleene algebras and applied the NU Strong Duality Corollary 3.3.9 to obtain a strong duality for this class. The schizophrenic object is

$$\underline{\mathbf{K}} = \langle \{0, a, 1\}; \vee, \wedge, \neg, 0, 1 \rangle \quad \text{and} \quad \underset{\sim}{\mathbf{K}} = \langle \{0, a, 1\}; \preccurlyeq, \sim, K_0; \mathcal{T} \rangle$$

where

$$0 < a < 1, \quad \neg 0 = 1, \quad \neg 1 = 0 \quad \text{and} \quad \neg a = a,$$

and $\preccurlyeq$ is the order with $0 \prec a$ and $1 \prec a$ while $\sim\, = K^2\backslash\{(0,1),(1,0)\}$ and $K_0 = \{0, 1\}$. We shall now see that not only is this duality optimal, but in a very natural sense it is the unique duality using algebraic relations of minimal arity. Both $\preccurlyeq$ and $\sim$ turn out to be absolutely unavoidable in $\Omega = \mathbb{S}(\underline{\mathbf{K}}) \cup \mathbb{S}(\underline{\mathbf{K}}^2)$ while K_0 is contained in a unique globally minimal failset U and is the only unary relation in U.

4.1 The Duality Again We begin by proving again that $\underset{\sim}{\mathbf{K}}$ yields a duality on $\mathcal{K}$. This time round we shall try to emphasise the choices which are available along the way. By the NU Duality Theorem 2.3.4 the set $\Omega := \mathbb{S}(\underline{\mathbf{K}}) \cup \mathbb{S}(\underline{\mathbf{K}}^2)$ yields a duality on $\mathcal{K}$. As in the proof of 4.3.10, it is easy to show by hand that $\underline{\mathbf{K}}^2$ has 11 subalgebras:

$$\Delta_K, \quad \Delta_{K_0}, \quad \preccurlyeq, \quad \succcurlyeq := \preccurlyeq^\smile, \quad \sim, \quad K \times K, \quad K \times K_0, \quad K_0 \times K,$$

$$K_0 \times K_0, \quad \preccurlyeq \cap (K_0 \times K) \quad \text{and} \quad \succcurlyeq \cap (K \times K_0).$$

The lattice of subalgebras of $\underline{\mathbf{K}}^2$ is as shown in Figure 8.1: only the relations $\preccurlyeq \cap (K_0 \times K)$ and $\succcurlyeq \cap (K \times K_0)$ are unlabelled. The meet-irreducible elements of $\mathbb{S}(\underline{\mathbf{K}}^2)$ are $K \times K_0$, $K_0 \times K$, $\preccurlyeq$, $\succcurlyeq$, and $\sim$. Any set which entails these relations will entail their intersections, and hence entail all subalgebras of $\underline{\mathbf{K}}^2$. Consequently $R := \{\preccurlyeq, \sim, K_0\}$ entails all subalgebras of $\underline{\mathbf{K}}^2$ and so yields a duality on $\mathcal{K}$.

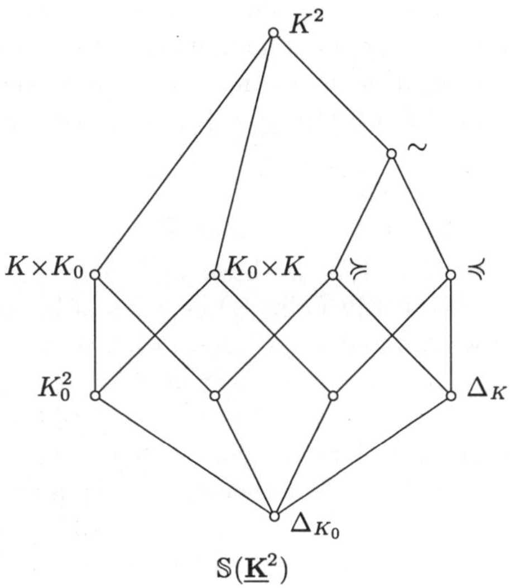

Figure 8.1 the subalgebras of $\underline{\mathbf{K}}^2$

How can we modify R without destroying the duality? It is easily seen that each member of the set

$$U^{K_0} := \{K_0, \Delta_{K_0}, K_0^2, K_0 \times K, K \times K_0, \preccurlyeq \cap (K_0 \times K), \succcurlyeq \cap (K \times K_0)\}$$

entails K_0. Thus, we could replace K_0 with any other member of U^{K_0}, but this would have the disadvantage of replacing a unary relation with a binary one. The only other change which comes to mind would be the trivial one of replacing $\preccurlyeq$ with its converse $\succcurlyeq$. This certainly *feels* like an optimal duality. To see that it is we need to find the minimal unavoidable sets, or equivalently, the global minimal failsets within $\Omega := \mathbb{S}(\underline{\mathbf{K}}) \cup \mathbb{S}(\underline{\mathbf{K}}^2)$.

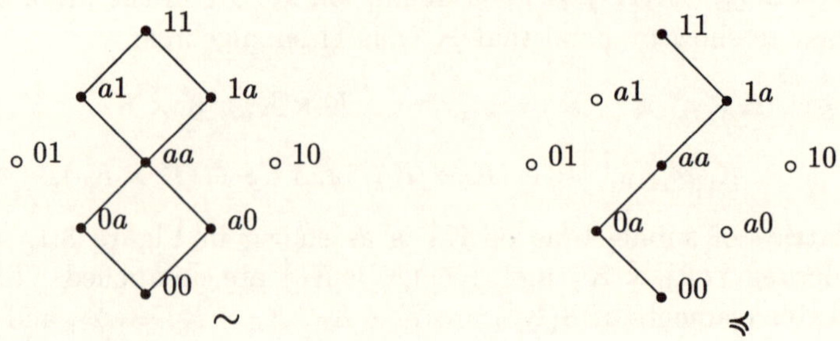

Figure 8.2 the subalgebras $\sim$ and $\preccurlyeq$ of $\underline{\mathbf{K}}^2$

4.2 The Globally Minimal Failsets First, consider the relation $\sim$: see Figure 8.2. Let $x\colon \sim \to \underline{\mathbf{K}}$ be a homomorphism. Because the fixpoint (a, a) of the Kleene negation must map to a, it is very easy to show that x is a projection. Thus $D(\sim) = \{\rho_1, \rho_2\}$. Define $\gamma\colon D(\sim) \to K$ by $\gamma(\rho_1) = 0$ and $\gamma(\rho_2) = 1$. Since $(0, 1) \notin \sim$, and, by Lemma 8.1.1, $(\rho_1, \rho_2) \in \sim^{D(\sim)}$, we conclude that $\sim \in \mathrm{Fail}_{\sim}(\gamma)$, that is, $\mathrm{Fail}_{\sim}(\gamma)$ is a failset of $\sim$. We shall now show that $\mathrm{Fail}_{\sim}(\gamma) = \{\sim\}$. Note that $K_0^{D(\sim)} = \varnothing$ since, for $i = 1, 2$, we have $\rho_i((a, a)) = a \notin K_0$. Thus γ preserves K_0 and consequently $\mathrm{Fail}_{\sim}(\gamma)$ consists of binary relations. Let $r \in \mathrm{Fail}_{\sim}(\gamma)$. Then we can find $x, y \in D(\sim)$ such that $(x, y) \in r^{D(\sim)}$ and $(\gamma(x), \gamma(y)) \notin r$. As $(0, 0), (1, 1) \in r$, we must have $x \neq y$. Consequently, as r is closed under the Kleene negation, $(0, 1), (1, 0) \notin r$, that is, $r \subseteq \sim$. By the **r**-on-**s** Lemma 8.1.2, there is a homomorphism, namely $\varphi = \rho_1 \sqcap \rho_2$ or $\varphi = \rho_2 \sqcap \rho_1$, from $\sim$ to $\mathbf{r}$. In either case, φ is one-to-one and hence $\varphi(\sim) \subseteq r \subseteq \sim$ implies that $r = \sim$. We have proved that $\mathrm{Fail}_{\sim}(\gamma) = \{\sim\}$, as claimed, and this failset is clearly minimal. Since this is also a minimal unavoidable set, $\sim$ is absolutely unavoidable.

(As subsets of K^2, the relation $\sim$ and the relational product $\succcurlyeq \cdot \preccurlyeq$ coincide. This identification fails on arbitrary spaces $D(\mathbf{A})$ for $\mathbf{A} \in \mathcal{K}$. Indeed, in a typically schizophrenic way, the identification fails on the dual $D(\sim)$ of the corresponding test algebra, that is, as we shall verify in 9.1.3,

$$\succcurlyeq^{D(\sim)} \cdot \preccurlyeq^{D(\sim)} \neq (\succcurlyeq \cdot \preccurlyeq)^{D(\sim)}.$$

Since γ preserves both $\succcurlyeq$ and $\preccurlyeq$, but does not preserve $\sim$, this shows that, in general, relational product cannot be an admissible construct.)

An almost identical argument applies to the relation $\preccurlyeq$. We leave it to the reader to check that $D(\preccurlyeq) = \{\rho_1, \rho_2\}$ and that, with γ defined by $\gamma(\rho_1) = a$ and $\gamma(\rho_2) = 0$, we have $\mathrm{Fail}_{\preccurlyeq}(\gamma) = \{\preccurlyeq, \succcurlyeq\}$, whence this failset is clearly minimal and $\preccurlyeq$ is also absolutely unavoidable within Ω.

Notice that any failset intersecting either $\{\preccurlyeq, \succcurlyeq\}$ or $\{\sim\}$ must contain that failset, and that the absolutely avoidable relations K, Δ_K and K^2 belong to no failset. Thus any other minimal failset must be a subset of $\Omega\backslash\{\preccurlyeq, \succcurlyeq, \sim, K, \Delta_K, K^2\} = U^{K_0}$. Thus we can conclude our search for the minimal unavoidable sets within Ω by showing that U^{K_0} itself is a minimal unavoidable set. First we prove that U^{K_0} is a failset of K_0. The dual of $\mathbf{K}_0$ is simply $D(\mathbf{K}_0) = \{\rho_1\}$, where $\rho_1 \colon \mathbf{K}_0 \to \underline{\mathbf{K}}$ is the inclusion map. Define $\gamma \colon D(\mathbf{K}_0) \to K$ by $\gamma(\rho_1) = a$. Again, the reader should show that the only unary relation in $\mathrm{Fail}_{\mathbf{K}_0}(\gamma)$ is K_0 and that a binary relation r is in $\mathrm{Fail}_{\mathbf{K}_0}(\gamma)$ if and only if $(a, a) \notin r$. It follows that $\mathrm{Fail}_{\mathbf{K}_0}(\gamma) = U^{K_0}$. It is easy to check by hand, using the Constructs for Entailment 2.4.5, that $\{\preccurlyeq, s\} \vdash r$ for all $r, s \in U^{K_0}$, whence U^{K_0} is a globally minimal failset by the equivalence of (ii) and (iv) in the Globally Minimal Failset Theorem 8.3.8.

This shows that there are exactly three minimal failsets within Ω, namely

$$\{\preccurlyeq, \succcurlyeq\}, \quad \{\sim\} \quad \text{and} \quad U^{K_0}.$$

By the Optimal Duality Theorem 8.3.10, the subsets of Ω which yield optimal dualities on $\mathcal{K}$ are the transversals of this family. Therefore, as expected, the only choices available are whether to use $\preccurlyeq$ or its converse and which member of U^{K_0} to use. Thus, as claimed, there is an essentially unique optimal duality for $\mathcal{K}$ using relations of minimal arity, namely $R = \{\preccurlyeq, \sim, K_0\}$.

4.3 The Lattice Λ When proving that U^{K_0} was a minimal failset we used condition (iv) of the Globally Minimal Failset Theorem 8.3.8. With more complex examples the aid of a computer is essential and it is more convenient to use condition (iii) and get the computer to check whether a given failset is a locally minimal failset of each of its members. The computer output for the Kleene example is given in Table 8.1. Here we see that $U^{K_0} = \{1, 2, 4, 5, 6, 7, 12\}$ is a locally minimal failset of each of its members, as the equivalence of (ii) and (iii) in the Globally Minimal Failset Theorem 8.3.8 predicts it will be. We also see that $\{6, 7, 9, 10, 11\}$ is a locally minimal failset of two different relations but is not a globally minimal failset.

The computer output enables us to find the lattice Λ should we wish to do so. In the case of Kleene algebras, the set $\mathcal{F}$ of minimal failsets has the following elements:

$$A = \{1, 2, 4, 5, 6, 7, 12\}, \qquad C = \{9, 10\},$$
$$B = \{6, 7, 9, 10, 11\}, \qquad D = \{11\}.$$

Subalgebra	Label	Minimal failsets
K_0^2	1	1 2 4 5 6 7 12
Δ_{K_0}	2	1 2 4 5 6 7 12
Δ_K	3	
$K \times K_0$	4	1 2 4 5 6 7 12
$K_0 \times K$	5	1 2 4 5 6 7 12
$\succcurlyeq \cap (K \times K_0)$	6	1 2 4 5 6 7 12 and 6 7 9 10 11
$\preccurlyeq \cap (K_0 \times K)$	7	1 2 4 5 6 7 12 and 6 7 9 10 11
K^2	8	
$\succcurlyeq$	9	9 10
$\preccurlyeq$	10	9 10
$\sim$	11	11
K_0	12	1 2 4 5 6 7 12
K	13	

Table 8.1 the minimal failsets within Ω

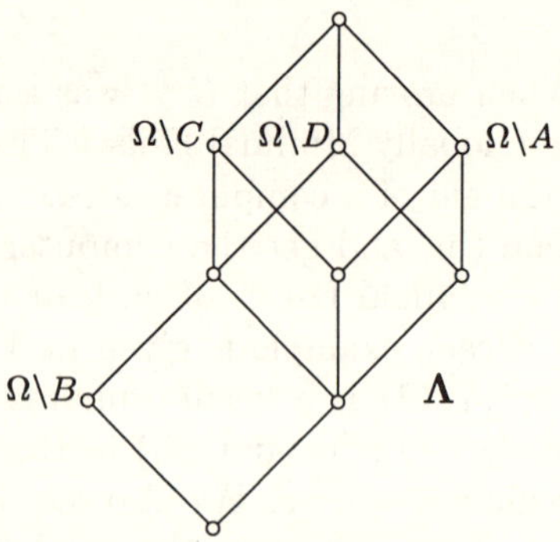

Figure 8.3 the lattice $\mathbf{\Lambda}$ of entailment-closed sets

Since the meet-irreducibles of $\mathbf{\Lambda}$ are precisely the complements of A, B, C and D in Ω, the lattice $\mathbf{\Lambda}$ is obtained by taking the $\cap$-semilattice generated by the family of sets $\{\Omega, \Omega\backslash A, \Omega\backslash B, \Omega\backslash C, \Omega\backslash D\}$. We obtain the lattice shown in Figure 8.3.

8.5 The structure of failsets

While it is important to be able to recognise absolutely unavoidable binary relations, like $\preccurlyeq$ and $\sim$ in the preceding section, such relations are relatively rare. It is therefore natural to study minimal unavoidable sets of relations. Indeed, since every optimal duality is given by a transversal of the minimal unavoidable sets, or equivalently, by Theorem 8.3.9, globally minimal failsets, an understanding of the structure of these sets more generally enables us to discover just how different optimal dualities for the same class $\mathcal{A}$ relate to one another.

The study of minimal failsets is in its infancy. Even amongst algebras with an underlying lattice structure, where we know that $\Omega = \mathbb{S}(\underline{\mathbf{M}}^2)$ yields a duality, there are very few examples for which all the globally minimal failsets in $\mathbb{S}(\underline{\mathbf{M}}^2)$ are known. Since, by Theorem 8.3.8, a set is a globally minimal failset if and only if it is a locally minimal failset of each of its members, we shall begin by searching for general results which characterise certain sets as locally minimal failsets within Ω. In the following section, we shall show how these results were applied in [DP96a] to describe the optimal dualities for varieties of distributive p-algebras.

As every n-ary algebraic relation in $\mathcal{B}$ can be constructed via intersection from the meet-irreducible elements of the lattice $\mathbb{S}(\underline{\mathbf{M}}^n)$, we need to characterise the meet-irreducibles and to relate them to failsets.

Let $a \in A$. A subalgebra $\mathbf{B}$ of an algebra $\mathbf{A}$ will be called a **value** of $\mathbf{A}$ at a if $\mathbf{B}$ is maximal with respect to not containing a.

5.1 Lemma *A subalgebra* $\mathbf{B}$ *of* $\mathbf{A}$ *is a value of* $\mathbf{A}$ *if and only if* $\mathbf{B}$ *is completely meet-irreducible in the lattice of subalgebras of* $\mathbf{A}$.

Proof Clearly, if $\mathbf{B}$ is a value of $\mathbf{A}$ at a, then the subalgebra generated by $B \cup \{a\}$ is the unique upper cover of $\mathbf{B}$ in $\mathbb{S}(\mathbf{A})$. Conversely, if $\mathbf{C}$ is the unique upper cover of $\mathbf{B}$ in $\mathbb{S}(\mathbf{A})$, then $\mathbf{B}$ is a value at each $a \in C \backslash B$. ■

Although a failset may include relations of different arities, we can still order the elements of a failset by set inclusion. This order helps us to identify subalgebras which are values. Further examples of the role this order has to play in the analysis of failsets are given in Exercise 8.9. If $\gamma\colon D(\mathbf{s}) \to M$ and $x_1, \dots, x_n \in D(\mathbf{s})$ with $(x_1, \dots, x_n) \in r^{D(\mathbf{s})}$ but $(\gamma(x_1), \dots, \gamma(x_n)) \notin r$ we say that $x_1, \dots, x_n$ **witness** $r \in \mathrm{Fail}_{\mathbf{s}}(\gamma)$.

5.2 Lemma *Let* $U = \mathrm{Fail}_{\mathbf{s}}(\gamma)$ *be a weak failset.*

(i) *Let* $r \subseteq M^n$ *be maximal in* U. *Then* $\mathbf{r}$ *is a value of* $\underline{\mathbf{M}}^n$; *explicitly, if* $x_1, \dots, x_n \in D(\mathbf{s})$ *witness* $r \in \mathrm{Fail}_{\mathbf{s}}(\gamma)$, *then* $\mathbf{r}$ *is a value of* $\underline{\mathbf{M}}^n$ *at* $(\gamma(x_1), \dots, \gamma(x_n))$.

(ii) *Every element of* U *is contained in a value in* U. *In particular, every failset contains a value.*

Proof Let $x_1, \ldots, x_n \in D(\mathbf{s})$ witness $r \in \mathrm{Fail}_{\mathbf{s}}(\gamma)$. Then we have $(x_1 \sqcap \cdots \sqcap x_n)(s) \subseteq r$, by Lemma 8.1.1, and $(\gamma(x_1), \ldots, \gamma(x_n)) \notin r$. Assume that $r \subseteq t \subseteq M^n$ with $(\gamma(x_1), \ldots, \gamma(x_n)) \notin t$. Then $(x_1 \sqcap \cdots \sqcap x_n)(s) \subseteq t$ and hence $x_1, \ldots, x_n \in D(\mathbf{s})$ witness $t \in \mathrm{Fail}_{\mathbf{s}}(\gamma)$. Since r is maximal in $\mathrm{Fail}_{\mathbf{s}}(\gamma)$, we conclude that $r = t$. Thus r is a value at $(\gamma(x_1), \ldots, \gamma(x_n))$. Hence (i) holds and (ii) is an immediate consequence of (i). ■

In the previous section we saw that U^{K_0} was the unique minimal failset containing $K_0 := \{0, 1\}$. Note that $\mathbf{K}_0$ is a value of $\underline{\mathbf{K}}$ at a. We now reveal the general structure of such unique minimal failsets determined by values of $\underline{\mathbf{M}}$.

Let $\mathbf{N} \leqslant \underline{\mathbf{M}}$ and define

$$\Delta_N := \{(a, a) \in M^2 \mid a \in N\},$$

and, more generally, for $n \geqslant 1$, define

$$\Delta_N^n := \{(a, \ldots, a) \in M^n \mid a \in N\}.$$

Now define

$$U^N := \{\, s \in \Omega \mid s \text{ is } n\text{-ary and } s \cap \Delta_M^n = \Delta_N^n \text{ for some } n \geqslant 1 \,\}.$$

The reader should verify that U^{K_0} is as claimed in the previous section and that it satisfies the conditions of the following theorem. Applications of this result to finitely generated varieties of distributive p-algebras, in general, may be found in Davey and Priestley [DP96a] (see the discussion after Proposition 8.6.3 below) and to Stone algebras, in particular, in Exercise 8.8. Recall that $\Psi_N := \{\, s \in \Omega \mid s \vdash N \,\}$.

5.3 First Failset Theorem *Let* $\mathbf{N} \leqslant \underline{\mathbf{M}}$ *and assume that* Ω *contains* $\mathbb{S}(\underline{\mathbf{M}})$ *and the (graphs of) all partial endomorphisms of* $\underline{\mathbf{M}}$. *Then the following are equivalent:*

(i) U^N *is a weak failset;*

(ii) U^N *is a failset of* N*;*

(iii) $U^N = \Psi_N$ *and hence is the unique minimal failset of* N*;*

(iv) $\Omega \backslash U^N$ *is completely meet-prime in* $\mathbf{\Lambda}$*;*

(v) *the inclusion* inc_N *of* N *into* M *is the only homomorphism from* $\mathbf{N}$ *to* $\underline{\mathbf{M}}$ *(that is,* $|D(\mathbf{N})| = 1$*) and* $\mathbf{N}$ *is a value of* $\underline{\mathbf{M}}$.

Proof Assume that U^N is a weak failset. Since, by hypothesis, $N \in \Omega$, we have $N \in U^N$ and consequently, by Corollary 8.3.6, U^N contains a locally minimal failset V of N. Since Ψ_N is contained in every failset of N, by Lemma 8.3.7, we have $N \in \Psi_N \subseteq V \subseteq U^N$. But if $s \in U^N$, then s entails N (via intersection with a trivial relation and repetition removal) and so $U^N \subseteq \Psi_N$. Thus $V = U^N = \Psi_N$, and we have proved the equivalence of (i), (ii) and (iii).

We now prove the equivalence of (i) and (iv). Assume that U^N is a weak failset. Then $\Omega\backslash U^N \in \Lambda$ by Lemma 8.2.4. For $i \in I$ let $R_i \in \Lambda$ with $R_i \nsubseteq \Omega\backslash U^N$. Hence, for each $i \in I$, there exists $r_i \in R_i \cap U^N$. Since $r_i \vdash N$ and since R_i is entailment-closed, we conclude that $N \in R_i$ for each $i \in I$. Thus $N \in \bigcap_{i\in I} R_i$ and consequently $\bigcap_{i\in I} R_i \nsubseteq \Omega\backslash U^N$. Hence $\Omega\backslash U^N$ is completely meet-prime in $\mathbf{\Lambda}$. Conversely, if $\Omega\backslash U^N$ is completely meet-prime, then it is CMI, whence $U^N = \Omega\backslash(\Omega\backslash U^N)$ is a failset by the Locally Minimal Failset Theorem 8.3.5.

To complete the proof, we establish the equivalence of (ii) and (v). Assume that U^N is a failset of N, say $U^N = \mathrm{Fail}_{\mathbf{N}}(\gamma)$. Since N is the only unary relation in U^N, it is maximal in U^N and so is a value of M by Lemma 8.5.2(i). Let $e\colon \mathbf{N} \to \underline{\mathbf{M}}$ be a homomorphism and suppose that $e \neq \mathrm{inc}_N$. If $\mathrm{graph}(e) \in U^N$, then $\mathrm{graph}(e) \cap \Delta_M = \Delta_N$. Since $|\,\mathrm{graph}(e)| = |N| = |\Delta_N|$, it follows that $\mathrm{graph}(e) = \Delta_N$, whence $e = \mathrm{inc}_N$, a contradiction. Hence $\mathrm{graph}(e) \notin U^N = \mathrm{Fail}_{\mathbf{N}}(\gamma)$. As $\mathrm{graph}(e) \in \Omega$, by hypothesis, we conclude that γ preserves $\mathrm{graph}(e)$ and so preserves e. As $N \in \mathrm{Fail}_{\mathbf{N}}(\gamma)$, there exists $x \in D(\mathbf{N})$ such that $x \in N^{D(\mathbf{N})}$ and $\gamma(x) \notin N$, that is, $x(N) \subseteq N$ and $\gamma(x) \notin N$. Now $x(N) \subseteq N$ implies that $x \in \mathrm{dom}(e^{D(\mathbf{N})})$ and consequently, as γ preserves e, we conclude that $\gamma(e \circ x) = e(\gamma(x))$, whence $\gamma(x) \in \mathrm{dom}(e) = N$, our final contradiction. Thus $e = \mathrm{inc}_N$.

Conversely, assume that $D(\mathbf{N}) = \{\mathrm{inc}_N\}$ and that $\mathbf{N}$ is a value of $\underline{\mathbf{M}}$ at d. Define $\gamma\colon D(\mathbf{N}) \to M$ by $\gamma(\mathrm{inc}_N) = d$. We claim that $\mathrm{Fail}_{\mathbf{N}}(\gamma) = U^N$. Let $s \in \Omega$ be n-ary. Then

$$
\begin{aligned}
s \in \mathrm{Fail}_{\mathbf{N}}(\gamma) &\Longrightarrow (\mathrm{inc}_N, \dots, \mathrm{inc}_N) \in s^{D(\mathbf{N})} \text{ and } (d, \dots, d) \notin s \\
&\Longrightarrow \Delta_N^n \subseteq s \text{ and } (d, \dots, d) \notin s \\
&\Longrightarrow \Delta_N^n \subseteq s \cap \Delta_M^n \text{ and } (d, \dots, d) \notin s \cap \Delta_M^n.
\end{aligned}
$$

Since $\mathbf{N}$ is a value of $\underline{\mathbf{M}}$ at d, it follows that Δ_N^n is a value of Δ_M^n at $(d, \dots, d)$. Consequently, $\Delta_N^n = s \cap \Delta_M^n$, that is, $s \in U^N$. Thus $\mathrm{Fail}_{\mathbf{N}}(\gamma) \subseteq U^N$. Since

$$N \in U^N \subseteq \Psi_N \subseteq \mathrm{Fail}_{\mathbf{N}}(\gamma)$$

holds trivially, we have $N \in U^N = \mathrm{Fail}_{\mathbf{N}}(\gamma)$. ■

In practice, the set U^N often turns out to be a globally minimal failset but, at this stage, we have no useful sufficient conditions which guarantee that this is the case. It is probably too much to hope for a complete understanding of globally minimal failsets within Ω, even when $\Omega = \mathbb{S}(\underline{\mathbf{M}}^2)$. Instead, we shall focus on particular relations $s \in \Omega$ and try to describe their (locally) minimal failsets. The First Failset Theorem above is a result of this type.

By Lemma 8.5.2, each failset contains a value. If the failset is globally minimal, then, by the Globally Minimal Failset Theorem 8.3.8, it is a locally minimal failset of the value. Thus, a natural objective would be to identify the relations lying in a minimal failset U of a given value $\mathbf{s}$ of $\underline{\mathbf{M}}^n$. Hopefully, the maximality condition in the definition of value will provide us with a handle on both the relation s and the failset U. To do this we need some means of constructing maps $\gamma\colon D(\mathbf{s}) \to M$ which fail to preserve s. In order to define a map on $D(\mathbf{s})$ we want a workable description of the homomorphisms from $\mathbf{s}$ to $\underline{\mathbf{M}}$. Of course, the maps $\rho_i := \pi_i{\restriction}_s$ $(i = 1,\dots,n)$ belong to $D(\mathbf{s})$, as do any well-defined composites of these with partial endomorphisms. The following lemma, which was foreshadowed in the proof of (ii) implies (iv) in the Unary Structure Theorem 6.2.2, shows that the complexity of the partial clone of algebraic partial operations on $\underline{\mathbf{M}}$ is reflected in the congruence lattices of the subalgebras of $\underline{\mathbf{M}}$.

5.4 Lemma *If there is a subalgebra* $\mathbf{N}$ *of* $\underline{\mathbf{M}}$ *such that the trivial congruence* $\mathbf{0^N}$ *is an irredundant meet of* k *non-trivial congruences, in particular if* $\mathrm{Irr}(\underline{\mathbf{M}}) = k$, *then* $\mathcal{P}$ *contains a* k*-ary partial map which depends upon all of its coordinates.*

Proof Assume that there exist a subalgebra $\mathbf{N}$ of $\underline{\mathbf{M}}$ and non-trivial congruences $\theta_1,\dots,\theta_k$ on $\mathbf{N}$ such that $\theta_1 \wedge \dots \wedge \theta_k = \mathbf{0^N}$ with this meet being irredundant. Define a $(k{+}1)$-ary algebraic relation r by

$$r = \{(a_1,\dots,a_k,a_{k+1}) \mid a_i\theta_i a_{k+1} \text{ for all } i\,\}.$$

It follows easily from the fact that the meet of the congruences is $\mathbf{0^N}$ that r is the graph of a partial operation $h := \mathrm{graph}^{-1}(r)$. Suppose that h does not depend upon its first variable. We shall show that $\theta_2 \wedge \dots \wedge \theta_k = \mathbf{0^N}$, contradicting the irredundancy of the original meet. Let $a, b \in M$; then

$$\begin{aligned} &a(\theta_2 \wedge \dots \wedge \theta_k)b \\ &\implies (a,\dots,a,a) \in r \text{ and } (b,a,\dots,a,b) \in r \\ &\implies h(a,\dots,a) = a \text{ and } h(b,a,\dots,a) = b. \end{aligned}$$

Consequently, $a = b$, since h does not depend upon its first variable. This contradiction shows that h does depend upon its first variable. By symmetry, h depends upon each of its variables. ■

Let $\mathbf{s} \leqslant M^n$ and let $h\colon \mathbf{s} \to \underline{\mathbf{M}}$ be a partial operation. We say that h is **essentially unary** if it is of the form $e \circ \rho_i$ for some $i \in \{1,\dots,n\}$ where e

is a partial endomorphism of $\underline{\mathbf{M}}$ such that $\operatorname{im}(\rho_i) \subseteq \operatorname{dom}(e)$. There is one important class of examples, which includes all the lattice-based examples from Chapter 4, for which every finitary algebraic operation is essentially unary. These examples correspond to the 'good' dualities described by the Unary Structure Theorem 6.2.2.

5.5 Lemma *Assume that $\underline{\mathbf{M}}$ generates a congruence distributive variety. Then the following are equivalent:*

(i) $\operatorname{Irr}(\underline{\mathbf{M}}) = 1$, *that is, every non-trivial subalgebra of $\underline{\mathbf{M}}$ is subdirectly irreducible;*

(ii) *every finitary algebraic partial operation on $\underline{\mathbf{M}}$ is essentially unary;*

(iii) *every binary algebraic partial operation on $\underline{\mathbf{M}}$ is essentially unary.*

Proof Assume (i), let $\mathbf{s} \leqslant \underline{\mathbf{M}}^n$ and let $x\colon \mathbf{s} \to \underline{\mathbf{M}}$ be a non-constant homomorphism. Since $\operatorname{im}(x)$ is subdirectly irreducible, Jónsson's Lemma implies that $\ker(\rho_i) \leqslant \ker(x)$ for some i (see Exercise A.9). Hence there is a partial endomorphism $e\colon \mathbf{N} \to \underline{\mathbf{M}}$, where $N = \operatorname{im}(\rho_i) \subseteq M$, such that $x = e \circ \rho_i$, as required to prove (ii). Part (ii) trivially implies (iii), and (iii) implies (i) by Lemma 8.5.4. ∎

Given that every homomorphism from s into $\underline{\mathbf{M}}$ is essentially unary, it is natural to attempt to define a map $\gamma\colon D(\mathbf{s}) \to M$ by choosing $(a_1, \dots, a_n) \in M^n$ and setting $\gamma(x) = e(a_i)$ whenever $x = e \circ \rho_i$. Note that we can restrict our attention here to non-extendable partial endomorphisms e of $\underline{\mathbf{M}}$. This has the advantage of raising the probability that γ will be well defined. It turns out that a dichotomy occurs. Either a well-defined map γ can be found, and $\operatorname{Fail}_{\mathbf{s}}(\gamma)$ can be described explicitly, or s is one of a number of types of relation built via admissible constructs solely from the non-extendable partial endomorphisms.

The following technical lemma does most of our work for us.

5.6 Lemma *Let $\mathbf{s} \leqslant \underline{\mathbf{M}}^n$ and assume that every homomorphism from s into $\underline{\mathbf{M}}$ is essentially unary. Let s^+ be the intersection of all n-ary algebraic relations r such that $s \subseteq r$ and r has one of the following forms:*

(i) *r is a trivial expansion of the domain of a non-extendable proper partial endomorphism,*

(ii) *r is a trivial expansion of the equaliser $\operatorname{eq}(e, f)$ of two non-extendable partial endomorphisms e and f of $\underline{\mathbf{M}}$,*

(iii) *r is a trivial expansion of the joint kernel $\ker(e, f)$ of two non-extendable partial endomorphisms of $\underline{\mathbf{M}}$.*

Choose $(a_1, \dots, a_n) \in M^n$ and 'define' $\gamma\colon D(\mathbf{s}) \to M$ by $\gamma(x) = e(a_i)$ whenever $x = e \circ \rho_i$ for some non-extendable partial endomorphism e with $\operatorname{im}(\rho_i) \subseteq \operatorname{dom}(e)$. Then γ is well defined if and only if $(a_1, \dots, a_n) \in s^+$.

Moreover, if $(a_1, \ldots, a_n) \in s^+$ *and* $r \in \mathrm{Fail}_{\mathbf{s}}(\gamma)$, *then there exist a map* $\varepsilon\colon \{1, \ldots, m\} \to \{1, \ldots, n\}$ *and non-extendable partial endomorphisms* $e_1, \ldots, e_m$ *of* $\underline{\mathbf{M}}$ *such that*

$$\begin{aligned} s &\subseteq ((e_1 \times \cdots \times e_m)^{-1}(r))^{\varepsilon} \\ &= \{(c_1, \ldots, c_n) \in (\mathrm{dom}(e_1) \times \cdots \times \mathrm{dom}(e_m))^{\varepsilon} \mid \\ &\qquad\qquad\qquad (e_1(c_{\varepsilon(1)}), \ldots, e_m(c_{\varepsilon(m)})) \in r\} \end{aligned}$$

and

$$(a_1, \ldots, a_n) \notin ((e_1 \times \cdots \times e_m)^{-1}(r))^{\varepsilon}.$$

Proof Note that γ is not a well-defined map if and only if one of the following three situations occurs.

(i) For some i there exists a proper non-extendable partial endomorphism e such that $a_i \notin \mathrm{dom}(e)$, whence $\gamma(e \circ \rho_i)$ is undefined. In this case we have $s \subseteq r := M \times \cdots \times M \times \mathrm{dom}(e) \times M \times \cdots \times M$ but $(a_1, \ldots, a_n) \notin r$.

(ii) For some i we have non-extendable partial endomorphisms e and f such that $e \circ \rho_i = f \circ \rho_i$ but $e(a_i) \neq f(a_i)$. Define

$$r := \{(c_1, \ldots, c_n) \mid c_i \in \mathrm{dom}(e) \cap \mathrm{dom}(f) \text{ and } e(c_i) = f(c_i)\}.$$

Then r is a trivial expansion of $\mathrm{eq}(e, f)$ and $e \circ \rho_i = f \circ \rho_i$ implies that $s \subseteq r$ while $e(a_i) \neq f(a_i)$ implies $(a_1, \ldots, a_n) \notin r$.

(iii) For some $i \neq j$ we have non-extendable partial endomorphisms e and f such that $e \circ \rho_i = f \circ \rho_j$ but $e(a_i) \neq f(a_j)$. Clearly,

$$r := \{(c_1, \ldots, c_n) \mid c_i \in \mathrm{dom}(e) \text{ and } c_j \in \mathrm{dom}(f) \text{ and } e(c_i) = f(c_j)\}$$

is a trivial expansion of $\ker(e, f)$ and $(a_1, \ldots, a_n) \notin r$.

Hence γ is a well-defined map if and only if $(a_1, \ldots, a_n) \in s^+$. Now let $(a_1, \ldots, a_n) \in s^+$ and assume that $r \in \mathrm{Fail}_{\mathbf{s}}(\gamma)$. Thus $(x_1, \ldots, x_m) \in r^{D(\mathbf{s})}$ and $(\gamma(x_1), \ldots, \gamma(x_m)) \notin r$ for some $x_1, \ldots, x_m \in D(\mathbf{s})$. Since x_j is essentially unary, we have $x_j = e_j \circ \rho_{\varepsilon(j)}$ where e_j is a non-extendable partial endomorphism of $\underline{\mathbf{M}}$ satisfying $\mathrm{im}(\rho_{\varepsilon(j)}) \subseteq \mathrm{dom}(e_j)$ and $\varepsilon\colon \{1, \ldots, m\} \to \{1, \ldots, n\}$. Hence, by the **r**-on-**s** Lemma,

$$(e_1 \circ \rho_{\varepsilon(1)}) \sqcap \cdots \sqcap (e_m \circ \rho_{\varepsilon(m)})(s) \subseteq r$$

and

$$(\gamma(e_1 \circ \rho_{\varepsilon(1)}), \ldots, \gamma(e_m \circ \rho_{\varepsilon(m)})) \notin r.$$

Define $D := (\mathrm{dom}(e_1) \times \cdots \times \mathrm{dom}(e_m))^{\varepsilon}$, or more explicitly,

$$D := \{(c_1, \ldots, c_n) \in M^n \mid (\forall j \in \{1, \ldots, m\})\ c_{\varepsilon(j)} \in \mathrm{dom}(e_j)\}.$$

Let $\overline{\rho}_i := \pi_i\restriction_D$ for all $i \in \{1,\dots,n\}$. Clearly $\mathrm{im}(\overline{\rho}_{\varepsilon(j)}) \subseteq \mathrm{dom}(e_j)$ and hence the composite $e_j \circ \overline{\rho}_{\varepsilon(j)} \colon D \to M$ is well defined. Since $\mathrm{im}(\rho_{\varepsilon(j)}) \subseteq \mathrm{dom}(e_j)$ for each j, the composite $e_j \circ \rho_{\varepsilon(j)} \colon s \to M$ is defined and $s \subseteq D$, whence $e_j \circ \overline{\rho}_{\varepsilon(j)}$ is an extension of $e_j \circ \rho_{\varepsilon(j)}$. Also, since $\gamma(e_j \circ \rho_{\varepsilon(j)})$ is defined, and equals $e_j(a_{\varepsilon(j)})$, we know that $a_{\varepsilon(j)} \in \mathrm{dom}(e_j)$ and hence $(a_1,\dots,a_n) \in D$. We thus have

$$(e_1 \circ \overline{\rho}_{\varepsilon(1)} \sqcap \dots \sqcap e_m \circ \overline{\rho}_{\varepsilon(m)})(s) \subseteq r \text{ and } (e_1(a_{\varepsilon(1)}),\dots,e_m(a_{\varepsilon(m)})) \notin r,$$

which implies that

$$s \subseteq ((e_1 \circ \rho_{\varepsilon(1)}) \sqcap \dots \sqcap (e_m \circ \rho_{\varepsilon(m)}))^{-1}(r) \text{ and}$$
$$(a_1,\dots,a_n) \notin ((e_1 \circ \overline{\rho}_{\varepsilon(1)}) \sqcap \dots \sqcap (e_m \circ \overline{\rho}_{\varepsilon(m)}))^{-1}(r).$$

Finally,

$$\begin{aligned} s &\subseteq ((e_1 \circ \overline{\rho}_{\varepsilon(1)}) \sqcap \dots \sqcap (e_m \circ \overline{\rho}_{\varepsilon(m)}))^{-1}(r) \\ &= \{(c_1,\dots,c_n) \in (\mathrm{dom}(e_1) \times \dots \times \mathrm{dom}(e_n))^{\varepsilon} \mid \\ &\qquad\qquad (e_1(c_{\varepsilon(1)}),\dots,e_m(c_{\varepsilon(m)})) \in r\} \\ &= ((e_1 \times \dots \times e_m)^{-1}(r))^{\varepsilon}, \end{aligned}$$

as required. ■

Now let $\mathfrak{T}_s$ be the set of all algebraic relations $r \in \mathcal{B}$ such that if r is m-ary then there exist a map $\varepsilon \colon \{1,\dots,m\} \to \{1,\dots,n\}$ and non-extendable partial endomorphisms $e_1,\dots,e_m$ of $\underline{\mathbf{M}}$ with

$$\begin{aligned} s &= ((e_1 \times \dots \times e_m)^{-1}(r))^{\varepsilon} \\ &= \{ (c_1,\dots,c_n) \in (\mathrm{dom}(e_1) \times \dots \times \mathrm{dom}(e_m))^{\varepsilon} \mid \\ &\qquad\qquad (e_1(c_{\varepsilon(1)}),\dots,e_m(c_{\varepsilon(m)})) \in r\}. \end{aligned}$$

Further details on the relations in $\mathfrak{T}_s$ are given in Exercise 8.17.

5.7 Second Failset Theorem *Let* $\mathbf{s} \leqslant \underline{\mathbf{M}}^n$ *be a value and assume that every homomorphism from* s *into* $\underline{\mathbf{M}}$ *is essentially unary. Then either*

(D) *s is a trivial expansion of a unary relation which is a value of $\underline{\mathbf{M}}$ and is the domain of a proper non-extendable partial endomorphism e of $\underline{\mathbf{M}}$, or*

(E) *s is a trivial expansion of a unary relation which is a value of $\underline{\mathbf{M}}$ and is the equaliser, $\mathrm{eq}(e,f)$, of non-extendable partial endomorphisms e and f of $\underline{\mathbf{M}}$, or*

(K) *s is a trivial expansion of a binary relation which is a value of $\underline{\mathbf{M}}^2$ and is the joint kernel, $\ker(e,f)$, of non-extendable partial endomorphisms e and f of $\underline{\mathbf{M}}$, or*

(F) *$\mathfrak{T}_s$ is a failset of s within $\mathcal{B}$, where $\mathfrak{T}_s$ is as defined above.*

Proof Assume that s is a value at $(a_1, \ldots, a_n)$ and define s^+ as in Lemma 8.5.6. If $s \subseteq s'$, where s' is a trivial expansion of the domain of a non-extendable partial endomorphism e and $(a_1, \ldots, a_n) \notin s'$, then, since s is a value at $(a_1, \ldots, a_n)$, we have $s = s'$ and so (D) holds: the domain of e is clearly a value of $\underline{\mathbf{M}}$ since s is a value of $\underline{\mathbf{M}}^n$. A similar argument shows that if $s \subseteq s'$ and $(a_1, \ldots, a_n) \notin s'$, where s' is a trivial expansion of either the equaliser or the joint kernel of two non-extendable partial endomorphisms, then (E) or (K) will hold. Consequently, if (D), (E) and (K) do not apply to s, then $(a_1, \ldots, a_n) \in s^+ \backslash s$. Thus, by Lemma 8.5.6, we have a well-defined map $\gamma \colon D(\mathbf{s}) \to M$ such that if $r \in \mathrm{Fail}_{\mathbf{s}}(\gamma)$, then there exist a map $\varepsilon \colon \{1, \ldots, m\} \to \{1, \ldots, n\}$ and non-extendable partial endomorphisms $e_1, \ldots, e_m$ of $\underline{\mathbf{M}}$ such that $s \subseteq ((e_1 \times \cdots \times e_m)^{-1}(r))^\varepsilon$ and $(a_1, \ldots, a_n) \notin ((e_1 \times \cdots \times e_m)^{-1}(r))^\varepsilon$. As $\mathbf{s}$ is a value of $\underline{\mathbf{M}}^n$ we have $s = ((e_1 \times \cdots \times e_m)^{-1}(r))^\varepsilon$. Note that $s \in \mathrm{Fail}_{\mathbf{s}}(\gamma)$ since $(\rho_1, \ldots, \rho_n) \in s^{D(\mathbf{s})}$, by Lemma 8.1.1(i), but $(\gamma(\rho_1), \ldots, \gamma(\rho_n)) = (a_1, \ldots, a_n) \notin s$. This proves that $\mathrm{Fail}_{\mathbf{s}}(\gamma) \subseteq \mathfrak{T}_s$.

Let $r \in \mathfrak{T}_s$. Thus $s = ((e_1 \times \cdots \times e_m)^{-1}(r))^\varepsilon$, for appropriate $e_1, \ldots, e_m$ and ε. It follows (see the proof of Lemma 8.5.6) that

$$(e_1 \circ \rho_{\varepsilon(1)}) \sqcap \cdots \sqcap (e_m \circ \rho_{\varepsilon(m)}) \colon s \to r$$

is well defined. Thus, by the **r**-on-**s** Lemma,

$$(e_1 \circ \rho_{\varepsilon(1)}, \ldots, e_m \circ \rho_{\varepsilon(m)}) \in r^{D(\mathbf{s})},$$

while $(\gamma(e_1 \circ \rho_{\varepsilon(1)}), \ldots, \gamma(e_m \circ \rho_{\varepsilon(m)})) \notin r$, since

$$\begin{aligned}(a_1, \ldots, a_n) \notin s = ((e_1 \times \cdots \times e_m)^{-1}(r))^\varepsilon \\ &\implies (a_{\varepsilon(1)}, \ldots, a_{\varepsilon(m)}) \notin (e_1 \times \cdots \times e_m)^{-1}(r) \\ &\implies (e_1(a_{\varepsilon(1)}), \ldots, e_m(a_{\varepsilon(m)})) \notin r \\ &\implies (\gamma(e_1 \circ \rho_{\varepsilon(1)}), \ldots, \gamma(e_m \circ \rho_{\varepsilon(m)})) \notin r.\end{aligned}$$

Thus $r \in \mathrm{Fail}_{\mathbf{s}}(\gamma)$ and hence $\mathfrak{T}_s \subseteq \mathrm{Fail}_{\mathbf{s}}(\gamma)$. ■

In order to show that $\mathfrak{T}_s$ is a locally minimal failset of s we shall need to know that $\mathfrak{T}_s$ contains no graph of a non-extendable partial endomorphism of $\underline{\mathbf{M}}$. The following two lemmas yield sufficient conditions for this to hold.

5.8 Lemma *Let e_1, e_2 and e be partial endomorphisms of $\underline{\mathbf{M}}$.*

(i) $(e_1 \times e_2)^{-1}(\mathrm{graph}(e)) = \ker(e \circ e_1, e_2)$.

(ii) *If $\varepsilon \colon \{1, 2\} \to \{1, \ldots, n\}$, then $((e_1 \times e_2)^{-1}(\mathrm{graph}(e)))^\varepsilon$ is a trivial expansion of* $\mathrm{eq}(e \circ e_1, e_2)$ *when ε is constant, and is a trivial expansion of* $\ker(e \circ e_1, e_2)$ *when ε is non-constant.*

Proof Let e_1, e_2 and e be partial endomorphisms of $\underline{\mathbf{M}}$. Then, using Exercise 8.15,

$$\begin{aligned}
&(e_1{\times}e_2)^{-1}(\mathrm{graph}(e)) \\
&\quad = (e_2 \cdot (e_1 \cdot \mathrm{graph}(e))^{\smile})^{\smile} && \text{as } (e_1 \times e_2)^{-1}(r) = (e_2 \cdot (e_1 \cdot r)^{\smile})^{\smile} \\
&\quad = (e_2 \cdot (\mathrm{graph}(e \circ e_1))^{\smile})^{\smile} && \text{as } e_1 \cdot \mathrm{graph}(e) = \mathrm{graph}(e \circ e_1) \\
&\quad = \ker(e_2, e \circ e_1)^{\smile} && \text{as } \ker(f,g) = f \cdot \mathrm{graph}(g)^{\smile} \\
&\quad = \ker(e \circ e_1, e_2) && \text{as } \ker(f,g)^{\smile} = \ker(g,f).
\end{aligned}$$

This proves (i), and (ii) follows easily. ■

We shall assume that '*s is not a trivial expansion of a unary relation*' rather than '*s is not a trivial expansion of* $\mathrm{eq}(e,f)$ *for some pair of partial endomorphisms e and f*' since every unary relation s can be written as $s = \mathrm{eq}(\mathrm{id}_s, \mathrm{id}_M)$.

5.9 Lemma *Assume that* $\mathbf{s} \leqslant \underline{\mathbf{M}}^n$ *and that* s *is not a trivial expansion of a unary relation and is not the trivial expansion of the joint kernel of two partial endomorphisms of* $\underline{\mathbf{M}}$. *Then* $\mathfrak{T}_s$ *contains no graph of a partial endomorphism of* $\underline{\mathbf{M}}$.

Proof Assume that $\mathfrak{T}_s$ contains the graph of the partial endomorphism e. Thus $s = ((e_1 \times e_2)^{-1}(\mathrm{graph}(e)))^{\varepsilon}$ for some map $\varepsilon\colon \{1,2\} \to \{1,\dots,n\}$ and non-extendable partial endomorphisms e_1 and e_2. Hence, by the previous lemma, either s is a trivial expansion of a unary relation or s is a trivial expansion of the joint kernel of two partial endomorphisms of $\underline{\mathbf{M}}$. ■

Although our results hold more generally, for the remainder of the section we shall concentrate on the case where s is binary and Ω is either $\mathbb{S}(\underline{\mathbf{M}}) \cup \mathbb{S}(\underline{\mathbf{M}}^2)$ or $\mathbb{S}(\underline{\mathbf{M}}^2)$.

5.10 Binary Minimal Failset Theorem *Let* $\Omega = \mathbb{S}(\underline{\mathbf{M}}) \cup \mathbb{S}(\underline{\mathbf{M}}^2)$ *or* $\Omega = \mathbb{S}(\underline{\mathbf{M}}^2)$. *Assume that* $\mathbf{s}$ *is a value of* $\underline{\mathbf{M}}^2$ *such that every homomorphism from* s *into* $\underline{\mathbf{M}}$ *is essentially unary,* $\mathbf{s}$ *is not of the form* $N \times M$ *nor of the form* $M \times N$ *for any subalgebra* $\mathbf{N}$ *of* $\underline{\mathbf{M}}$ *and* $\mathbf{s}$ *is not the joint kernel of two partial endomorphisms of* $\underline{\mathbf{M}}$. *Then*

$$V^s := \{\, r \mid \mathbf{r} \leqslant \underline{\mathbf{M}}^2 \text{ and } s = (e \times f)^{-1}(r) \text{ or } s = (e \times f)^{-1}(r^{\smile}) \text{ for some non-extendable partial endomorphisms } e \text{ and } f \text{ of } \underline{\mathbf{M}} \,\}$$

is a locally minimal failset of s *within* Ω *(and is the unique minimal failset of* s *within* Ω *which avoids the graphs of non-extendable partial endomorphisms of* $\underline{\mathbf{M}}$*).*

Proof Assume that s has the properties listed. Our first claim is that $V^s := \mathfrak{T}_s \cap \Omega$: the proof is left as an easy but worthwhile exercise in unravelling subscript manipulation—see Exercise 8.18. By the Second Failset Theorem, the assumptions on s guarantee that V^s is a failset within Ω and the lemma above guarantees that V^s contains no graph of a partial endomorphism of $\underline{\mathbf{M}}$.

Let U be a failset within Ω which contains s but contains no graph of a partial endomorphism of $\underline{\mathbf{M}}$. If

$$s = (e \times f)^{-1}(r) \quad \text{or} \quad s = (e \times f)^{-1}(r^{\smile})$$

for some non-extendable partial endomorphisms e, f and binary relation $r \in \Omega$, then $\{e, f, r\} \vdash s$, which may be shown directly or via Exercise 8.16. Since U contains s and does not intersect $\{e, f\}$, it follows that $r \in U$, as $\Omega \backslash U$ is entailment-closed. Hence $V^s \subseteq U$. The case where $s \in U \subseteq V^s$ shows that V^s is a locally minimal failset of s, while the case in which U is not assumed to be contained in V^s shows that V^s is the unique minimal failset of s which contains no graph of a partial endomorphism of $\underline{\mathbf{M}}$. ∎

We close this section by considering absolutely unavoidable relations in the at-most-binary case. Since we wish to use relations of minimum arity, we shall consider both $\Omega = \mathbb{S}(\underline{\mathbf{M}}) \cup \mathbb{S}(\underline{\mathbf{M}}^2)$ and $\Omega = \mathbb{S}(\underline{\mathbf{M}}^2)$. As far as absolute unavoidability is concerned, $\mathbb{S}(\underline{\mathbf{M}})$ plays no role.

5.11 Lemma *A unary relation is never absolutely unavoidable within $\mathbb{S}(\underline{\mathbf{M}}) \cup \mathbb{S}(\underline{\mathbf{M}}^2)$, and a binary relation of the form $N \times M$ or $M \times N$ for $\mathbf{N} \leqslant \underline{\mathbf{M}}$ is neither absolutely unavoidable in $\mathbb{S}(\underline{\mathbf{M}}^2)$ nor absolutely unavoidable in $\mathbb{S}(\underline{\mathbf{M}}) \cup \mathbb{S}(\underline{\mathbf{M}}^2)$.*

Proof Let N be a unary algebraic relation and define t to be N in the unary case and either $N \times M$ or $M \times N$ in the binary case. As t is entailed by the binary relation Δ_N, any failset containing t also contains Δ_N. ∎

In order to apply the Second Failset Theorem to show that $\mathfrak{T}_s \cap \Omega$ is a failset of s within Ω, we need to know that s is not the joint kernel of two non-extendable partial endomorphisms. If e is a partial endomorphism, then the graph of e equals $\ker(e, \mathrm{id}_M)$. Thus we shall assume that s is not the graph of a non-extendable partial endomorphism. When applied to the three-element Kleene algebra, the following theorem shows at once that both $\sim$ and $\preccurlyeq$ are absolutely unavoidable.

5.12 Unavoidable Binary Relation Theorem *Let $\mathbf{s}$ be a binary algebraic relation on $\underline{\mathbf{M}}$. Assume that every homomorphism from s into $\underline{\mathbf{M}}$ is essentially unary, that neither s nor $s^{\smile}$ is the graph of a non-extendable partial endomorphism of $\underline{\mathbf{M}}$, and that s is not the joint kernel* *of two non-extendable partial endomorphisms of* $\underline{\mathbf{M}}$. *Then the following conditions are equivalent:*

(i) $\{s, s^{\smile}\}$ *is a (minimal) failset in* $\mathbb{S}(\underline{\mathbf{M}}) \cup \mathbb{S}(\underline{\mathbf{M}}^2)$;

(ii) $\{s, s^{\smile}\}$ *is a (minimal) failset in* $\mathbb{S}(\underline{\mathbf{M}}^2)$;

(iii) s *is a value of* $\underline{\mathbf{M}}^2$ *and is not a product of* M *with a subalgebra of* $\underline{\mathbf{M}}$, *and* $\mathbb{S}(\underline{\mathbf{M}}^2)\backslash\{s, s^{\smile}\}$ *is closed under the action of the non-extendable partial endomorphisms of* $\underline{\mathbf{M}}$.

Proof Since s is binary, the equivalence of (i) and (ii) is trivial. Assume that $\{s, s^{\smile}\}$ is a (necessarily minimal) failset in $\mathbb{S}(\underline{\mathbf{M}}^2)$. The subalgebra $\mathbf{s}$ is a value of $\underline{\mathbf{M}}^2$, by Lemma 8.5.2, and is not a product of $\underline{\mathbf{M}}$ with a subalgebra of $\underline{\mathbf{M}}$, by the previous lemma. Assume that $e \cdot r \in \{s, s^{\smile}\}$ for some binary relation r and some non-extendable partial endomorphism e. Thus $\{e, r\} \vdash s$ and, since $\{s, s^{\smile}\}$ is a failset, we conclude that either $e \in \{s, s^{\smile}\}$ or $r \in \{s, s^{\smile}\}$. If neither s nor $s^{\smile}$ is the graph of a non-extendable partial endomorphism, then $e \in \{s, s^{\smile}\}$ is impossible. Thus (ii) implies (iii).

Finally assume that (iii) holds. By the Binary Minimal Failset Theorem, V^s is a locally minimal failset of s. If $r \in V^s$ then r is a binary relation such that $s = (e \times f)^{-1}(r)$ or $s^{\smile} = (e \times f)^{-1}(r)$ for some non-extendable partial endomorphisms e and f of $\underline{\mathbf{M}}$. Since $(e \times f)^{-1}(r) = (f \cdot (e \cdot r)^{\smile})^{\smile}$ and since $\mathbb{S}(\underline{\mathbf{M}}^2)\backslash\{s, s^{\smile}\}$ is closed under the action of the non-extendable partial endomorphisms of $\underline{\mathbf{M}}$, a simple calculation shows that $(e \times f)^{-1}(r) \in \{s, s^{\smile}\}$ implies $r \in \{s, s^{\smile}\}$. Hence $\{s, s^{\smile}\} = V^s$ is a (minimal) failset in $\mathbb{S}(\underline{\mathbf{M}}^2)$. Thus (iii) implies (ii). ■

8.6 Optimal dualities for varieties of distributive p-algebras

In the previous chapter, we applied the Piggyback Duality Theorem to obtain a duality for the variety $\mathcal{B}_n$ of distributive p-algebras generated by the algebra $\underline{\mathbf{B}}_n$, where $\underline{\mathbf{B}}_n$ is obtained by adjoining a new top, 1, to the Boolean lattice with 2^n elements, whose top we denote by a and whose bottom we denote by 0. The lattice homomorphism $\omega = \chi_{\{1\}}$ from $\underline{\mathbf{B}}_n$ onto the two-element bounded distributive lattice plays a vital role in the piggyback duality. Recall that the piggyback duality is given by $\underset{\sim}{\mathbf{B}}_n = \langle B_n; G, T, \mathcal{T}\rangle$, where G is a (three-element) generating set for the endomorphism monoid of $\underline{\mathbf{B}}_n$ and T is the set of all subalgebras of $\underline{\mathbf{B}}_n^2$ which are maximal in

$$\omega^{-1}(\leqslant) := \{\, (b, c) \in B_n^2 \mid \omega(b) \leqslant \omega(c)\}.$$

By considering actions of the endomorphisms on T we were able to reduce T to a set of relations, the left-packed relations in T, in one-to-one correspondence with the partitions of the natural number n. In this section we will reveal how the results of the previous section may be applied to determine many of the globally minimal failsets of $\Omega = G \cup T$, thereby allowing us to further optimise this duality. Recall from Exercise 7.10 that every $s \in T$ is of the form

$$E(g) := \{(b, g(b)) \mid b \in {\downarrow}a\} \cup \{(b, 1) \mid g(b) = a\} \cup \{(1, 1)\},$$

where $g\colon {\downarrow}a \to {\downarrow}a$ is a Boolean endomorphism. A simple consequence of this description is that if $s \in T$ then $(a, a), (a, 1), (1, 1) \in s$ but $(1, a) \notin s$, and that $|s| \geqslant 2^n + 2$. In fact, s is a value at $(1, a)$.

6.1 Lemma *A subalgebra* $\mathbf{s}$ *of* $\underline{\mathbf{B}}_n^2$ *is a value at* $(1, a)$ *if and only if* $s = B_n \times \{0, 1\}$ *or* s *is maximal in* $\omega^{-1}(\leqslant)$.

Proof Assume that $s \in T$ and assume that $\mathbf{r}$ is a subalgebra of $\underline{\mathbf{B}}_n^2$ with $s \subsetneqq r \subseteq 2$. Then $r \nsubseteq \omega^{-1}(\leqslant)$, whence there exists $(b, c) \in r$ such that $(\omega(b), \omega(c)) = (1, 0)$. Since $\omega(b) = 1$ implies that $b = 1$, we have $(1, c) \in r$. Also $c < 1$ so $(1, a) = (1, c) \vee (a, a) \in r$ (since $(a, a) \in s$). Thus any $s \in T$ is indeed a value at $(1, a)$. Since $\{0, 1\}$ is a value of $\underline{\mathbf{B}}_n$ at a it follows that $\underline{\mathbf{B}}_n \times \{0, 1\}$ is also a value at $(1, a)$.

Conversely, assume that $\mathbf{s}$ is a value of $\underline{\mathbf{B}}_n^2$ at $(1, a)$. If $\pi_2(s) = \{0, 1\}$ then, since s is maximal with respect to not containing $(1, a)$, we have $s = B_n \times \{0, 1\}$. Now assume that $\pi_2(s) \supsetneqq \{0, 1\}$, i.e. $a \in \pi_2(s)$. Thus there exists $b < 1$ such that $(b, a) \in s$. If $s \nsubseteq \omega^{-1}(\leqslant)$, then there exists $c \leqslant a$ such that $(1, c) \in s$. Then $(1, a) = (1, c) \vee (b, a) \in s$, contrary to the assumption that $(1, a) \notin s$. Thus $s \subseteq \omega^{-1}(\leqslant)$. No subalgebra of $\omega^{-1}(\leqslant)$ contains $(1, a)$, so that the assumption that $\mathbf{s}$ is a value at $(1, a)$ forces s to be maximal in $\omega^{-1}(\leqslant)$. ■

Define $T^{\smile} := \{\, s^{\smile} \mid s \in T\,\}$. Our aim now is to show that the isomorphism classes partition the set $T \cup T^{\smile}$ into failsets which are globally minimal within both $\mathbb{S}(\underline{\mathbf{B}}_n^2)$ and $\mathbb{S}(\underline{\mathbf{B}}_n) \cup \mathbb{S}(\underline{\mathbf{B}}_n^2)$. We shall need the following lemma which gives valuable information on the partial endomorphisms of $\underline{\mathbf{B}}_n$ and how they act on the set T. For any subset r of B_n^2, let $\mathrm{conv}\colon r \to r^{\smile}$ be the natural map which sends (b, c) to (c, b).

6.2 Lemma

(i) *Let* e *be a partial endomorphism of* $\underline{\mathbf{B}}_n$. *Then either*

(a) *e is one-to-one, or*

(b) $e{\restriction}_{({\downarrow}a \cap \mathrm{dom}(e))}$ *is a Boolean homomorphism and the range of* e *is* $\{0, 1\}$.

(ii) *Let* $s \in T$ *and let* r *be a binary algebraic relation on* $\underline{\mathbf{B}}_n$. *Assume that* e *and* f *are partial endomorphisms of* $\underline{\mathbf{B}}_n$ *such that* $(e \times f)^{-1}(r) = s$. *Then*

(a) *e and f are one-to-one, and*

(b) *r belongs to T and $(e \times f)\restriction_{\mathbf{s}}\colon \mathbf{s} \to \mathbf{r}$ is an isomorphism.*

(iii) *Let $s \in T$ and let r be a binary algebraic relation on $\underline{\mathbf{B}}_n$. Assume that $u\colon \mathbf{s} \to \mathbf{r}$ is an isomorphism. Then*

(a) *there are one-to-one partial endomorphisms e and f of $\underline{\mathbf{B}}_n$ such that $u = (e \times f)\restriction_{\mathbf{s}}$, in which case $s = (e \times f)^{-1}(r)$, or $u = \mathrm{conv} \circ (e \times f)\restriction_{\mathbf{s}}$, in which case $s = (e \times f)^{-1}(r^{\smallsmile})$, and*

(b) *r belongs to $T \cup T^{\smallsmile}$.*

Proof Part (i) is left as an exercise. Assume that $s \in T$. Let e and f be partial endomorphisms of $\underline{\mathbf{B}}_n$ such that $s = (e \times f)^{-1}(r)$ for some binary algebraic relation r. Since $(a, a) \in s$, we have $a \in \mathrm{dom}(e) \cap \mathrm{dom}(f)$. By (i), in order to prove that f is one-to-one it suffices to show that if $b \in \mathrm{dom}(f)$ and $f(b) = 0$, then $b = 0$. But

$$b \in \mathrm{dom}(f) \text{ and } f(b) = 0 \implies (e(0), f(b)) = (0, 0) \in r \implies (0, b) \in s.$$

Since $s \subseteq \omega^{-1}(\leqslant)$, we have $(1, c) \in s$ implies $c = 1$. Thus

$$(0, b) \in s \implies (1, b^*) = (0, b)^* \in s \implies b^* = 1 \implies b = 0.$$

Hence f is one-to-one, whence in particular $f(a) = a$. If e is not one-to-one, then $e(a) = e(1) = 1$, whence

$$(a, a) \in s \implies (e(a), f(a)) \in r \implies (e(1), f(a)) \in r \implies (1, a) \in s,$$

which contradicts the fact that $s \subseteq \omega^{-1}(\leqslant)$. Therefore, e is also one-to-one. Consequently, (ii)(a) holds and we turn to (ii)(b).

Since $a \in \mathrm{dom}(f)$ and f is one-to-one, we have $f(a) = a$. Hence, $(1, a) \in \mathrm{dom}(e \times f)$ and $(1, a)$ is a fixpoint of $e \times f$. As $(1, a) \notin s$ we conclude that $(1, a) \notin r$. Let $\mathbf{t}$ be a value of $\underline{\mathbf{B}}_n^2$ at $(1, a)$ with $r \subseteq t$. Thus, by (ii)(a), the map $(e \times f)\restriction_{\mathbf{s}}$ embeds $\mathbf{s}$ into $\mathbf{r}$ and hence into $\mathbf{t}$. By Lemma 8.6.1, either $t \in T$ or $t = B_n \times \{0, 1\}$. The latter is impossible since $(a, a) \in r \subseteq t$ as $(a, a) \in s$ and (a, a) is a fixpoint of $e \times f$. Hence $t \in T$. Since $s \subseteq (e \times f)^{-1}(t)$ and $(1, a) \notin (e \times f)^{-1}(t)$ and since s is a value of $\underline{\mathbf{B}}_n^2$ at $(1, a)$, we must have $s = (e \times f)^{-1}(t)$. Consequently, $(e \times f)\restriction_{\mathbf{s}}\colon \mathbf{s} \to \mathbf{t}$ is an isomorphism, whence $r = t \in T$ and $(e \times f)\restriction_{\mathbf{s}}\colon \mathbf{s} \to \mathbf{r}$ is an isomorphism, as required.

Now consider (iii). Let $s \in T$ and let $u\colon \mathbf{s} \to \mathbf{r}$ be an isomorphism for some binary algebraic relation r. Denote the restrictions of the projections to r and s by ρ_1^r, ρ_2^r and ρ_1^s, ρ_2^s, respectively. By Lemma 8.5.5, there are partial endomorphisms e_1 and e_2 and a map $\varepsilon\colon \{1, 2\} \to \{1, 2\}$ such that $\rho_i^r \circ u = e_i \circ \rho_{\varepsilon(i)}^s$, for $i = 1, 2$, that is,

$$u((b, c)) = (e_1(\rho_{\varepsilon(1)}^s((b, c))), e_2(\rho_{\varepsilon(2)}^s((b, c)))).$$

If ε is constant with value 1, then u depends only on the first coordinate, whence $u(a,a) = u(a,1)$, contradicting the fact that u is an isomorphism. Similarly, if ε is constant with value 2, then we obtain the contradiction $u(a,1) = u(1,1)$. Hence, either $\varepsilon = \mathrm{id}_{\{1,2\}}$ or ε is the 2-cycle (1 2).

Assume that $\varepsilon = \mathrm{id}_{\{1,2\}}$. Then $u = (e_1 \times e_2){\restriction}_{\mathbf{s}}$. If e_1 is not one-to-one, then $e_1(a) = e_1(1) = 1$, which yields the contradiction

$$u((a,1)) = (e_1(a), e_2(1)) = (1,1) = u((1,1)).$$

In the same way, if e_2 is not one-to-one, we obtain the contradiction

$$u((a,a)) = (e_1(a), e_2(a)) = (e_1(a), e_2(1)) = u((a,1)).$$

Thus both e_1 and e_2 are one-to-one from which it follows that $s = (e \times f)^{-1}(r)$. Finally, assume that $\varepsilon = (1\ 2)$. Then $u = \mathrm{conv} \circ (e_2 \times e_1){\restriction}_{\mathbf{s}}$. A little symmetry applied to the calculations above shows that both e_1 and e_2 are one-to-one and consequently $s = (e_2 \times e_1)^{-1}(r^{\smallsmile})$. Hence, (iii)(a) holds and (iii)(b) follows at once from (ii)(b). ■

The set V^s was defined in the Binary Minimal Failset Theorem 8.5.10.

6.3 Proposition *Let $\Omega = \mathbb{S}(\underline{\mathbf{B}}_n) \cup \mathbb{S}(\underline{\mathbf{B}}_n^2)$ or $\Omega = \mathbb{S}(\underline{\mathbf{B}}_n^2)$. The globally minimal failsets in Ω which avoid the graphs of non-extendable partial endomorphisms of $\underline{\mathbf{B}}_n$ partition $T \cup T^{\smallsmile}$ into isomorphism classes. Indeed,*

(i) *if $s \in T$, then $V^s = \{\, r \mid \mathbf{r} \leqslant \underline{\mathbf{B}}_n^2 \text{ and } \mathbf{r} \cong \mathbf{s} \,\}$ and $V^s \subseteq T \cup T^{\smallsmile}$, and*

(ii) *a subset U of Ω is a globally minimal failset which avoids the graphs of non-extendable partial endomorphisms of $\underline{\mathbf{B}}_n$ if and only if $U = V^s$ for some $s \in T$.*

Proof Let $s \in T$. By definition, a binary algebraic relation r is in V^s if and only if $s = (e \times f)^{-1}(r)$ or $s = (e \times f)^{-1}(r^{\smallsmile})$ for some non-extendable partial endomorphisms e and f of $\underline{\mathbf{B}}_n$. Hence (i) follows immediately from the previous lemma. In order to prove (ii), we must first show that the assumptions of the Binary Minimal Failset Theorem 8.5.10 hold. By Lemma 8.5.5, every homomorphism from $\mathbf{s}$ into $\underline{\mathbf{M}}$ is essentially unary. Since $(1,1)$ is join-irreducible in $\mathbf{s}$, we conclude that s is not a product. Suppose that $s = \ker(e,f)$ for some partial endomorphisms e and f. Then

$$\begin{aligned}(a,a),(a,1) \in s = \ker(e,f) &\implies e(a) = f(a)\ \&\ e(a) = f(1) = 1 = e(1)\\ &\implies e(1) = f(a) = 1 \implies (1,a) \in \ker(e,f) = s,\end{aligned}$$

a contradiction. Thus the Binary Minimal Failset Theorem applies and consequently V^s is the unique minimal failset of s which avoids the graphs of non-extendable partial endomorphisms of $\underline{\mathbf{B}}_n$. By the Globally Minimal Failset Theorem 8.3.8, in order to prove that V^s is a globally minimal failset, it suffices to show that it is a locally minimal failset of each $r \in V^s$. Let $r \in V^s$. Then either $r \in T$ or $r^{\smile} \in T$. Since r and $r^{\smile}$ have exactly the same failsets, there is no loss of generality in assuming that $r \in T$. Applying the argument above to r, we see that $V^s = V^r$ is a locally minimal failset of r. Thus V^s is a globally minimal failset which avoids the graphs of non-extendable partial endomorphisms of $\underset{\sim}{\mathbf{B}}_n$.

To prove the converse, assume that U is a globally minimal failset in Ω which avoids the graphs of non-extendable partial endomorphisms of $\underset{\sim}{\mathbf{B}}_n$. Since $\mathrm{End}(\underset{\sim}{\mathbf{B}}_n) \cup T$ yields a duality on $\mathcal{B}_n$ and since U contains no graph of an endomorphism, the Optimal Duality Theorem 8.3.10 implies that there exists $s \in U \cap T$. As $s \in U$, the set U is a minimal failset of s which avoids the graphs of non-extendable partial endomorphisms. As $s \in T$, the Binary Minimal Failset Theorem guarantees that V^s is the unique such failset, whence $U = V^s$, as claimed. ∎

We conclude at once from this result that

(1) any (optimal) duality for $\mathcal{B}_n$ must include a transversal R of the isomorphism classes in $T \cup T^{\smile}$, and

(2) if R is any such transversal, G is a generating set for the monoid $\mathrm{End}(\underset{\sim}{\mathbf{B}}_n)$ and H is the set of non-extendable proper partial endomorphisms of $\underset{\sim}{\mathbf{B}}_n$, then $G \cup H \cup R$ yields a duality on $\mathcal{B}_n$.

In fact, much more is true. In Davey and Priestley [DP93b], the restricted Priestley duality is used to prove that if G is a generating set for $\mathrm{End}(\underset{\sim}{\mathbf{B}}_n)$ and R is a transversal of the isomorphism classes in $T \cup T^{\smile}$, then $G \cup R$ yields a duality on $\mathcal{B}_n$, that is, the set H of non-extendable proper partial endomorphisms can be dispensed with. This is further refined in Davey and Priestley [DP96a] where all the globally minimal failsets in $\mathbb{S}(\mathbf{B}_n^2)$ and in $\mathbb{S}(\mathbf{B}_n) \cup \mathbb{S}(\mathbf{B}_n^2)$ are described. It is proved, for example, that the non-extendable proper partial endomorphisms of $\underset{\sim}{\mathbf{B}}_n$ are absolutely avoidable and that the set U^N of the First Failset Theorem 8.5.3 (with $N = \{0,1\}$) is a globally minimal failset and contains all the endomorphisms of $\underset{\sim}{\mathbf{B}}_n$ with codomain $\{0,1\}$.

Exercises 8

Exercises from the text. Complete the proof of 8.1.5. Prove the claim, made in the proof of 8.3.2, that $\mathrm{Fail}_{\mathbf{N}}(\gamma) = \{N\}$. Complete the proofs of Proposition 8.3.4 and Corollary 8.3.6. Verify all the unproved assertions in 8.4.2. Show that U^{K_0} as defined in 8.4.1 equals U^{K_0} with U and K_0 as defined in the preamble to 8.4.1. Fill in the missing details in the proof of Lemma 8.5.5. Prove the claim in 8.6.2(i)—either do this algebraically or appeal to the restricted Priestley duality as done in Chapter 7.

8.1 Prove that if $\mathbf{s} \leqslant \underline{\mathbf{M}}^n$ and $\mathbf{s}$ is a retract of $\underline{\mathbf{M}}$, then $\mathrm{End}(\underline{\mathbf{M}}) \vdash s$. [Hint: Use Exercise 2.4(i), Exercise 2.3 and Corollary 8.1.4.]

8.2 Use Theorem 7.3.3 and Exercise 8.1 to prove that the Hetying algebra $\mathbf{2}^2 \oplus \mathbf{1}$ is endodualisable.

8.3 [DP00] Prove that every finite Heyting algebra in $\mathbb{ISP}\underline{\mathbf{C}}_n$, for each $n \in \mathbb{N}$, and every finite Heyting algebra in $\mathbb{ISP}(\mathbf{2}^2 \oplus \mathbf{1})$ is endodualisable. [Hint: Use Theorem 7.8.2 and the fact that $\underline{\mathbf{C}}_n$ and $\mathbf{2}^2 \oplus \mathbf{1}$ are projective Heyting algebras (see Balbes and Horn [BH70]).]

8.4 Use the Injectivity and Boolean Powers Theorem 5.5.15 and the fact that $\underline{\mathbf{C}}_n$ and $\mathbf{2}^2 \oplus \mathbf{1}$ are endodualisable to describe the injectives, weak injectives and absolute subretracts in the (quasi-)varieties they generate. (The case of $\underline{\mathbf{C}}_n$, which is due to Davey [Dav76], led the authors to the general theorem presented in Chapter 5.)

8.5 [DHP95b] Let $\underline{\mathbf{L}} = \langle L; \vee, \wedge, 0, 1\rangle$ be a finite non-trivial Boolean lattice and let R be a set of finitary algebraic relations on $\underline{\mathbf{L}}$. Prove that if R yields a duality on the variety of bounded distributive lattices, then R includes a non-Boolean relation. [Hint: Let

$$U := \{\, r \subseteq L^n \mid n \in \mathbb{N} \text{ and } \mathbf{r} \leqslant \underline{\mathbf{L}}^n \text{ is non-Boolean}\,\}.$$

Let $D = \{0,1\} \subseteq L$ and define $\gamma\colon D(\leqslant_D) \to L$ by declaring $\gamma(x)$ to be the complement of $x((0,1))$ in $\underline{\mathbf{L}}$. Prove that $U = \mathrm{Fail}_{\leqslant_D}(\gamma)$ and that U is a locally minimal failset of each of its members, whence U is a minimal unavoidable set in $\mathcal{B}$.]

8.6 Let Ω be a set and let $\vdash$ be a relation between $\wp(\Omega)$ and Ω. For $R \subseteq \Omega$, define $\overline{R} := \{\, s \in \Omega \mid R \vdash s\,\}$. Show that this defines a closure operation on Ω if and only if for all $R, S \subseteq \Omega$ and all $s, t \in \Omega$,

(a) $\{s\} \vdash s$,

(b) if $R \vdash s$ and $R \subseteq S$, then $S \vdash s$, and

(c) if $R \vdash s$ for all $s \in S$ and $S \vdash t$, then $R \vdash t$.

8.7 Let $\underline{\mathbf{D}} = \langle\{0,1\}; \vee, \wedge\rangle$. Prove that $\leqslant$ is absolutely unavoidable within $\mathbb{S}(\underline{\mathbf{D}}) \cup \mathbb{S}(\underline{\mathbf{D}}^2)$. [Hint: Find a map $\gamma\colon D(\leqslant) \to \{0,1\}$ such that $\mathrm{Fail}_{\leqslant}(\gamma) = \{\leqslant, \geqslant\}$.]

8.8 Let $\underline{\mathbf{S}} = \langle\{0,a,1\}; \vee, \wedge, {}^*, 0, 1\rangle$ be the three-element Stone algebra. In 4.3.6 we showed that $\underset{\sim}{\mathbf{S}} = \langle\{0,a,1\}; d, \preccurlyeq, \mathcal{T}\rangle$ yields a strong duality on the quasi-variety $\mathcal{S} = \mathbb{ISP}\underline{\mathbf{S}}$, where d is the endomorphism of $\underline{\mathbf{S}}$ given by $0 \mapsto 0$ and $1, a \mapsto 1$ and $\preccurlyeq\, = \{(0,0),(a,a),(1,1),(1,a)\}$ is an order on S. Let $\Omega = \mathbb{S}(\underline{\mathbf{S}}) \cup \mathbb{S}(\underline{\mathbf{S}}^2)$ and $S_0 = \{0,1\}$. Prove each of the following claims.

(i) The lattice $\mathbb{S}(\underline{\mathbf{S}}^2)$ is isomorphic to the corresponding lattice, given in Figure 8.1 (p.247), for the three-element Kleene algebra.

(ii) The order $\preccurlyeq$ is absolutely unavoidable within Ω.

(iii) The set $\{ S_0, \Delta_{S_0}, S_0^2, S_0 \times S, S \times S_0, \text{graph}(d), \text{graph}(d)^{\smile} \}$ is a minimal unavoidable set within Ω.

(iv) The relation $\ker(d)$ is absolutely avoidable within Ω. (It is interesting to observe that the corresponding element of the lattice $\mathbb{S}(\underline{\mathbf{K}}^2)$ in the Kleene case is absolutely unavoidable.)

(v) Up to replacing a binary relation by its converse, there is a unique optimal duality for $\mathcal{S}$ using relations of minimal arity, namely the duality given by $\underset{\sim}{\mathbf{S}}' = \langle \{0, a, 1\}; \preccurlyeq, S_0, \mathcal{T} \rangle$.

(vi) $\underset{\sim}{\mathbf{S}}'$ does not yield a full duality on $\mathcal{S}$.

(vii) Up to replacing a binary relation by its converse, there is a unique optimal strong duality for $\mathcal{S}$ using operations and relations of minimal arity, namely the duality given by $\underset{\sim}{\mathbf{S}} = \langle \{0, a, 1\}; d, \preccurlyeq, \mathcal{T} \rangle$.

8.9 Let $r, s \in \Omega \subseteq \mathcal{B}$. Prove each of the following claims. In parts (iii) to (vi), assume that Ω is $\mathbb{S}$-closed.

(i) If $u \colon \mathbf{s} \to \mathbf{r}$ is a homomorphism and $\gamma \colon D(\mathbf{s}) \to M$ is any map, then $\text{Fail}_{\mathbf{r}}(\lambda) \subseteq \text{Fail}_{\mathbf{s}}(\gamma)$, where $\lambda := \gamma \circ D(u)$.

(ii) Let $\mathbf{s}$ be a retract of $\mathbf{r}$, with associated maps $u \colon \mathbf{s} \to \mathbf{r}$ and $v \colon \mathbf{r} \to \mathbf{s}$ with $v \circ u = \text{id}_s$. Then, for every map $\gamma \colon D(\mathbf{s}) \to M$, we have $\text{Fail}_{\mathbf{r}}(\lambda) = \text{Fail}_{\mathbf{s}}(\gamma)$, where $\lambda := \gamma \circ D(u)$.

(iii) Let $x_1, \dots, x_n \in D(\mathbf{s})$ witness $r \in \text{Fail}_{\mathbf{s}}(\gamma)$ and define $r' := (x_1 \sqcap \dots \sqcap x_n)(s)$. Then $r' \subseteq r$ and $r' \in \text{Fail}_{\mathbf{s}}(\gamma)$.

(iv) Let $x_1, \dots, x_n \in D(\mathbf{s})$ witness $r \in \text{Fail}_{\mathbf{s}}(\gamma)$ and assume that r is minimal in $\text{Fail}_{\mathbf{s}}(\gamma)$. Then $x_1 \sqcap \dots \sqcap x_n \colon \mathbf{s} \to \mathbf{r}$ is surjective.

(v) Let $x_1, \dots, x_n \in D(\mathbf{s})$ witness $s \in \text{Fail}_{\mathbf{s}}(\gamma)$ and assume that s is minimal in $\text{Fail}_{\mathbf{s}}(\gamma)$. Then $x_1 \sqcap \dots \sqcap x_n \colon \mathbf{s} \to \mathbf{s}$ is an automorphism of $\mathbf{s}$.

(vi) Assume that s is minimal in $\text{Fail}_{\mathbf{s}}(\gamma)$. Then there exists a map $\lambda \colon D(\mathbf{s}) \to M$ such that $s \in \text{Fail}_{\mathbf{s}}(\lambda) \subseteq \text{Fail}_{\mathbf{s}}(\gamma)$ and $s \in \text{Fail}_{\mathbf{s}}(\lambda)$ is witnessed by $\rho_1, \dots, \rho_n \in D(\mathbf{s})$.

8.10 Let $r, s \in \Omega \subseteq \mathcal{B}$. Show that the following are equivalent:

(i) $r \vdash s$;

(ii) every failset containing s also contains r;

(iii) $r \in \bigcap \mathcal{F}_s$;

(iv) every minimal failset of s contains a minimal failset of r;

(v) every failset of s contains a failset of r.

8.11 In a complete lattice $\mathbf{L}$, an element x is **completely meet-prime** (CMP) if $\bigwedge\{ y_i \mid i \in I\} \leqslant x$ implies $y_i \leqslant x$ for some $i \in I$. **Completely join-prime** (CJP) is defined dually. Completely meet-prime elements and completely join-prime elements occur in matched pairs: if x is CMP, then $y := \bigwedge L \backslash {\downarrow} x$ is CJP, and if y is CJP, $x := \bigvee L \backslash {\uparrow} y$ is CMP. If $L = {\downarrow} x \,\dot{\cup}\, {\uparrow} y$ then x is CMP and y is CJP. Under these circumstances we refer to (x, y) as a **completely prime pair**. Let $\boldsymbol{\Lambda}$ be the lattice of closed sets of some closure operator on a set Ω and define $R \vdash s$ to mean $s \in \overline{R}$.

(i) Show that a closed set S is completely join-irreducible in $\boldsymbol{\Lambda}$ if and only if $S = \overline{\{s\}}$ for some $s \in \Omega$ and

$$S_* := \overline{\{s\}} \cap (\Omega \backslash \Psi_s) = \{ r \in \Omega \mid s \vdash r \text{ and } r \nvdash s \}$$

is closed.

(ii) Show that if (R, S) is a completely prime pair, then $R = \Omega \backslash \Psi_s$ and $S = \overline{\{s\}}$ for some $s \in \Omega$. Conversely, if $\Omega \backslash \Psi_s$ is closed, then $(\Omega \backslash \Psi_s, \overline{\{s\}})$ is a completely prime pair.

8.12 Let $s \in \Omega \subseteq \mathcal{B}$. Use the previous exercise to show that the following are equivalent:

(i) Ψ_s is a failset of s;

(ii) Ψ_s is the unique minimal failset of s;

(iii) $\Omega \backslash \Psi_s$ is closed;

(iv) $\overline{\{s\}}$ is CJP in $\boldsymbol{\Lambda}$;

(v) $\Omega \backslash \Psi_s$ is CMP in $\boldsymbol{\Lambda}$;

(vi) $\Omega \backslash \Psi_s$ is CMI in $\boldsymbol{\Lambda}$;

(vii) Ψ_s is a (weak) failset.

8.13 Let U be a subset of Ω. Show that the following are equivalent:

(i) U is the unique minimal failset of s for some $s \in \Omega$;

(ii) U is a failset and $U = \Psi_s$ for some $s \in \Omega$;

(iii) $\Omega \backslash U$ is CMP in $\boldsymbol{\Lambda}$.

8.14 (This exercise generalises the First Failset Theorem.) Let $\mathbf{N} \leqslant \underline{\mathbf{M}}$ and assume that Ω contains $\mathbb{S}(\underline{\mathbf{M}})$ and the (graphs of) all partial endomorphisms of $\underline{\mathbf{M}}$. Assume that $\mathbf{N}$ has exactly ℓ upper covers in $\mathbb{S}(\underline{\mathbf{M}})$, *viz.* $[N \cup \{d_1\}], \dots, [N \cup \{d_\ell\}]$. Then the following are equivalent:

(i) the inclusion map inc_N of N into M is the only homomorphism from $\mathbf{N}$ into $\underline{\mathbf{M}}$ (that is, $|D(\mathbf{N})| = 1$);

(ii) $\Psi_N = U^N$ and N has exactly ℓ minimal failsets, namely the failsets $\mathrm{Fail}_{\mathbf{N}}(\gamma_1), \dots, \mathrm{Fail}_{\mathbf{N}}(\gamma_\ell)$, where $\gamma_i \colon D(\mathbf{N}) \to M$ is the constant map onto d_i.

8.15 Let f and g be partial endomorphisms of $\underline{\mathbf{M}}$ and let $r \in \mathcal{B}$ be binary. Verify each of the following equalities which were used in the proof of Lemma 8.5.8.

(i) $(f \times g)^{-1}(r) = (g \cdot (f \cdot r)^{\smile})^{\smile}$.

(ii) $f \cdot \mathrm{graph}(g) = \mathrm{graph}(g \circ f)$.

(iii) $\ker(f, g) = f \cdot \mathrm{graph}(g)^{\smile}$.

8.16 Let $r \in \mathcal{B}$ be m-ary, let $e_1, \dots, e_m$ be partial endomorphisms of $\underline{\mathbf{M}}$ and let σ be the cycle $(m \dots 2\ 1)$. Prove that

$$e_1 \cdot r = (e_1 \times \mathrm{id}_M \times \cdots \times \mathrm{id}_M)^{-1}(r)$$

and that

$$(e_1 \times \cdots \times e_m)^{-1}(r) = (e_m \cdot \ldots \cdot (e_2 \cdot (e_1 \cdot r)^{\sigma})^{\sigma} \ldots)^{\sigma}.$$

8.17 Let $s \in \mathcal{B}$ be n-ary. Prove the following claims concerning $\mathfrak{T}_s$.

(i) If an m-ary relation r belongs to $\mathfrak{T}_s$, then s can be obtained from r by application of the action of non-extendable partial endomorphisms alternating with m applications of permutation of subscripts followed by a single application of m-to-n subscript manipulation.

(ii) Assume that $\mathfrak{T}_s$ is a failset of s (within $\mathcal{B}$) and $\mathfrak{T}_s$ contains no non-extendable partial endomorphism of $\underline{\mathbf{M}}$. Then a relation $r \in \mathcal{B}$ is in $\mathfrak{T}_s$ if and only if s can be obtained from r via a finite number of applications of the action of non-extendable partial endomorphisms of $\underline{\mathbf{M}}$ and subscript manipulation.

8.18 Given partial endomorphisms e and f of $\underline{\mathbf{M}}$ and $r \subseteq M^2$, define

$$(e \boxtimes f)^{-1}(r) := \{\, c \in M \mid c \in \mathrm{dom}(e) \cap \mathrm{dom}(f) \text{ and } (e(c), f(c)) \in r \,\}.$$

(i) Let s be unary. Prove that a unary relation r belongs to $\mathfrak{T}_s$ if and only if $s = e^{-1}(r)$ for some non-extendable partial endomorphism e of $\underline{\mathbf{M}}$. A binary relation r is in $\mathfrak{T}_s$ if and only if $s = (e \boxtimes f)^{-1}(r)$ for some non-extendable partial endomorphisms e and f of $\underline{\mathbf{M}}$.

(ii) Let s be binary. Prove that a unary relation r belongs to $\mathfrak{T}_s$ if and only if $s = e^{-1}(r) \times M$ or $s = M \times e^{-1}(r)$ for some non-extendable partial endomorphism e of $\underline{\mathbf{M}}$. A binary relation r belongs to $\mathfrak{T}_s$ if and only if

$$s \in \{(e \boxtimes f)^{-1}(r) \times M, M \times (e \boxtimes f)^{-1}(r), (e \times f)^{-1}(r), (e \times f)^{-1}(r)^{\smile}\}$$

for some non-extendable partial endomorphisms e and f.

9

Completeness Theorems for Entailment

In Chapter 8 we developed an algorithmic method to find all restrictions of a given duality to an optimal duality. Even in relatively small examples, we saw that this method can be computationally tedious enough to require computer assistance. The alternative remains to use constructs for entailment, as we did in Chapter 4, to eliminate unnecessary relations from a given duality until it becomes optimal. Although this bare-hands method is not algorithmic, it continues to be preferable in many cases where the given dualising set of relations is not too large.

If we are to rely upon sets of constructs for either entailment or hom-entailment, we must be certain in advance of two facts. Our constructs must be **sound** in the sense that the relations and partial operations they produce are appropriately entailed by the given structures. They must also be **complete** in the sense that they do indeed yield all appropriately entailed relations and partial operations. Our primary aims in this chapter are to give a complete set of constructs for entailment as promised in Chapter 2, and to give a complete set of constructs for hom-entailment as promised in Chapter 3. It turns out that establishing completeness of our constructs for entailment is considerably more involved than for hom-entailment, and requires that we add several new constructs beyond those listed in Constructs for Entailment 2.4.5. We will give a purely syntactic description of the constructs for the relations entailed by $G \cup H \cup R$ in terms of first-order formulæ of type $\langle G, H, R \rangle$. This syntactic description will prove to be very useful in its own right. The results of this chapter are based on Davey, Haviar and Priestley [DHP95a].

9.1 The syntax of entailment

By a **construct** of type $\langle G, H, R \rangle$ we mean a rule which associates each structure $\underset{\sim}{\mathrm{M}} = \langle M; G, H, R, \mathcal{T} \rangle$ with an operation, partial operation or relation s on M. By an **admissible construct** we mean a construct associating $\underset{\sim}{\mathrm{M}}$ with s in such a way that

(1) s is algebraic over $\underline{\mathbf{M}}$ whenever $\underset{\sim}{\mathrm{M}}$ is algebraic over $\underline{\mathbf{M}}$ and

(2) $G \cup H \cup R$ entails s.

For example, in Constructs of Entailment 2.4.5 we listed fifteen different admissible constructs.

A simple method of generating admissible constructs uses first-order formulæ of type $\langle G, H, R \rangle$. In the presence of partial operations, satisfaction of formulæ has the usual meaning with the added proviso that each term is actually defined under the given substitution. Any formula $\Phi(x_1, x_2, \dots, x_n)$ with n free variables determines the construct which associates the structure $\utilde{\mathbf{M}}$ with the relation

$$\Phi^{\utilde{\mathbf{M}}} := \{(c_1, c_2, \dots, c_n) \in M^n \mid \utilde{\mathbf{M}} \text{ satisfies } \Phi(c_1, c_2, \dots, c_n)\}.$$

We would like to find restrictions on the form of $\Phi(x_1, x_2, \dots, x_n)$ guaranteeing that the construct it determines is admissible. A **primitive positive formula** $\Phi(x_1, \dots, x_n)$ of type $\langle G, H, R\rangle$ is an existential conjunct of atomic formulæ of type $\langle G, H, R\rangle$, *viz.*

$$(\exists y_1, \dots, y_m)\ [\beta_1(x_1, \dots, x_n, y_1, \dots, y_m) \wedge \dots \wedge \beta_k(x_1, \dots, x_n, y_1, \dots, y_m)],$$

where each $\beta_i(x_1, \dots, x_n, y_1, \dots, y_m)$ is an atomic formula whose variables are among $x_1, \dots, x_n, y_1, \dots, y_m$. If $\Phi(x_1, x_2, \dots, x_n)$ is a primitive positive formula, then we refer to $\utilde{\mathbf{M}} \mapsto \Phi^{\utilde{\mathbf{M}}}$ as a **primitive positive construct**.

We ask the reader to verify that, in the context of duality, primitive positive constructs always satisfy condition (1) for admissibility.

1.1 Lemma *If $\utilde{\mathbf{M}} = \langle M; G, H, R, \mathcal{T}\rangle$ is algebraic over $\underline{\mathbf{M}}$ and the formula $\Phi(x_1, \dots, x_n)$ is primitive positive, then $\Phi^{\utilde{\mathbf{M}}}$ is algebraic over $\underline{\mathbf{M}}$.*

Our syntactic description of entailment is very similar to a fundamental result in clone theory: the description of the relations in $\mathrm{Inv}(\mathrm{Pol}(R))$ for some set R of finitary relations on a finite set M. Indeed, we shall show in Exercises 9.3–9.5 that our entailment theorem and the clone-theoretic result have a common core.

1.2 Syntax of Entailment Theorem *Let s be an n-ary algebraic relation on $\underline{\mathbf{M}}$. The following are equivalent:*

(i) *$G \cup H \cup R$ entails s;*

(ii) *$s = \{(\alpha(\rho_1), \dots, \alpha(\rho_n)) \mid \alpha\colon D(\mathbf{s}) \to M$ preserves $G \cup H \cup R\}$;*

(iii) *there is a primitive positive formula $\Phi(x_1, x_2, \dots, x_n)$ having type $\langle G, H, R\rangle$ such that $s = \Phi^{\utilde{\mathbf{M}}}$ and $D(\mathbf{s})$ satisfies $\Phi(\rho_1, \rho_2, \dots, \rho_n)$.*

Proof Assume (i). For each $c \in s$ the evaluation map $e_{\mathbf{s}}(c)\colon D(\mathbf{s}) \to M$ preserves $G \cup H \cup R$ and hence

$$c = (c_1, \dots, c_n) = (\rho_1(c), \dots, \rho_n(c)) = (e_{\mathbf{s}}(c)(\rho_1), \dots, e_{\mathbf{s}}(c)(\rho_n)) \in s^\flat,$$

where

$$s^\flat := \{(\alpha(\rho_1), \dots, \alpha(\rho_n)) \mid \alpha\colon D(\mathbf{s}) \to M \text{ preserves } G \cup H \cup R\}.$$

Thus $s \subseteq s^\flat$. For the reverse inclusion, let $\alpha\colon D(\mathbf{s}) \to M$ preserve $G \cup H \cup R$, so that $(\alpha(\rho_1), \dots, \alpha(\rho_n)) \in s^\flat$. But $(\rho_1, \dots, \rho_n) \in s^{D(\mathbf{s})}$

and consequently $(\alpha(\rho_1), \dots, \alpha(\rho_n)) \in s$ as $G \cup H \cup R$ entails s, by (i). Thus $s = s^\flat$, whence (ii) holds.

Now assume (ii). Since there is only a finite number of maps from $D(\mathbf{s})$ to M which do not preserve $G \cup H \cup R$, there is a finite subset Υ of $G \cup H \cup R$ such that $\alpha\colon D(\mathbf{s}) \to M$ preserves $G \cup H \cup R$ if and only if α preserves Υ. Let $D(\mathbf{s}) = \{\rho_1, \dots, \rho_n, \tau_1, \dots, \tau_m\}$. For each relation $r \in R \cup \{=\}$, there is a formula $\Phi_r(x_1, \dots, x_n, y_1, \dots, y_m)$ which is a conjunct of atomic formulæ and which precisely describes the relation r on $D(\mathbf{s})$ via the correspondence $x_i \leftrightarrow \rho_i$ and $y_j \leftrightarrow \tau_j$. In a similar way, there is a formula $\Phi_e(x_1, \dots, x_n, y_1, \dots, y_m)$ describing the graph of each $e \in G \cup H$ on $D(\mathbf{s})$. Define $\Phi(x_1, \dots, x_n)$ to be

$$(\exists y_1 \dots y_m) \,\&\{\, \Phi_k(x_1, \dots, x_n, y_1, \dots, y_m) \mid k \in \Upsilon \cup \{=\}\,\}.$$

(Note that we need to include '=' since it is possible, for example, that $\rho_1 = \rho_2$: let $n = 2$ and let $s = \Delta_M$ be the diagonal.) By construction, $D(\mathbf{s})$ satisfies $\Phi(\rho_1, \dots, \rho_n)$.

We shall show that $s = \Phi^{\underset{\sim}{\mathbf{M}}}$. Let $(\alpha(\rho_1), \dots, \alpha(\rho_n)) \in s^\flat$ with $\alpha\colon D(\mathbf{s}) \to M$ a $(G \cup H \cup R)$-preserving map. Since $D(\mathbf{s})$ satisfies $\Phi(\rho_1, \dots, \rho_n)$, it follows at once that $\underset{\sim}{\mathbf{M}}$ satisfies $\Phi(\alpha(\rho_1), \dots, \alpha(\rho_n))$, whence $(\alpha(\rho_1), \dots, \alpha(\rho_n)) \in \Phi^{\underset{\sim}{\mathbf{M}}}$. Hence $s^\flat \subseteq \Phi^{\underset{\sim}{\mathbf{M}}}$. Let $(c_1, \dots, c_n) \in \Phi^{\underset{\sim}{\mathbf{M}}}$. Thus there exist $d_1, \dots, d_m \in M$ such that $\underset{\sim}{\mathbf{M}}$ satisfies the formula $\Phi_k(c_1, \dots, c_n, d_1, \dots, d_m)$ for each $k \in \Upsilon \cup \{=\}$. It follows that $\rho_i \mapsto c_i$ and $\tau_j \mapsto d_j$ yields a well-defined Υ-preserving and therefore $(G \cup H \cup R)$-preserving map from $D(\mathbf{s})$ to M, whence $(c_1, \dots, c_n) \in s^\flat$. Thus $\Phi^{\underset{\sim}{\mathbf{M}}} \subseteq s^\flat$. Hence, by (ii), we have $s = s^\flat = \Phi^{\underset{\sim}{\mathbf{M}}}$, whence (iii) holds.

Finally, assume (iii). By the Test Algebra Lemma, to prove (i) we must prove that $G \cup H \cup R$ entails s on $D(\mathbf{s})$. Let $\alpha\colon D(\mathbf{s}) \to M$ be a $(G \cup H \cup R)$-preserving map. We shall show that α preserves s. Let $z_1, \dots, z_n \in D(\mathbf{s})$, with $(z_1, \dots, z_n) \in s^{D(\mathbf{s})}$. By Lemma 8.1.1, there exists a $(G \cup H \cup R)$-preserving map $\varphi\colon D(\mathbf{s}) \to D(\mathbf{s})$ with $\varphi(\rho_i) = z_i$ for all i. Thus $\psi := \alpha \circ \varphi\colon D(\mathbf{s}) \to M$ is a $(G \cup H \cup R)$-preserving map satisfying $\psi(\rho_i) = \alpha(z_i)$ for all i. Then $\underset{\sim}{\mathbf{M}}$ satisfies $\Phi(\psi(\rho_1), \dots, \psi(\rho_n))$ since $D(\mathbf{s})$ satisfies $\Phi(\rho_1, \dots, \rho_n)$, whence (by (iii)), $(\alpha(z_1), \dots, \alpha(z_n)) \in s$. Hence α preserves s and consequently $G \cup H \cup R$ entails s on $D(\mathbf{s})$. ∎

1.3 Kleene algebras once more Consider once again the absolutely unavoidable relations $\preccurlyeq$ and $\sim$ which arise in the duality for Kleene algebras: see Section 5 of Chapter 8. It is straightforward to check that $(c_1, c_2) \in \sim$ if and only if there exists $d \in K$ such that $d \preccurlyeq c_1$ and $d \preccurlyeq c_2$. This says precisely that

$$\sim\, = \{(c_1, c_2) \in K^2 \mid \underset{\sim}{\mathbf{K}} \text{ satisfies } \Phi(c_1, c_2)\},$$

where

$$\Phi(x_1, x_2) := (\exists y)\,(y \preccurlyeq x_1 \wedge y \preccurlyeq x_2).$$

Consequently, $\sim$ is determined by the formula $\Phi(x_1, x_2)$. It follows that we can define at least one map $u\colon \sim \to K$ such that $u((c_1, c_2)) \preccurlyeq c_1$ and $u((c_1, c_2)) \preccurlyeq c_2$ for all $(c_1, c_2) \in \sim$. Thus, in the power $\underset{\sim}{\mathbf{K}}^{\sim}$, we have $u \preccurlyeq \rho_1$ and $u \preccurlyeq \rho_2$, which tells us that

$$\underset{\sim}{\mathbf{K}}^{\sim} \text{ satisfies } \Phi(\rho_1, \rho_2).$$

It turns out that the map u cannot be chosen in such a way that it is a homomorphism from $\sim$ to $\underset{\sim}{\mathbf{K}}$, that is, we must have $u \notin D(\sim)$. Indeed, in 8.4.2 we showed that $D(\sim) = \{\rho_1, \rho_2\}$. Since $(a, 0)$ and $(0, a)$ are in $\sim$ and

$$\rho_1((a,0)) = a \not\preccurlyeq 0 = \rho_2((a,0)) \text{ and } \rho_2((0,a)) = a \not\preccurlyeq 0 = \rho_1((0,a)$$

it follows that $\rho_1 \not\preccurlyeq \rho_2$ and $\rho_2 \not\preccurlyeq \rho_1$ in $D(\sim)$. Consequently, $D(\sim)$ does not satisfy $\Phi(\rho_1, \rho_2)$. Since the formula $\Phi(x_1, x_2)$ describes the relational product, $\succcurlyeq \cdot \preccurlyeq$, it follows, as was discussed parenthetically in The Globally Minimal Failsets 8.4.2, that relational product is not in general an admissible construct.

As we have just seen, existential quantifiers can prevent the construct determined by a primitive positive formula $\Phi(x_1, \ldots, x_n)$ from being admissible. Quantifier-free formulæ are much better behaved.

1.4 Lemma *Every primitive positive construct which is determined by a conjunct of atomic formulæ is admissible.*

Proof Let $\Phi(x_1, \ldots, x_n)$ be a conjunct of atomic formulæ which determines the relation $\Phi^{\underset{\sim}{\mathbf{M}}} \subseteq M^n$. Let $c = (c_1, \ldots, c_n) \in s$; then $\underset{\sim}{\mathbf{M}}$ satisfies $\Phi(\rho_1(c), \ldots, \rho_n(c))$, by the definition of s. Since $\Phi(x_1, \ldots, x_n)$ is a conjunct of atomic formulæ, it follows that $D(\mathbf{s})$ satisfies $\Phi(\rho_1, \ldots, \rho_n)$. By the Syntax of Entailment Theorem 9.1.2, $\underset{\sim}{\mathbf{M}}$ entails s. ■

A construct determined by a conjunct of atomic formulæ of type $\langle G, H, R\rangle$ defines a relation on $\underset{\sim}{\mathbf{M}}$ which depends only upon the local properties of the elements in question and not their global properties within $\underset{\sim}{\mathbf{M}}$. For this reason all constructs of this type are referred to as **local constructs**. We shall say that a construct, $\underset{\sim}{\mathbf{M}} \mapsto h$, which produces a (partial) operation, is local if the corresponding relation-producing construct, $\underset{\sim}{\mathbf{M}} \mapsto \text{graph}(h)$, is local. All of the constructs introduced in Chapter 2 are of this special type.

9.2 Completeness of constructs for entailment

In this section we introduce three new admissible constructs, and then prove that our resulting list of constructs for entailment is complete.

2.1 Constructs for Entailment Revisited As examples of local constructs, we exhibit conjuncts of atomic formulæ required to determine six of the constructs given in 2.4.5. The numbering below corresponds to the numbering of the constructs in 2.4.5.

(1) **Trivial relations** If θ is an equivalence relation on $\{1,\ldots,n\}$ then the trivial relation $\Delta^\theta := \{(c_1,\ldots,c_n) \in M^n \mid i\,\theta\,j \Rightarrow c_i = c_j\}$ is determined by the formula $\&_{(i,j)\in\theta}(x_i \approx x_j)$.

(5) **Repetition removal** If r is an n-ary relation and, for fixed i and j, we have $c_i = c_j$ for all $(c_1,\ldots,c_n) \in r$, then the $(n-1)$-ary relation

$$r'_j := \{(c_1,\ldots,c_{j-1},c_{j+1},\ldots,c_n) \in M^{n-1} \mid (c_1,\ldots,c_n) \in r\}$$

is determined by the formula $r(x_1,\ldots,x_{j-1},x_i,x_{j+1},\ldots,x_n)$.

(6) **Intersection** If r and s are n-ary relations, then $r \cap s$ is determined by the formula $r(x_1,\ldots,x_n) \wedge s(x_1,\ldots,x_n)$.

(7) **Product** Let r be n-ary and s be m-ary. Then $r \times s$ is determined by the formula $r(x_1,\ldots,x_n) \wedge s(x_{n+1},\ldots,x_{n+m})$.

(11) **Composition** If g is a k-ary partial operation and $h_1,\ldots,h_k$ are n-ary partial operations, then (the graph of) $g(h_1,\ldots,h_k)$ is defined by the formula $g(h_1(x_1,\ldots,x_n),\ldots,h_k(x_1,\ldots,x_n)) \approx y$.

(12) **Domain** If h is an n-ary partial operation, then $\mathrm{dom}(h)$ is determined by the formula $h(x_1,\ldots,x_n) \approx h(x_1,\ldots,x_n)$ since the equality $h(c_1,\ldots,c_n) = h(c_1,\ldots,c_n)$ says just that $h(c_1,\ldots,c_n)$ is defined.

In general, if $G \cup H$ is non-empty we require one further local construct which was not introduced in Chapter 2. We leave as an exercise for the reader to check that it yields, as special cases, each of the remaining relation-producing constructs of Chapter 2 and that these constructs are therefore also local.

(16) **Term manipulation** Given an m-ary relation $r \in R\cup\{=\}$ and n-ary terms $t_1,\ldots,t_m$ we may take $\Phi(x_1,\ldots,x_n)$ to be the corresponding atomic formula

$$r(t_1(x_1,\ldots,x_n),\ldots,t_m(x_1,\ldots,x_n)).$$

We call the associated construct **term manipulation**.

To complete our list of admissible constructs we introduce two non-local constructs.

(17) **Retractive projection** Given an m-ary relation r and $n \leqslant m$, let $p\colon \mathbf{r} \to \underline{\mathbf{M}}^n$ be the projection

$$p(c_1, \ldots, c_n, c_{n+1}, \ldots, c_m) = (c_1, \ldots, c_n).$$

We call $s = p(r)$ a **retractive projection of** r if $p\colon \mathbf{r} \to \mathbf{s}$ is a retraction, that is, there exists a homomorphism $q\colon \mathbf{s} \to \mathbf{r}$ such that $p \circ q = \mathrm{id}_s$. The relation s is called a **bijective projection of** r if $p\colon \mathbf{r} \to \mathbf{s}$ is a bijection. We claim that s is a retractive projection of r if and only if $D(\mathbf{s})$ satisfies $\Phi(\rho_1^s, \ldots, \rho_n^s)$ where the formula $\Phi(x_1, \ldots, x_n)$ is given by $(\exists y_{n+1} \cdots y_m)\; r(x_1, \ldots, x_n, y_{n+1}, \ldots, y_m)$. To prove the claim, first assume that s is a retractive projection of r. Define homomorphisms $\tau_i := \rho_i^r \circ q$ for $i = n+1, \ldots, m$. Then $D(\mathbf{s})$ satisfies $r(\rho_1^s, \ldots, \rho_n^s, \tau_{n+1}, \ldots, \tau_m)$. For the converse assume there exist homomorphisms $\tau_i\colon \mathbf{s} \to \underline{\mathbf{M}}$ $(i = n+1, \ldots, m)$ such that $D(\mathbf{s})$ satisfies $r(\rho_1^s, \ldots, \rho_n^s, \tau_{n+1}, \ldots, \tau_m)$. Then the map q required for s to be a retractive projection of r is

$$q := \rho_1^s \sqcap \cdots \sqcap \rho_n^s \sqcap \tau_{n+1} \sqcap \cdots \sqcap \tau_m.$$

(18) **Homomorphic relational product** Given an n-ary relation r and an m-ary relation s, construct the relational product

$$r \cdot s := \{\, (c_1, \ldots, c_{n+m-2}) \in M^{n+m-2} \mid (\exists c \in M) \\ (c_1, \ldots, c_{n-1}, c) \in r \text{ and } (c, c_n, \ldots, c_{n+m-2}) \in s \,\}.$$

We say $r \cdot s$ is a **homomorphic relational product** if there exists an $(n+m-2)$-ary homomorphism $u\colon \mathbf{r} \cdot \mathbf{s} \to \underline{\mathbf{M}}$ which satisfies $(c_1, \ldots, c_{n-1}, c) \in r$ and $(c, c_n, \ldots, c_{n+m-2}) \in s$, where $c := u(c_1, \ldots, c_{n+m-2})$, for all $(c_1, \ldots, c_{n+m-2}) \in r \cdot s$. Observe that $r \cdot s$ is a homomorphic relational product precisely when

$$D(\mathbf{r} \cdot \mathbf{s}) \text{ satisfies } (\exists \tau)(r(\rho_1, \ldots, \rho_{n-1}, \tau) \wedge s(\tau, \rho_n, \ldots, \rho_{n+m-2})).$$

2.2 Soundness of Entailment Theorem *Each of the constructs for entailment* (1) *to* (18) *is admissible.*

It remains to show that our listed constructs suffice to build any relation s entailed by $G \cup H \cup R$. We do this in three stages. Firstly, we show that every local construct may be built up from our listed local constructs. Secondly, we show that the second dual $ED(\mathbf{s})$ can be locally constructed from $G \cup H \cup R$, whether or not $G \cup H \cup R$ entails s. Thirdly, we show that if $G \cup H \cup R$ entails s, then s can be obtained from $ED(\mathbf{s})$ by a retractive projection, which is a bijective projection in the case that $G \cup H \cup R$ yields a duality on $\mathbf{s}$.

2.3 Proposition

(i) *Assume that s can be obtained from $G \cup H \cup R$ via a local construct. Then s can be obtained from $G \cup H \cup R$ by term manipulation and intersection.*

(ii) *Assume that s can be obtained from R via a local construct. Then s can be obtained from R by a finite number of applications of product, intersection, trivial relations and repetition removal.*

Proof Assume that s can be obtained from $G \cup H \cup R$ via a local construct given by the formula

$$\beta_1(x_1, \ldots, x_n) \wedge \cdots \wedge \beta_k(x_1, \ldots, x_n)$$

where each atomic formula β_i has the form

$$r(t_1(x_1, \ldots, x_n), \ldots, t_m(x_1, \ldots, x_n))$$

for some $r \in R \cup \{=\}$ and some terms $t_1, \ldots, t_m$. Then s is an intersection of k relations, each of which is obtained via term manipulation.

To prove (ii), let $G \cup H = \varnothing$. Since the only terms are the coordinate projections, term manipulation reduces to subscript manipulation. More explicitly, let the formula $\Phi(x_1, \ldots, x_n)$ which determines s be the conjunct of atomic formulæ,

$$r_1(x_{i_{11}}, \ldots, x_{i_{1n_1}}) \wedge \cdots \wedge r_k(x_{i_{k1}}, \ldots, x_{i_{kn_k}}),$$

where $r_i \in R$ for all i. Then $s = \bigcap_{j=1}^{k} r_j^{\sigma_j}$ where $\sigma_j\colon \{1, \ldots, n_j\} \to \{1, \ldots, n\}$ is given by $\sigma_j(\ell) := i_{j\ell}$. Thus we get s from R via intersection and subscript manipulation applied to relations in R. We may now appeal to Exercise 2.7(iv). ■

Let Z be a finite subset of M^T, for some non-empty set T, let $Z = \{z_1, \ldots, z_k\}$ be a fixed labelling of Z and let C be a subset of M^Z. By the **induced labelling of** C we shall mean the relation

$$[C]^\ell := \{(\alpha(z_1), \ldots, \alpha(z_k)) \in M^k \mid \alpha \in C\}.$$

2.4 Lemma *Let $Z = \{z_1, \ldots, z_k\}$ be a finite substructure of $\underset{\sim}{M}^T$ for some non-empty set T. Then the relation*

$$[E(Z)]^\ell := \{(\alpha(z_1), \ldots, \alpha(z_k)) \mid \alpha\colon Z \to M \text{ preserves } G \cup H \cup R\}$$

can be locally constructed from $G \cup H \cup R$.

Proof As above, write $Z = \{z_1, \ldots, z_k\}$. Then, exactly as in the proof of the Syntax of Entailment Theorem 9.1.2, there is a primitive positive formula $\Psi(x_1, \ldots, x_k)$ such that

$$[E(Z)]^\ell = \{\, (c_1, \ldots, c_k) \in M^k \mid M \models \Psi(c_1, \ldots, c_k) \,\}.$$

Since $z_i \leftrightarrow x_i$ is a bijection between the elements of Z and the variables in Ψ, the formula Ψ is quantifier-free and the proof is complete. ■

2.5 Lemma *Let $D(\mathbf{s}) = \{\rho_1, \ldots, \rho_n, \tau_1, \ldots, \tau_m\}$ be an enumeration of the elements of $D(\mathbf{s})$.*

(i) *If $G \cup H \cup R$ entails s, then s is a retractive projection of the induced labelling $[ED(\mathbf{s})]^\ell$ of $ED(\mathbf{s})$.*

(ii) *If $G \cup H \cup R$ yields a duality on s, then s is a bijective projection of the induced labelling $[ED(\mathbf{s})]^\ell$ of $ED(\mathbf{s})$.*

Proof Let $D(\mathbf{s}) = \{\rho_1, \ldots, \rho_n, \tau_1, \ldots, \tau_m\}$ and let

$$r = [ED(\mathbf{s})]^\ell = \{(\alpha(\rho_1), \ldots, \alpha(\rho_n), \alpha(\tau_1), \ldots, \alpha(\tau_m)) \mid \alpha \in ED(\mathbf{s})\}$$

be the induced labelling of $ED(\mathbf{s})$. Take

$$\eta\colon \{1, \ldots, n\} \to \{1, \ldots, n, n+1, \ldots, n+m\}$$

to be the inclusion map. Then

$$\begin{aligned} r_\eta &= \{(c_1, \ldots, c_n) \in M^n \mid (\exists (d_1, \ldots, d_{n+m}) \in r)\ d_1 = c_1, \ldots, d_n = c_n\} \\ &= \{(\alpha(\rho_1), \ldots, \alpha(\rho_n)) \mid \alpha \in ED(\mathbf{s})\} \\ &= s, \end{aligned}$$

by the Syntax of Entailment Theorem. For all $c \in s$,

$$\begin{aligned} &(\rho_1(c), \ldots, \rho_n(c), \tau_1(c), \ldots, \tau_m(c)) \\ &\qquad = (e_{\mathbf{s}}(c)(\rho_1), \ldots, e_{\mathbf{s}}(c)(\rho_n), e_{\mathbf{s}}(c)(\tau_1), \ldots, e_{\mathbf{s}}(c)(\tau_n)) \\ &\qquad \in [ED(\mathbf{s})]^\ell = r. \qquad \text{(Take } \alpha = e_{\mathbf{s}}(c).) \end{aligned}$$

Thus $(\rho_1, \ldots, \rho_n, \tau_1, \ldots, \tau_m) \in r^{D(\mathbf{s})}$, that is, $D(\mathbf{r}_\eta) = D(\mathbf{s})$ satisfies $\Phi(\rho_1, \ldots, \rho_n)$ where $\Phi(x_1, \ldots, x_n)$ is

$$(\exists y_{n+1}, \ldots, y_{n+m})\, r(x_1, \ldots, x_n, y_{n+1}, \ldots, y_{n+m}).$$

Hence $s = r_\eta$ is a retractive projection of $[ED(\mathbf{s})]^\ell = r$ onto its first n coordinates: see (17) in 9.2.1. If $G \cup H \cup R$ yields a duality on $\mathbf{s}$, then $|r| = |ED(\mathbf{s})| = |s| = |r_\eta|$ and hence the projection p of r onto r_η must be a bijection. ∎

The next theorem, which provides the complete set of admissible constructs for entailment as promised in Chapter 2, is now immediate from Proposition 9.2.3, Lemma 9.2.4, applied with $T = s$ and $Z = D(\mathbf{s})$, and Lemma 9.2.5.

2.6 Completeness of Entailment Theorem *If $G \cup H \cup R$ entails a relation s, then s can be obtained from $G \cup H \cup R$ by term manipulation and intersection followed by one application of retractive projection. If $G \cup H = \varnothing$, then product, intersection, trivial relations, repetition removal* *and retractive projection suffice to produce s. If $G \cup H \cup R$ yields a duality on* **s**, *then retractive projection may be replaced by retractive bijection. If $G \cup H \cup R$ entails a (partial) operation h, then h can be obtained from $G \cup H \cup R$ by term manipulation and intersection followed by one application of retractive projection and one application of* graph^{-1}.

9.3 Local entailment

In this section we seek conditions under which local constructs alone suffice to determine every relation s entailed by $G \cup H \cup R$, and in such circumstances to show quite explicitly how entailment operates. We shall do this in the context of the Unary Structure Theorem 6.2.2, leaving open the more general problem of determining when local constructs are sufficient.

Recall that $\mathcal{P}$ denotes the set of all finitary, algebraic, partial or total operations on $\underline{\mathbf{M}}$ and that $[P]$ denotes the partial clone generated by a subset P of $\mathcal{P}$. The set $\mathcal{P}$ is partially ordered by defining $t_1 \leqslant t_2$ in $\mathcal{P}$ if and only if t_2 is an extension of t_1 (that is, $\mathrm{dom}(t_1) \subseteq \mathrm{dom}(t_2)$ and $t_1 = t_2{\restriction}_{\mathrm{dom}(t_1)}$, or equivalently, $\mathrm{graph}(t_1) \subseteq \mathrm{graph}(t_2)$).

3.1 Proposition *Assume that $G \cup H \cup R$ entails s and that every element of $D(\mathbf{s})$ has an extension in $[G \cup H]$; then s can be constructed locally from $G \cup H \cup R$. If the maximal elements of $\mathcal{P}$ lie in $[G \cup H]$, then any s entailed by $G \cup H \cup R$ can be locally constructed from $G \cup H \cup R$.*

Proof Let s be n-ary and assume that $\tau \in D(\mathbf{s})$ has an extension t in $[G \cup H]$, that is, $s \subseteq \mathrm{dom}(t)$ and $t{\restriction}_s = \tau$. For $x_1, \dots, x_n \in D(\mathbf{s})$ we have $(x_1, \dots, x_n) \in \mathrm{dom}(t^{D(\mathbf{s})})$ if and only if

$$(\forall c \in s)\ (x_1(c), \dots, x_n(c)) \in \mathrm{dom}(t).$$

But

$$(\forall c \in s)\ (\rho_1(c), \dots, \rho_n(c)) = c \in s \subseteq \mathrm{dom}(t),$$

whence $(\rho_1, \dots, \rho_n) \in \mathrm{dom}(t^{D(\mathbf{s})})$. Since

$$(\forall c \in s)\ t^{D(\mathbf{s})}(\rho_1, \dots, \rho_n)(c) = t(\rho_1(c), \dots, \rho_n(c)) = t(c) = \tau(c),$$

it follows that $t(\rho_1, \dots, \rho_n) = \tau$ in $D(\mathbf{s})$.

Now assume that $G \cup H \cup R$ entails s. Then by the Syntax of Entailment Theorem, there is a primitive positive formula $\Phi(x_1, \dots, x_n)$ which determines s and such that $D(\mathbf{s})$ satisfies $\Phi(\rho_1, \dots, \rho_n)$. Let $\Phi(x_1, \dots, x_n)$ be the formula

$$(\exists y_1, \dots, y_m)\ \Psi(x_1, \dots, x_n, y_1, \dots, y_m),$$

where Ψ is a conjunct of atomic formulæ. Thus there exist $\tau_1, \ldots, \tau_m \in D(\mathbf{s})$ such that $D(\mathbf{s})$ satisfies $\Psi(\rho_1, \ldots, \rho_n, \tau_1, \ldots, \tau_m)$. By the discussion above, each τ_i has an extension $t_i \in [G \cup H]$ such that $t_i(\rho_1, \ldots, \rho_n) = \tau_i$ in $D(\mathbf{s})$. It is immediate that $D(\mathbf{s})$ satisfies

$$\Psi(\rho_1, \ldots, \rho_n, t_1(\rho_1, \ldots, \rho_n), \ldots, t_m(\rho_1, \ldots, \rho_n)).$$

Define $\Phi'(x_1, \ldots, x_n)$ to be

$$\Psi(x_1, \ldots, x_n, t_1(x_1, \ldots, x_n), \ldots, t_m(x_1, \ldots, x_n)).$$

Since Φ determines s, it follows easily that Φ' also determines s. As Φ' is quantifier-free, s can be constructed locally from $G \cup H \cup R$. This proves the first assertion and the second follows at once. ∎

Recall that if $\underline{\mathbf{M}}$ generates a congruence distributive variety and every non-trivial subalgebra of $\underline{\mathbf{M}}$ is subdirectly irreducible, then every $h \in \mathcal{P}$ is essentially unary: see Lemma 8.5.5. Thus the following corollary applies to a large class of algebras. Note that if $e \in G \cup H$ is a (partial) endomorphism of $\underline{\mathbf{M}}$, then the n-ary (partial) map e_i^n, determined by $e_i^n = e \circ \rho_i$ and $\operatorname{dom}(e_i^n) = M \times \cdots \times M \times \operatorname{dom}(e) \times M \times \cdots \times M$, belongs to $[G \cup H]$ and moreover e_i^n is maximal in $[G \cup H]$ if and only if e is non-extendable.

3.2 Corollary *Assume that every homomorphism from* $\mathbf{s}$ *into* $\underline{\mathbf{M}}$ *is essentially unary and that* $G \cup H \cup R$ *entails* s. *If* $[G \cup H]$ *contains the set of all non-extendable partial endomorphisms of* $\underline{\mathbf{M}}$, *then* s *can be locally constructed from* $G \cup H \cup R$.

Proof By hypothesis, every element of $x \in D(\mathbf{s})$ is of the form $e \circ \rho_i$, where e is a non-extendable partial endomorphism and $\operatorname{im} \rho_i \subseteq \operatorname{dom} e$. Thus, if $e \in [G \cup H]$, then e_i^n is an extension of x in $[G \cup H]$. ∎

We now reveal how to translate this corollary into an explicit semantic description when each homomorphism from $\mathbf{s}$ into $\underline{\mathbf{M}}$ is essentially unary. The first step, when $G \cup H$ is empty, is provided by another application of Lemma 8.5.6.

3.3 Proposition *Assume that* $\mathbf{s} \leqslant \underline{\mathbf{M}}^n$. *If each homomorphism from* $\mathbf{s}$ *into* $\underline{\mathbf{M}}$ *is essentially unary and* R *entails* s, *then* s *is a finite intersection of relations of the following forms:*

(D) *the trivial expansion of the domain of a non-extendable partial endomorphism of* $\underline{\mathbf{M}}$,

(E) *the trivial expansion of the equaliser of two non-extendable partial endomorphisms of* $\underline{\mathbf{M}}$,

(K) *the trivial expansion of the joint kernel of two non-extendable partial endomorphisms of* $\underline{\mathbf{M}}$,

(T) *a relation* t, *where*

$$\begin{aligned} t &= ((e_1 \times \cdots \times e_m)^{-1}(r))^{\varepsilon} \\ &= \{(c_1, \ldots, c_n) \in (\mathrm{dom}(e_1) \times \cdots \times \mathrm{dom}(e_m))^{\varepsilon} \mid \\ &\qquad\qquad (e_1(c_{\varepsilon(1)}), \ldots, e_m(c_{\varepsilon(m)})) \in r\} \end{aligned}$$

for some m*-ary relation* $r \in R$, *map* $\varepsilon\colon \{1, \ldots, m\} \to \{1, \ldots, n\}$ *and non-extendable partial endomorphisms* $e_1, \ldots, e_m$ *of* $\underline{\mathbf{M}}$.

Proof Define s^+ as in Lemma 8.5.6. If $s^+ = s$ there is nothing to prove. If there exists $(a_1, \ldots, a_n) \in s^+ \backslash s$, then $\gamma(e \circ \rho_i) := e(a_i)$ gives a well-defined map. Since $(\rho_1, \ldots, \rho_n) \in s^{D(\mathbf{s})}$, and since $(\gamma(\rho_1), \ldots, \gamma(\rho_n)) = (a_1, \ldots, a_n) \notin s$, we have $s \in \mathrm{Fail}_{\mathbf{s}}(\gamma)$. As $R \vdash s$, there exists $r \in R$ with $r \in \mathrm{Fail}_{\mathbf{s}}(\gamma)$. Hence, by the last part of Lemma 8.5.6, there exist non-extendable partial endomorphisms $e_1, \ldots, e_m$ such that

$$s \subseteq ((e_1 \times \cdots \times e_m)^{-1}(r))^{\varepsilon} \text{ and } (a_1, \ldots, a_n) \notin ((e_1 \times \cdots \times e_m)^{-1}(r))^{\varepsilon}.$$

Since $((e_1 \times \cdots \times e_m)^{-1}(r))^{\varepsilon}$ includes s and excludes $(a_1, \ldots, a_n)$ and is a relation of the type described in (T), it is now clear that the intersection of all relations of the types (D), (E), (K) and (T) which contain s is equal to s, as required. ∎

Each of the constructs (D), (E), (K) and (T) will be an admissible (local) construct provided $[G \cup H]$ includes all non-extendable partial endomorphisms of $\underline{\mathbf{M}}$.

3.4 Local Entailment Theorem *Assume that every map in* $G \cup H$ *is nullary or unary and that* $[G \cup H]$ *includes all non-extendable partial endomorphisms of* $\underline{\mathbf{M}}$. *Let* $\mathbf{s} \leqslant \underline{\mathbf{M}}^n$. *If each homomorphism from* $\mathbf{s}$ *into* $\underline{\mathbf{M}}$ *is essentially unary then* $G \cup H \cup R$ *entails* s *if and only if* s *is a finite intersection of relations of the form* (D) *or* (T) *along with relations which are*

(E)′ *the trivial expansion of the equaliser of* $e \circ e_1$ *and* e_2 *for some* $e \in G \cup H \cup \{\mathrm{id}_M\}$ *and non-extendable partial endomorphisms* e_1 *and* e_2 *of* $\underline{\mathbf{M}}$, *or*

(K)′ *the trivial expansion of the joint kernel of* $e \circ e_1$ *and* e_2 *for some* $e \in G \cup H \cup \{\mathrm{id}_M\}$ *and non-extendable partial endomorphisms* e_1 *and* e_2 *of* $\underline{\mathbf{M}}$.

Proof If $[G \cup H]$ includes all non-extendable partial endomorphisms of $\underline{\mathbf{M}}$, then (D), (E)′, (K)′ and (T) will be admissible (local) constructs, whence any relation s which is a finite intersection of such relations will be entailed by $G \cup H \cup R$. Note that, since we allow $e = \mathrm{id}_M$, cases (E) and (K) are subcases of (E)′ and (K)′, respectively. For the converse, let $R' := R \cup \{\, \mathrm{graph}(e) \mid e \in G \cup H \,\}$. If $G \cup H \cup R$ entails s, then so does R'. Hence by the previous proposition, s is a finite intersection of relations of the form (D), (E), (K), or (T) with $r \in R'$. Consider relations t of the form (T) with $r \in R'$. If $r \in R$, then t is covered by (T). If $r = \mathrm{graph}(k)$ for some nullary operation k, then $r = \{k\}$ and hence $t = (e_1^{-1}(\{k\}))^\varepsilon = \mathrm{dom}(e_1)^\varepsilon$. In this case, t is covered by (D). If $r = \mathrm{graph}(e)$ for some unary (partial) operation, then there exist non-extendable partial endomorphisms e_1 and e_2 of $\underline{\mathbf{M}}$ and a map $\varepsilon\colon \{1,2\} \to \{1,\dots,n\}$ such that $t = ((e_1 \times e_2)^{-1}(\mathrm{graph}(e)) = \ker(e \circ e_1, e_2))^\varepsilon$. By Lemma 8.5.8, if $e \circ e_1 = e_2$, then t is covered by (D), and if $e \circ e_1 \neq e_2$, then t is covered by (E)′ or (K)′. ■

3.5 Corollary *Assume that every map in $G \cup H$ is nullary or unary and that $[G \cup H]$ includes all non-extendable partial endomorphisms of $\underline{\mathbf{M}}$. Let $\mathbf{s} \leqslant \underline{\mathbf{M}}^n$. If each homomorphism from $\mathbf{s}$ into $\underline{\mathbf{M}}$ is essentially unary then $G \cup H \cup R$ entails s if and only if s can be obtained from $G \cup H \cup R$ via the local constructs intersection, equaliser, joint kernel, composition, endomorphism action, trivial expansion, trivial relations, repetition removal and product.*

Proof This follows at once from the fact that relations of type (T) may be obtained via endomorphism action and subscript manipulation (see Exercise 8.16) while subscript manipulation can be obtained via trivial expansion, trivial relations, repetition removal and product (see Exercise 2.7(iv)). ■

The possible forms of the relation s as listed in Exercise 8.18 provide an explicit description of the at-most-binary relations of the form (T) which arise when r is at most binary.

9.4 Optimal strong dualities and hom-entailment

Hom-entailment was introduced briefly back in Chapter 3. Recall that if $P \subseteq \mathcal{P}$ and $k \in \mathcal{P}$, then P **hom-entails** k if, for all non-empty sets S, each topologically closed subset of M^S which is closed under the partial operations in P is also closed under k. Note, in particular, that a nullary operation hom-entails the corresponding constant endomorphism but not conversely. Define

$$\widehat{P} := \{\, k \in \mathcal{P} \mid P \text{ hom-entails } k \,\}.$$

Then $P \mapsto \widehat{P}$ is a closure operator on $\mathcal{P}$ and we refer to $\widehat{P}$ as the **hom-entailment closure** of P. As we pointed out in the Brute Force Strong

Duality Theorem 3.2.2, $G \cup H$ must be **hom-entailment dense**, that is, $\widehat{G \cup H} = \mathcal{P}$, in order for $\underset{\sim}{\mathbf{M}}$ to yield a strong duality on $\mathcal{A}$.

Assume that $\underset{\sim}{\mathbf{M}} = \langle M; G, H, R, \mathcal{T} \rangle$ yields a strong duality on $\mathcal{A}$. Since the relations in R play no role in considerations of hom-entailment, in order to refine $\underset{\sim}{\mathbf{M}}$ to a structure $\underset{\sim}{\mathbf{M}}^{\mathrm{o}}$ which yields an optimal strong duality on $\mathcal{A}$, we must

(1) first replace G and H by sets G^{o} and H^{o} such that $G^{\mathrm{o}} \cup H^{\mathrm{o}}$ is minimal with respect to hom-entailing every map in $G \cup H$, and then

(2) apply the theory of the previous chapter in order to replace R by a set R^{o} which is minimal with respect to the property that $G^{\mathrm{o}} \cup H^{\mathrm{o}} \cup R^{\mathrm{o}}$ entails every $s \in G \cup H \cup R$.

The resulting structure $\underset{\sim}{\mathbf{M}}^{\mathrm{o}} := \langle M; G^{\mathrm{o}}, H^{\mathrm{o}}, R^{\mathrm{o}}, \mathcal{T} \rangle$ will yield an **optimal strong duality** on $\mathcal{A}$, that is, $\underset{\sim}{\mathbf{M}}^{\mathrm{o}}$ will yield a strong duality on $\mathcal{A}$ but removal of any element in $G^{\mathrm{o}} \cup \widetilde{H}^{\mathrm{o}} \cup H^{\mathrm{o}}$ will result in a structure that will yield a duality which is not strong or will not yield a duality at all.

The key to step (1) is a complete list of constructs for hom-entailment. By comparison with the problem of giving a sufficient set of constructs to describe the entailment closure of $G \cup H \cup R$, it is surprisingly easy to give an internal description of the hom-entailment closure of $G \cup H$. Our next lemma, which does all the work for us, is to maps what the Test Algebra Lemma is to relations.

4.1 Lemma *Let $P \subseteq \mathcal{P}$ and let $k \in \mathcal{P}$ with $k\colon D \to M$ where $\mathbf{D} \leqslant \underline{\mathbf{M}}^n$.*

(i) *The following are equivalent:*

 (a) *P hom-entails k;*

 (b) *P hom-entails k on subsets of M^D;*

 (c) *there exists $p \in [P]$ such that $k = p{\restriction}_D$.*

 In particular, if k is nullary, then P hom-entails k if and only if P contains a nullary and k belongs to the substructure of $\underset{\sim}{\mathbf{M}} = \langle M; P, \mathcal{T} \rangle$ generated by the nullaries in P.

(ii) $\widehat{P} = \{ p{\restriction}_D \mid p \in [P] \text{and } \mathbf{D} \leqslant \mathbf{dom}(p) \}$.

Proof We shall prove (i) from which (ii) follows. Only (c) $\Rightarrow$ (a) and (b) $\Rightarrow$ (c) require proof. It is easy to see that if $p \in [P]$ then P hom-entails p and that p hom-entails $p{\restriction}_D$ for each subalgebra $\mathbf{D}$ of $\mathbf{dom}(p)$. Thus (c) implies (a). Now assume (b) and consider the subset

$$X := \{ p{\restriction}_D\colon D \to M \mid p \text{ is } n\text{-ary}, p \in [P] \text{ and } \mathbf{D} \leqslant \mathbf{dom}(p) \}$$

of M^D. If $p_1{\restriction}_D, \dots, p_m{\restriction}_D \in X$ and $h \in P$ is an m-ary partial map such that $h(p_1, \dots, p_m)$ is defined, then $h(p_1, \dots, p_m){\restriction}_D$ is defined and $h(p_1{\restriction}_D, \dots, p_m{\restriction}_D) = h(p_1, \dots, p_m){\restriction}_D$. Thus X is closed under the partial operations in P. By assumption, P hom-entails k on X, whence X is closed under k. Now $\rho_1 = \pi_1\restriction_D, \ldots, \rho_n = \pi_n\restriction_D \in X$ and moreover $(\rho_1, \ldots, \rho_n) \in \mathrm{dom}(k^X)$, whence $k = k(\rho_1, \ldots, \rho_n) \in X$. Thus $k = p\restriction_D$ for some $p \in [P]$. Hence (b) implies (c). ∎

We may regard the projections $\pi_i \colon M^n \to M$ as nullary constructs for hom-entailment just as the trivial relations are nullary constructs for entailment. The next result is an immediate consequence of Lemma 9.4.1. The reader should verify that the projections, restriction of domain and composition are valid constructs for hom-entailment.

4.2 Completeness of Hom-entailment Theorem *The projections $\pi_i \colon M^n \to M$ for each $n \geqslant 1$, restriction of domain (viz. $p \mapsto p\restriction_D$ for $\mathbf{D} \lessdot \mathbf{dom}(p)$) and composition of partial functions (viz. $(h, g_1, \ldots, g_n) \mapsto h(g_1, \ldots, g_n)$, where defined) form a complete set of constructs for hom-entailment.*

An important corollary of this result tells us what we are required to include in $G \cup H$ if we wish to obtain a strong duality.

4.3 Corollary *If $\underset{\sim}{\mathbf{M}}$ yields a strong duality on $\mathcal{A}$, then the enriched partial clone of $\underset{\sim}{\mathbf{M}}$ must include an extension of every algebraic partial operation on $\underline{\mathbf{M}}$.*

Proof If $\underset{\sim}{\mathbf{M}}$ yields a strong duality on $\mathcal{A}$, then every substructure of a power of $\underset{\sim}{\mathbf{M}}$ is closed under h (by the definition of strong). Hence $G \cup H$ hom-entails h, whence h has an extension in $[G \cup H]$ by the completeness theorem. ∎

Since a partial operation is hom-entailed by each of its extensions, when striving for an optimal strong duality we may restrict our attention to the non-extendable members of $\mathcal{P}$. Lemma 9.4.1 yields a characterisation of minimal hom-entailment dense subsets P of $\mathcal{P}$ consisting of non-extendable maps.

4.4 Proposition *Let P be a subset of $\mathcal{P}$. If P is a minimal generating set for the partial clone $\mathcal{P}_{\mathrm{ne}}$ generated by the set of all non-extendable members of $\mathcal{P}$, then P is a minimal hom-entailment dense set. The converse holds provided P consists of non-extendable members of $\mathcal{P}$.*

Proof Let P be a minimal generating set for the partial clone $\mathcal{P}_{\mathrm{ne}}$. Since P generates $\mathcal{P}_{\mathrm{ne}}$, every member p of $\mathcal{P}$ has an extension in $[P]$ and hence P is hom-entailment dense in $\mathcal{P}$. Now assume that P consists of non-extendable members of $\mathcal{P}$. If $k \in P$ were hom-entailed by $P\backslash\{k\}$, then by Lemma 9.4.1 we would have $k = p\restriction_D$ for some $p \in [P\backslash\{k\}]$. Since k is non-extendable, we would have $k = p \in [P\backslash\{k\}]$, contradicting the minimality assumption on P. Thus P is a minimal hom-entailment dense subset of $\mathcal{P}$. Conversely, assume that P is a minimal hom-entailment dense subset of $\mathcal{P}$. By Lemma 9.4.1, $[P]$ contains every non-extendable member of $\mathcal{P}$ and hence $[P] = \mathcal{P}_{\text{ne}}$ as $P \subseteq \mathcal{P}_{\text{ne}}$, by assumption. If k were in the partial clone generated by $P\backslash\{k\}$ for some $k \in P$, then $P\backslash\{k\}$ would hom-entail k, contradicting the minimality assumption on P. Hence, P is a minimal generating set for the partial clone $\mathcal{P}_{\text{ne}}$. ∎

We shall close this section by looking at the case where the irreducibility index of $\underline{\mathbf{M}}$ is 1. Although our next result is not the most general we could state, it is by far the most useful special case. Recall that $\mathbf{K}$ denotes the (possibly empty) substructure of $\underset{\sim}{\mathbf{M}}$, consisting of those elements of M which determine one-element subalgebras of $\underline{\mathbf{M}}$, and that, by Lemma 3.1.2, a necessary condition for full (and therefore for strong) duality is that $\mathbf{K}$ is the substructure of $\underset{\sim}{\mathbf{M}}$ generated by its distinguished elements.

4.5 Special Optimal Strong Duality Theorem *Assume that $\underline{\mathbf{M}}$ has a majority term and that* $\operatorname{Irr}(\underline{\mathbf{M}}) = 1$. *Assume that*

(i) *G_0 is a subset of K,*

(ii) *G_1 is a set of endomorphisms of $\underline{\mathbf{M}}$,*

(iii) *H is a set of proper non-extendable partial endomorphisms of $\underline{\mathbf{M}}$, and*

(iv) *R is a set of binary algebraic relations on $\underline{\mathbf{M}}$.*

Then $\underset{\sim}{\mathbf{M}} := \langle M; G_0 \cup G_1, H, R, \mathcal{T}\rangle$ yields an optimal strong duality on $\mathcal{A}$ if and only if

(1) *G_0 is a minimal generating set for the substructure $\mathbf{K}$ of $\underset{\sim}{\mathbf{M}}$,*

(2) *$\overline{G}_0 \cup G_1 \cup H$ is a minimal generating set for the monoid $\mathcal{M}_{\text{ne}}$ generated by the endomorphisms and the proper non-extendable partial endomorphisms of $\underline{\mathbf{M}}$, where $\overline{G}_0$ is the set of constant endomorphisms corresponding to elements of G_0, and*

(3) *R is a transversal of the family of the globally minimal failsets within $\mathbb{S}(\underline{\mathbf{M}}^2)$ which do not intersect* $\{\operatorname{graph}(p) \mid p \in \overline{G}_0 \cup G_1 \cup H\}$.

Proof Assume that $\underline{\mathbf{M}}$ has a majority term, that $\operatorname{Irr}(\underline{\mathbf{M}}) = 1$ and that (i)–(iv) hold. Since $\operatorname{Irr}(\underline{\mathbf{M}}) = 1$, Lemma 8.5.5 guarantees that every element of $\mathcal{P}$ is either nullary or essentially unary. Now assume that $\underset{\sim}{\mathbf{M}}$ satisfies (1)–(3). A simple application of the NU Strong Duality Corollary 3.3.9 shows that $\underset{\sim}{\mathbf{M}}$ yields a strong duality on $\mathcal{A}$. By Proposition 9.4.4 and the Optimal Duality Theorem 8.3.10, the minimality assumptions in conditions (1)–(3) guarantee that $\underset{\sim}{\mathbf{M}}$ yields an optimal strong duality. Conversely, assume that $\underset{\sim}{\mathbf{M}}$ yields an optimal strong duality on $\mathcal{A}$. Since each $p \in G_0 \cup G_1 \cup H$ is non-extendable in $\mathcal{P}$ and since every member of $\mathcal{P}$

is nullary or essentially unary, Proposition 9.4.4 and the Optimal Duality Theorem 8.3.10 show that $\underset{\sim}{\mathbf{M}}$ satisfies (1)–(3). ■

We now obtain the following description of the optimal strong dualities for the variety $\mathcal{B}_n$ of distributive p-algebras as an immediate consequence of Proposition 8.6.3 and the Special Optimal Strong Duality Theorem. We refer the reader to Section 6 of Chapter 8 for a discussion of the relations in the set T.

4.6 Optimal Strong Dualities for $\mathcal{B}_n$ *Let $n \geqslant 1$ and assume that*

(i) *G is a set of endomorphisms of $\underline{\mathbf{B}}_n$,*

(ii) *H is a set of proper non-extendable partial endomorphisms of $\underline{\mathbf{B}}_n$, and*

(iii) *R is a set of binary algebraic relations on $\underline{\mathbf{B}}_n$.*

Then $\underset{\sim}{\mathbf{M}} := \langle M; G, H, R, \mathcal{T}\rangle$ yields an optimal strong duality on $\mathcal{B}_n$ if and only if $G \cup H$ is a minimal generating set for the monoid $\mathcal{M}_{\text{ne}}$ generated by the endomorphisms and the proper non-extendable partial endomorphisms of $\underline{\mathbf{B}}_n$ and R is a transversal of the isomorphism classes of $T \cup T^{\smile}$.

The description of the relations in T via the restricted Priestley duality given in Chapter 7 shows that, when considering $\underline{\mathbf{B}}_n$, there are exactly n isomorphism classes within $T \cup T^{\smile}$. Consequently, once we have chosen G and H such that $G \cup H$ is a minimal generating set for the monoid $\mathcal{M}_{\text{ne}}$, if $|R| = n$ and $\underset{\sim}{\mathbf{M}} := \langle M; G, H, R, \mathcal{T}\rangle$ yields a duality on $\mathcal{B}_n$, then it is automatically an optimal strong duality. We now apply this observation in the special cases $n = 1, 2, 3$.

4.7 Optimal Strong Dualities for $\mathcal{B}_1$, $\mathcal{B}_2$ and $\mathcal{B}_3$ For $n = 1$ (Stone algebras) and for $n = 2$, the algebra $\underline{\mathbf{B}}_n$ is injective and hence the Total Structure Theorem 6.1.2 tells us that no partial operations are required. For $n = 1$, there is only one non-identity endomorphism, namely d (see 4.3.6), and consequently the duality given by d and $\preccurlyeq$ (see 4.3.7) is an optimal strong duality. In fact (up to replacing the order $\preccurlyeq$ by its converse), this is the unique optimal strong duality for $\mathcal{S} = \underline{\mathbf{B}}_1$ using operations and relations of minimal arity—see Exercise 8.8. For $n = 2$, the endomorphism monoid of $\underline{\mathbf{B}}_n$ is generated by the non-trivial automorphism f and any $\{0,1\}$-valued endomorphism. Consequently the duality for $\mathcal{B}_n$ given by the two maps f and d and two binary relations $\leqslant$ and $\preccurlyeq$ described in Theorem 4.3.18 is an optimal strong duality. For $n \geqslant 3$, the endomorphism monoid of $\underline{\mathbf{B}}_n$ is generated by two automorphisms and any $\{0,1\}$-valued endomorphism. Since $\underline{\mathbf{B}}_n$ is not injective in this case, we must add partial operations to the type of $\underset{\sim}{\mathbf{M}}$. For $n = 3$, it suffices to add any one of the nine proper non-extendable partial endomorphisms as it along with the automorphisms generates the other eight. Thus there is an optimal strong duality for $\mathfrak{B}_3$ given by three endomorphisms, one proper non-extendable partial endomorphism and three binary relations.

In the final section of this chapter we shall consider a small example whose irreducibility index is 2.

9.5 An example: the three-element chain

We shall take $\underline{\mathbf{M}}$ to be $\langle \{0, d, 1\}; \vee, \wedge, 0, 1 \rangle$, the three-element chain $0 < d < 1$. Then $\mathbb{ISP}\underline{\mathbf{M}}$ is the variety $\mathfrak{D}_{01}$ of bounded distributive lattices. The only proper subalgebra of $\underline{\mathbf{M}}$ has $D := \{0, 1\}$ as underlying set; as in Section 5 of Chapter 8, we denote $\{(0,0), (1,1)\}$ by Δ_D. The NU Duality Theorem 2.3.4 ensures that $R = \mathbb{S}(\underline{\mathbf{M}}^2)$ yields a duality on $\mathfrak{D}_{01}$. The lattice $\mathbb{S}(\underline{\mathbf{M}}^2)$ has 49 elements (which was checked by computer). The meet-irreducibles are the following subalgebras:

$$V_1 = D \times M, \qquad V_2 = \{(0,0), (0,d), (0,1), (d,1), (1,1)\},$$
$$V_3 = M^2 \backslash \{(1,0), (1,d)\}, \quad V_4 = M^2 \backslash \{(d,0), (1,0)\},$$
$$V_5 = M^2 \backslash \{(1,0)\},$$

and their converses.

5.1 Duality via Endomorphisms In Chapter 7, we showed that the endomorphism monoid, $\mathrm{End}(\underline{\mathbf{L}})$, of a finite distributive lattice $\underline{\mathbf{L}}$ together with the usual order $\leqslant_D := \{(0,0), (0,1), (1,1)\}$ on $\mathbf{D} \leqslant \underline{\mathbf{L}}$ always yields a duality on $\mathfrak{D}_{01}$ (7.7.6), and, further, that $\mathrm{End}(\underline{\mathbf{L}})$ yields a duality on $\mathfrak{D}_{01}$ if and only if $\underline{\mathbf{L}}$ is non-Boolean (7.8.5). Thus, in particular, $\mathrm{End}(\underline{\mathbf{M}})$ yields a duality on $\mathfrak{D}_{01}$. The endomorphism monoid of $\underline{\mathbf{M}}$ consists of the identity map and two retractions, e and f, onto $\mathbf{D}$, given by $e^{-1}(1) = \{1\}$ and $f^{-1}(0) = \{0\}$. Thus it follows from the general theory that $\{e, f\}$ entails $\leqslant_D$. (See Exercise 9.2.) We will show directly how to obtain $\leqslant_D$ from $\{e, f\}$ using our constructs, and then show that local constructs alone are not sufficient to do this.

Note that $D(\leqslant_D) = \{\rho_1, \rho_2, \tau\}$, where $\tau((0,1)) = d$. Since $e(\tau) = e \circ \tau = \rho_1$ and $f(\tau) = f \circ \tau = \rho_2$, it follows that $D(\leqslant_D)$ satisfies the formula $(\exists y)\,(e(y) \approx \rho_1 \mathbin{\&} f(y) \approx \rho_2)$ and, moreover,

$$\leqslant_D = \{(c_1, c_2) \in M^2 \mid (\exists c_3 \in M)\,(e(c_3) = c_1 \mathbin{\&} f(c_3) = c_2)\}.$$

Thus, $\leqslant_D = \mathrm{graph}(e)^{\smile} \cdot \mathrm{graph}(f)$, and this is a homomorphic relational product, whence $\{e, f\}$ entails $\leqslant_D$.

We now show that $\leqslant_D$ cannot be expressed in terms of the endomorphisms e and f using local constructs only. If $\leqslant_D$ were given locally from e and f, then there would be a (quantifier-free) conjunct of atomic formulæ, $\Phi(x_1, x_2)$, such that $\leqslant_D = \{(a,b) \in M^2 \mid \underset{\sim}{\mathbf{M}} \text{ satisfies } \Phi(a,b)\}$ where $\underset{\sim}{\mathbf{M}} = \langle\{0,d,1\}; e, f, \mathcal{T}\rangle$. In particular, $\Phi(0,1)$ would hold. Since $0 \leftrightarrows 1$ is an automorphism of the substructure $\langle\{0,1\}; e, f, \mathcal{T}\rangle$, it would also satisfy $\Phi(1,0)$. This contradicts the fact that $1 \nleqslant_D 0$.

5.2 Duality via a Single Relation A computer analysis, based on the theory in Section 6 of Chapter 8, allowed the authors of [DHP95a] to determine all optimal dualities for $\mathfrak{D}_{01}$ using subsets of $\mathbb{S}(\underline{\mathbf{M}}) \cup \mathbb{S}(\underline{\mathbf{M}}^2)$. This analysis revealed that the single relation

$$t = \{(0,0), (0,d), (d,1), (1,1)\}$$

yields a duality on $\mathfrak{D}_{01}$! The intersection $t \cap t^{\smile}$ is Δ_D. Thus V_1 may be obtained from t by local constructs (as D may be obtained from Δ_D via repetition removal). The following formulæ express the remaining meet-irreducibles in $\mathbb{S}(\underline{\mathbf{M}}^2)$ in terms of t. In each case the relational products involved are homomorphic relational products.

$$\begin{aligned} V_2 &= t \cdot t, \\ V_3 &= t^{\smile} \cdot (t \cdot t), \\ V_4 &= t \cdot (t \cdot t^{\smile}), \\ V_5 &= (t^{\smile} \cdot (t \cdot t)) \cdot t^{\smile}. \end{aligned}$$

These formulæ are not unique: for example, V_5 is also given by

$$V_5 = t^{\smile} \cdot (t \cdot (t \cdot t^{\smile})).$$

We observe also that the order on the three-element chain is $V_3 \cap V_4$.

Not all relational products in $\mathbb{S}(\underline{\mathbf{M}}^2)$ are homomorphic. Take $r_1 = \{(0,0), (0,d), (d,d), (1,1)\}$ and $r_2 = M^2 \backslash \{(0,1)\}$. Then $M^2 = r_1 \cdot r_2$. However, this is not a homomorphic relational product. To see this, first note that $D(\underline{\mathbf{M}}^2)$ has six elements, $g \circ \pi_i$ ($g \in \operatorname{End}(\underline{\mathbf{M}})$ and $i = 1,2$). On $D(\underline{\mathbf{M}}^2)$ the relation $\underline{\mathbf{M}}^2$ is $D(\underline{\mathbf{M}}^2)^2$. We need to show that there is no $z \in D(\underline{\mathbf{M}}^2)$ for which $(\rho_1, z) \in r_1$ and $(z, \rho_2) \in r_2$ on $D(\underline{\mathbf{M}}^2)$. A simple calculation shows that the only possible choice for z if we are to have $(\rho_1, z) \in r_1$ is $z = \rho_1$. However, $(\rho_1, \rho_2) \notin r_2$ on $D(\underline{\mathbf{M}}^2)$.

5.3 A strong duality How can we upgrade the duality for $\mathfrak{D}_{01}$ given by e and f to a strong duality? Since $\underline{\mathbf{M}}$ is not injective in $\mathfrak{D}_{01}$, the Total Structure Theorem 6.1.2 tells us that if $\underset{\sim}{\mathbf{M}}'$ yields a strong duality on $\mathfrak{D}_{01}$, then the type of $\underset{\sim}{\mathbf{M}}'$ must include some proper partial operations. Since the irreducibility index of $\underline{\mathbf{M}}$ is 2, the NU Strong Duality Corollary 3.3.9

implies that if $\underset{\sim}{\mathbf{M}}$ is known to yield a duality then a strong duality may be obtained by adding the unary and binary algebraic partial operations on $\underline{\mathbf{M}}$ to the type of $\underset{\sim}{\mathbf{M}}$. Moreover, by Lemma 8.5.4 unary algebraic partial operations will not suffice. Proposition 9.4.4 now tells us that it is sufficient to add to the type of $\underset{\sim}{\mathbf{M}}$ a set P of partial unary and binary algebraic operations such that the partial clone generated by P includes the non-extendable partial unary and binary algebraic operations on $\underline{\mathbf{M}}$.

Let θ_e and θ_f be the kernels of e and f, respectively. Since $\theta_e \cap \theta_f = \mathbf{0}^{\underline{\mathbf{M}}}$ in $\mathrm{Con}(\underline{\mathbf{M}})$ and since $\theta_e \cdot \theta_f = V_5$ we may define $h\colon V_5 \to M$ by declaring $h(a,b)$ to be the unique element $c \in M$ such that $a\,\theta_e\,c\,\theta_f\,b$. Note that h is an extension of the map $\tau\colon \leqslant_D \to M$ defined above. We claim that $\underset{\sim}{\mathbf{M}}' = \langle\{0,d,1\}; e, f, h, \mathcal{T}\rangle$ yields a strong duality on $\mathfrak{D}_{01}$. By the remarks above, it remains to see that every non-extendable partial binary algebraic operation on $\underline{\mathbf{M}}$ is in the partial clone generated by $\{e, f, h\}$. The authors of [DHP95a] carried out a computer-assisted check which revealed that there are 308 partial binary homomorphisms on $\underline{\mathbf{M}}$ and that, up to symmetry, of these only 7 are non-extendable. These 7 non-extendable partial binary homomorphisms are the total maps π_1, $e\circ\pi_1$, $f\circ\pi_1$ and the proper partial maps h, $h(\mathrm{id}, e)$, $h(f, \mathrm{id})$, and $h(f,e)$, all of which belong to the partial clone generated by $\{e, f, h\}$.

Exercises 9

Exercises from the text. Prove Lemma 9.1.1. Show that each of the relation-producing constructs (2), (3), (4), (5), (10) (graph), (12), (13), (14) and (15) from 2.4.5 is a special case of term manipulation (16) and is therefore local. Show that for each of the (partial-) operation-producing constructs (8), (10) (graph^{-1}), and (11), the corresponding relation-producing construct is a special case of term manipulation (16) and is therefore local. Prove the Soundness Theorem 9.2.2 by checking that the remaining operation-producing construct (9) is admissible. Verify that the projections, restriction of domain and composition are valid constructs for hom-entailment (see 9.4.2). Verify each of the formulæ, given in 9.5.2, which express the relations V_2–V_5 in terms of the relation t and show that each is a homomorphic relational product. Complete the details of the argument at the end of 9.5.2 which shows that $r_1 \cdot r_2$ is not a homomorphic relational product. Confirm that each of the seven maps specified at the end of 9.5.3 is non-extendable.

9.1 [Dav96] Let $\mathbf{s} \leqslant \underline{\mathbf{M}}^n$. Prove that the following are related by (i) $\Leftrightarrow$ (ii) $\Rightarrow$ (iii). (Use Exercise 8.1.)

(i) $\mathbf{s}$ is a retract of $\underline{\mathbf{M}}$;

(ii) there exist endomorphisms $e_1, \ldots, e_n$ of $\underline{\mathbf{M}}$ such that

$$s = \{(c_1, \ldots, c_n) \in M^n \mid M \text{ satisfies } (\exists a) \mathop{\&}_{i=1}^{n} e_i(a) = c_i\}$$

and $D(\mathbf{s})$ satisfies $(\exists y) \mathop{\&}_{i=1}^{n} e_i(y) = \rho_i$;

(iii) there exist $e_1, \ldots, e_n \in \mathrm{End}(\underline{\mathbf{M}})$ such that $\{e_1, \ldots, e_n\} \vdash s$.

Moreover, if $\mathbf{D} \leqslant \underline{\mathbf{M}}$ and $\mathbf{s} \leqslant \mathbf{D}^n \leqslant \underline{\mathbf{M}}^n$, then the endomorphisms in (i) and (ii) can be chosen so that $\mathrm{im}(e_i) \subseteq D$ for all i.

9.2 Use the previous exercise to prove that if $\underline{\mathbf{L}} = \langle L; \vee, \wedge, 0, 1\rangle$ is a non-Boolean finite bounded distributive lattice, and $\leqslant_D$ is the induced order on $D := \{0, 1\}$, then $\mathrm{End}(\underline{\mathbf{L}}) \vdash \leqslant_D$.

9.3 Let M be a finite set, let $s \subseteq M^n$, let $Z \subseteq M^s$ and let $\rho_i := \pi_i\restriction_s$. Assume that Z satisfies

(i) $\rho_1, \ldots, \rho_n \in Z$ and

(ii) if $(z_1, \ldots, z_n) \in s^Z$ and $w \in Z$, then $w \circ (z_1 \sqcap \cdots \sqcap z_n) \in Z$.

Then $\rho_1, \ldots, \rho_n$ satisfies the following universality condition within Z (cf. Lemma 8.1.1):

$(\mathcal{U}_s)$ for any $z_1, \ldots, z_n \in Z$ with $(z_1, \ldots, z_n) \in s^Z$ there exists a unique map $\varphi\colon Z \to Z$ which preserves every finitary relation on M and satisfies $\varphi(\rho_i) = z_i$ for $i = 1, \ldots, n$.

Conclude that if $\mathbf{s} \leqslant \underline{\mathbf{M}}^n$, then both M^s and $D(\mathbf{s})$ satisfy $(\mathcal{U}_s)$.

9.4 Let G be a family of finitary operations, let H be a family of finitary partial operations, let R be a family of finitary relations on a finite set M and let $\underset{\sim}{\mathbf{M}} = \langle M; G, H, R, \mathcal{T}\rangle$. Let $s \subseteq M^n$ and assume that $\mathbf{Z}$ is a substructure of $\underset{\sim}{\mathbf{M}}^s$ which contains $\rho_1, \ldots, \rho_n$ and satisfies $(\mathcal{U}_s)$. Prove that the following are equivalent:

(i) $G \cup H \cup R$ entails s on Z, that is, every map $\alpha\colon Z \to M$ which preserves $G \cup H \cup R$ also preserves s;

(ii) $s = \{(\alpha(\rho_1), \ldots, \alpha(\rho_n)) \mid \alpha\colon Z \to M \text{ preserves } G \cup H \cup R\}$;

(iii) there exists a primitive positive formula $\Phi(x_1, \ldots, x_n)$ of type $\langle G, H, R\rangle$ such that

(a) $s = \{(c_1, \ldots, c_n) \in M^n \mid \underset{\sim}{\mathbf{M}} \text{ satisfies } \Phi(c_1, \ldots, c_n)\}$, and

(b) $\mathbf{Z}$ satisfies $\Phi(\rho_1, \ldots, \rho_n)$.

9.5 (Clearly Exercises 9.3 and 9.4, applied with $\mathbf{Z} = D(\mathbf{s})$, combine with the Test Algebra Lemma 8.1.3 to yield a two-line proof of the Syntax of Entailment Theorem. We now apply Exercises 9.3 and 9.4 to the case $\mathbf{Z} = \underset{\sim}{\mathbf{M}}^s$ to prove an important result from clone theory. The usual case considered in clone theory arises when G and H are empty.) Let $\underset{\sim}{\mathbf{M}}$ be as in the previous exercise. Define

$$\mathrm{Pol}(G \cup H \cup R) := \bigcup\{\,\mathrm{Hom}(\underset{\sim}{\mathbf{M}}^n, \underset{\sim}{\mathbf{M}}) \mid n \in \mathbb{N}\}.$$

If F is a family of finitary operations on M, then define

$$\operatorname{Inv}(F) := \bigcup\{\, \mathbb{S}(\underline{\mathbf{M}}^n) \mid n \in \mathbb{N}\},$$

where $\underline{\mathbf{M}} := \langle M; F\rangle$. Let $s \subseteq M^n$. Prove that the following are equivalent:

(i) $s \in \operatorname{Inv}(\operatorname{Pol}(G \cup H \cup R))$;

(ii) $G \cup H \cup R$ entails s on M^s;

(iii) $s = \{(\alpha(\rho_1), \dots, \alpha(\rho_n)) \mid \alpha\colon M^s \to M \text{ preserves } G \cup H \cup R\}$;

(iv) there are some finite structure $\mathbf{Z}$ of the same type as $\underset{\sim}{\mathbf{M}}$ and elements $z_1, \dots, z_n \in Z$ such that

$$s = \{(\alpha(z_1), \dots, \alpha(z_n)) \mid \alpha\colon Z \to M \text{ preserves } G \cup H \cup R\};$$

(v) $s = \{(c_1, \dots, c_n) \in M^n \mid \underset{\sim}{\mathbf{M}} \text{ satisfies } \Phi(c_1, \dots, c_n)\}$ for some primitive positive formula $\Phi(x_1, \dots, x_n)$ of type $\langle G, H, R\rangle$.

9.6 Prove the following claims, where $\underset{\sim}{\mathbf{M}}$ is as in the previous exercises.

(i) If $s \in \operatorname{Inv}(\operatorname{Pol}(G \cup H \cup R))$, then s can be obtained from $G \cup H \cup R$ by a finite number of applications of coordinate projection, composition, term manipulation, equaliser, intersection and projection.

(ii) If $s \in \operatorname{Inv}(\operatorname{Pol}(R))$, then s can be obtained from R by a finite number of applications of product, intersection, trivial relations, repetition removal and projection.

9.7 Assume that $e\colon \operatorname{dom}(e) \to \underline{\mathbf{M}}$ is a partial endomorphism and s an m-ary relation. Show that $\operatorname{graph}(e) \cdot s$ is a homomorphic relational product. (This illustrates the fact that changing the type of $\underset{\sim}{\mathbf{M}}$ by replacing a (partial) operation by its graph can convert a local construct, namely endomorphic action, into a non-local construct, namely homomorphic relational product.)

9.8 [CISSW] Show that the following are equivalent:

(i) for all $S \neq \varnothing$, each closed substructure of $\underset{\sim}{\mathbf{M}}^S$ is term closed;

(ii) for all $\mathbf{A} \in \mathbb{ISP}\underline{\mathbf{M}}$, the set $\mathcal{A}(\mathbf{A}, \underline{\mathbf{M}})$ is the only closed substructure of $\mathcal{A}(\mathbf{A}, \underline{\mathbf{M}}) \leqslant \underset{\sim}{\mathbf{M}}^A$ which separates A;

(iii) for all I and all $\mathbf{A} \leqslant \underline{\mathbf{M}}^I$, the closed substructure of $\underset{\sim}{\mathbf{M}}^A$ generated by $\{\, \rho_i \mid i \in I\}$ is $\mathcal{A}(\mathbf{A}, \underline{\mathbf{M}})$;

(iv) $\widehat{G \cup H} = \mathcal{P}$ and, for all $\mathbf{A} \in \mathbb{ISP}\underline{\mathbf{M}}$, the only topologically closed subset of $\mathcal{A}(\mathbf{A}, \underline{\mathbf{M}}) \subseteq M^A$ which is closed under all $k \in \mathcal{P}$ and separates A is $\mathcal{A}(\mathbf{A}, \underline{\mathbf{M}})$ itself.

[Hint: For (iii) $\Rightarrow$ (i), use (iv) $\Rightarrow$ (i) of the Closure Theorem 3.1.3.]

10

Dualisable Algebras

In Chapters 8 and 9 we concentrated on the important practical problem of how we refine the structure $\underset{\sim}{\mathbf{M}}$ without destroying the duality which it yields on the quasi-variety $\mathcal{A} = \mathbb{ISP}\underline{\mathbf{M}}$. In Chapter 7 we found a way to transfer a duality for $\underline{\mathbf{M}}$ to a duality for closely related algebras. This presupposes that the algebra $\underline{\mathbf{M}}$ is dualisable in the first place. We now turn from the practical to the theoretical and address the most fundamental question in the theory of natural dualities.

10.1 Dualisability

Recall that an algebra $\underline{\mathbf{M}}$ admits a duality or is **dualisable** if there is some choice of $\underset{\sim}{\mathbf{M}} = \langle M; G, H, R, \mathcal{T} \rangle$ which yields a duality on the quasi-variety $\mathcal{A}$ generated by $\underline{\mathbf{M}}$. By Lemma 2.1.3, this is equivalent to asking that the brute force structure $\underset{\sim}{\mathbf{M}} = \langle M; \mathcal{B}, \mathcal{T} \rangle$ yields a duality on $\mathcal{A}$.

1.1 Dualisablity Problem *Which finite algebras are dualisable?*

Since this problem seems to be particularly intractable, it is natural to consider various modifications of it.

- *Characterise the dualisable algebras in a given class $\mathcal{C}$ of algebras.*

Of course, this modification of the dualisability problem is only of interest if the class $\mathcal{C}$ has some intrinsic appeal. For example, $\mathcal{C}$ might be the class of all finite tgroups, or more generally, the class of all finite algebras which possess a Mal'cev congruence permutability term (see A.11). In the next section we shall answer this question in the case that $\mathcal{C}$ is the class of all algebras which generate a congruence distributive variety.

When we don't have a complete solution to the previous question we can

- *give sufficient conditions for an algebra in $\mathcal{C}$ to be dualisable*,

thereby extending the list of algebras known to be dualisable, or

- *give necessary conditions for an algebra in $\mathcal{C}$ to be dualisable*,

thereby extending the list of algebras known to be non-dualisable. In Chapter 7 we considered the sufficiency question in the case that $\mathcal{C}$ is the class of all finite algebras in the quasi-variety generated by a finite dualisable algebra.

Turning the question on its head, we can

- *give necessary or sufficient conditions for a finite algebra* $\underline{\mathbf{M}}$ *to be dualised by a particular structure* $\underset{\sim}{\mathbf{M}} = \langle M; G, H, R, \mathcal{T}\rangle$ *definable in terms of* $\underline{\mathbf{M}}$.

For example, in Chapter 7 we gave sufficient conditions for a finite algebra $\underline{\mathbf{M}}$ to be dualised by $\underset{\sim}{\mathbf{M}} = \langle M; \mathrm{End}(\underline{\mathbf{M}}), \mathcal{T}\rangle$ and thereby characterised the endoprimal algebras in several familiar classes of algebras.

10.2 Near-unanimity as an obstacle to duality

In this section we shall discover that near-unanimity is a two-edged sword. One of the most used results in the theory of natural dualities is the NU Duality Theorem 2.3.4. It tells us that a finite algebra which has a near-unanimity term is dualisable (via a structure of finite type) and so yields a plethora of dualisable algebras. Nevertheless, near-unanimity is actually an obstacle to the existence of a natural duality: in the presence of even a small degree of congruence distributivity, a dualisable algebra must possess a near-unanimity term. As a consequence, we shall find large classes of algebras which are not dualisable just because they do not possess a near-unanimity term. Our two NU Obstacle Theorems come from Davey, Heindorf and McKenzie [DHM95].

If $\underline{\mathbf{M}}$ possesses a majority term, then the NU Duality Theorem provides us with a particularly well-behaved duality: $\underset{\sim}{\mathbf{M}}$ is of finite type (that is, $G \cup H \cup R$ is finite), the operations in G and the partial operations in H are at most unary and the relations in R are at most binary, and $\underset{\sim}{\mathbf{M}}$ is injective in $\mathfrak{X}$. Our first theorem provides a partial converse.

2.1 Small NU Obstacle Theorem *Assume that* $\underset{\sim}{\mathbf{M}}$ *is of finite type and that the (partial) operations in* $G \cup H$ *are at most unary. Let* m *be the maximum of the arities of the relations in* R. *If* $\underset{\sim}{\mathbf{M}}$ *yields a duality on* $\mathcal{A}$ *with* $\underset{\sim}{\mathbf{M}}$ *injective in* $\mathfrak{X}_{\mathrm{fin}}$, *then* $\underline{\mathbf{M}}$ *has a* $(k{+}1)$*-ary near-unanimity term where* $k = \max\{2, m\}$.

Proof Let $\underset{\sim}{\mathbf{M}}$ and k be as described in the statement of the theorem. Define $X := \dot{\bigcup}\{\, X_{ab} \mid a, b \in M\,\}$, where

$$X_{ab} := \{(a, \dots, a, b), (a, \dots, b, a), \dots, (b, a, \dots, a)\} \subseteq M^{k+1}.$$

Since the (partial) operations in $G \cup H$ are at most unary, $\mathbf{X}$ is a substructure of $\underset{\sim}{\mathbf{M}}^{k+1}$. Define $\alpha \colon X \to M$ by $\alpha(x) = a$ if $x \in X_{ab}$. On any k-or-fewer-element subset of X, the map α is a projection. Hence α preserves all the relations in R since they are at most k-ary, and clearly α preserves the (partial) operations in $G \cup H$ as they are at most unary. Thus $\alpha \in \mathfrak{X}(\mathbf{X}, \underset{\sim}{\mathbf{M}})$. Since $\underset{\sim}{\mathbf{M}}$ is injective in $\mathfrak{X}_{\mathrm{fin}}$, there is a morphism

$t\colon \underset{\sim}{\mathbf{M}}^{k+1} \to \underset{\sim}{\mathbf{M}}$ which extends α. As $\underset{\sim}{\mathbf{M}}$ yields a duality on $\mathcal{A}$, the map t is a $(k+1)$-ary term function on $\underline{\mathbf{M}}$ (by the First Duality Theorem 2.2.2) and $t{\restriction}_X = \alpha$ says exactly that t is a near-unanimity function. ∎

While this theorem shows that properties of $\underset{\sim}{\mathbf{M}}$ can force $\underline{\mathbf{M}}$ to have a near-unanimity term, the Big NU Obstacle Theorem shows that, in the presence of dualisability, simple properties of $\underline{\mathbf{M}}$ can force it to have a near-unanimity term. A lattice which satisfies the quasi-equation

$$x \vee y = x \vee z \implies x \vee (y \wedge z) = x \vee y$$

is called **join-semidistributive**.

2.2 Big NU Obstacle Theorem *The following are equivalent:*

(i) $\underline{\mathbf{M}}$ *has a near-unanimity term;*

(ii) $\underline{\mathbf{M}}$ *is dualisable and generates a congruence distributive variety;*

(iii) $\underline{\mathbf{M}}$ *is dualisable and* $\mathrm{Con}(\mathbf{A})$ *is join-semidistributive for every finite algebra* $\mathbf{A}$ *in* $\mathcal{A}$.

Furthermore, if $\underline{\mathbf{M}}$ *is dualisable and* $\mathrm{Con}_{\mathcal{A}}(\mathbf{A})$ *is join-semidistributive for every finite algebra* $\mathbf{A}$ *in* $\mathcal{A}$, *then* $\underline{\mathbf{M}}$ *has a near-unanimity term.*

We present a proof of this result which is due to Zádori [Zád95]. In fact, Zádori proves a little more—see Exercise 10.1. First observe that (i) implies (ii) by the NU Duality Theorem 2.3.4 and Theorem A.14, and that (ii) trivially implies (iii). It remains to prove that, if $\underline{\mathbf{M}}$ is dualisable and either $\mathrm{Con}(\mathbf{A})$ or $\mathrm{Con}_{\mathcal{A}}(\mathbf{A})$ is join-semidistributive for every finite algebra $\mathbf{A}$ in $\mathcal{A}$, then $\underline{\mathbf{M}}$ has a near-unanimity term. We will present the proof via three intermediate results.

Define

$$N_4 := N_4^{(1)} \cup N_4^{(2)} \cup N_4^{(3)} \cup N_4^{(4)},$$

where $N_4^{(1)} := \{(b,a,a,a) \mid a,b \in M\}$, $N_4^{(2)} := \{(a,b,a,a) \mid a,b \in M\}$, and $N_4^{(3)}$ and $N_4^{(4)}$ are defined similarly. Recall that for $X \subseteq M^S$, the term closure of X in M^S is denoted by X^{tm}. (See Exercise 3.3.)

2.3 Lemma *Assume that* $\mathrm{Con}(\mathbf{A})$ *or* $\mathrm{Con}_{\mathcal{A}}(\mathbf{A})$ *is join-semidistributive for every finite algebra* $\mathbf{A} \in \mathcal{A}$. *Then* $(c,c,d,d) \notin N_4^{\mathrm{tm}}$ *for all* $c,d \in M$.

Proof We shall prove the $\mathrm{Con}_{\mathcal{A}}(\mathbf{A})$ version. The proof for the $\mathrm{Con}(\mathbf{A})$ version is obtained by simply removing the '$\mathcal{A}$-' prefix wherever it occurs.

Assume that $\mathrm{Con}_{\mathcal{A}}(\mathbf{A})$ is join-semidistributive for every finite algebra $\mathbf{A} \in \mathcal{A}$ and, by way of contradiction, suppose that there exist $c \neq d$ in M such that $(c,c,d,d) \in N_4^{\mathrm{tm}}$. Denote the subalgebra $\mathbf{F}_{\underline{\mathbf{M}}}(4){\restriction}_{N_4}$ of $\underline{\mathbf{M}}^{M^4}$ by $\mathbf{A}$. By Exercise 3.3, there exists a homomorphism $u\colon \mathbf{A} \to \underline{\mathbf{M}}$ with

$u(\rho_1) = u(\rho_2) = c$ and $u(\rho_3) = u(\rho_4) = d$, where $\rho_i := \pi_i{\restriction}_{N_4} \in A$ is the restriction to N_4 of the ith projection. Let θ be the least $\mathcal{A}$-congruence on $\mathbf{A}$ containing (ρ_1, ρ_2) and (ρ_3, ρ_4) and let ψ be the least $\mathcal{A}$-congruence on $\mathbf{A}$ which identifies ρ_1, ρ_2, ρ_3 and ρ_4. Note that $\theta \leqslant \ker(u)$. For $i = 1, 2, 3, 4$, let θ_i be the kernel of the projection $\pi_{N_4^{(i)}}$ which restricts each α in A to the set $N_4^{(i)}$.

Since N_4 is the union of the sets $N_4^{(i)}$, we have $\theta_1 \wedge \theta_2 \wedge \theta_3 \wedge \theta_4 = \mathbf{0}^{\mathbf{A}}$ in $\mathrm{Con}_{\mathcal{A}}(\mathbf{A})$. We claim that $\theta \vee \theta_i = \psi$ for all i. Certainly, $\theta \leqslant \psi$. We now show that $\theta_1 \leqslant \psi$. Let $\alpha, \beta \in A$. Since $\mathbf{A}$ is generated by ρ_1, ρ_2, ρ_3 and ρ_4, there exist 4-ary term functions t_1 and t_2 such that $\alpha = t_1(\rho_1, \rho_2, \rho_3, \rho_4)$ and $\beta = t_2(\rho_1, \rho_2, \rho_3, \rho_4)$. Thus

$$\begin{aligned}\alpha \equiv \beta \pmod{\theta_1} &\iff \alpha(b,a,a,a) = \beta(b,a,a,a) \text{ for all } a, b \in M\\ &\implies \alpha(a,a,a,a) = \beta(a,a,a,a) \text{ for all } a \in M\\ &\implies t_1(\rho_1,\rho_1,\rho_1,\rho_1) = t_2(\rho_1,\rho_1,\rho_1,\rho_1), \text{ in } \mathbf{A},\end{aligned}$$

since, for all $(a, b, c, d) \in N_4$, we have

$$\begin{aligned}t_1(\rho_1,\rho_1,\rho_1,\rho_1)(a,b,c,d) &= t_1(a,a,a,a)\\ &= t_1(\rho_1,\rho_2,\rho_3,\rho_4)(a,a,a,a) = \alpha(a,a,a,a),\end{aligned}$$

and similarly for t_2 and β. Hence $\alpha \equiv \beta \pmod{\theta_1}$ implies that

$$\begin{aligned}\alpha &= t_1(\rho_1,\rho_2,\rho_3,\rho_4) \equiv t_1(\rho_1,\rho_1,\rho_1,\rho_1)\\ &= t_2(\rho_1,\rho_1,\rho_1,\rho_1) \equiv t_2(\rho_1,\rho_2,\rho_3,\rho_4) = \beta \pmod{\psi},\end{aligned}$$

that is, $\theta_1 \leqslant \psi$, and consequently $\theta \vee \theta_1 \leqslant \psi$ in $\mathrm{Con}_{\mathcal{A}}(\mathbf{A})$. To show that $\theta \vee \theta_1 \geqslant \psi$, we must prove that $\theta \vee \theta_1$ identifies ρ_1, ρ_2, ρ_3 and ρ_4. Since $\rho_2(b,a,a,a) = \rho_3(b,a,a,a) = \rho_4(b,a,a,a) = a$, the maps ρ_2, ρ_3 and ρ_4 agree on $N_4^{(1)}$ and thus they are identified by the congruence θ_1 and therefore by $\theta \vee \theta_1$. But θ identifies ρ_1 and ρ_2, so $\theta \vee \theta_1$ identifies all four projections. Thus $\theta \vee \theta_1 = \psi$, and, similarly, we have $\theta \vee \theta_i = \psi$ for all i. By the join-semidistributivity of $\mathrm{Con}_{\mathcal{A}}(\mathbf{A})$, we have

$$\psi = \theta \vee (\theta_1 \wedge \theta_2 \wedge \theta_3 \wedge \theta_4) = \theta \vee \mathbf{0}^{\mathbf{A}} = \theta.$$

As $\theta \leqslant \ker(u)$ we have $\psi \leqslant \ker(u)$, whence $c = u(\rho_1) = u(\rho_3) = d$, a contradiction. Hence, $(c, c, d, d) \notin N_4^{\mathrm{tm}}$, as claimed. ∎

We define an infinite version of N_4 in the obvious way: for $n \in \mathbb{N}$ and $a, b \in M$, define $a_n^b \colon \mathbb{N} \to M$ by

$$a_n^b(k) = \begin{cases} a & \text{if } k \neq n,\\ b & \text{if } k = n,\end{cases}$$

and let

$$N_\omega := \{\, a_n^b \colon \mathbb{N} \to M \mid n \in \mathbb{N} \text{ and } a, b \in M\}.$$

The following extension of N_ω plays a vital role:

$$C := \{x \colon \mathbb{N} \to M \mid (\exists a \in M)(\forall b \in M \backslash \{a\})\ |x^{-1}(b)| \leqslant 1\,\}.$$

2.4 Lemma *If $(c,c,d,d) \notin N_4^{\mathrm{tm}}$ for all $c,d \in M$, then $N_\omega^{\mathrm{tm}} \subseteq C$.*

Proof If t is a 4-ary term and T is a four-element subset of $\mathbb{N}$, then we shall denote the induced $\mathbb{N}$-ary term function on $\underline{\mathbf{M}}$ by t^T, thus $t^T := t \circ \pi_T$, where $\pi_T \colon M^{\mathbb{N}} \to M^T$ is the obvious projection. Since $\pi_T(N_\omega) = N_4$, we have

$$(1) \qquad t_1{\restriction}_{N_4} = t_2{\restriction}_{N_4} \implies t_1^T{\restriction}_{N_\omega} = t_2^T{\restriction}_{N_\omega},$$

for every four-element subset T of $\mathbb{N}$ and for all 4–ary terms t_1 and t_2.

Assume that $(c,c,d,d) \notin N_4^{\mathrm{tm}}$ for all $c,d \in M$. Let $y \in M^{\mathbb{N}} \backslash C$. We shall prove that $y \notin N_\omega^{\mathrm{tm}}$. As $y \notin C$, there exists a four-element subset $T = \{a_1, a_2, b_1, b_2\}$ of $\mathbb{N}$ such that

$$(2) \qquad c := y(a_1) = y(a_2) \neq d := y(b_1) = y(b_2).$$

As $(c,c,d,d) \notin N_4^{\mathrm{tm}}$, there exist 4–ary terms t_1 and t_2 such that $t_1{\restriction}_{N_4} = t_2{\restriction}_{N_4}$ but $t_1(c,c,d,d) \neq t_2(c,c,d,d)$. By (1), we have $t_1^T{\restriction}_{N_\omega} = t_2^T{\restriction}_{N_\omega}$ and, since $t_1^T(y) \neq t_2^T(y)$, we conclude that $y \notin N_\omega^{\mathrm{tm}}$. ∎

The next lemma completes the proof of the Big NU Obstacle Theorem.

2.5 Lemma *Assume that $\underline{\mathbf{M}}$ satisfies $N_\omega^{\mathrm{tm}} \subseteq C$. Then $\underline{\mathbf{M}}$ is dualisable if and only if it has a near-unanimity term.*

Proof If $\underline{\mathbf{M}}$ has a near-unanimity term, then it is dualisable by the NU Duality Theorem 2.3.4. Assume that $N_\omega^{\mathrm{tm}} \subseteq C$ and define $\alpha \colon N_\omega^{\mathrm{tm}} \to M$ by $\alpha(x) := \lim\limits_{n\to\infty} x(n)$. To see that α is continuous, let $a \in M$ and let $\alpha(x) = a$. Choose $m, n \in \mathbb{N}$ with $m \neq n$ and $x(m) = x(n) = a$. Consider the clopen subset $V := N_\omega^{\mathrm{tm}} \cap (U_{m,a} \cap U_{n,a})$ of N_ω^{tm}. If $y \in V$, then since $N_\omega^{\mathrm{tm}} \subseteq C$, we have $\alpha(y) = \lim\limits_{n\to\infty} y(n) = a$, whence $x \in V \subseteq \alpha^{-1}(a)$. Thus α is continuous.

Next we check that α preserves every ℓ-ary algebraic relation r on $\underline{\mathbf{M}}$. On any finite subset $\{x_1, \dots, x_\ell\}$ of N_ω^{tm}, the map α is given by evaluation at some $k \in \mathbb{N}$. Indeed, define

$$k(x) := \min\{\, n \in \mathbb{N} \mid x(m) = x(n) \text{ for all } m \geqslant n\,\}$$

and let $k := \max\{k(x_1), \dots, k(x_\ell)\}$. Clearly, $\alpha(x_i) = x_i(k)$ for $i = 1, \dots, \ell$. Thus $(x_1, \dots, x_\ell) \in r^{N_\omega^{\mathrm{tm}}}$ implies that $(\alpha(x_1), \dots, \alpha(x_\ell)) =$

$(x_1(k), \ldots, x_\ell(k)) \in r^M$. Consequently, α preserves every algebraic relation on $\underline{\mathbf{M}}$. If $\underline{\mathbf{M}}$ is dualisable, then brute force yields a duality on $\mathcal{A}$ and consequently α extends to an $\mathbb{N}$-ary term function τ on $\underline{\mathbf{M}}$, by the Third Duality Theorem 3.1.6.

For some $n \geqslant 3$, there exists an n-ary term t such that $\tau(x) = t(x(1), \ldots, x(n))$ for all $x \in M^{\mathbb{N}}$. Hence, for all $a, b \in M$,

$$t(b, a, \ldots, a) = t(a_1^b(1), \ldots, a_1^b(n)) = \tau(a_1^b) = \alpha(a_1^b) = \lim_{n \to \infty} a_1^b(n) = a,$$

and similarly, $t(b, a, \ldots, a) = t(a, b, a, \ldots, a) = \cdots = t(a, a, \ldots, a, b) = a$. Thus, t is a near-unanimity term on $\underline{\mathbf{M}}$. ■

10.3 Some applications of the NU obstacle theorems

It seems highly unlikely that there is an algorithm which, given a finite algebra $\underline{\mathbf{M}}$ of finite type, would decide whether or not $\underline{\mathbf{M}}$ is dualisable. The closely related problem of whether there is an algorithm which will decide whether or not $\underline{\mathbf{M}}$ has a near-unanimity term may be more tractable. The Big NU Obstacle Theorem shows that a positive answer to the dualisability question yields a positive answer to the near-unanimity question. (First decide whether or not $\underline{\mathbf{M}}$ generates a congruence distributive variety: since $\mathbf{F}_{\underline{\mathbf{M}}}(3)$ is finite, Jónsson's First Theorem A.13 shows that this is decidable. If $\underline{\mathbf{M}}$ does not generate a congruence distributive variety, then $\underline{\mathbf{M}}$ has no near-unanimity term by Mitschke's Theorem A.14. If $\underline{\mathbf{M}}$ generates a congruence distributive variety, then, by the NU Duality Theorem and the Big NU Obstacle Theorem, $\underline{\mathbf{M}}$ has a near-unanimity term if and only if $\underline{\mathbf{M}}$ is dualisable, so, in this case, the decision algorithm for dualisability also decides the near-unanimity question.) McKenzie's [McK96] groundbreaking decidability results may yield the techniques required for a negative solution to the following problem.

3.1 Decidability Problem *Is the dualisability of a finite algebra of finite type decidable? In particular, is the existence of a near-unanimity term decidable?*

The NU Obstacle Theorems yield large classes of non-dualisable algebras. The following important result due to Daniels [Dan97] not only provides a very easy proof that certain algebras do not admit a duality, it also provides us with new examples of algebras which generate relatively congruence distributive quasi-varieties. A class $\mathcal{A}$ is called **relatively congruence distributive** if $\mathrm{Con}_{\mathcal{A}}(\mathbf{A})$ is a distributive lattice for all $\mathbf{A} \in \mathcal{A}$.

3.2 Proposition *Given the finite algebra $\underline{\mathbf{M}}$, suppose that there is a set G of at most unary total operations on M such that $\underset{\sim}{\mathbf{M}} = \langle M; G, \mathcal{T}\rangle$ satisfies* (FTC). *Then*

(i) $\underline{\mathbf{M}}$ *is injective in the quasi-variety $\mathcal{A}$ generated by $\underline{\mathbf{M}}$,*

(ii) *the quasi-variety $\mathcal{A}$ is relatively congruence distributive, and*

(iii) $\underline{\mathbf{M}}$ *is dualisable if and only if $\underline{\mathbf{M}}$ has a near-unanimity term.*

Proof Assume that $\underset{\sim}{\mathbf{M}} = \langle M; G, \mathcal{T}\rangle$ satisfies (FTC) and let $\widehat{\underset{\sim}{\mathbf{M}}} := \langle M; G, \mathcal{B}_\infty, \mathcal{T}\rangle$ be the amalgam of $\underset{\sim}{\mathbf{M}}$ with the ultimate brute force structure $\underset{\sim}{\mathbf{M}}_\infty := \langle M; \mathcal{B}_\infty\rangle$, where $\mathcal{B}_\infty$ denotes the class of all subalgebras of $\underline{\mathbf{M}}^\kappa$ where κ ranges over all non-zero cardinals. Since $\underset{\sim}{\mathbf{M}}_\infty$ yields a duality on $\mathcal{A}$ (see Theorem 2.3.1 and the discussion following it) and since $\underset{\sim}{\mathbf{M}}$ satisfies (FTC), it follows, by Theorem 3.2.9, that $\widehat{\underset{\sim}{\mathbf{M}}}$ yields a strong duality on $\mathcal{A}$. Since $\widehat{\underset{\sim}{\mathbf{M}}}$ is a total structure, the Total Structure Theorem 6.1.2 implies that $\underline{\mathbf{M}}$ is injective in $\mathcal{A}$, and by Theorem 3.2.1, $\mathrm{Con}_{\mathcal{A}}(\mathbf{A})$ is isomorphic to the lattice of closed substructures of $D(\mathbf{A})$ for all $\mathbf{A} \in \mathcal{A}$. As the operations in G are at most unary, the set of closed substructures of $D(\mathbf{A})$ is closed under union and intersection, whence $\mathrm{Con}_{\mathcal{A}}(\mathbf{A})$ is distributive. Thus $\mathcal{A}$ is relatively congruence distributive. The Big NU Obstacle Theorem now implies that $\underline{\mathbf{M}}$ is dualisable if and only if $\underline{\mathbf{M}}$ has a near-unanimity term. ■

If the type of $\underset{\sim}{\mathbf{M}}$ includes partial operations, we may use brute force rather than ultimate force to prove the following slight modification of the previous result.

3.3 Proposition *Assume that $\underset{\sim}{\mathbf{M}} = \langle M; G, H, \mathcal{T}\rangle$, that each $h \in G \cup H$ is at most unary and that $\underset{\sim}{\mathbf{M}}$ satisfies* (FTC). *Then $\mathcal{A}_{\text{fin}}$ is relatively congruence distributive and, therefore, $\underline{\mathbf{M}}$ is dualisable if and only if $\underline{\mathbf{M}}$ has a near-unanimity term.*

Recall from Chapter 5 that a non-trivial algebra $\underline{\mathbf{M}}$ is called **order primal** if some order $\leqslant$ on M determines the clone of term functions of $\underline{\mathbf{M}}$, that is, for all $n \in \mathbb{N}$, a map $t\colon M^n \to M$ is a term function of $\underline{\mathbf{M}}$ if and only if t is order-preserving with respect to $\leqslant$. The following result was first proved in Davey, Quackenbush and Schweigert [DQS90].

3.4 Dualisability of Order Primal Algebras *Let $\underline{\mathbf{M}}$ be a finite order primal algebra. Then*

(i) $\underline{\mathbf{M}}$ *generates a relatively congruence distributive quasi-variety in which it is injective, and*

(ii) $\underline{\mathbf{M}}$ *is dualisable if and only if it possesses a near-unanimity term.*

Proof Assume that $\underline{\mathbf{M}}$ is order primal with respect to $\leqslant$. If $\langle M; \leqslant\rangle$ is an antichain, then label any pair of elements of M with 0 and 1. If $\langle M; \leqslant\rangle$

is not an antichain, choose a pair of comparable elements of M and label them 0 and 1 with $0 < 1$. Let X be a subset of an ordered set $\langle Y; \leqslant\rangle$ and let $y \in Y\backslash X$. Define $t_1\colon Y \to M$ to be the characteristic function of $\uparrow y$ and $t_2\colon Y \to M$ to be the characteristic function of $(\uparrow y)\backslash\{y\}$. Then t_1 and t_2 are order-preserving, $t_1{\restriction}_X = t_2{\restriction}_X$ and $t_1(y) \neq t_2(y)$. It follows that $\underset{\sim}{\mathrm{M}} := \langle M; \mathcal{T}\rangle$ satisfies (FTC) and the result follows from Proposition 10.3.2 with $G = \varnothing$. ■

We shall say that a clone C on a finite set M is **dualisable** if the algebra $\underline{\mathbf{M}}_C := \langle M; C\rangle$ is dualisable. Hence, an algebra $\underline{\mathbf{M}}$ is dualisable if and only if its clone of term functions $\mathrm{Clo}(\underline{\mathbf{M}})$ is dualisable. Thus we obtain a useful reformulation of the Dualisability Problem.

3.5 Dualisable Clones Problem *Which clones on a finite set are dualisable?*

Reworded in this way, the previous result says that *the clone of monotone functions on a finite ordered set is dualisable if and only if the clone includes an n-ary near-unanimity function for some $n \geqslant 3$.*

We shall look at clones on a two-element set later in the chapter. For the moment, we shall consider certain clones on a finite set M with at least three elements. Recall that a function $f\colon M^n \to M$ is **essentially unary** if $f = \lambda \circ \pi_i$ for some projection $\pi_i\colon M^n \to M$ and some unary map $\lambda \in M^M$. Assume that $|M| = m$ and, for $1 \leqslant k \leqslant m$, consider the following sets of functions on M:

$$V_k := \{\, f\colon M^n \to M \mid n \in \mathbb{N} \text{ and } f \text{ is a projection or } |f(M^n)| \leqslant k \,\},$$

$$U_k := V_k \cup \{\, f\colon M^n \to M \mid n \in \mathbb{N} \text{ and } f \text{ is essentially unary} \,\}.$$

The sets U_k and V_k are clones on M for each k. The clone U_{m-1} is known as the **Słupecki clone** on M. When $k = m$, both V_k and U_k coincide with the clone of all finitary functions on M. Thus $\underline{\mathbf{M}}_{U_m}$ is primal and therefore is dualisable (see Theorem 4.1.1). At the other end of the spectrum, $\underline{\mathbf{M}}_{V_1}$ is equivalent to the algebra $\underline{\mathbf{M}}^0 = \langle M; a_1, \dots, a_m\rangle$ in which every element of M is the value of a nullary operation and there are no other operations. This algebra is dualisable—see Exercise 3.7. The algebra $\underline{\mathbf{M}}_{U_1}$ is equivalent to the unary algebra $\underline{\mathbf{M}}^u = \langle M; M^M\rangle$. We shall see a little later that $\underline{\mathbf{M}}^u$ is not dualisable—see Corollary 10.4.5. It is easy to see that the remaining clones U_k and V_k are non-dualisable.

3.6 Non-dualisable Słupecki-Type Clones *Let M be a finite set with $|M| = m \geqslant 3$. For $1 < k < m$, each of the algebras $\underline{\mathbf{M}}_{U_k}$ and $\underline{\mathbf{M}}_{V_k}$ generates a relatively congruence distributive quasi-variety and each of the clones U_k and V_k is non-dualisable.*

Proof We claim that $\underset{\sim}{\mathbf{M}} = \langle M; \mathcal{T}\rangle$ satisfies (FTC). Without loss of generality, assume that $\{0, 1, 2\} \subseteq M$. Let $X \subseteq M^n$ and let $y \in M^n \backslash X$. Let $t_1 \colon M^n \to M$ be the constant map onto $\{1\}$ and let $t_2 \colon M^n \to M$ be the characteristic function of X. Clearly, $t_1{\restriction}_X = t_2{\restriction}_X$ and $t_1(y) \neq t_2(y)$. As $|M| \geqslant 3$ and $k \geqslant 2$, we have $t_1, t_2 \in V_k \subseteq U_k$, whence (FTC) holds. By Proposition 10.3.2, it remains to see that U_{m-1} does not contain a near-unanimity function. We leave this as an easy exercise. ∎

10.4 The balanced subalgebra method

In this section we present a technique due to L. Heindorf which allows us to construct a multitude of non-dualisable algebras. Consider the subalgebra $\mathbf{A}^\#$ of $\underline{\mathbf{M}}^{\mathcal{A}(\mathbf{A},\underline{\mathbf{M}})}$ whose underlying set is the set

$$A^\# := \{\, e_{\mathbf{A}}(a) \colon \mathcal{A}(\mathbf{A}, \underline{\mathbf{M}}) \to M \mid a \in A \,\}$$

of all evaluation maps from $\mathcal{A}(\mathbf{A}, \underline{\mathbf{M}})$ to M. Since the evaluation maps are continuous, $\mathbf{A}^\#$ is a subalgebra of the algebra $C(\mathcal{A}(\mathbf{A}, \underline{\mathbf{M}}), \underline{\mathbf{M}})$ of all continuous maps from the $\mathcal{A}(\mathbf{A}, \underline{\mathbf{M}})$ into $\underline{\mathbf{M}}$ (with the discrete topology) and $e_{\mathbf{A}} \colon \mathbf{A} \to \mathbf{A}^\#$ is an isomorphism. Let $u \colon \mathbf{A}^\# \to \underline{\mathbf{M}}$ be a homomorphism, and define $x := u \circ e_{\mathbf{A}}$. Then $x \in \mathcal{A}(\mathbf{A}, \underline{\mathbf{M}})$ and, for all $a \in A$,

$$u(e_{\mathbf{A}}(a)) = (u \circ e_{\mathbf{A}})(a) = x(a) = (e_{\mathbf{A}}(a))(x) = \rho_x(e_{\mathbf{A}}(a)),$$

where $\rho_x \colon \mathbf{A}^\# \to \underline{\mathbf{M}}$ is the restriction to the algebra $\mathbf{A}^\#$ of the projection $\pi_x \colon \underline{\mathbf{M}}^{\mathcal{A}(\mathbf{A},\underline{\mathbf{M}})} \to \underline{\mathbf{M}}$. We have proved that every homomorphism from $\mathbf{A}^\#$ to $\underline{\mathbf{M}}$ is the restriction of a projection. Since it is trivial that $A^\#$ separates the points of $\mathcal{A}(\mathbf{A}, \underline{\mathbf{M}})$, this motivates our next definition and proves the following lemma.

Let $\mathbf{Y}$ be a Boolean space. A subalgebra $\mathbf{B}$ of the Boolean power $C(\mathbf{Y}, \underline{\mathbf{M}})$ is called **balanced** if B separates the points of $\mathbf{Y}$ (that is, if $x, y \in Y$ and $b(x) = b(y)$ for all $b \in B$, then $x = y$) and each homomorphism $u \colon \mathbf{B} \to \underline{\mathbf{M}}$ is given by evaluation at some point $y \in Y$ (that is, $u = \rho_y := \pi_y{\restriction}_B$, where $\pi_y \colon M^Y \to M$ is the natural projection).

4.1 Lemma *For all $\mathbf{A} \in \mathcal{A}$ the algebra $\mathbf{A}^\#$ is a balanced subalgebra of $C(\mathcal{A}(\mathbf{A}, \underline{\mathbf{M}}), \underline{\mathbf{M}})$ which is isomorphic to $\mathbf{A}$, and hence every algebra in $\mathcal{A}$ is isomorphic to a balanced subalgebra of $C(\mathbf{Y}, \underline{\mathbf{M}})$ for some Boolean space $\mathbf{Y}$.*

If $\mathbf{B}$ is a balanced subalgebra of $C(\mathbf{Y}, \underline{\mathbf{M}})$, then the connection between $\mathbf{Y}$ and $\mathbf{B}$ is very tight. We leave the proof of the following lemma to the reader.

4.2 Lemma *Let* $\mathbf{Y}$ *be a Boolean space and let* $\mathbf{B}$ *be a subalgebra of* $C(\mathbf{Y}, \underline{\mathbf{M}})$. *Then* $\mathbf{B}$ *is balanced precisely when the map* $\varphi\colon Y \to \mathcal{A}(\mathbf{B}, \underline{\mathbf{M}})$, *defined by* $\varphi(y) := \rho_y = \pi_y{\restriction}_B$, *is a homeomorphism.*

As usual, we put the discrete topology on M and then endow M^Y with the product topology. We refer to the induced subspace topology on $C(\mathbf{Y}, \underline{\mathbf{M}})$ as the **pointwise topology**. The following result of L. Heindorf raises Exercise 2.1 to theorem status. (See [Hei93] and [DHM95].)

4.3 Balanced Subalgebra Theorem *Let* $\underline{\mathbf{M}}$ *be a finite non-trivial algebra and let* $\mathbf{A} \in \mathcal{A} := \mathbb{ISP}\underline{\mathbf{M}}$. *The following are equivalent:*

(i) *there is some structure* $\underset{\sim}{\mathbf{M}}$ *which yields a duality on* $\mathbf{A}$;

(ii) $A^{\#}$ *is closed in the pointwise topology on* $C(\mathcal{A}(\mathbf{A}, \underline{\mathbf{M}}), \underline{\mathbf{M}})$;

(iii) $\mathbf{A}$ *is isomorphic to a closed and balanced subalgebra of a Boolean power of* $\underline{\mathbf{M}}$;

(iv) *if a balanced subalgebra* $\mathbf{B}$ *of a Boolean power of* $\underline{\mathbf{M}}$ *is isomorphic to* $\mathbf{A}$, *then* B *is closed in the pointwise topology.*

Proof The equivalence of (i) and (ii) is outlined in Exercise 2.1, and Lemma 10.4.1 shows that (ii) implies (iii) and that (iv) implies (ii) since $\mathbf{A}^{\#} \cong \mathbf{A}$. Thus it remains to prove that (iii) implies (iv).

Assume that $\mathbf{B}$ and $\mathbf{C}$ are balanced subalgebras of the Boolean powers $C(\mathbf{Y}, \underset{\sim}{\mathbf{M}})$ and $C(\mathbf{X}, \underset{\sim}{\mathbf{M}})$ respectively and that $u\colon \mathbf{C} \to \mathbf{B}$ is an isomorphism. We claim that if C is closed in $C(\mathbf{X}, \underset{\sim}{\mathbf{M}})$, then B is closed in $C(\mathbf{Y}, \underset{\sim}{\mathbf{M}})$. That (iii) implies (iv) then follows easily.

The discretely topologised set $\underset{\sim}{\mathbf{M}}_0 = \langle M; \mathcal{T}\rangle$ yields a preduality between $\mathcal{A}$ and the category $\mathcal{Z}$ of Boolean spaces, so $D(u)\colon C(\mathbf{B}, \underline{\mathbf{M}}) \to C(\mathbf{C}, \underline{\mathbf{M}})$ is a homeomorphism. Composing $D(u)$ with the two isomorphisms given by Lemma 10.4.2,

$$\mathbf{Y} \xrightarrow{\varphi_{\mathbf{Y}}} \mathcal{A}(\mathbf{B}, \underline{\mathbf{M}}) \xrightarrow{D(u)} \mathcal{A}(\mathbf{C}, \underline{\mathbf{M}}) \xrightarrow{\varphi_{\mathbf{X}}^{-1}} \mathbf{X},$$

gives us the isomorphism $\psi := \varphi_{\mathbf{X}}^{-1} \circ D(u) \circ \varphi_{\mathbf{Y}}\colon \mathbf{Y} \to \mathbf{X}$. For $y \in Y$, the point $\psi(y)$ is the unique point $x \in X$ such that $\rho_x = \rho_y \circ u$, that is, $c(\psi(y)) = u(c)(y)$ for each $c \in C$. Since ψ is a bijection, u extends to an isomorphism $v\colon \underline{\mathbf{M}}^X \to \underline{\mathbf{M}}^Y$ defined by $v(c)(y) = c(\psi(y))$ for all $c \in M^X$ and $y \in Y$. Clearly v is also a homeomorphism with respect to the pointwise topology. Consequently, if C is closed in $C(\mathbf{X}, \underset{\sim}{\mathbf{M}})$, then B is closed in $C(\mathbf{Y}, \underset{\sim}{\mathbf{M}})$. ■

This theorem provides a technique for proving that $\underline{\mathbf{M}}$ is not dualisable: if we can find a Boolean space $\mathbf{X}$ and a subalgebra $\mathbf{B}$ of $C(\mathbf{X}, \underline{\mathbf{M}})$ which is balanced but not closed in the pointwise topology, then $\underline{\mathbf{M}}$ is not dualisable. The following example, which is adapted from Heindorf [Hei93], illustrates this **balanced subalgebra method**.

4.4 Non-dualisable Sub-Słupecki Clones *Let $M = \{0, 1, \dots, m-1\}$ for some $m \geqslant 3$ and let $\underline{\mathbf{M}} = \langle M; F\rangle$, where F contains all $\{0,1\}$-valued unary maps on M and is a subset of the Słupecki clone on M. Then $\underline{\mathbf{M}}$ is not dualisable. Indeed, let $\mathbf{X}_0, \dots, \mathbf{X}_{m-1}$ be infinite Boolean spaces, let $\mathbf{X}$ be the disjoint union of $\mathbf{X}_0, \dots, \mathbf{X}_{m-1}$ and define A to be the set of all continuous maps $a\colon \mathbf{X} \to \underline{\mathbf{M}}$ such that the partition of X associated with $\ker(a)$ is not $\{X_0|X_1|\dots|X_{m-1}\}$; then $\mathbf{A}$ is a balanced subalgebra of $C(\mathbf{X}, \underline{\mathbf{M}})$ which is not closed in the pointwise topology.*

Proof Let $\underline{\mathbf{M}}$, $\mathbf{X}_i$, $\mathbf{X}$ and A be as described. Note that A is just the set of all continuous maps from $\mathbf{X}$ into $\underline{\mathbf{M}}$ with $m!$ elements deleted. On any finite subset of X, each of the deleted elements agrees with some function in A and consequently A is not closed in $C(\mathbf{X}, \underline{\mathbf{M}})$. The fact that F is a subset of the Słupecki clone on M guarantees that $\mathbf{A}$ is a subalgebra of $C(\mathbf{X}, \underline{\mathbf{M}})$. We now show that the assumption that F contains all $\{0,1\}$-valued unary maps on M implies that $\mathbf{A}$ is balanced. Let $B := C(\mathbf{X}, \{0,1\})$. Note that B is a subset of A and that while B is not necessarily a subalgebra of $\mathbf{A}$, the set B is closed under the usual Boolean operations $\vee$, $\wedge$ and $\neg$ on $\{0,1\}$. It is clear that A separates the points of X; in fact, the subset B does.

Let $u\colon \mathbf{A} \to \underline{\mathbf{M}}$ be a homomorphism. We must find $x \in X$ such that $u = \rho_x$. As a first step we prove that

$$X_u := \bigcap \{\, a^{-1}(1) \mid a \in B \text{ and } u(a) = 1 \,\}$$

is non-empty. Since $\mathbf{X}$ is compact, by Exercise B.1, it suffices to show that $\{\, a^{-1}(1) \mid a \in B \text{ and } u(a) = 1 \,\}$ satisfies the Finite Intersection Property. We achieve this via a sequence of claims. We shall constantly use the fact that u commutes with every unary map $\lambda\colon M \to \{0,1\}$.

(i) *For all $a \in B$, we have $u(a) \in \{0,1\}$, $u(\neg a) = \neg(u(a))$, $u(\hat{0}) = 0$ and $u(\hat{1}) = 1$, where $\hat{0}$ and $\hat{1}$ are the constant maps.*

Define $\lambda_1\colon M \to \{0,1\}$ by $\lambda_1 = \begin{bmatrix} 0 & 1 & * \\ 1 & 0 & 0 \end{bmatrix}$, which is to be read vertically, that is, λ_1 maps 0 to 1, and 1 to 0, and every other element of M (hence the wild card $*$) to 0, and define $\lambda_2\colon M \to \{0,1\}$ by $\lambda_2 = \begin{bmatrix} 0 & 1 & * \\ 0 & 1 & 0 \end{bmatrix}$. Since $a = \lambda_2 a$, we have $u(a) = u(\lambda_2 a) = \lambda_2 u(a) \in \{0,1\}$. Moreover, $u(\neg a) = u(\lambda_1 a) = \lambda_1 u(a) = \neg(u(a))$. Similar arguments, using the constant unary maps $\underline{0}, \underline{1}\colon M \to \{0,1\}$, yield $u(\hat{0}) = 0$ and $u(\hat{1}) = 1$.

(ii) *If $a, b \in B$ and $u(a) = u(b) = 1$, then $a \wedge b \neq 0$.*

Let $a, b \in B$ with $u(a) = u(b) = 1$ and suppose that $a \wedge b = 0$, that is, $a^{-1}(1) \cap b^{-1}(1) = \varnothing$. Define $c \in C(\mathbf{X}, \underline{\mathbf{M}})$ by

$$c(x) = \begin{cases} 0 & \text{if } x \in a^{-1}(1), \\ 1 & \text{if } x \in b^{-1}(1), \\ 2 & \text{otherwise.} \end{cases}$$

`Case (a):` $c \in A$. In this case, $\lambda_1 c = a$ and $\lambda_2 c = b$, whence

$$\lambda_1 u(c) = u(\lambda_1 c) = u(a) = 1 = u(b) = u(\lambda_2 c) = \lambda_2 u(c),$$

from which we obtain the contradiction $0 = u(c) = 1$.

`Case (b):` $c \notin A$. Thus the partition associated with $\ker(c)$ must be $\{X_0, \ldots, X_{m-1}\}$. Hence $m = 3$ and we may assume without loss of generality that $a^{-1}(1) = X_0$, $b^{-1}(1) = X_1$ and $X \backslash (a^{-1}(1) \cup b^{-1}(1)) = X_2$. Let $X_2 = X_{20} \dot\cup X_{21}$ be a non-trivial clopen partition of X_2 and let $d \in B$ be the characteristic function of $X_1 \cup X_{21}$. By (i), $\{u(d), u(\neg d)\} = \{0, 1\}$. If $u(d) = 1$, then Case (a) applies to a and d and we obtain a contradiction. If $u(\neg d) = 1$, then Case (a) applies to b and $\neg d$ and again we obtain a contradiction. Each case lead to a contradiction; consequently, $a \wedge b \neq 0$.

(iii) *If $a, b \in B$ with $u(a) = 1$ and $a \leqslant b$, then $u(b) = 1$.*

If $a, b \in B$ with $u(a) = 1$ and $u(b) = 0$, then $u(\neg b) = 1$ whence, by (ii), $a^{-1}(1) \cap b^{-1}(0) = a^{-1}(1) \cap (\neg b)^{-1}(1) \neq \varnothing$. Let $x \in a^{-1}(1) \cap b^{-1}(0)$, then $a(x) = 1 \nleqslant 0 = b(x)$, and consequently, $a \nleqslant b$. Thus $u(a) = 1$ and $a \leqslant b$ imply $u(b) = 1$.

(iv) *If $a, b \in B$ and $u(a) = u(b) = 1$, then $u(a \wedge b) = 1$.*

Consider $c_1, c_2 \colon \mathbf{X} \to \underline{\mathbf{M}}$ defined by

$$c_1(x) = \begin{cases} 1 & \text{if } x \in a^{-1}(1) \cap b^{-1}(1), \\ 2 & \text{if } x \in a^{-1}(1) \backslash b^{-1}(1), \\ 0 & \text{otherwise,} \end{cases}$$

and

$$c_2(x) = \begin{cases} 1 & \text{if } x \in a^{-1}(1) \cap b^{-1}(1), \\ 2 & \text{if } x \in b^{-1}(1) \backslash a^{-1}(1), \\ 0 & \text{otherwise.} \end{cases}$$

We shall distinguish three cases.

`Case (a):` $c_1 \in A$. Define $\lambda_3, \lambda_4 \colon M \to \{0, 1\}$ by $\lambda_3 = \begin{bmatrix} 0 & 1 & * \\ 0 & 1 & 1 \end{bmatrix}$ and $\lambda_4 = \begin{bmatrix} 0 & 1 & * \\ 1 & 1 & 0 \end{bmatrix}$. Then $\lambda_3 c_1 = a$ and therefore $\lambda_3 u(c_1) = u(\lambda_3 c_1) =$

$u(a) = 1$, whence $u(c_1) \neq 0$. But $b \leqslant \lambda_4 c_1$ and $u(b) = 1$ so, by (iii), $1 = u(\lambda_4 c_1) = \lambda_4 u(c_1)$, whence $u(c_1) \in \{0, 1\}$. Consequently, $u(c_1) = 1$. As $a \wedge b = \lambda_2 c_1$, we conclude that $u(a \wedge b) = u(\lambda_2 c_1) = \lambda_2 u(c_1) = \lambda_2(1) = 1$.

`Case (b):` $c_2 \in A$. By symmetry, this case also yields $u(a \wedge b) = 1$.

`Case (c):` $c_1, c_2 \notin A$. In this case, we may assume without loss of generality that $X_0 = a^{-1}(1) \cap b^{-1}(1)$, $X_1 = a^{-1}(1) \backslash b^{-1}(1)$ and $X_2 = b^{-1}(1) \backslash a^{-1}(1)$. Let $X_2 = X_{20} \,\dot\cup\, X_{21}$ be a non-trivial, clopen partition of X_2 and let d be the characteristic function of $X_0 \cup X_1 \cup X_{21}$. Thus $a < d < \underline{1}$. As $a \leqslant d$ and $u(a) = 1$, we conclude, by (iii), that $u(d) = 1$. As $b^{-1}(1) \cap d^{-1}(1) = X_0 \cup X_{21}$, Case (a) applies to b and d and hence $u(b \wedge d) = 1$. As $X \backslash (b \wedge d)^{-1}(1) = X_1 \cup X_{20}$, Case (a) applies to $b \wedge d$ and a and hence $u((b \wedge d) \wedge a) = 1$. But $(b \wedge d) \wedge a = a \wedge b$, whence $u(a \wedge b) = 1$.

It follows from (ii) and (iv) that $\{\, a^{-1}(1) \mid a \in B \text{ and } u(a) = 1 \,\}$ satisfies the Finite Intersection Property and so its intersection, X_u, is non-empty. Let $x \in X_u$. We claim that $u = \rho_x$. Let $a \in B$. If $u(a) = 1$ then $\rho_x(a) = a(x) = 1 = u(a)$ as $x \in X_u$. If $u(a) = 0$, then $u(\neg a) = 1$, by (i), whence $\neg a(x) = 1$ (as $x \in X_u$), and so $\rho_x(a) = a(x) = \neg\neg a(x) = \neg 1 = 0 = u(a)$. Hence, $u(a) = \rho_x(a)$ for all $a \in B$. For all $\lambda \in \{0,1\}^M$ and $a \in A$, we have $\lambda a \in B$ and thus $\lambda u(a) = u(\lambda a) = \rho_x(\lambda a) = \lambda \rho_x(a)$. Since such maps λ separate the points of M, it follows that $u(a) = \rho_x(a)$. Thus $u = \rho_x$ and consequently $\mathbf{A}$ is a balanced subalgebra of $C(\mathbf{X}, \underline{\mathbf{M}})$.■

4.5 Corollary *Let M be a finite set with at least three elements and assume that $\{0,1\}$ is a subset of M. Then both $\underline{\mathbf{M}}^{01} := \langle M; \{0,1\}^M \rangle$ and $\underline{\mathbf{M}}^{u} := \langle M; M^M \rangle$ are non-dualisable.*

The second corollary of the Non-dualisable Sub-Słupecki Clones result tells us that if M has at least three elements then non-dualisable clones on M abound. This is in stark contrast to the case where $|M| = 2$: only eight of the countably many clones on a two-element set are non-dualisable (see the Dualisable Two-Element Algebras Theorem 10.7.1).

4.6 Theorem (Heindorf [Hei93]) *If M is a finite set with at least three elements, then the number of non-dualisable clones on M is uncountable.*

Proof Let $M = \{0, 1, \dots, k{-}1\}$ with $k \geqslant 3$ and for $n \geqslant 2$ define an n-ary function on M by

$$f(a_1, \dots, a_n) := \begin{cases} 1 & \text{if } (a_1, \dots, a_n) \in X_{21}, \\ 0 & \text{otherwise,} \end{cases}$$

where $X_{21} := \{(2, \dots, 2, 1), (2, \dots, 1, 2), \dots, (1, 2, \dots, 2)\}$. For each subset S of $\{2, 3, \dots\}$, let C_S be the clone generated by

$$F_S := \{0,1\}^M \cup \{\, f_n \mid n \in S \,\}.$$

By the Non-dualisable Sub-Słupecki Clones result, the algebra $\underline{\mathbf{M}}_S := \langle M; F_S \rangle$ is non-dualisable and hence the clone C_S is non-dualisable. It remains to prove that if $S \neq T$, then $C_S \neq C_T$. We leave it to the reader to verify that if $n \notin S$, then $f_n \notin C_S$. ■

While the size of the set of non-dualisable clones on M is known, the size of the set of dualisable clones is not.

4.7 Countability Problem *Is there only a countable number of dualisable clones on a finite set?*

The most natural way to settle the Countability Problem in the affirmative would be to give an affirmative answer to the following problem, which is a clone-theoretic reformulation of the Finite Type Problem 2.2.13. A clone C on M is said to be of **finite degree** if there is some finite set R of finitary relations on M such that C is the clone of functions which preserve all of the relations in R.

4.8 Finite Degree Problem *Is every dualisable clone on a finite set M necessarily of finite degree?*

In Exercise 10.13 we ask the reader to show, with the aid of the theory presented in Section 7, that the answer to the Finite Degree Problem is 'yes' in the case that $|M| = 2$.

10.5 The ghost element method

We now present another approach to proving that a finite algebra $\underline{\mathbf{M}}$ is not dualisable. This approach dates back to Davey and Werner [DW83] where it was used to prove the first non-dualisability result, namely that the two-element implication algebra is not dualisable—see the Dualisable Two-Element Algebras Theorem 10.7.1 below. It was also used in R. McKenzie's original (unpublished) proof of the Big NU Obstacle Theorem.

The idea is quite simple. The algebra $\underline{\mathbf{M}}$ is dualisable provided that, for all $\mathbf{A} \in \mathcal{A}$ and each continuous map $\alpha\colon \mathcal{A}(\mathbf{A}, \underline{\mathbf{M}}) \to M$ which preserves every algebraic relation on $\underline{\mathbf{M}}$, there is an element $a_\alpha \in A$ such that α is the evaluation map $e_{\mathbf{A}}(a_\alpha)$. Since $\mathbf{A} \in \mathcal{A}$, we may assume it to be a subalgebra of a power $\underline{\mathbf{M}}^S$. In this case we can identify a_α in advance by taking $\rho_s := \pi_s{\restriction}_A$ for each $s \in S$ and computing

$$a_\alpha(s) = \rho_s(a_\alpha) = e_{\mathbf{A}}(a_\alpha)(\rho_s) = \alpha(\rho_s).$$

This tells us that if α is an evaluation at all, then it must be the evaluation at the element $a_\alpha \in M^S$ defined by $a_\alpha(s) := \alpha(\rho_s)$. When a_α fails to be in A, we refer to it as a **ghost element** of $\mathbf{A}$. This gives us a different method to demonstrate that $\underline{\mathbf{M}}$ is *not* dualisable: find a set S, a subalgebra $\mathbf{A}$ of $\underline{\mathbf{M}}^S$, and a continuous map $\alpha\colon \mathcal{A}(\mathbf{A},\underline{\mathbf{M}}) \to M$ which preserves all algebraic relations on $\underline{\mathbf{M}}$ such that the element a_α is a ghost element of $\mathbf{A}$. Later, in Theorem 10.6.3, we shall see that the algebra $\mathbf{A}$ can always be chosen to be countable.

We now show how to refine the ideas in the previous paragraph to produce a step-by-step guide for the ghost element method.

5.1 Ghost Element Theorem *Let $\underline{\mathbf{M}}$ be a finite non-trivial algebra, let $\mathbf{A}$ be a subalgebra of $\underline{\mathbf{M}}^S$ for some set S and let $\alpha\colon \mathcal{A}(\mathbf{A},\underline{\mathbf{M}}) \to M$ be a map satisfying*

(i) *there exists a finite subset T of A such that, for all $x, y \in \mathcal{A}(\mathbf{A},\underline{\mathbf{M}})$, we have $x{\restriction}_T = y{\restriction}_T$ implies $\alpha(x) = \alpha(y)$ (that is, α has finite support),*

(ii) *if Z is a finite subset of $\mathcal{A}(\mathbf{A},\underline{\mathbf{M}})$, then there exists $a \in A$ such that $\alpha{\restriction}_Z = e_{\mathbf{A}}(a){\restriction}_Z$ (that is, α is locally an evaluation), and*

(iii) *the map $a_\alpha \in \underline{\mathbf{M}}^S$ is a ghost element of $\mathbf{A}$.*

Then $\underline{\mathbf{M}}$ is not dualisable. Indeed, (i) *holds if and only if α is continuous,* (ii) *holds if and only if α preserves every finitary algebraic relation on $\underline{\mathbf{M}}$, while* (iii) *guarantees that α is not an evaluation map.*

Proof That (i) is equivalent to the continuity of α is just a reformulation of Lemma B.6 in this context. We now show that (ii) holds if and only if α preserves every relation in $\mathcal{B}$. Assume that $\alpha\colon \mathcal{A}(\mathbf{A},\underline{\mathbf{M}}) \to M$ is locally an evaluation. Let $r \in \mathcal{B}$ be n-ary and assume that $(x_1, \dots, x_n) \in r^{\mathcal{A}(\mathbf{A},\underline{\mathbf{M}})}$. Since α agrees with some evaluation map on the set $Z := \{x_1, \dots, x_n\}$ and since evaluation maps preserve every relation in $\mathcal{B}$, it follows that $(\alpha(x_1), \dots, \alpha(x_n)) \in r$. Conversely, assume that α preserves every relation $r \in \mathcal{B}$. Let $Z = \{x_1, \dots, x_n\}$ be a finite subset of $\mathcal{A}(\mathbf{A},\underline{\mathbf{M}})$. Since each x_i is a homomorphism, $r := \{(x_1(a), \dots, x_n(a)) \mid a \in A\}$ is an algebraic relation on $\underline{\mathbf{M}}$ and by construction $(x_1, \dots, x_n) \in r^{\mathcal{A}(\mathbf{A},\underline{\mathbf{M}})}$. Since α preserves r we have $(\alpha(x_1), \dots, \alpha(x_n)) \in r$, that is, there exists $a \in A$ such that $(\alpha(x_1), \dots, \alpha(x_n)) = (x_1(a), \dots, x_n(a))$, which says precisely that $\alpha{\restriction}_Z = e_{\mathbf{A}}(a){\restriction}_Z$. Thus α is locally an evaluation. Finally, if $\underline{\mathbf{M}}$ were dualisable then, by Lemma 2.1.3, α would be an evaluation which (iii) guarantees it is not. Thus $\underline{\mathbf{M}}$ is not dualisable. ∎

While a ghost element a_α of $\mathbf{A}$ in $\underline{\mathbf{M}}^S$ is, by definition, not an element of A, it must look locally like an element of A.

5.2 Lemma *If $\mathbf{A}$ is a subalgebra of $\underline{\mathbf{M}}^S$ and $\alpha\colon \mathbf{A} \to M$ preserves every finitary algebraic relation on $\underline{\mathbf{M}}$, then the induced map $a_\alpha \in M^S$ is* *locally in A, that is, if T is a finite subset of S, then there exists $a \in A$ such that $a_\alpha{\restriction}_T = a{\restriction}_T$.*

Proof Let $\mathbf{A}$ be a subalgebra of $\underline{\mathbf{M}}^S$ and assume that α preserves every $r \in \mathcal{B}$. Let T be a finite subset of S and consider $Z := \{\, \rho_t \mid t \in T \,\} \subseteq \mathcal{A}(\mathbf{A}, \underline{\mathbf{M}})$. By the Ghost Element Theorem, there exists $a \in A$ such that $\alpha{\restriction}_Z = e_{\mathbf{A}}(a){\restriction}_Z$. Hence, $a_\alpha(t) = \alpha(\rho_t) = e_{\mathbf{A}}(a)(\rho_t) = \rho_t(a) = a(t)$, for all $t \in T$, that is $a_\alpha{\restriction}_T = a{\restriction}_T$, whence a_α is locally in A. ■

When applying the Ghost Element Theorem it is common to begin with a specific potential ghost element $g \in M^S \backslash A$ and use it to define the map α as described in the guide below.

5.3 Ghost Element Guide *A finite algebra $\underline{\mathbf{M}}$ will be non-dualisable provided we are able to carry out the following ghost element construction.*

1. *Let S be a countable set and, for each $s \in S$, let $a_s \in M^S$. Let $\mathbf{A}$ be a subalgebra of $\underline{\mathbf{M}}^S$ with $\{\, a_s \mid s \in S \,\} \subseteq A$.*
2. *Choose (the ghost element) $g \in M^S \backslash A$ such that g is locally in A.*
3. *Let $\mathcal{A}(\mathbf{A}, \underline{\mathbf{M}}) = \{\, \rho_s \mid s \in S \,\} \cup X_0$, where X_0 consists of the elements of $\mathcal{A}(\mathbf{A}, \underline{\mathbf{M}})$ which are not projections. Give an explicit description of the maps in X_0.*
4. *Define $\alpha\colon \{\, \rho_s \mid s \in S \,\} \to M$ by $\alpha(\rho_s) := g(s)$ for all $s \in S$ and use the explicit description of the maps in X_0 from (3) to define α on the set X_0. The map $\alpha\colon \mathcal{A}(\mathbf{A}, \underline{\mathbf{M}}) \to M$ so obtained should be such that*
   - (A) *there exist $n \in \mathbb{N}$ and $s_1, \dots, s_n \in S$ such that $\{\, a_{s_1}, \dots, a_{s_n} \}$ is a finite support for α, and*
   - (B) *if Z is a finite subset of $\mathcal{A}(\mathbf{A}, \underline{\mathbf{M}})$, then, for some $n \in \mathbb{N}$, there is an n-ary term $t(v_1, \dots, v_n)$ and elements $s_1, \dots, s_n$ of S such that $\alpha(z) = t(z(a_{s_1}), \dots, z(a_{s_n}))$ for all $z \in Z$.*

   (In practice, for (B) it often occurs that there exists $s \in S$ such that $\alpha(z) = z(a_s)$ for all $z \in Z$, that is, $n = 1$ and $t(v_1) = v_1$.)

Proof Assume that we have defined the maps a_s, the algebra $\mathbf{A}$ and the ghost element g as described in (1) and (2), that we have described the set X_0 as required in (3) and that we are able to define the map $\alpha\colon \mathcal{A}(\mathbf{A}, \underline{\mathbf{M}}) \to M$ satisfying (A) and (B) in (4). Condition (A) guarantees that α is continuous. We claim that (B) guarantees that α is locally an evaluation. To see this, let Z be a finite subset of $\mathcal{A}(\mathbf{A}, \underline{\mathbf{M}})$. By (B), there exist an n-ary term t and $s_1, \dots, s_n \in S$ such that $\alpha(z) = t(z(a_{s_1}), \dots, z(a_{s_n})))$ for all $z \in Z$. Define $a := t(a_{s_1}, \dots, a_{s_n}) \in A$. For all $z \in Z$, we have

$$\alpha(z) = t(z(a_{s_1}), \dots, z(a_{s_n})) = z(t(a_{s_1}, \dots, a_{s_n})) = z(a) = e_{\mathbf{A}}(a)(z).$$

This proves that α is locally an evaluation. By construction, the map a_α induced by α equals g and hence is a ghost element of $\mathbf{A}$, whence $\underline{\mathbf{M}}$ is not dualisable. ∎

When using the Ghost Element Guide, it is sometimes convenient to replace $\mathcal{A}(\mathbf{A}, \underline{\mathbf{M}})$ by an isomorphic copy within M^S. Define a map $\eta\colon \mathcal{A}(\mathbf{A}, \underline{\mathbf{M}}) \to M^S$ by $\eta(x)(s) := x(a_s)$ and let $Y \subseteq M^S$ be the image of η. Then $\eta\colon D(\mathbf{A}) \to \mathbf{Y}$ is an isomorphism with respect to the brute force construction. (Consider the homomorphism $u\colon \mathbf{F}_{\underline{\mathbf{M}}}(S) \to \mathbf{A}$ determined by $u(\pi_s) = a_s$, for all $s \in S$. The map η is precisely $\rho_S \circ D(u)$, where $\rho_S\colon D(\mathbf{F}_{\underline{\mathbf{M}}}(S)) \to M^S$ is the isomorphism given in Lemma 2.2.1. Since u is surjective, $\eta = \rho_S \circ D(u)$ is an embedding.) For all $t \in S$, define $\widehat{a}_t \in M^S$ by $\widehat{a}_t(s) := a_s(t)$ for all $s \in S$. Thus $\widehat{a}_t = \eta(\rho_t)$ and consequently $Y = \{\, \widehat{a}_t \mid t \in S \,\} \cup Y_0$, where Y_0 consists of the elements of Y which do not correspond under η to projection maps. We may now work with $\mathbf{Y}$ instead of $\mathcal{A}(\mathbf{A}, \underline{\mathbf{M}})$.

In the previous sections we used the NU Obstacle Theorems and the Balanced Subalgebra Method to produce an assortment of non-dualisable algebras on sets with at least three elements. Our next theorem, which illustrates the Ghost Element Method, will be applied in the last section to determine all non-dualisable algebras on the two-element set $\{0, 1\}$.

5.4 Theorem *Let $\underline{\mathbf{M}} = \langle \{0,1\}; F \rangle$ be a two-element algebra such that*

(i) *$d(a, b, c) := a \vee (b \wedge c)$ is a term function of $\underline{\mathbf{M}}$, and*

(ii) *for each $n \geqslant 0$ and each n-ary operation $f \in F$ there is an index k such that $\underline{\mathbf{M}}$ satisfies $f(x_1, x_2, \dots, x_n) \approx 0 \implies x_k \approx 0$.*

Then $\underline{\mathbf{M}}$ is not dualisable.

Proof We shall carry out the four steps indicated in the Ghost Element Guide.

`Step (1).` Define $e_n\colon \mathrm{N} \to \{0, 1\}$ to be the characteristic function of $\{n\}$ and let $\mathbf{A}$ be the subalgebra of $\underline{\mathbf{M}}^{\mathrm{N}}$ generated by the set $\{\, e_n \mid n \in \mathrm{N} \}$.

`Step (2).` Choose the constant map $g\colon S \to \{0, 1\}$ onto $\{0\}$ as the potential ghost element. (Note that g is locally in A.) It follows from (ii) that, for all $n \geqslant 0$, we have $t(e_1, \dots, e_n) \neq g$ for every n-ary term t (see Exercise 10.7) and hence $g \notin A$.

`Step (3).` We first claim that $\mathcal{A}(\mathbf{A}, \underline{\mathbf{M}}) = \{\, \rho_n \mid n \in \mathrm{N} \} \,\dot\cup\, X_0$, where $1 \leqslant |X_0| \leqslant 2$. (Condition (ii) guarantees that $\{1\}$ forms a one-element subalgebra of $\underline{\mathbf{M}}$, whence the constant map u_1 onto $\{1\}$ belongs to $X_0 \subseteq \mathcal{A}(\mathbf{A}, \underline{\mathbf{M}})$. If the constant map from $\{\, e_n \mid n \in \mathrm{N} \}$ onto $\{0\}$ extends to a homomorphism u_0 from $\mathbf{A}$ to $\underline{\mathbf{M}}$, then u_0 also belongs to X_0.) Let $x\colon \mathbf{A} \to \underline{\mathbf{M}}$ be a homomorphism. Assume that $x(e_k) = x(e_\ell) = 1$ for $k \neq \ell$. As $k \neq \ell$, we have $d(e_n, e_k, e_\ell) = e_n \vee (e_k \wedge e_\ell) = e_n$ for all $n \in \mathrm{N}$.

Thus $x(e_n) = x(d(e_n, e_k, e_\ell)) = d(x(e_n), x(e_k), x(e_\ell)) = d(x(e_n), 1, 1) = x(e_n) \vee (1 \wedge 1) = 1$ for all $n \in \mathbb{N}$. We conclude that $x = u_1$. Thus x is either u_0 or u_1 or x maps exactly one of the generators, say e_k, to 1, in which case $x = \rho_k$.

Step (4). Define $\alpha\colon \mathcal{A}(\mathbf{A}, \underline{\mathbf{M}}) \to \{0,1\}$ by $\alpha(u_1) = 1$ and $\alpha(\rho_n) = g(n) = 0$, for all $n \in \mathbb{N}$, and, when $u_0 \in \mathcal{A}(\mathbf{A}, \underline{\mathbf{M}})$, define $\alpha(u_0) = 0$. Since $\{e_1, e_2\}$ is a finite support for α, and since $\alpha(z) = z(e_{n+1})$ for all $z \in \{u_0, u_1, \rho_1, \rho_2, \dots, \rho_n\}$, it follows that α satisfies (A) and (B).

The Ghost Element Guide now guarantees that $\underline{\mathbf{M}}$ is not dualisable. ■

In Section 7 we shall apply this result to prove that there are exactly eight non-dualisable clones on a two-element set.

The Ghost Element Method can be used to prove that some algebras, in particular the two-element algebras to which Theorem 10.5.4 applies, enjoy a very strong form of non-dualisability. A finite algebra $\underline{\mathbf{D}}$ is called **inherently non-dualisable** provided $\underline{\mathbf{M}}$ is non-dualisable whenever $\underline{\mathbf{M}}$ is a finite algebra with $\underline{\mathbf{D}} \in \mathbb{ISP}\underline{\mathbf{M}}$. The idea behind the following theorem is simple: we first construct the algebra $\mathbf{A} \leqslant \underline{\mathbf{D}}^S$ which shows that $\underline{\mathbf{D}}$ is non-dualisable and then attempt to prove that $\mathbf{A}$ also yields the non-dualisability of $\underline{\mathbf{M}}$. Since a homomorphism from $\mathbf{A}$ to $\underline{\mathbf{M}}$ has a finite range, we may obtain information on the set $\mathcal{A}(\mathbf{A}, \underline{\mathbf{M}})$ by focusing on the congruences θ on $\mathbf{A}$ which are of **finite index**, that is, such that $\mathbf{A}/\theta$ is finite. The following result is due to Davey, Idziak, Lampe and McNulty [DILM97].

5.5 Inherently Non-dualisable Algebra Theorem *Assume that $\underline{\mathbf{D}}$ is a finite algebra, that $\mathbf{A}$ is a subalgebra of $\underline{\mathbf{D}}^S$ and that B is an infinite subset of A such that*

(i) *there is a function $u\colon \mathbb{N} \to \mathbb{N}$ such that if θ is a congruence on $\mathbf{A}$ of finite index at most n, then $\theta{\restriction}_B$ has only one class with more than $u(n)$ elements,*

(ii) *$g \notin A$ where g is the element of D^S such that $g(s) := \rho_s(b)$, for each $s \in S$, with b any element of the block of $\ker(\rho_s){\restriction}_B$ which has size greater than $u(|D|)$.*

Then $\underline{\mathbf{D}}$ is inherently non-dualisable.

Proof Let $\underline{\mathbf{M}}$ be a finite algebra with $\underline{\mathbf{D}} \in \mathbb{ISP}\underline{\mathbf{M}}$. We may assume that $\underline{\mathbf{D}} \leqslant \underline{\mathbf{M}}^I$, for some finite set I. Let $S' := S \times I$. So $\underline{\mathbf{D}}^S$ is isomorphic to a subalgebra of $\underline{\mathbf{M}}^{S'}$ via the embedding σ with $\sigma(a)((s,i)) = a(s)(i)$. Let $\mathbf{A}' := \sigma(\mathbf{A})$. Thus $\mathbf{A} \cong \mathbf{A}' \leqslant \underline{\mathbf{M}}^{S'}$. Define $B' := \sigma(B) \subseteq A'$, so that assumption (i) is now valid with $\mathbf{A}$ and B replaced by $\mathbf{A}'$ and B'. Let θ be a congruence on $\mathbf{A}'$ of finite index n. By assumption, $\theta{\restriction}_{B'}$ has one infinite class and the remaining classes are of size bounded by $u(n)$, whence the complement of the infinite class has size at most $(n-1)u(n)$. In particular, the infinite class is cofinite. Let $x\colon \mathbf{A}' \to \underline{\mathbf{M}}$ be a homomorphism; then $\ker(x)$ is a congruence on $\mathbf{A}'$ of finite index at most $|M|$. Hence $\ker(x){\restriction}_{B'}$ has one cofinite class and its complement has size at most $(|M|-1)u(|M|)$.

Define $\alpha\colon \mathcal{A}(\mathbf{A}', \underline{\mathbf{M}}) \to M$ by setting $\alpha(x)$ to be the value of x at any element of the cofinite class of $\ker(x){\restriction}_{B'}$. It is easy to see that any subset of B' of size $N := |M|u(|M|)+1$ is a finite support for α. Since the intersection of a finite number of cofinite subsets of B' is a cofinite subset of B', it is easy to see that α is locally an evaluation. Define $g' \in \underline{\mathbf{M}}^{S'}$ by $g'(t) := x(\rho_t)$, for each $t \in S'$. Then $g' = \sigma(g)$, whence $g' \notin A'$ as $g \notin A$. By the Ghost Element Theorem, $\underline{\mathbf{M}}$ is not dualisable. Hence, $\underline{\mathbf{D}}$ is inherently non-dualisable. ■

Many applications of this theorem are to situations in which the function u takes the constant value 1 (see Exercises 10.9–10.11). In [DILM97] it was applied to prove that every finite non-finitely-based graph algebra is inherently non-dualisable. We do not have an example which shows that inherent non-dualisability is actually a stronger notion than non-dualisability. The following problem was raised in [DILM97].

5.6 Inherent Non-dualisability Problem *Is there a finite algebra which is non-dualisable but not inherently non-dualisable? In particular, is there a dualisable algebra with a non-dualisable subalgebra? Does there exist a pair of finite algebras, one dualisable and the other not, which generate the same quasi-variety?*

10.6 The term-closed subset method

As $\underline{\mathbf{M}}$ is dualisable if and only if brute force yields a duality on $\mathcal{A}$, the Third Duality Theorem 3.1.6 yields a characterisation of dualisability.

6.1 Third Duality Theorem Revisited *$\underline{\mathbf{M}}$ is dualisable if and only if, for every non-empty set S and term-closed subset X of M^S, each continuous map $\alpha\colon X \to M$ which preserves every finitary algebraic relation on $\underline{\mathbf{M}}$ extends to an S-ary term function on $\underline{\mathbf{M}}$.*

We have already seen this form of the theorem at work in Zádori's proof of the Big NU Obstacle Theorem (see Lemma 10.2.5). In this section we shall present two further applications: a technical but powerful result due to Zádori, and a proof that every finite subdirectly irreducible pseudocomplemented semilattice with more than two elements is non-dualisable.

If we have a sequence $I_0 \subseteq I_1 \subseteq \cdots \subseteq I_n \subseteq \cdots \subseteq I$ of subsets of a set I, we shall denote the natural projections π_{I_n} from M^I to M^{I_n}

and $\pi_{I_m I_n}$ from M^{I_m} to M^{I_n}, for $m \geqslant n$, by π_n and π_{mn}, respectively, provided the meaning is unambiguous. As usual, if $Z \subseteq M^I$, then the restriction of π_n to Z will be denoted by ρ_n and if $Z_m \subseteq M^{I_m}$, then the restriction of π_{mn} to Z_m will be denoted by ρ_{mn}. The following theorem gives a characterisation of non-dualisability in terms of an inverse limit diagram and shows that the set S in the Third Duality Theorem Revisited can be assumed to be countable.

6.2 Non-dualisability Limit Theorem (Zádori [Zád95]) *The following conditions are equivalent:*

(i) $\underline{\mathbf{M}}$ *is not dualisable;*

(ii) *there exist a sequence of finite sets* $\varnothing \neq S_0 \subseteq \cdots \subseteq S_n \subseteq \dots$, *non-empty subsets* $X_n \subseteq M^{S_n}$, *for each* $n \geqslant 0$, *and a map* $\alpha_0 \colon X_0 \to M$ *such that, for all* $n \geqslant 1$,

   (a) $\pi_{nn-1}(X_n) = \pi_{nn-1}(X_n^{\mathrm{tm}}) = X_{n-1}$, *and*

   (b) *the map* $\alpha_n := \alpha_0 \circ \rho_{n0} \colon X_n \to M$ *preserves every* n*-ary algebraic relation on* $\underline{\mathbf{M}}$ *but does not extend to an* S_n*-ary term function on* $\underline{\mathbf{M}}$*;*

(iii) *there exist a sequence of finite sets* $\varnothing \neq T_0 \subseteq \cdots \subseteq T_n \subseteq \dots$, *non-empty term-closed subsets* $Y_n \subseteq M^{T_n}$, *for each* $n \geqslant 0$, *and a map* $\beta_0 \colon Y_0 \to M$ *such that, for all* $n \geqslant 1$,

   (a) $\pi_{nn-1}(Y_n) \subseteq Y_{n-1}$, *and*

   (b) *the map* $\beta_n := \beta_0 \circ \rho_{n0} \colon Y_n \to M$ *preserves every* n*-ary algebraic relation on* $\underline{\mathbf{M}}$ *but does not extend to a* T_n*-ary term function on* $\underline{\mathbf{M}}$*;*

(iv) *there are a countable set* T, *a term-closed subset* Y *of* M^T *and a continuous map* $\beta \colon Y \to M$ *which preserves every finitary algebraic relation on* $\underline{\mathbf{M}}$ *but does not extend to a* T*-ary term function on* $\underline{\mathbf{M}}$.

Proof We first prove that (i) implies (ii). Assume that $\underline{\mathbf{M}}$ is not dualisable. By the Third Duality Theorem Revisited, there are a set S, a term-closed subset X of M^S and a continuous map $\alpha \colon X \to M$ which preserves every finitary algebraic relation on $\underline{\mathbf{M}}$ but does not extend to an S-ary term function on $\underline{\mathbf{M}}$. Since α is continuous, it has a finite support S_0, where $\varnothing \neq S_0 \subseteq S$, whence there exists a map $\alpha_0 \colon X_0 \to M$, where $X_0 := \pi_0(X)$, such that $\alpha = \alpha_0 \circ \rho_0$. We now give an inductive definition of S_n and X_n for $n \geqslant 1$.

Assume that S_{n-1} and X_{n-1} have been successfully defined and assume that in addition to the properties listed in (ii)(a),(b) we also have $\pi_{n-1}(X) = X_{n-1}$ (which certainly holds at the base of the induction when $n = 1$). We shall define S_n such that the properties listed in (ii)(a),(b) hold with $X_n := \pi_n(X)$. By Lemma 2.2.6 (with $R = \mathcal{B}_n$), there is a finite subset E_n of S such that α factors as $\alpha = \alpha_1 \circ \pi_{E_n}$ where $\alpha_1 \colon \pi_{E_n}(X) \to M$ preserves every n-ary algebraic relation on $\underline{\mathbf{M}}$. (Thus the set $S_{n-1} \cup E_n$ is a first approximation to S_n. We need to extend $S_{n-1} \cup E_n$ so as to force $\pi_{nn-1}(X_n^{\text{tm}}) = \pi_{nn-1}(X_n)$, where $X_n := \pi_n(X)$.)

Let $Z := \pi_{n-1}^{-1}(X_{n-1}^{\text{tm}} \backslash X_{n-1})$; then Z is closed in M^S, and $Z \cap X = \varnothing$ since $\pi_{n-1}(X) = X_{n-1}$ by our inductive assumption. Since X is term-closed, for each $z \in Z$, there exist S-ary term functions τ_z and τ'_z which agree on X but differ at z. The set $V_z := M^S \backslash \operatorname{eq}(\tau_z, \tau'_z)$ is an open set containing z. Since Z is closed in M^S, it is compact and hence there exist $z_1, \dots, z_m \in Z$ such that $Z \subseteq V_{z_1} \cup \cdots \cup V_{z_m}$. Let F_n be a finite subset of S which is a support for each of the functions $\tau_{z_1}, \tau'_{z_1}, \dots, \tau_{z_m}, \tau'_{z_m}$.

Define $S_n := S_{n-1} \cup E_n \cup F_n$ and $X_n := \pi_n(X)$. We must now verify the conditions given in (ii)(a),(b). Certainly,

$$\pi_{nn-1}(X_n) = \pi_{nn-1}(\pi_n(X)) = \pi_{nn-1} \circ \pi_n(X) = \pi_{n-1}(X) = X_{n-1}.$$

It follows that $X_{n-1} \subseteq \pi_{nn-1}(X_n^{\text{tm}})$. We claim that $\pi_{nn-1}(X_n^{\text{tm}}) \subseteq X_{n-1}^{\text{tm}}$. To establish this, we must show that if two S_{n-1}-ary term functions τ and τ' agree on X_{n-1} then they agree on $\pi_{nn-1}(X_n^{\text{tm}})$. Assume that τ and τ' agree on X_{n-1} and let $y \in X_n^{\text{tm}}$. The S_n-ary term functions $\tau \circ \pi_{nn-1}$ and $\tau' \circ \pi_{nn-1}$ agree on X_n and hence must agree at y, that is, τ and τ' agree at $\pi_{nn-1}(y)$, as required. Thus

$$X_{n-1} \subseteq \pi_{nn-1}(X_n^{\text{tm}}) \subseteq X_{n-1}^{\text{tm}}.$$

In order to prove that $\pi_{nn-1}(X_n^{\text{tm}}) \subseteq X_{n-1}$, suppose, by way of contradiction, that there exists $y \in \pi_{nn-1}(X_n^{\text{tm}}) \backslash X_{n-1}$. Thus $y \in X_{n-1}^{\text{tm}} \backslash X_{n-1}$ and there exists $w \in X_n^{\text{tm}}$ such that $\pi_{nn-1}(w) = y$. Let $z \in M^S$ with $\pi_n(z) = w$ and therefore $\pi_{n-1}(z) = y$. Hence, $z \in \pi_{n-1}^{-1}(X_{n-1}^{\text{tm}} \backslash X_{n-1}) = Z$. Since $F_n \subseteq S_n$, there exist S-ary term functions τ and τ' which have S_n as support and which agree on X but differ at z. Thus the induced S_n-ary term functions τ_0 and τ'_0 agree on $\pi_n(X) = X_n$ and differ at $\pi_n(z) = w$. Thus $w \notin X_n^{\text{tm}}$, contradicting the choice of w. Hence $\pi_{nn-1}(X_n^{\text{tm}}) = X_{n-1}$.

The map $\alpha_n := \alpha_0 \circ \rho_{n0}$ preserves every n-ary algebraic relation on $\underline{\mathbf{M}}$ since $E_n \subseteq S_n$. If α_n were to extend to an S_n-ary term function τ, then $\pi_n \circ \tau$ would be an S-ary term function extending α. Hence, α_n does not extend to an S_n-ary term function. This completes the proof that (i) implies (ii).

Now assume (ii) to prove (iii). Let $T_n := S_{n+1}$ and $Y_n := X_{n+1}^{\text{tm}}$ for all $n \geqslant 0$ and let $\beta_0 := \alpha_0 \circ \rho_{S_1 S_0} \colon X_1^{\text{tm}} \to M$. Since $\pi_{S_{n+1}S_n}(X_{n+1}^{\text{tm}}) = X_n$ for all $n \geqslant 0$, we have $\pi_{T_n T_{n-1}}(Y_n) \subseteq Y_{n-1}$ for all $n \geqslant 1$. Moreover, $\pi_{S_{n+1}S_n}(X_{n+1}^{\text{tm}}) = X_n$ implies that the projection $\pi_{S_{n+1}S_n}$ maps the domain of β_n onto the domain of α_n. Thus the composite $\alpha_n \circ (\pi_{S_{n+1}S_n}{\restriction_{Y_n}})$

is well defined. A simple calculation shows that $\alpha_n \circ (\pi_{S_{n+1}S_n}{\restriction_{Y_n}}) = \beta_n$, whence β_n preserves every n-ary algebraic relation on $\underline{\mathbf{M}}$ since both α_n and $\pi_{S_{n+1}S_n}{\restriction_{Y_n}}$ do. An extension of β_n to a T_n-ary term function would also be an extension of α_{n+1} to an S_{n+1}-ary term function, whence no such extension of β_n exists. Thus (ii) implies (iii).

We now show that (iii) implies (iv). Assume (iii) and let $T := \bigcup_{n=0}^{\infty} T_n$ and $Y := \bigcap\{\, \pi_n^{-1}(Y_n) \mid n \geqslant 0 \,\}$. Since $\pi_{nn-1}(Y_n) \subseteq Y_{n-1}$ and since π_n is continuous, the sets $\pi_n^{-1}(Y_n)$, for $n \geqslant 0$, form a decreasing sequence of non-empty closed subsets of M^T. It follows, as M^T is compact, that Y is a non-empty closed subset of M^T (see Exercise B.1), and we claim that Y is term-closed in M^T. Let $x \in M^T \backslash Y$. Then there exists $n \geqslant 0$ such that $x \notin \pi_n^{-1}(Y_n)$. Since Y_n is term-closed in M^{T_n}, there exist T_n-ary term functions τ_1 and τ_2 which agree on Y_n but differ at $\pi_n(x)$. Consequently, since $\pi_n(Y) \subseteq Y_n$, the T-ary term functions $\tau_1 \circ \pi_n$ and $\tau_2 \circ \pi_n$ agree on Y but differ at x. Hence, Y is term-closed in M^T.

Define $\beta\colon Y \to M$ by $\beta := \beta_0 \circ \rho_0$. Certainly, β is continuous. For all $n \geqslant 1$, we have $\beta = \beta_0 \circ \rho_{n0} \circ \rho_n = \beta_n \circ \rho_n$. Since both β_n and ρ_n preserve all the n-ary algebraic relations on $\underline{\mathbf{M}}$, it follows that β preserves every finitary algebraic relation on $\underline{\mathbf{M}}$. It remains to prove that β does not extend to a T-ary term function τ on $\underline{\mathbf{M}}$. Suppose that such an extension does exist. This means that if $y \in M^T$ satisfies $\pi_n(y) \in Y_n$ for all $n \geqslant 0$, then $\tau(y) = \beta(y)$. As τ is a term function, it must have a finite support which has to be a subset of T_m for some $m \geqslant 0$. Hence there exists a T_m-ary term function t such that $t(\pi_m(y)) = \tau(y) = \beta(y) = \beta_m(\pi_m(y))$, for all $y \in Y$. Thus, $t(x) = \beta_m(x)$ for all $x \in \pi_m(Y)$. Suppose, for the moment, that there exists $k \geqslant m$ such that $\pi_{km}(Y_k) \subseteq \pi_m(Y)$, that is, for each $z \in Y_k$, there exists $y \in Y$ such that $\pi_{km}(z) = \pi_m(y)$. Then we would have $t(\pi_{km}(z)) = t(\pi_m(y)) = \beta_m(\pi_m(y)) = \beta_m(\pi_{km}(z)) = \beta_k(z)$, that is, the T_k-ary term function $t \circ \pi_{km}$ would extend β_k, yielding the desired contradiction. Thus, it remains to show that there exists $k \geqslant m$ with $\pi_{km}(Y_k) \subseteq \pi_m(Y)$.

As $\pi_{nn-1}(Y_n) \subseteq Y_{n-1}$ for all $n \geqslant 1$, we have a decreasing sequence of subsets of Y_m, namely,

$$Y_m \supseteq \pi_{m+1m}(Y_{m+1}) \supseteq \pi_{m+2m}(Y_{m+2}) \supseteq \cdots \supseteq \pi_m(Y).$$

As Y_m is finite, this sequence is eventually constant, that is, there exists $k \geqslant m$ such that $\pi_{\ell m}(Y_\ell) = \pi_{km}(Y_k)$ for all $\ell \geqslant k$. We claim that this is the required k. To see this, let $x \in \pi_{km}(Y_k)$ and, for all $\ell \geqslant k$, define

$$Z_\ell := \{\, z \in Y_\ell \mid \pi_{\ell m}(z) = x \,\}.$$

Since $\ell \geqslant k$, the set Z_ℓ is non-empty. Thus the family $\{\, Z_\ell \mid \ell \geqslant k \,\}$ forms an inverse system of non-empty finite sets with the map $\varphi_{n\ell}\colon Z_n \to Z_\ell$ for

$n \geqslant \ell \geqslant k$ given by $\varphi_{n\ell} = \pi_{n\ell}{\restriction}_{Z_n}$. By the Inverse Limit Theorem 1.3.3, for all $\ell \geqslant k$ there exists $z_\ell \in Z_\ell$ such that $\varphi_{n\ell}(z_n) = z_\ell$ for all $n \geqslant \ell \geqslant k$. Thus, we can unambiguously define $y \in Y$ by $y(t) := z_\ell(t)$ where ℓ is any natural number greater than k such that $t \in T_\ell$. By construction, $\pi_\ell(y) \in Z_\ell$ for all $\ell \geqslant k$. In particular, $\pi_k(y) \in Z_k$, so $\pi_m(y) = \pi_{km}(\pi_k(y)) = x$, whence $\pi_{km}(Y_k) \subseteq \pi_m(Y)$, as claimed. This completes the proof of (iii) $\Rightarrow$ (iv) and, since (iv) $\Rightarrow$ (i) follows immediately from the Third Duality Theorem Revisited, it also completes the proof of the theorem. ∎

This result has two very important consequences.

6.3 Theorem *$\underline{\mathbf{M}}$ is non-dualisable if and only if there is a countable algebra $\mathbf{A}$ in $\mathbb{ISP}\underline{\mathbf{M}}$ such that brute force does not yield a duality on $\mathbf{A}$.*

Proof Assume that $\underline{\mathbf{M}}$ is non-dualisable and let $\underset{\sim}{\mathbf{M}} = \langle M; \mathcal{B}, \mathcal{T}\rangle$ be the brute force structure. Then, by the Non-dualisability Limit Theorem, there are a countable set T, a term-closed subset Y of M^T and a morphism $\beta \in \mathcal{X}(\mathbf{Y}, \underset{\sim}{\mathbf{M}})$ which does not extend to a T-ary term function on $\underline{\mathbf{M}}$. The argument given in the proof of the Third Duality Theorem 3.1.6 shows that $\underset{\sim}{\mathbf{M}}$ does not yield a duality on $\mathbf{A} := \mathbf{F}_{\underline{\mathbf{M}}}(T){\restriction}_Y$. Since $\mathbf{A}$ is generated by the countable set $\{\, \pi_t \colon M^T \to M \mid t \in T \,\}$ of projections, it is countable. This establishes the forward implication. The converse is trivial. ∎

The second consequence of the Non-dualisability Limit Theorem is another proof of the Duality Compactness Theorem 2.2.11.

6.4 Duality Compactness Theorem Revisited *If $\underset{\sim}{\mathbf{M}}$ is of finite type and yields a duality on $\mathcal{A}_{\text{fin}}$, then $\underset{\sim}{\mathbf{M}}$ yields a duality on $\mathcal{A}$.*

Proof (Zádori [Zád95]) Assume that $\underset{\sim}{\mathbf{M}}$ yields a duality on $\mathcal{A}_{\text{fin}}$. Assume that we have a sequence of finite sets T_n and non-empty term-closed subsets $Y_n \subseteq M^{T_n}$ such that $\pi_{nn-1}(Y_n) \subseteq Y_{n-1}$ for all $n \geqslant 1$. Let $\beta_0 \colon Y_0 \to M$ be a map such that, for all $n \geqslant 1$, the map $\beta_n := \beta_0 \circ \rho_{n0} \colon Y_n \to M$ preserves every n-ary algebraic relation on $\underline{\mathbf{M}}$. Choose $k \in \mathbb{N}$ to be an upper bound for the arities of the relations in $R \cup \{\, \text{graph}(h) \mid h \in G \cup H \,\}$. Then $\beta_k \colon \mathbf{Y}_k \to \underset{\sim}{\mathbf{M}}$ is a morphism and hence, by the finite version of the Third Duality Theorem 3.1.6, β_k extends to a T_n-ary term function on $\underline{\mathbf{M}}$. Hence, by the Non-dualisability Limit Theorem, $\underline{\mathbf{M}}$ is dualisable. ∎

6.5 Pseudocomplemented Semilattices A meet semilattice $\mathbf{A} = \langle A; \wedge, {}^*, 0, 1\rangle$ equipped with an operation * of pseudocomplementation defined, as in the case of lattices, by

$$a^* = \max\{\, b \in A \mid a \wedge b = 0 \,\}$$

is referred to as a **pseudocomplemented semilattice** or, more succinctly, a p-**semilattice**. The class of all p-semilattices forms a variety $\mathcal{S}_\omega$. G.T. Jones [Jon:PS] proved that the structure of the subdirectly irreducibles in $\mathcal{S}_\omega$ is surprisingly similar to the structure of the subdirectly irreducibles in the variety $\mathcal{B}_\omega$ of distributive p-algebras (see 4.3.17): they are precisely the p-semilattices constructed by adding a new top element 1 to a Boolean lattice (whose top and bottom we once again denote by a and 0 respectively). For all $n \geqslant 1$, let $\underline{\mathbf{S}}_n$ be the subdirectly irreducible p-semilattice constructed by adding a new top element to the n-atom Boolean lattice. Thus, $\underline{\mathbf{S}}_n$ is the reduct of the distributive p-algebra $\underline{\mathbf{B}}_n$ obtained by deleting the operation $\vee$ from its type.

As far as duality theory is concerned, the analogy ends here. Distributive p-algebras have a majority term and hence the distributive p-algebra $\underline{\mathbf{B}}_n$ is dualisable for all n. In contrast, the p-semilattice $\underline{\mathbf{S}}_n$ is non-dualisable for all $n \geqslant 1$ (an unpublished result of D.M. Clark). Our proof, which involves two preliminary lemmas of independent interest, is closely related to the proof of Lemma 10.2.5.

6.6 Lemma *Let S and T be non-empty sets and let Σ be a subset of S^T. Given a subset X of M^T define a subset $X^{(\Sigma)}$ of M^S by*

$$X^{(\Sigma)} := \{\, y \in M^S \mid (\forall \sigma \in \Sigma)\ y \circ \sigma \in X \,\}.$$

Then $X^{(\Sigma)}$ is term-closed in M^S provided X is term-closed in M^T.

Proof Assume that X is term-closed in M^T and let $z \in M^S \backslash X^{(\Sigma)}$. Thus there exists $\sigma \in \Sigma$ such that $z \circ \sigma \notin X$. As X is term-closed in M^T, there are T-ary term functions t_1 and t_2 such that $t_1{\restriction}_X = t_2{\restriction}_X$ and $t_1(z \circ \sigma) \neq t_2(z \circ \sigma)$. Consequently, the S-ary term functions t_1' and t_2', given by $t_i'(y) := t_i(y \circ \sigma)$, agree on $X^{(\Sigma)}$ and differ at z, whence $X^{(\Sigma)}$ is term-closed in M^S. ∎

6.7 Lemma *For $X \subseteq M^2$ define $D_X := \{\, a \in M \mid (a, a) \in X \,\}$ and*

$$Y := \{\, y \in M^{\mathbb{N}} \mid i \neq j \implies (y(i), y(j)) \in X \,\}.$$

Assume that there exists $\ell \in \mathbb{N}$ such that, for every $y \in Y$, there is a unique $a \in M$ such that $|y^{-1}(a)| > \ell$. If X is term-closed in M^2 and $\underline{\mathbf{M}}$ is dualisable, then there exist $k \geqslant 3$ and a k-ary term function t on $\underline{\mathbf{M}}$ such that

(i) *$t(a, \ldots, a) = a$ for all $a \in D_X$, and*

(ii) *$t(b, a, \ldots, a) = t(a, b, a, \ldots, a) = \cdots = t(a, a, \ldots, a, b) = a$ for all $a \in D_X$ and all $b \in M$ such that $(a, b), (b, a) \in X$.*

Proof Assume that X is term-closed in M^2 and define Σ to be the set of one-to-one maps from $\{1,2\}$ into $\mathbb{N}$. Observe that Y is exactly $X^{(\Sigma)}$. By the previous lemma, Y is term-closed in $M^{\mathbb{N}}$. We may define $\alpha\colon Y \to M$ by $\alpha(y) := \lim_{n\to\infty} y(n)$. Since each element of M, except the limiting value of y, occurs at most ℓ times as a value of $y \in Y$, it follows that $\{1, 2, \ldots, m{+}1\}$ is a support of α, where $m = \ell|M|$; whence α is continuous. On any finite subset of Y, the map α agrees with some projection $\pi_n\colon M^{\mathbb{N}} \to M$; whence α preserves every relation $r \in \mathcal{B}$. Consequently, by the Third Duality Theorem Revisited, if $\underline{\mathbf{M}}$ is dualisable there exists an $\mathbb{N}$-ary term function $t'\colon M^{\mathbb{N}} \to M$ which extends α. Assume that $\{1, \ldots, k\}$ is a support for t', with $k \geqslant 3$, and let t be a k-ary term function such that $t'(y) = t(y(1), \ldots, y(k))$, for all $y \in M^{\mathbb{N}}$. If $a \in D_X$, then $\overline{a} \in Y$, and hence

$$t(a, \ldots, a) = t'(\overline{a}) = \alpha(\overline{a}) = a.$$

Recall that $a_\ell^b := (a, \ldots, a, b, a, \ldots)$, with b in the ℓth position. If $a \in D_X$ and $(a, b), (b, a) \in X$, then a_ℓ^b belongs to Y, whence, for all $\ell \leqslant n$,

$$t(a, \ldots, a, b, a, \ldots, a) = t'(a_\ell^b) = \alpha(a_\ell^b) = a,$$

as required. ■

Exercise 10.4 shows that $\mathbf{S}_n$ is not congruence join-semidistributive (for $n \geqslant 1$) and not relatively congruence join-semidistributive (for $n \geqslant 2$). Thus the non-dualisability of the p-semilattice $\underline{\mathbf{S}}_n$ cannot be obtained from the Big NU Obstacle Theorem.

6.8 Theorem *For all $n \geqslant 1$, the p-semilattice $\underline{\mathbf{S}}_n$ is non-dualisable.*

Proof Let $n \geqslant 1$ and suppose that $\underline{\mathbf{S}}_n$ is dualisable. Define $X \subseteq S_n^2$ by

$$X := \{(1,1), (a,a)\} \cup X_{b1} \cup X_{b1}^{\smile} \cup X_{ba} \cup X_{ba}^{\smile} \cup X_{bc},$$

where $X_{b1} = \{(b, 1) \mid 0 \leqslant b < a\}$, $X_{ba} = \{(b, a) \mid 0 < b < a\}$ and $X_{bc} = \{(b, c) \mid b, c < a \text{ and } b \vee c = a\}$. Note that X consists of all pairs in S_n^2 other than

$$(0, a), (a, 0), (1, a), (a, 1) \text{ and } (b, c) \text{ with } b \vee c < a.$$

Table 10.1 shows the behaviour of two pairs and one triple of binary term functions on $\underline{\mathbf{S}}_n^2$. The top half of the table shows the agreement of the term functions on X. From the bottom half we are able to read off, for each element (u, v) of $S_n^2 \backslash X$, a pair of term functions which agree on X but differ at (u, v). It follows that X is term-closed in S_n^2. (A few easy calculations are required to verify some of the values given in the table. If

	$x^* \wedge y$	x^*	$x \wedge y^*$	y^*	$x \wedge y$	$x \wedge y^{**}$	$x^{**} \wedge y$
$(1,1)$	0	0	0	0	1	1	1
(a,a)	0	0	0	0	a	a	a
$(b,1)$	b^*	b^*	0	0	b	b	b
$(1,b)$	0	0	b^*	b^*	b	b	b
(b,a)	b^*	b^*	0	0	b	b	b
(a,b)	0	0	b^*	b^*	b	b	b
(b,c)	b^*	b^*	c^*	c^*	$b \wedge c$	$b \wedge c$	$b \wedge c$
$(0,a)$	a	1					
$(a,0)$			a	1			
$(1,a)$					a	1	
$(a,1)$					a		1
(b,c)			$b \wedge c^*$	c^*			

Table 10.1 proof that X is term-closed in S_n^2

$b \vee c = a$, then $b \geqslant c^*$ and hence $b \wedge c^* = c^*$. Similarly, $b \vee c = a$ implies that $b^* \wedge c = b^*$. If $b \vee c < a$, then $b \ngeqslant c^*$ and hence $b \wedge c^* < c^*$.)

We have $D_X = \{a, 1\}$ and $(0,1), (1,0) \in X$. Let

$$Y := \{\, y \in M^{\mathbb{N}} \mid i \neq j \implies (y(i), y(j)) \in X \,\}.$$

If $y \in Y$, then only a and 1 can occur more than once as values of y and both cannot occur simultaneously. Thus $|y^{-1}(b)| \leqslant 1$ either for all $b \in S_n$ where $b \neq a$ or for all $b \in S_n$ where $b \neq 1$. By Lemma 10.6.7, with $\ell = 1$, there is a k-ary term function t on $\underline{\mathbf{S}}_n$, for some $k \geqslant 3$, such that

(i) $t(a, \dots, a) = a$ and $t(1, \dots, 1) = 1$, and

(ii) $t(0, 1, \dots, 1) = t(1, 0, 1, \dots, 1) = \dots = t(1, \dots, 1, 0) = 1$.

It is easy to check that

$$r := \{0,1\}^{k+1} \cup \{(a_1, \dots, a_{k+1}) \in \{0, a, 1\}^{k+1} \mid (\exists i)\, a_i = a \Rightarrow (\exists j)\, a_j = 0\}$$

is a $(k+1)$-ary algebraic relation on $\underline{\mathbf{S}}_n$. Hence r must be closed under the term function t. But the $(k+1)$-tuples

$$x_1 := (a, 0, 1, \dots, 1),\ x_2 := (a, 1, 0, 1, \dots, 1), \dots, x_k := (a, 1, \dots, 1, 0)$$

are all in r, while

$$\begin{aligned} t(x_1, \dots, x_k) &= (t(a, \dots, a), t(0, 1, \dots, 1), \dots, t(1, \dots, 1, 0)) \\ &= (a, 1, \dots, 1) \end{aligned}$$

is not in r. This contradiction shows that $\underline{\mathbf{S}}_n$ is non-dualisable. ∎

10.7 Dualisable clones on a two-element set

E.L. Post [Pos21, Pos41] gave a complete description of the lattice $\mathfrak{L}_2$ of all clones on the set $\{0,1\}$—see Figure 10.1 and Figure 10.2. The labels used in these lattice diagrams are explained in Tables 10.2–10.4. Wherever possible, for each clone C on $\{0,1\}$, we have given a small generating set F for C so that the algebra $\underline{\mathbf{M}} = \langle\{0,1\}; F\rangle$ is familiar. The sets F are essentially those given in Berman [Ber80].

In each table, the first column gives the notation for the algebra $\underline{\mathbf{M}}$, the second column gives Post's original notation for the clone C, the third column gives the set F and the fourth column gives the usual name for the quasi-variety generated by $\underline{\mathbf{M}}$ (when there is one). If g is an operation on $\{0,1\}$, then we define $\underline{\mathbf{M}}^g := \langle\{0,1\}; F \cup \{g\}\rangle$.

The operations $\vee$, $\wedge$, $\rightarrow$, $'$ are the usual Boolean operations and $+$ is the usual group addition. For every function f, we obtain an **opposite** function f^∂ by interchanging 0 and 1. More precisely, f^∂ is defined by

$$f^\partial(a_1, \ldots, a_n) := f(a_1', \ldots, a_n')'.$$

The opposite of $\rightarrow$ is denoted by $-$. We require six ternary operations: m, p, d, d^∂, q and q^∂. The operations m and p are the usual median operation, $m(x, y, z) := (x \vee y) \wedge (y \vee z) \wedge (z \vee x)$, and Mal'cev operation, $p(x, y, z) := x+y+z$. The operation d, defined by $d(x, y, z) := x \vee (y \wedge z)$, is the operation we met in Theorem 10.5.4, while q is defined by $q(x, y, z) = (x \vee y \vee z) \wedge (x \vee y' \vee z')$ and incorporates both join (since $x \vee y = q(x, x, y)$) and relative complement (if $a \leqslant x \leqslant b$, then $q(a, b, x) = a \vee (b \wedge x')$ is the complement of x in the interval $[a, b]$). For each $n \geqslant 4$, the n-ary median operations

$$m_n(x_1, \ldots, x_n) := \bigwedge_i (\bigvee_{j \neq i} x_j) \text{ and } m_n^\partial(x_1, \ldots, x_n) := \bigvee_i (\bigwedge_{j \neq i} x_j)$$

will also be required. There is no commonly used notation for the algebras which involve these functions. Since m_n and m_n^∂ are near-unanimity functions, we denote these algebras by $\underline{\mathbf{N}}$ or by $\underline{\mathbf{U}}$ with appropriate subscripts and superscripts.

We have not necessarily given an irredundant set of operations for the algebra $\underline{\mathbf{M}}$. In particular, we have included all definable constants as nullary operations. For example, we write the implication algebra as $\underline{\mathbf{I}} = \langle\{0,1\}; \rightarrow, 1\rangle$ rather than $\underline{\mathbf{I}} = \langle\{0,1\}; \rightarrow\rangle$ and the Boolean lattice as $\underline{\mathbf{L}} = \langle\{0,1\}; q, \vee, \wedge\rangle$ rather than $\underline{\mathbf{L}} = \langle\{0,1\}; q, \wedge\rangle$.

Note that the lattice $\mathfrak{L}_2$ is countable, in contrast to the lattice $\mathfrak{L}_n$ of clones on a finite set with $n \geqslant 3$ elements which is uncountable—see 10.4.6. (For a discussion of the lattice of clones and a wealth of references to the

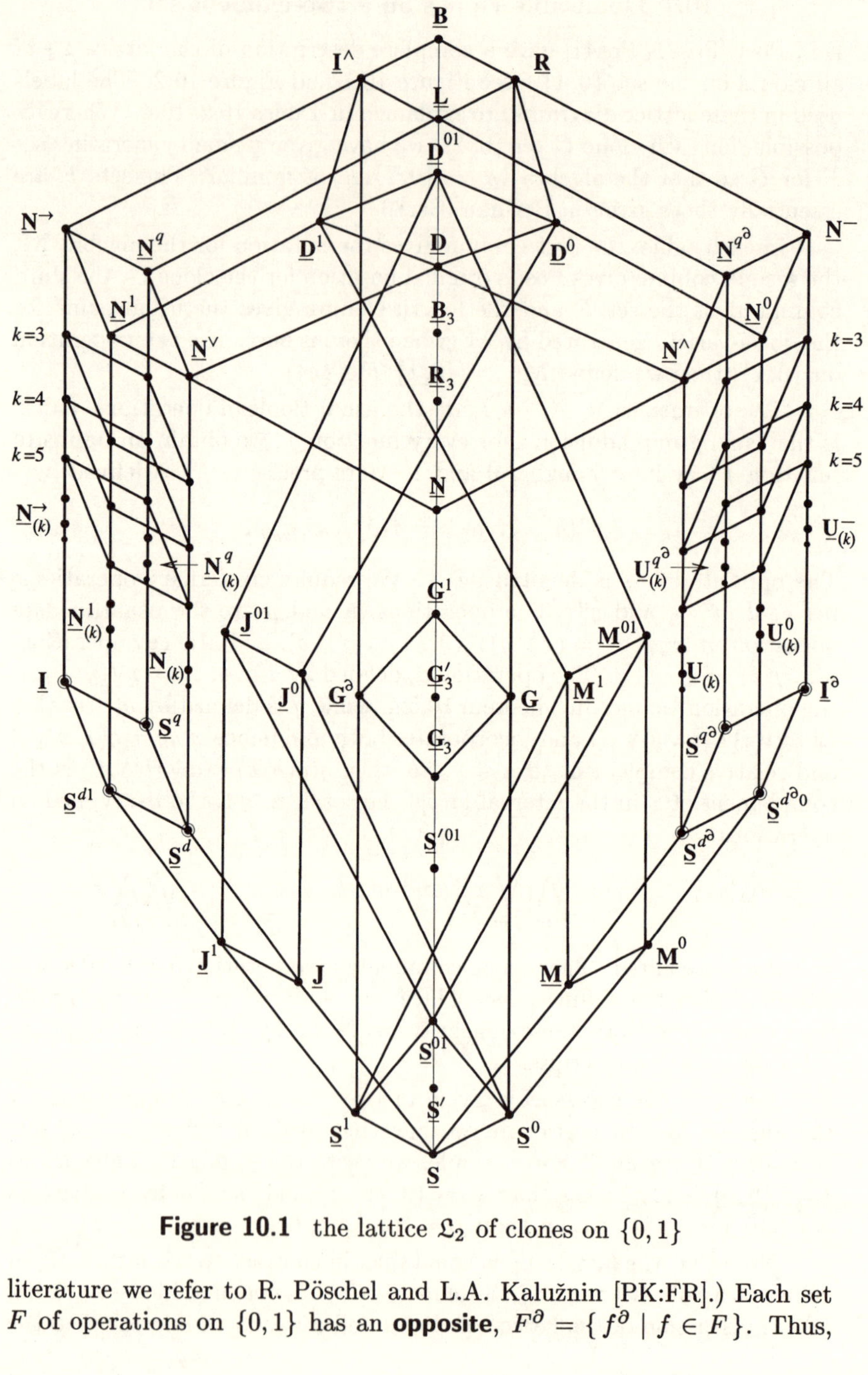

Figure 10.1 the lattice $\mathfrak{L}_2$ of clones on $\{0,1\}$

literature we refer to R. Pöschel and L.A. Kalužnin [PK:FR].) Each set F of operations on $\{0,1\}$ has an **opposite**, $F^\partial = \{\, f^\partial \mid f \in F \,\}$. Thus,

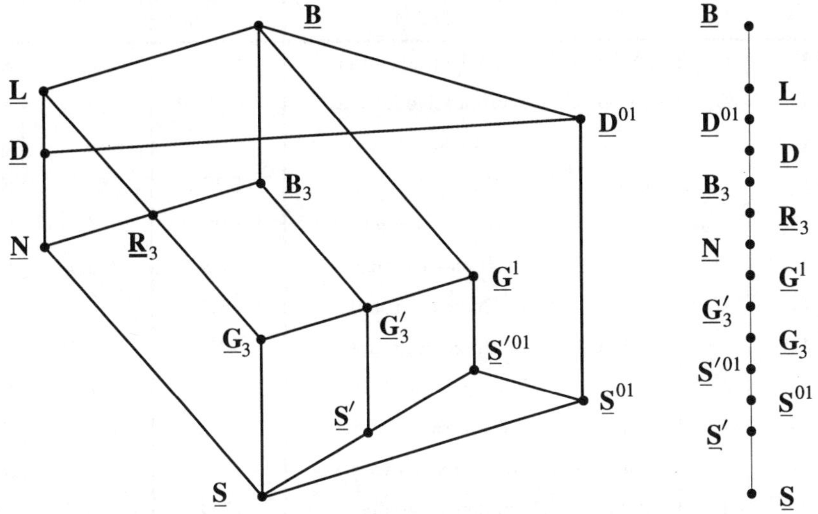

Figure 10.2 fixed points of $\underline{\mathbf{M}} \mapsto \underline{\mathbf{M}}^{\partial}$ in $\mathfrak{L}_2$

for each algebra $\underline{\mathbf{M}} = \langle\{0,1\}; F\rangle$ and clone C there is a corresponding opposite algebra $\underline{\mathbf{M}}^{\partial} := \langle\{0,1\}; F^{\partial}\rangle$, which is isomorphic to the algebra $\underline{\mathbf{M}} = \langle\{0,1\}; F\rangle$, and an opposite clone C^{∂}. In Figure 10.1, the opposite of each clone is drawn horizontally opposite to it. Along the thin vertical centreline are the 14 clones which are the fixpoints of the automorphism $\underline{\mathbf{M}} \mapsto \underline{\mathbf{M}}^{\partial}$ of $\mathfrak{L}_2$. The centreline linearly orders these 14 clones by an order which is an extension of their lattice order that is illustrated in Figure 10.2.

Since the implication operation $x \to y := \neg x \vee y$ satisfies the quasi-equation $(x \to y \approx 0) \implies (y \approx 0)$, Theorem 10.5.4 tells us that each of the clones $\underline{\mathbf{S}}^{q}$, $\underline{\mathbf{S}}^{d}$, $\underline{\mathbf{S}}^{d1}$ and $\underline{\mathbf{I}}$, which are contained in the clone $\underline{\mathbf{I}}$ generated by $\{\to, 1\}$ and which contain d, is non-dualisable. Adding their opposites, $\underline{\mathbf{S}}^{q^{\partial}}$, $\underline{\mathbf{S}}^{d^{\partial}}$, $\underline{\mathbf{S}}^{d^{\partial}0}$ and $\underline{\mathbf{I}}^{\partial}$, gives a total of eight non-dualisable clones on $\{0,1\}$ each of which, by Exercise 10.10, is actually inherently non-dualisable. In Tables 10.2 and 10.3, we indicate how a strong duality can be obtained for every other clone on $\{0,1\}$. Table 10.2 lists the 25 dualities which are given by total algebras. Column 5 of Table 10.2 indicates the theorems from which the indicated duality may be obtained: the NU Strong Duality Theorem (NUS), the Semi-primal Strong Duality Theorem (SPS), the Two-for-One Strong Duality Theorem (TFO), and the Strong Self-Duality Theorem (SSD). The other dualisable clones are listed in Table 10.3: in each case the clone includes a near-unanimity function and the duality is obtained from the NU Strong Duality Theorem. The dualities for two-element algebras with a near-unanimity term function

$\underline{\mathbf{M}}$	Post	F	$\mathbb{ISP}\underline{\mathbf{M}}$	Theorems	G	$\underline{\mathbf{M}}'$
$\underline{\mathbf{B}}$	C_1	$\vee, \wedge, ', 0, 1$	Boolean algebras	SPS	$\varnothing$	$\underline{\mathbf{S}}$
$\underline{\mathbf{I}}^{\wedge}$	C_2	$\vee, \wedge, \to, 1$	implicative lattices	SPS	1	$\underline{\mathbf{S}}^1$
$\underline{\mathbf{R}}$	C_3	$+, \wedge, 0$	Boolean rings	SPS	0	$\underline{\mathbf{S}}^0$
$\underline{\mathbf{L}}$	C_4	$q, \vee, \wedge$	Boolean lattices	SPS	$0, 1$	$\underline{\mathbf{S}}^{01}$
$\underline{\mathbf{B}}_3$	D_3	$m, '$	Boolean 3-algebras	SPS	$'$	$\underline{\mathbf{S}}'$
$\underline{\mathbf{R}}_3$	D_1	m, p	Boolean 3-rings	SPS	$', 0, 1$	$\underline{\mathbf{S}}'^{01}$
$\underline{\mathbf{G}}$	L_3	$+, 0$	Boolean groups	SSD	$+, 0$	self
$\underline{\mathbf{G}}^{\partial}$	L_2	$+^{\partial}, 1$	Boolean groups	SSD	$+^{\partial}, 1$	self
$\underline{\mathbf{G}}^1$	L_1	$+, 0, 1$	Bool. groups with one	TFO	p	$\underline{\mathbf{G}}_3$
$\underline{\mathbf{G}}_3$	L_4	p	Boolean 3-groups	TFO	$+, 0, 1$	$\underline{\mathbf{G}}^1$
$\underline{\mathbf{G}}'_3$	L_5	$p, '$	Bool. 3-groups with inv.	SSD	$p, '$	self
$\underline{\mathbf{M}}^{01}$	P_6	$\wedge, 0, 1$	bounded $\wedge$-semilattices	TFO	$\wedge$	$\underline{\mathbf{M}}$
$\underline{\mathbf{M}}^1$	P_4	$\wedge, 1$	$\wedge$-semilattices with one	SSD	$\wedge, 1$	self
$\underline{\mathbf{M}}^0$	P_5	$\wedge, 0$	$\wedge$-semilattices with zero	SSD	$\wedge, 0$	self
$\underline{\mathbf{M}}$	P_2	$\wedge$	$\wedge$-semilattices	TFO	$\wedge, 0, 1$	$\underline{\mathbf{M}}^{01}$
$\underline{\mathbf{J}}^{01}$	S_6	$\vee, 0, 1$	bounded $\vee$-semilattices	TFO	$\vee$	$\underline{\mathbf{J}}$
$\underline{\mathbf{J}}^1$	S_4	$\vee, 1$	$\vee$-semilattices with one	SSD	$\vee, 1$	self
$\underline{\mathbf{J}}^0$	S_5	$\vee, 0$	$\vee$-semilattices with zero	SSD	$\vee, 0$	self
$\underline{\mathbf{J}}$	S_2	$\vee$	$\vee$-semilattices	TFO	$\vee, 0, 1$	$\underline{\mathbf{J}}^{01}$
$\underline{\mathbf{S}}'^{01}$	R_{13}	$', 0, 1$	doubly pointed inv. sets	NUS, TFO	m, p	$\underline{\mathbf{R}}_3$
$\underline{\mathbf{S}}'$	R_4	$'$	involution sets	NUS, TFO	$m, '$	$\underline{\mathbf{B}}_3$
$\underline{\mathbf{S}}^{01}$	R_{11}	$0, 1$	doubly pointed sets	NUS, TFO	$q, \wedge$	$\underline{\mathbf{L}}$
$\underline{\mathbf{S}}^1$	R_6	1	1-pointed sets	NUS, TFO	$\vee, \wedge, \to, 1$	$\underline{\mathbf{I}}^{\wedge}$
$\underline{\mathbf{S}}^0$	R_8	0	0-pointed sets	NUS, TFO	$+, \wedge, 0$	$\underline{\mathbf{R}}$
$\underline{\mathbf{S}}$	R_1	$\varnothing$	sets	NUS, TFO	$\vee, \wedge, ', 0, 1$	$\underline{\mathbf{B}}$

Table 10.2 strong dualities given by total algebras

were first described in Deneke [Den86] and Deneke and Lüders [DL89]. The eight non-dualisable clones are described in Table 10.4 and are circled in Figure 10.1.

With many details left for the reader, we have established the following result. Recall that two algebras $\underline{\mathbf{M}}$ and $\underline{\mathbf{M}}'$ on the same set M are **term equivalent** if they have the same clones of term functions.

$\underline{\mathbf{M}}$	Post	F	$\mathbb{ISP}\underline{\mathbf{M}}$	$G \cup R$
$\underline{\mathbf{D}}^{01}$	A_1	$\vee, \wedge, 0, 1$	bounded dist. lattices	$\leqslant$
$\underline{\mathbf{D}}^{1}$	A_2	$\vee, \wedge, 1$	dist. lattices with one	$1, \leqslant$
$\underline{\mathbf{D}}^{0}$	A_3	$\vee, \wedge, 0$	dist. lattices with zero	$0, \leqslant$
$\underline{\mathbf{D}}$	A_3	$\vee, \wedge$	distributive lattices	$0, 1, \leqslant$
$\underline{\mathbf{N}}$	D^2	m	median algebras	$', 0, 1, \leqslant$
$\underline{\mathbf{N}}^{q}$	F_1^2	m, q		$0, 1, \{0,1\}^2 \backslash \{(0,0)\}$
$\underline{\mathbf{N}}^{\vee}$	F_2^2	$m, \vee$		$0, 1, \leqslant, \{0,1\}^2 \backslash \{(0,0)\}$
$\underline{\mathbf{N}}^{1}$	F_3^2	$m, 1$		$1, \leqslant, \{0,1\}^2 \backslash \{(0,0)\}$
$\underline{\mathbf{N}}^{\rightarrow}$	F_4^2	$m, \rightarrow$		$1, \{0,1\}^2 \backslash \{(0,0)\}$
$\underline{\mathbf{N}}^{q}_{(k)}$	F_1^k	m_{k+1}, q		$0, 1, \{0,1\}^k \backslash \{(0,\dots,0)\}$
$\underline{\mathbf{N}}_{(k)}$	F_2^k	m_{k+1}		$0, 1, \leqslant, \{0,1\}^k \backslash \{(0,\dots,0)\}$
$\underline{\mathbf{N}}^{1}_{(k)}$	F_3^k	$m_{k+1}, 1$		$1, \leqslant, \{0,1\}^k \backslash \{(0,\dots,0)\}$
$\underline{\mathbf{N}}^{\rightarrow}_{(k)}$	F_4^k	$m_{k+1}, \rightarrow$		$1, \{0,1\}^k \backslash \{(0,\dots,0)\}$
$\underline{\mathbf{N}}^{q^\partial}$	F_5^2	m, q^∂		$0, 1, \{0,1\}^2 \backslash \{(1,1)\}$
$\underline{\mathbf{N}}^{\wedge}$	F_6^2	$m, \wedge$		$0, 1, \leqslant, \{0,1\}^2 \backslash \{(1,1)\}$
$\underline{\mathbf{N}}^{0}$	F_7^2	$m, 0$		$0, \leqslant, \{0,1\}^2 \backslash \{(1,1)\}$
$\underline{\mathbf{N}}^{-}$	F_8^2	$m, -$		$0, \{0,1\}^2 \backslash \{(1,1)\}$
$\underline{\mathbf{U}}^{q^\partial}_{(k)}$	F_5^k	$m^\partial_{k+1}, q^\partial$		$0, 1, \{0,1\}^k \backslash \{(1,\dots,1)\}$
$\underline{\mathbf{U}}_{(k)}$	F_6^k	m^∂_{k+1}		$0, 1, \leqslant, \{0,1\}^k \backslash \{(1,\dots,1)\}$
$\underline{\mathbf{U}}^{0}_{(k)}$	F_7^k	$m^\partial_{k+1}, 0$		$0, \leqslant, \{0,1\}^k \backslash \{(1,\dots,1)\}$
$\underline{\mathbf{U}}^{-}_{(k)}$	F_8^k	$m^\partial_{k+1}, -$		$0, \{0,1\}^k \backslash \{(1,\dots,1)\}$

Table 10.3 NU strong dualities involving relations

7.1 Dualisable Two-Element Algebras Theorem *Up to term equivalence, there are precisely* 8 *non-dualisable algebras on a two-element set, namely,* $\underline{\mathbf{S}}^q$, $\underline{\mathbf{S}}^d$, $\underline{\mathbf{S}}^{d1}$ *and* $\underline{\mathbf{I}}$, *and their opposites,* $\underline{\mathbf{S}}^{q^\partial}$, $\underline{\mathbf{S}}^{d^\partial}$, $\underline{\mathbf{S}}^{d^\partial 0}$ *and* $\underline{\mathbf{I}}^\partial$.

It is interesting to note that, although the two-element implication algebra $\underline{\mathbf{I}}$ is non-dualisable, the algebra $\underline{\mathbf{I}}^0$ obtained from $\underline{\mathbf{I}}$ by adding 0 as a nullary operation is term equivalent to the two-element Boolean algebra (look in the lattice $\mathfrak{L}_2$) and is therefore dualisable. The non-dualisable algebra $\underline{\mathbf{S}}^d$ is perhaps more perverse. The algebra $\underline{\mathbf{S}}^{d1}$, obtained by adding 1 as a nullary operation, is non-dualisable while the algebra $\underline{\mathbf{S}}^{d0}$, obtained by adding 0 is term equivalent to $\underline{\mathbf{D}}^0$ and hence is dualisable!

$\underline{\mathbf{M}}$	Post	F	$\mathbb{ISP}\underline{\mathbf{M}}$
$\underline{\mathbf{S}}^q$	F_1^∞	q	
$\underline{\mathbf{S}}^d$	F_2^∞	d	
$\underline{\mathbf{S}}^{d1}$	F_3^∞	$d, 1$	
$\underline{\mathbf{I}}$	F_4^∞	$\to, 1$	implication algebras
$\underline{\mathbf{S}}^{q^\partial}$	F_5^∞	q^∂	
$\underline{\mathbf{S}}^{d^\partial}$	F_6^∞	d^∂	
$\underline{\mathbf{S}}^{d^\partial 0}$	F_7^∞	$d^\partial, 0$	
$\underline{\mathbf{I}}^\partial$	F_8^∞	$-, 0$	

Table 10.4 the eight non-dualisable clones

Exercises 10

Exercises from the text. Complete the proof of 10.3.6 by showing that if $|M| = m \geqslant 3$ then the Słupecki clone U_{m-1} contains no near-unanimity function. Prove Lemma 10.4.2. Complete the proof of 10.4.6. Show that the relation r, defined in the proof of 10.6.8, is algebraic over $\underline{\mathbf{S}}_n$.

10.1 [Zád95] Define $N_m := N_m^{(1)} \cup \cdots \cup N_m^{(m)} \subseteq M^m$, where

$$N_m^{(1)} := \{(b, a, a, \dots, a) \mid a, b \in M\},$$
$$N_m^{(2)} := \{(a, b, a, \dots, a) \mid a, b \in M\},$$

and $N_m^{(3)}, \dots, N_m^{(m)}$ are defined similarly. For each $\ell \geqslant 2$, define

$$C_\ell := \{x \in M^{\mathbb{N}} \mid \text{all but one block of } \ker(x) \text{ has size at most } \ell{-}1\}.$$

(i) Let $(c, \dots, c, d, \dots, d)$ be the $2k$-tuple whose first k entries are c and whose last k are d. Prove the following are equivalent:
   (a) there exists $k \geqslant 2$ such that the term closure of N_{2k} does not contain $(c, \dots, c, d, \dots, d)$ for all $c \neq d$ in M;
   (b) the term closure of N_ω is a subset of C_ℓ for some $\ell \geqslant 2$.

(ii) Assume that the equivalent conditions of (i) hold. Show that $\underline{\mathbf{M}}$ is dualisable if and only if $\underline{\mathbf{M}}$ has a near-unanimity term.

10.2 [Dan97] Let $\underset{\sim}{\mathbf{M}}$ be algebraic over $\underline{\mathbf{M}}$. We say that $\underline{\mathbf{M}}$ yields a **duality** on $\mathcal{X} := \mathbb{IS}_c\mathbb{P}^+\underset{\sim}{\mathbf{M}}$ if $\varepsilon_{\mathbf{X}} \colon \mathbf{X} \to DE(\mathbf{X})$ is an isomorphism for all $\mathbf{X} \in \mathcal{X}$. Let $\underset{\sim}{M} = \{0, 1, \dots, m-1\}$ for some $m \geqslant 3$ and let $\underline{\mathbf{M}} = \langle M; F\rangle$, where F contains all $\{0,1\}$-valued unary maps on M and is a subset of the Słupecki clone on M. Let $\underset{\sim}{\mathbf{M}} = \langle M; \mathcal{T}\rangle$. Show that $\underline{\mathbf{M}}$ yields a duality on the category $\mathcal{Z} = \mathbb{IS}_c\mathbb{P}^+\underset{\sim}{\mathbf{M}}$ of

Boolean spaces. [Hint: Let $\mathbf{X} \in \mathcal{Z}$; then $\mathbf{A} := E(\mathbf{X}) = C(\mathbf{X}, \underline{\mathbf{M}})$. Let $u\colon \mathbf{A} \to \underline{\mathbf{M}}$ be a homomorphism. Show that a simplified version of the proof of 10.4.4 yields $x \in X$ with $\varepsilon_{\mathbf{X}}(x) = u$.]

10.3 [Dan97] Let $M = \{0, 1, \dots, m-1\}$ for some $m \geqslant 3$ and let $\underline{\mathbf{M}} = \langle M; \varnothing\rangle$. Show that $\underset{\sim}{\mathbf{M}} = \langle M; \{0,1\}^M, \mathcal{T}\rangle$ yields a duality on the quasi-variety $\mathbb{ISP}\underline{\mathbf{M}}$ of non-empty sets. (This provides an alternative proof that every set with at least three elements is endodualisable—see 7.8.5.) [Hint: Apply the previous exercise to the algebra $\underline{\mathbf{M}}' := \langle M; \{0,1\}^M\rangle$ and then appeal to Theorem 2.2.11.]

10.4 Show that the p-semilattice $\underline{\mathbf{S}}_1^2$ has exactly seven congruences and that $\mathrm{Con}(\underline{\mathbf{S}}_1^2)$ is not join-semidistributive. Show that $\underline{\mathbf{S}}_1^2/\theta \in \mathbb{ISP}\underline{\mathbf{S}}_2$, for all $\theta \in \mathrm{Con}(\underline{\mathbf{S}}_1^2)$ and conclude that $\mathcal{S}_2$ is not relatively congruence join-semidistributive. [Hint: Find all seven semilattice congruences on the meet semilattice $\uparrow(a, a) \subseteq S_1^2$ and show that each has a unique extension to a p-semilattice congruence on $\underline{\mathbf{S}}_1^2$.]

10.5 Verify the strong duality claims made in Tables 10.2 and 10.3: use the indicated theorems to show that the given structure $\underset{\sim}{\mathbf{M}}$ yields a strong duality on $\mathbb{ISP}\underline{\mathbf{M}}$. [First look in Chapter 4.]

10.6 Show that the ternary Boolean function $d\colon \{0,1\}^3 \to \{0,1\}$ given by the term $x \vee (y \wedge z)$ is contained in the clone of term functions of the two-element implication algebra $\underline{\mathbf{I}} = \langle \{0,1\}; \to, 1\rangle$ by verifying that $d(x, y, z) = [(y \to x) \vee (z \to x)] \to x$, where $x \vee y = (y \to x) \to x$.

10.7 Consider the quasi-equation (I_k): $f(x_1, \dots, x_n) \approx 0 \implies x_k \approx 0$.

(i) Let $\underline{\mathbf{M}} = \langle \{0,1\}; F\rangle$ and assume that, for every operation $f \in F$, there exists k such that f satisfies (I_k). Show that, for all $n \geqslant 0$, every n-ary term function on the algebra $\underline{\mathbf{M}}$ also satisfies (I_k) for some k. Conclude that $\underline{\mathbf{M}}$ has no near-unanimity term function.

(ii) Consider the operations $\to$, d and q defined in Section 7. Show that $\to$ satisfies (I_1), d satisfies (I_2), and q satisfies (I_1).

10.8 [Dan97] Use Proposition 10.3.2 and the previous exercise to show that the two-element implication algebra $\underline{\mathbf{I}}$ is not dualisable. [Hint: Show that $\underset{\sim}{\mathbf{I}} := \langle \{0,1\}; 1, \mathcal{T}\rangle$ satisfies (FTC). Let $\mathbf{X}$ be a substructure of $\underset{\sim}{\mathbf{I}}^n$ and let $y \in I^n \backslash X$. Argue by induction on $|X|$. Assume that t_1' and t_2' are term functions which agree on $X\backslash\{x\}$ and differ at y. Show that, for an appropriately chosen coordinate k, the term functions given below agree on X and differ at y:

$$t_1 := \begin{cases} v_k \to t_1' & \text{if } y_k = 1,\\ t_1' \to v_k & \text{otherwise,}\end{cases} \quad\text{and}\quad t_2 := \begin{cases} v_k \to t_2' & \text{if } y_k = 1,\\ t_2' \to v_k & \text{otherwise.]}\end{cases}$$

10.9 Prove the following are equivalent for an algebra $\mathbf{A}$ and $B \subseteq A$:

(i) for all $\theta \in \mathrm{Con}(\mathbf{A})$, only one class of $\theta{\restriction}_B$ is non-trivial;

(ii) $a \neq b$ and $c \neq d$ implies $a \equiv c\ (\theta(a,b) \vee \theta(c,d))$, for all $a,b,c,d \in B$., where $\theta(a,b)$ denotes the smallest congruence on $\mathbf{A}$ which identifies a and b;

(iii) for all $a,b,c,d \in B$ with $a \neq b$ and $c \neq d$ there exist $n \geqslant 1$ and 4-ary polynomial functions $p_1,\dots,p_n$ on $\mathbf{A}$ (that is, $(k{+}4)$-ary term functions on $\mathbf{A}$, for some k, with k variables replaced by fixed elements of A) such that

$$a = p_1(a,b,c,d), \quad p_1(b,a,d,c) = p_2(a,b,c,d), \dots$$
$$p_{n-1}(b,a,d,c) = p_n(a,b,c,d), \quad p_n(b,a,d,c) = c.$$

10.10 [DILM97] Show that every two-element algebra $\underline{\mathbf{D}} = \langle \{0,1\}; F\rangle$ which satisfies the assumptions of Theorem 10.5.4 is inherently non-dualisable. [Hint: Let $\mathbf{A}$ be the subalgebra of $\underline{\mathbf{D}}^{\mathbb{Z}}$ generated by $B = \{\, 0_m \mid m \in \mathbb{Z}\,\}$, where $0_m \colon \mathbb{Z} \to \{0,1\}$ is the characteristic function of $\{m\}$. Use the previous exercise with $n = 2$: given $m,n,k,l \in \mathbb{Z}$, define $p_1(x,y,u,v) := d(x,0_l,u)$ and $p_2(x,y,u,v) := d(v,0_n,y)$. Now apply 10.5.5 with $u(n) = 1$ for all n.]

10.11 [DILM97] Let $\underline{\mathbf{M}} = \langle\{0,1,2\};\cdot\rangle$ with $1\cdot 2 = 1,\ 2\cdot 1 = 2\cdot 2 = 2$ and all other products equal to 0. Show that $\underline{\mathbf{M}}$ is inherently non-dualisable. (This algebra is often referred to as Murskiĭ's algebra, see [Mur65], and is an example of a graph algebra, see [DILM97].) [Hint: Apply Exercise 10.9. Let $\mathbf{A}$ be the subalgebra of $\underline{\mathbf{M}}^{\mathbb{Z}}$ generated by $B = \{\, 2^1_m \mid m \in \mathbb{Z}\,\}$, where $2^1_m \colon \mathbb{Z} \to \{0,1,2\}$ takes the value 1 at m and the value 2 elsewhere. Let $m \neq n$ and $k \neq \ell$ and choose $s \notin \{m,n,k,\ell\}$. Define

$$p_1(x,y,u,v) := g(g(f_n(x), f_m(y)), g(f_\ell(u), f_k(v)))$$
$$p_2(x,y,u,v) := g(g(f_k(u), f_\ell(v)), g(f_m(x), f_n(y))),$$

where $g(x,y) := x\cdot(2^1_s\cdot y)$ and $f_n(x) := x \cdot 2^1_n$.]

10.12 Analyse the proof of the Big NU Obstacle Theorem in [DHM95] and show, via 10.5.5, that if $\underline{\mathbf{D}}$ has no near-unanimity term and $\mathrm{Con}(\mathbf{A})$ ($\mathrm{Con}_{\mathcal{A}}(\mathbf{A})$) is join-semidistributive for every finite algebra in $\mathbf{A} \in \mathbb{ISP}\underline{\mathbf{D}}$, then $\underline{\mathbf{D}}$ is inherently non-dualisable.

10.13 Use the structure of the lattice $\mathfrak{L}_2$ and the information given in Tables 10.2 and 10.3 to show that each of the eight non-dualisable clones on $\{0,1\}$ is not of finite degree.

Appendix A

Algebras

We shall assume the reader is familiar with the basic notions of **universal algebra** as found in the introductory chapters of standard texts such as Burris and Sankappanavar [BS:UA], Grätzer [Grä:UA] or McKenzie, McNulty and Taylor [MMT:UA]. Our purpose here is to fix the formulations of these basic notions as used in this book, and to give an exposition of a few more specialised topics that are needed in duality theory.

Let A be a non-empty set. For any set S, an S-**ary operation** on A is a function from A^S into A. If $a \in A^S$, then we denote the value of a at $s \in S$ by either $a(s)$ or a_s depending upon the context. When $S = \{1, 2, \dots, n\}$, for some natural number n, we refer to an S-ary operation as an n-**ary operation** on A. A 0-**ary** or **nullary** operation on A is a function from $A^{\varnothing} = \{\varnothing\}$ into A. A nullary operation on A serves to specify a **distinguished element** of A, namely, the value of $\varnothing$ under that nullary operation. We identify a nullary operation with the element of A which it distinguishes. A **finitary operation** on A is an n-ary operation on A for some natural number n.

If f is an n-ary operation on A and $t_1, \dots, t_n$ are S-ary operations on A, then the **composition** $f(t_1, \dots, t_n)$ is the S-ary operation on A whose value at $a \in A^S$ is $f(t_1(a), \dots, t_n(a))$. Notice that both $n = 0$ and $S = \varnothing$ are allowable in this definition. When $n = 0$, we obtain an n-ary constant operation corresponding to the nullary operation f.

Algebra is the study of sets and their finitary operations. Certain syntactic notions are necessary to provide a language for this study. By a **type** we mean a set F of **operation symbols**, each of which has a finite **arity** which is determined by an **arity function** from F into the non-negative integers. Given a set $\{v_s \mid s \in S\}$ of pairwise distinct **variables** indexed by the set S, we define an S-**ary term of type** F recursively:

(i) each variable, and each nullary operation symbol of F, is an S-ary term, and

(ii) if $t_1, t_2, \dots, t_n$ are S-ary terms and $f \in F$ is an n-ary operation symbol for some $n \geqslant 1$, then the string of symbols $f(t_1, t_2, \dots, t_n)$ is an S-ary term.

Since an S-ary term never contains more than finitely many variables, we normally refer to a term t as $t(v_{s_1}, v_{s_2}, \dots, v_{s_n})$ to indicate that the variables of t are among $v_{s_1}, v_{s_2}, \dots, v_{s_n}$. If n is a natural number, then an n-**ary term** is an S-ary term where $S = \{1, 2, \dots, n\}$. The S-ary terms for $S = \varnothing$ are called 0-ary or **nullary** terms. Note that if F contains no nullary operation symbols, then there are no nullary terms.

Given a set F of operation symbols, an **algebra $\mathbf{A}$ of type** F is a non-empty set A, the **underlying set** of $\mathbf{A}$, together with an n-ary operation $f^{\mathbf{A}}: A^n \to A$ for each n-ary operation symbol $f \in F$. Thus an algebra of type F provides an interpretation of the operation symbols in F. This interpretation extends in a natural way to the terms of type F. If t is an S-ary term of type F, then $t^{\mathbf{A}}$ is the S-ary operation on A whose value at $a \in A^S$ is obtained by substituting, for each $s \in S$, the element a_s for the variable v_s and substituting, for each $f \in F$, the operation $f^{\mathbf{A}}$ for the operation symbol f. If $t = t(v_{s_1}, v_{s_2}, \ldots, v_{s_n})$ is an S-ary term, then we normally denote $t^{\mathbf{A}}(a)$ by $t^{\mathbf{A}}(a_{s_1}, a_{s_2}, \ldots, a_{s_n})$.

An S-**ary term function** on $\mathbf{A}$ is an S-ary operation $t^{\mathbf{A}}$ defined by some S-ary term t. When $S = \{1, 2, \ldots, n\}$, for some natural number n, an S-ary term function is referred to as an n-**ary term function**. As with nullary operations, we identify a nullary term function with its value. Special S-ary term functions on $\mathbf{A}$ are the **projection functions** $\pi_s = v_s^{\mathbf{A}}$, where $s \in S$ and

$$\pi_s(a) = a_s \quad \text{for} \quad a \in A^S,$$

and the **fundamental operations**

$$f^{\mathbf{A}} = f(v_1, v_2, \ldots, v_n)^{\mathbf{A}}$$

for each n-ary operation symbol $f \in F$.

A set C of finitary operations on a set A is called a **clone** on A if C contains the projection functions $\{\pi_i: A^n \to A \mid i = 1, 2, \ldots, n\}$ for each natural number n and C is closed under composition. If $\mathbf{A}$ is an algebra, then the set $\mathrm{Clo}(\mathbf{A})$ of all operations which are n-ary term functions on A for some natural number n is an example of a clone on A. This clone, called the **clone of** $\mathbf{A}$, consists of all non-nullary operations on the set A obtainable from the structure on the algebra $\mathbf{A}$. In fact, an arbitrary clone C on A is the clone of an algebra with underlying set A, namely, the algebra $\mathbf{A} = \langle A, C\rangle$ having the operations of C as fundamental operations. Algebras $\mathbf{A} = \langle A; F\rangle$ and $\mathbf{A}' = \langle A; F'\rangle$ with the same underlying set A are said to be **term equivalent** if $\mathrm{Clo}(\mathbf{A}) = \mathrm{Clo}(\mathbf{A}')$.

Let $\mathbf{B}$ be an algebra having the same type F as $\mathbf{A}$. We say that $\mathbf{B}$ is a **subalgebra** of the algebra $\mathbf{A}$ if $B \subseteq A$ and $f^{\mathbf{B}}$ is the restriction of $f^{\mathbf{A}}$ to B^n for each n-ary operation symbol $f \in F$. Thus the subalgebras of $\mathbf{A}$ are determined by the non-empty subsets of A that are **closed under** the fundamental operations of $\mathbf{A}$, and therefore also under the term functions of $\mathbf{A}$. If $\{\, a_s \mid s \in S \,\}$ is a subset of A, then

$$[\{\, a_s \mid s \in S \,\}] := \{\, t^{\mathbf{A}}(a_{s_1}, a_{s_2}, \ldots, a_{s_n}) \mid t \text{ is an } S\text{-ary term} \,\}$$

is called the **subalgebra generated by** $\{ a_s \mid s \in S \}$ and is easily seen to be the smallest subalgebra of $\mathbf{A}$ which contains $\{ a_s \mid s \in S \}$ (Exercise A.1).

If $\mathbf{A}$ and $\mathbf{B}$ are algebras of the same type F, then a map $u\colon A \to B$ is called a **homomorphism** if it preserves the operations in F, that is, for each n-ary $f \in F$ and $a_1, \ldots, a_n \in A$, we have

$$u(f^{\mathbf{A}}(a_1, \ldots, a_n)) = f^{\mathbf{B}}(u(a_1), \ldots, u(a_n)).$$

We indicate that u is a homomorphism by writing $u\colon \mathbf{A} \to \mathbf{B}$. We say that the homomorphism u is an **embedding** if it is one-to-one and that it is an **isomorphism** if it is bijective.

The **Cartesian product** of an indexed collection of sets $\{A_s \mid s \in S\}$ is the set $\prod\{A_s \mid s \in S\}$ of all maps a with domain S such that $a(s) \in A_s$ for each $s \in S$. Associated with each $t \in S$ we have the **projection function** $\pi_t\colon \prod\{A_s \mid s \in S\} \to A_t$ defined by $\pi_t(a) = a(t)$. If $\mathbf{A}_s$ is an algebra of type F for each $s \in S$, we can form the **direct product** $\prod\{\mathbf{A}_s \mid s \in S\}$ over the Cartesian product by defining operations and distinguished elements coordinatewise. Each projection function π_t is then a homomorphism. In particular, the product indexed by the empty set of algebras is the one-element algebra (whose single element is $\varnothing$). A special case of the direct product is the **direct power** denoted by $\mathbf{A}^S$ and obtained by taking each $\mathbf{A}_s$ to be the same algebra $\mathbf{A}$. A subalgebra $\mathbf{B}$ of the direct product $\prod\{\mathbf{A}_s \mid s \in S\}$ is a **subdirect product** of the algebras $\{\mathbf{A}_s \mid s \in S\}$ if, for each $t \in S$, the image of $\mathbf{B}$ under the projection π_t is all of $\mathbf{A}_t$.

An **equation** (of type F) is an expression of the form

$$t(v_{s_1}, v_{s_2}, \ldots, v_{s_n}) \approx t'(v_{s_1}, v_{s_2}, \ldots, v_{s_n})$$

where t and t' are S-ary terms for some set S. An algebra $\mathbf{A}$ **satisfies** the equation $t \approx t'$ if $t^{\mathbf{A}} = t'^{\mathbf{A}}$. An algebra is a **model** of a set of equations provided that it satisfies every equation in the set. The **equational theory** of the algebra $\mathbf{A}$ is the set of all equations that are satisfied by $\mathbf{A}$. The **variety generated by** $\mathbf{A}$, denoted by $\mathcal{V}(\mathbf{A})$, is the class of all algebras $\mathbf{B}$ of the same type as $\mathbf{A}$ which are models of the equational theory of $\mathbf{A}$.

Let $\mathcal{M}$ be a class of algebras, let $\mathbf{F} \in \mathcal{M}$ and let $G \subseteq F$. We say that $\mathbf{F}$ is **$\mathcal{M}$-freely generated by** G if G generates $\mathbf{F}$ and, for every algebra $\mathbf{A} \in \mathcal{M}$ and every map $w\colon G \to A$, there is a (unique) homomorphism $u\colon \mathbf{F} \to \mathbf{A}$ which extends w. An alternate description of free generating sets is usually used to identify them.

A.1 Lemma *If G is a generating set for an algebra $\mathbf{F} \in \mathcal{M}$, then the following are equivalent:*

(i) *$\mathbf{F}$ is $\mathcal{M}$-freely generated by G;*

(ii) *for every natural number n, for every choice of two n-ary terms $t(v_1, \ldots, v_n)$ and $t'(v_1, \ldots, v_n)$ and for every list of distinct elements $g_1, \ldots, g_n \in G$, the equality $t^{\mathbf{F}}(g_1, \ldots, g_n) = t'^{\mathbf{F}}(g_1, \ldots, g_n)$ implies that every algebra $\mathbf{A} \in \mathcal{M}$ is a model of the equation $t \approx t'$.*

Proof Assume (i) and let $(a_1, a_2, \ldots, a_n) \in A^n$. By (i), there is a homomorphism $u\colon \mathbf{F} \to \mathbf{A}$ such that $u(g_i) = a_i$. Under the hypotheses of (ii) we have

$$t^{\mathbf{A}}(a_1, \ldots, a_n) = u(t^{\mathbf{F}}(g_1, \ldots, g_n)) = u(t'^{\mathbf{F}}(g_1, \ldots, g_n)) = t'^{\mathbf{A}}(a_1, \ldots, a_n).$$

Now assume (ii); let $\mathbf{A} \in \mathcal{M}$ and let $w\colon G \to A$. Then (ii) assures us that we can unambiguously define $u\colon F \to A$ by $u(t^{\mathbf{F}}(g_1, g_2, \ldots, g_n)) = t^{\mathbf{A}}(a_1, a_2, \ldots, a_n)$, where $a_i = u(g_i)$. ∎

Under some simple conditions we can guarantee the existence of free algebras in $\mathcal{M}$.

A.2 Theorem *If $\mathcal{M}$ is a class of algebras closed under $\mathbb{S}$ and $\mathbb{P}$, then for every non-empty (index) set S there is an algebra $\mathbf{F}(S)$ in $\mathcal{M}$ and a bijection g from S onto a subset G of $F(S)$ such that $\mathbf{F}(S)$ is $\mathcal{M}$-freely generated by G.*

Proof Use S to index a set $\{v_s \mid s \in S\}$ of distinct variables. Let I denote the set of pairs (t, t') of S-ary terms such that the equation $t \approx t'$ does not hold in every member of $\mathcal{M}$. For each $(t, t') \in I$, choose $\mathbf{A}_{tt'} \in \mathcal{M}$ and $a_{tt'} \in A_{tt'}^S$ such that $t(a_{tt'}(s_1), \ldots, a_{tt'}(s_n)) \neq t'(a_{tt'}(s_1), \ldots, a_{tt'}(s_n))$. For $s \in S$, define $g_s \in \prod\{A_{tt'} \mid (t, t') \in I\}$ by $g_s(t, t') = a_{tt'}(s)$. It follows from Lemma A.1 that the subalgebra $\mathbf{F}(S)$ of $\prod\{\mathbf{A}_{tt'} \mid (t, t') \in I\}$ generated by $G = \{g_s \mid s \in S\}$ is $\mathcal{M}$-freely generated by G. ∎

In fact we can give a more concrete construction of an $\mathcal{M}$-free algebra on S generators if $\mathcal{M} = \mathcal{V}(\mathbf{A})$, the variety generated by the single algebra $\mathbf{A}$. Recall that each S-ary term function on $\mathbf{A}$ is an element of the direct power $\mathbf{A}^{A^S}$ of $\mathbf{A}$. Moreover, the set of S-ary term functions is closed under the operations of $\mathbf{B} = \mathbf{A}^{A^S}$ as we can easily check that

$$(*) \qquad f^{\mathbf{B}}(t_1^{\mathbf{A}}, t_2^{\mathbf{A}}, \ldots, t_n^{\mathbf{A}}) = f(t_1, t_2, \ldots, t_n)^{\mathbf{A}}$$

for every n-ary operation symbol $f \in F$ and every choice, t_1, t_2, $\ldots$, t_n, of S-ary terms. Consequently the S-ary term functions on $\mathbf{A}$ form a subalgebra of $\mathbf{A}^{A^S}$ which we denote by $\mathbf{F_A}(S)$. Notice that $\mathbf{F_A}(S)$ is exactly the subalgebra of $\mathbf{A}^{A^S}$ generated by the set $G = \{\pi_s \mid s \in S\}$ of projections as we have, for any S-ary term $t(v_{s_1}, v_{s_2}, \ldots, v_{s_n})$,

$$(**) \qquad t(v_{s_1}, v_{s_2}, \ldots, v_{s_n})^{\mathbf{A}} = t^{\mathbf{F_A}(S)}(\pi_{s_1}, \pi_{s_2}, \ldots, \pi_{s_n})$$

(Exercise A.3). We ask the reader to use Lemma A.1 to verify that G freely generates $\mathbf{F_A}(S)$.

A.3 Proposition *If S is a non-empty set, then $\mathbf{F_A}(S)$ is $\mathcal{V}(\mathbf{A})$-freely generated by $G = \{\pi_s \mid s \in S\}$.*

Accordingly, we refer to $\mathbf{F_A}(S)$ as the **free algebra of S-ary term functions over $\mathbf{A}$**.

For a class $\mathcal{M}$ of algebras of the same type, we will denote by

$$\mathbb{I}\mathcal{M}, \quad \mathbb{S}\mathcal{M}, \quad \mathbb{P}\mathcal{M}, \quad \mathbb{H}\mathcal{M},$$

respectively, the classes of isomorphic copies, subalgebras, direct products (over possibly empty index sets) and homomorphic images of members of $\mathcal{M}$. Because these four constructions are used so often in algebra, it is of considerable importance to understand which classes of algebras are closed under all four of them. A classical result of general algebra, due to G. Birkhoff [Bir35], shows that these are exactly the equationally definable classes of algebras. We use free algebras of term functions to prove a slightly restricted version of this theorem which is suitable for our needs.

A.4 $\mathbb{HSP}$-Theorem *If $\mathbf{A}$ is an algebra, then $\mathcal{V}(\mathbf{A}) = \mathbb{HSP}\mathbf{A}$.*

Proof We ask the reader to verify that $\mathbb{HSP}\mathbf{A} \subseteq \mathcal{V}(\mathbf{A})$ by showing that satisfaction of equations is preserved by $\mathbb{H}$, $\mathbb{S}$ and $\mathbb{P}$ (Exercise A.4). Conversely, suppose $\mathbf{B} \in \mathcal{V}(\mathbf{A})$ and consider the free algebra $\mathbf{F_A}(B)$ of B-ary term functions over $\mathbf{A}$. Since $\mathbf{F_A}(B) \leq \mathbf{A}^{A^B}$, Exercise A.4 assures us that $\mathbf{F_A}(B) \in \mathcal{V}(\mathbf{A})$. We now apply Proposition A.3 to see that $\mathbf{B}$ is a homomorphic image of $\mathbf{F_A}(B)$, a subalgebra of a power of $\mathbf{A}$, by mapping π_b to b for each $b \in B$. ∎

A bounded lattice $\mathbf{L} = \langle L, \vee, \wedge, 0, 1\rangle$ is **complete** if every subset $S \subseteq L$ has a greatest lower bound $\bigwedge S$ and, consequently, a least upper bound $\bigvee S = \bigwedge\{a \in L \mid a \geqslant b \text{ for all } b \in S\}$. An element k of a complete lattice $\mathbf{L}$ is **compact** if $S \subseteq L$ and $k \leqslant \bigvee S$ implies that $k \leqslant \bigvee S_0$ for some finite subset S_0 of S. A complete lattice $\mathbf{L}$ is **algebraic** if each element of $\mathbf{L}$ is a join of compact elements. (See Exercise A.5.) An element k of a complete lattice is **completely meet-irreducible**, abbreviated **CMI**, if it is not the meet of the elements properly above it. The CMI elements of an algebraic lattice play a special role.

A.5 Theorem *Every element of an algebraic lattice is the meet of CMI elements.*

Proof Let $a \in L$ where $\mathbf{L}$ is an algebraic lattice, and let b be the meet of all CMI elements above a. Then $a \leqslant b$. Suppose $a < b$. Since $\mathbf{L}$ is algebraic, there is a compact element $k \in L$ such that $b \geqslant k$ but $a \ngeqslant k$. Let $D = \{d \in L \mid d \ngeqslant k, d \geqslant a\}$. Thus $a \in D$ but $b \notin D$. Using the compactness of k, we can apply Zorn's Lemma to obtain a maximal element e of D. Now let $S = \{s \in L \mid s > e\}$. Since each element of S is above k, we have $\bigwedge S \geqslant k$ so $\bigwedge S \neq e$. Thus e is CMI, and since it is above a it must also be above b. But then $e \geqslant b \geqslant k$, contradicting the fact that $e \in D$. ∎

Algebraic lattices can be viewed as arising from a special kind of operation on the power set $\mathcal{P}(X)$ of some base set X. By a **closure operation** on a set X we mean a map $C \colon \mathcal{P}(X) \to \mathcal{P}(X)$ such that, for every $U, V \in \mathcal{P}(X)$, we have

$$U \subseteq C(U), \quad CC(U) = C(U) \quad \text{and} \quad U \subseteq V \implies C(U) \subseteq C(V).$$

A subset $U \subseteq X$ is **closed** if $C(U) = U$.

A.6 Lemma *If C is a closure operation on a set X, then the closed subsets of X form a complete lattice $\mathbf{L}_C$ where, for $S \subseteq L_C$, we have*

$$\bigwedge S = \bigcap S \quad \text{and} \quad \bigvee S = \bigcap \{U \in L_C \mid \bigcup S \subseteq U\} = C(\bigcup S).$$

Proof We must show that, for $S \subseteq L_C$, the set $\bigcap S$ is closed. Notice that for each $U \in S$, we have $C(\bigcap S) \subseteq C(U)$. Thus

$$C(\bigcap S) \subseteq \bigcap_{U \in S} C(U) = \bigcap_{U \in S} U = \bigcap S \subseteq C(\bigcap S).$$

It follows that $\bigcap S$ is closed. ∎

A closure operation C on a set X is **algebraic** if the closure of each set is equal to the union of the closures of its finite subsets. (See Exercise A.6.)

A.7 Lemma *Let C be an algebraic closure operation on X and let $K \in L_C$. Then K is compact if and only if K is the closure of a finite subset of X.*

Proof Assume $K = C(F)$ is the closure of the finite set $F \subseteq X$ and that $K \subseteq \bigvee S$ where $S \subseteq L_C$. Then $F \subseteq K \subseteq \bigvee S = C(\bigcup S)$ which, since C is compact, is the union of the closures of the finite subsets of $\bigcup S$. It follows that there is a finite subset $S_0 \subseteq S$ such that $F \subseteq C(\bigcup S_0)$. Consequently $K \subseteq C(\bigcup S_0) = \bigvee S_0$, and we see that K is compact.

Conversely, assume that K is compact. Since K is the union of the closures of its finite subsets, it is the union of the closures of finitely many of its finite subsets: $K = C(G_1) \cup C(G_2) \cup \cdots \cup C(G_n)$. Let $F = G_1 \cup G_2 \cup \cdots \cup G_n$. Then we have $F \subseteq K \subseteq C(F)$ and, since K is closed, we conclude that $K = C(F)$. ∎

It follows immediately that the lattice $\mathbf{L}_C$ of closed sets under an algebraic closure operation C is an algebraic lattice. In fact, all algebraic lattices can be represented this way. The proof of the following theorem is outlined in Exercise A.7.

A.8 Theorem *A lattice* $\mathbf{L}$ *is algebraic if and only if it is isomorphic to the lattice* $\mathbf{L}_C$ *of closed sets under some algebraic closure operation* C.

An equivalence relation θ on the underlying set A of an algebra $\mathbf{A}$ is called a **congruence** on $\mathbf{A}$ if it is also a subalgebra of $\mathbf{A} \times \mathbf{A}$; that is, if the θ-classes of $\mathbf{A}$ form an algebra (denoted by $\mathbf{A}/\theta$) under unambiguously defined operations

$$f(a_1/\theta, a_2/\theta, \ldots, a_n/\theta) := f(a_1, a_2, \ldots, a_n)/\theta.$$

(Notice that here, and in most of this book, we omit the superscripts on operation symbols when it is clear from the context on what domain they are being interpreted.) The equality relation, $\Delta_A := \{ (a, a) \mid a \in A \}$, is the smallest congruence on $\mathbf{A}$ and is denoted by $\mathbf{0}^{\mathbf{A}}$ when it is thought of as a congruence. At the other extreme, the universal relation $A \times A$ is the largest congruence on $\mathbf{A}$ and is denoted by $\mathbf{1}^{\mathbf{A}}$. Congruences on $\mathbf{A}$ are closed under arbitrary intersection, and this fact allows us to define **join** and **meet** operations on the set of all congruences on $\mathbf{A}$ by taking, for congruences θ and ψ,

$$\theta \vee \psi := \bigcap \{ \lambda \mid \lambda \text{ is a congrence and } \theta \cup \psi \subseteq \lambda \} \quad \text{and} \quad \theta \wedge \psi := \theta \cap \psi.$$

Under these operations the congruences on $\mathbf{A}$ form a (complete) lattice $\langle \mathrm{Con}(\mathbf{A}); \vee, \wedge, \mathbf{0}^{\mathbf{A}}, \mathbf{1}^{\mathbf{A}} \rangle$. In fact, we can use Theorem A.8 to show that $\mathrm{Con}(\mathbf{A})$ is an algebraic lattice. (See Exercise A.8.)

An embedding $u \colon \mathbf{B} \to \prod\{ \mathbf{A}_s \mid s \in S \}$ of an algebra $\mathbf{B}$ into a direct product is a **subdirect representation** of $\mathbf{B}$ if the image of $\mathbf{B}$ under u is a subdirect product of the factors. Algebra abounds with representations of particular algebras as subdirect products of other algebras which are, in some special sense, more restricted. Since a natural duality is, among other things, an example of subdirect representation, we will examine this construction in a bit more detail. A subdirect representation u of $\mathbf{B}$ is **trivial** if, for some $s \in S$, the composition $\pi_s \circ u \colon \mathbf{B} \to \mathbf{A}_s$ is an isomorphism. We say that $\mathbf{B}$ is **subdirectly irreducible** if every subdirect representation of $\mathbf{B}$ is trivial. It turns out that these concepts are fully visible within the congruence lattice of $\mathbf{B}$. Part (i) of the following lemma follows from Theorem A.5 and the remainder is left as an exercise.

A.9 Lemma *For any algebra* $\mathbf{B}$ *and any congruence* $\psi \in \mathrm{Con}(\mathbf{B})$:

(i) ψ *is the intersection of* CMI *congruences;*
(ii) $\mathbf{B}/\psi$ *is subdirectly irreducible if and only if* ψ *is* CMI;
(iii) $\mathbf{B}$ *is subdirectly irreducible if and only if* $\mathbf{0}^{\mathbf{B}}$ *is* CMI;
(iv) *if* $\mathbf{B}/\psi$ *is finite, then* ψ *is* CMI *if and only if it is* MI.

These ideas were first formulated by G. Birkhoff [Bir44] who used them to prove the following fundamental representation theorem.

A.10 Birkhoff's Theorem *Every algebra is a subdirect product of its subdirectly irreducible homomorphic images.*

Proof Let $\mathbf{B}$ be an algebra and let Σ be the set of all CMI congruences in $\mathrm{Con}(\mathbf{B})$. By Lemma A.9(i), we have $\bigcap \Sigma = \mathbf{0^B}$ and consequently the canonical homomorphism

$$u\colon \mathbf{B} \to \prod\{\,\mathbf{B}/\theta \mid \theta \in \Sigma\,\} \quad \text{where} \quad u(b)_\theta = b/\theta$$

is a subdirect representation of $\mathbf{B}$. By part (ii) each factor is subdirectly irreducible. ∎

Because a given set of (subdirectly irreducible) algebras normally admits many different subdirect products, Birkhoff's Theorem invites us to discover more specific subdirect representations. For example, both sheaf representations and dualities view the index set S as an object in a particular category and then attempt to represent an algebra as the subdirect product consisting of the set of *all* morphisms from S into the factors.

Perhaps one of the most significant steps in the development of universal algebra was the realisation that, in many contexts, algebras are better understood when they are classified according to properties of their congruence lattices rather than according to their types and equational properties. We say that **congruences permute** on an algebra $\mathbf{B}$ if for any congruences θ and ψ on $\mathbf{B}$, we have $\theta\psi = \psi\theta$ (and this relational product is therefore equal to the join $\theta \vee \psi$ in $\mathrm{Con}(\mathbf{B})$). A variety $\mathcal{V}$ of algebras is said to be **congruence permutable** if congruences permute on every algebra in $\mathcal{V}$. We say that $\mathcal{V}$ is **congruence distributive** if the congruence lattice of every algebra in $\mathcal{V}$ is a distributive lattice. A variety $\mathcal{V}$ is said to be **arithmetic** if it is both permutable and distributive. We will review several landmark theorems which characterise these properties and reveal some of their important consequences. The following theorem, published by Mal'cev in 1954 [Mal54], initiated these new insights into algebra.

A.11 Mal'cev's Theorem *A variety is congruence permutable if and only if there is a ternary term p such that the equations*

$$x \approx p(x, y, y) \approx p(y, y, x)$$

are satisfied by every algebra in the variety.

We refer the reader to [BS:UA], Theorem 12.2, [Grä:UA], Theorem 26.4 or [MMT:UA], Theorem 4.142, for a proof. The following year, I. Fleischer discovered a very useful consequence of congruence permutability (Fleischer [Fle55]). Suppose that $\mathbf{P}_1$, $\mathbf{P}_2$ and $\mathbf{C}$ are algebras and $f_i\colon \mathbf{P}_i \to \mathbf{C}$, for $i = 1, 2$, are homomorphisms. Then the set $\ker(f_1, f_2) = \{(p_1, p_2) \in P_1 \times P_2 \mid f_1(p_1) = f_2(p_2)\}$ is called the **joint kernel** of f_1 and f_2. Notice that $\ker(f_1, f_2)$ is a subalgebra of $\mathbf{P}_1 \times \mathbf{P}_2$ provided that it is non-empty.

A.12 Fleischer's Theorem *Let* $\mathbf{A}_1$ *and* $\mathbf{A}_2$ *be algebras such that congruences on subalgebras of* $\mathbf{A}_1 \times \mathbf{A}_2$ *permute. Then every subalgebra* $\mathbf{B}$ *of* $\mathbf{A}_1 \times \mathbf{A}_2$ *is a joint kernel* $\ker(f_1, f_2)$ *for some algebra* $\mathbf{C}$*, some subalgebras* $\mathbf{P}_i \le \mathbf{A}_i$ *and some surjections* $f_i \colon \mathbf{P}_i \to \mathbf{C}$*.*

Proof For $i = 1, 2$ let θ_i be the kernel of the projection $\rho_i \colon \mathbf{B} \to \mathbf{A}_i$ and let $\mathbf{P}_i \le \mathbf{A}_i$ be its image. We take $\mathbf{C}$ to be the quotient $\mathbf{B}/\theta_1\theta_2$, and $h \colon \mathbf{B} \to \mathbf{C}$ to be the natural map. Since $\theta_i \subseteq \theta_1\theta_2$, there are surjections $f_i \colon \mathbf{P}_i \to \mathbf{C}$ such that $h = f_i \circ \rho_i$ for $i = 1, 2$.

To show that $\mathbf{B} = \ker(f_1, f_2)$, let $(b_1, b_2) \in B$. Then

$$f_1(b_1) = f_1\rho_1(b_1, b_2) = h(b_1, b_2) = f_2\rho_2(b_1, b_2) = f_2(b_2).$$

Conversely, let $(b_1, b_2) \in P_1 \times P_2$ where $f_1(b_1) = f_2(b_2)$. Then there are $a_i \in A_i$ such that $(b_1, a_2), (a_1, b_2) \in B$. Now we compute

$$h(b_1, a_2) = f_1\rho_1(b_1, a_2) = f_1(b_1) = f_2(b_2) = f_2\rho_2(a_1, b_2) = h(a_1, b_2)$$

from which we conclude that $(b_1, a_2)\theta_1\theta_2(a_1, b_2)$. The interpolating member of $\mathbf{B}$ must be (b_1, b_2). ■

In 1967 Jónsson [Jón67] offered a powerful analysis of congruence distributivity. His first theorem ([BS:UA]; Theorem 12.6, [Grä:UA], Theorem 60.2, or [MMT:UA], Theorem 4.144) gives the appropriate analogue of Mal'cev's Theorem.

A.13 Jónsson's First Theorem *A variety* $\mathcal{V}$ *is congruence distributive if and only if, for some natural number* m*, there are ternary terms* $d_0, d_1, \dots, d_m$ *such that the following equations are satisfied by every algebra in* $\mathcal{V}$*.*

(i) $d_0(x, y, z) \approx x$, $d_i(x, y, x) \approx x$ *for* $0 \le i \le m$, *and* $d_m(x, y, z) \approx z$.
(ii) $d_i(x, y, y) \approx d_{i+1}(x, y, y)$ *for all even* $i < m$.
(iii) $d_i(x, x, y) \approx d_{i+1}(x, x, y)$ *for all odd* $i < m$.

Near-unanimity terms, introduced by Baker and Pixley [BP75] in 1975, play a central role in duality theory. Mitschke [Mit78] quickly discovered that Jónsson terms can be derived from a near-unanimity term, establishing congruence distributivity.

A.14 Mitschke's Theorem *Suppose that there are a* $k \ge 2$ *and a* $(k+1)$*-ary term* n *such that the near-unanimity equations*

$$n(x, \dots, x, y) \approx n(x, \dots, y, x) \approx \cdots \approx n(y, x, \dots, x) \approx x$$

hold in a variety $\mathcal{V}$*. Then* $\mathcal{V}$ *is congruence distributive.*

Proof Let $d_0(x, y, z) = x$ and, for $1 \le i \le k$, let

$$d_{2i-1}(x, y, z) = n(x, \dots, x, x, z, \dots, z),$$
$$d_{2i}(x, y, z) = n(x, \dots, x, y, z, \dots, z),$$

where z occurs i times in each expression. It is now quick to check that the conditions of Jónsson's First Theorem are satisfied for $m = 2k$. ■

While Mitschke's Theorem will give us a useful sufficient condition for congruence distributivity, Jónsson [Jón67] also discovered that congruence distributivity has some powerful consequences. An **ultrafilter** on a set S is a non-empty collection U of subsets of S such that, for all $u, v \subseteq S$,

(a) $u, v \in U$ implies $u \cap v \in U$,
(b) $u \in U$ and $u \subseteq v$ imply $v \in U$,
(c) $u \cup v \in U$ implies $u \in U$ or $v \in U$, and
(d) $\varnothing \notin U$.

If U is an ultrafilter on S, then for any partition of S into finitely many subsets, exactly one of those subsets is in U. If $\mathbf{B} \le \prod\{\mathbf{A}_s \mid s \in S\}$, then each ultrafilter U on S induces a congruence θ_U on $\mathbf{B}$ where

$$b\theta_U c \quad \text{if and only if} \quad \{s \in S \mid b_s = c_s\} \in U.$$

The reader will find a well presented proof of the following result in [BS:UA], Corollary 6.10.

A.15 Jónsson's Lemma *If* $\mathbf{B} \le \prod\{\mathbf{A}_s \mid s \in S\}$ *is a subdirect product, the congruence lattice of* $\mathbf{B}$ *is distributive and* $\psi \in \mathrm{Con}(\mathbf{B})$ *is CMI, then there is an ultrafilter* U *on* S *such that* $\theta_U \subseteq \psi$.

From this lemma comes Jónsson's structure theorem for finitely generated congruence distributive varieties [Jón67].

A.16 Jónsson's Second Theorem *Suppose that the variety* $\mathcal{V}$ *generated by the finite algebra* $\underline{\mathbf{M}}$ *is congruence distributive. Then*

(i) $\mathcal{V}$ *contains (up to isomorphism) only finitely many subdirectly irreducibles, each of which is a homomorphic image of a subalgebra of* $\underline{\mathbf{M}}$, *and*
(ii) $\mathcal{V} = \mathbb{ISPHS}\underline{\mathbf{M}}$.

Proof We prove (i), from which (ii) will follow by Theorem A.10. Let $\mathbf{A} \in \mathcal{V} = \mathbb{HSP}\underline{\mathbf{M}}$ be subdirectly irreducible. Then there are a subdirect product $\mathbf{B} \le \prod\{\underline{\mathbf{M}}_s \mid s \in S\}$ of subalgebras of $\underline{\mathbf{M}}$ and a congruence $\psi \in \mathrm{Con}(\mathbf{B})$ such that $\mathbf{A} \cong \mathbf{B}/\psi$. By Lemma A.9, the congruence ψ is CMI. Let U be an ultrafilter on S such that $\theta_U \subseteq \psi$. Then $\mathbf{A}$ is a homomorphic image of $\mathbf{B}/\theta_U$.

It remains to show that $\mathbf{B}/\theta_U$ is isomorphic to a subalgebra of $\underline{\mathbf{M}}$. Let $M = \{a_1, a_2, \ldots, a_m\}$. Then $\{b^{-1}(a_1), b^{-1}(a_2), \ldots, b^{-1}(a_m)\}$ is a finite partition of S for every $b \in B$. Consequently there is a unique i, where $1 \le i \le m$, such that $b^{-1}(a_i) \in U$. Define $f\colon \mathbf{B} \to \underline{\mathbf{M}}$ where $f(b) = a_i$. We leave it as an exercise to verify that f is a homomorphism with kernel θ_U. ■

Also following in Mal'cev's footsteps, A. Pixley [Pix63] found conditions for a variety to be arithmetic. This proof is also well ensconced in the standard literature ([BS:UA], Theorem 12.5, or [MMT:UA], Theorem 4.143), and follows a pattern similar to that of Mal'cev's Theorem and Jónsson's First Theorem.

A.17 Pixley's Theorem *A variety is arithmetic if and only if there is a terary term p such that the equations*

$$p(x,y,y) \approx p(x,y,x) \approx p(y,y,x) \approx x$$

are satisfied by every algebra in the variety.

Exercises

Exercises from the text. Prove Lemma A.9. Complete the proofs of Mitschke's Theorem A.14 and of Jónsson's Second Theorem A.16.

A.1 Show that $[\{\, a_s \mid s \in S \,\}]$ is the smallest subalgebra of $\mathbf{A}$ containing the generating set $\{\, a_s \mid s \in S \,\}$.

A.2 Assume that F contains some nullary operation symbols. Show that the set of (values of the) nullary term functions on $\mathbf{A}$ is the subalgebra of $\mathbf{A}$ generated by the distinguished elements of $\mathbf{A}$.

A.3 Show that $\mathbf{F_A}(S)$ is a subalgebra of $\mathbf{A}^{A^S}$ by verifying equation $(*)$. Then show that it is generated by the projections by verifying equation $(**)$.

A.4 Let $t \approx t'$ be an equation.

(i) Show that if $\mathbf{A}$ satisfies the equation $t \approx t'$, then every subalgebra and every homomorphic image of $\mathbf{A}$ also satisfy $t \approx t'$.

(ii) Show that any direct product of algebras, each of which satisfies $t \approx t'$, also satisfies $t \approx t'$.

A.5 Let $\mathbf{L}$ be the lattice obtained by adjoining the empty set to the collection of subalgebras of the algebra $\mathbf{A}$. Prove the following.

(i) $\mathbf{L}$ is a complete lattice.

(ii) The compact elements of $\mathbf{L}$ are exactly the finitely generated subalgebras of $\mathbf{A}$.

(iii) $\mathbf{L}$ is an algebraic lattice.

A.6 Let $\mathbf{L}$ be as defined in Exercise A.5. For $U \subseteq A$, let $C(U)$ be the subalgebra generated by U. Show that C is an algebraic closure operation and that $\mathbf{L} = \mathbf{L}_C$.

A.7 In this exercise we outline a proof that every algebraic lattice $\mathbf{L}$ is isomorphic to $\mathbf{L}_C$ for some algebraic closure operation C, thereby completing the proof of Theorem A.8. First show that the set X of compact elements of $\mathbf{L}$ is closed under $\vee$ and therefore forms a join semilattice $\mathbf{X} = \langle X, \vee, 0 \rangle$ with 0. An **ideal** of $\mathbf{X}$ is a subsemilattice J such that $a \in J$ and $b \leqslant a$ implies $b \in J$. Let $\mathrm{Id}(\mathbf{X})$ denote the set of ideals of $\mathbf{X}$.

(i) Show that if $a \in L$, then ${\downarrow}a = \{x \in X \mid x \leqslant a\}$ is in $\mathrm{Id}(\mathbf{X})$.

(ii) Show that ${\downarrow}\colon \mathbf{L} \to \mathrm{Id}(\mathbf{X})$ is an order (and therefore lattice) isomorphism.

(iii) If $U \subseteq X$, define $C(U)$ to be $\{0\}$ if $U = \varnothing$; otherwise $C(U)$ is the set of all $x \in X$ such that $x \leqslant u_1 \vee u_2 \vee \ldots \vee u_n$ for some $u_1, u_2, \ldots, u_n \in U$. Show that C is an algebraic closure operation and that $\mathbf{L} \cong \mathrm{Id}(\mathbf{X}) = \mathbf{L}_C$.

A.8 Let $\mathbf{A}$ be an algebra, $\mathrm{Con}(\mathbf{A})$ its congruence lattice.

(i) Show that the intersection of any collection of congruences on $\mathbf{A}$ is again a congruence on $\mathbf{A}$, and that $\mathrm{Con}(\mathbf{A})$ forms a complete lattice.

(ii) For a subset S of $A \times A$, let $\Theta(S)$ be the intersection of all congruences on $\mathbf{A}$ containing S. Show that Θ is an algebraic closure operation by showing that $\Theta(S)$ is the subalgebra of $\mathbf{A} \times \mathbf{A}$ generated by the equivalence relation generated by S.

(iii) Show that $\mathrm{Con}(\mathbf{A})$ is exactly the algebraic lattice $\mathbf{L}_\Theta$ of Θ-closed sets.

A.9 Prove the following finitary version of A.16. Assume that $\underline{\mathbf{M}}$ generates a congruence distributive variety, $\mathbf{s} \leqslant \underline{\mathbf{M}}^n$ and the image of the homomorphism $x\colon \mathbf{s} \to \underline{\mathbf{M}}$ is subdirectly irreducible. Then $\ker(\rho_i) \leqslant \ker(x)$ for some $i \leqslant n$.

A.10 Given congruences θ_1 and θ_2 on $\mathbf{A}$ and $\mathbf{B}$ respectively, define (θ_1, θ_2) on $\mathbf{A} \times \mathbf{B}$ by $(a, b)\,(\theta_1, \theta_2)\,(c, d) \Leftrightarrow a\,\theta_1\,c \;\&\; b\,\theta_2\,d$. Show that if $\mathrm{Con}(\mathbf{A} \times \mathbf{B})$ is distributive, then every congruence on $\mathbf{A} \times \mathbf{B}$ is of this form.

Appendix B

Boolean Spaces

While this treatment is self contained, it is written with the assumption that our readers have some working knowledge of elementary topology as might be obtained from a beginning course or from the introductory chapter of any topology text. We present here a streamlined introduction to topology with a more detailed discussion of the specialised topics that are needed for natural dualities.

A **topological space** $\mathbf{X}$ consists of a set X and an associated collection $\mathcal{T}$ of subsets of X, the **open sets** of $\mathbf{X}$, such that $\mathcal{T}$ includes $\varnothing$ and X and is closed under finite intersections and arbitrary unions. The collection $\mathcal{T}$ is called the **topology** on $\mathbf{X}$. A subset of X is **closed** if its complement is open, and it is **clopen** if it is both closed and open. A subset Y of a topological space $\mathbf{X}$ forms a topological space $\mathbf{Y}$ itself in which the open sets are the intersections of open sets of $\mathbf{X}$ with the set Y. This is called the **relative topology** on Y. A collection $\mathcal{U}$ of open sets **covers** a subset Z of the space $\mathbf{X}$ if the union of the members of $\mathcal{U}$ contains Z. We say that Z is **compact** if every open cover contains a finite subcover. The topological space $\mathbf{X}$ is said to be **Hausdorff** if for any pair of distinct points x and y of $\mathbf{X}$ there are disjoint open sets U containing x and V containing y. If U and V can always be chosen to be clopen, then $\mathbf{X}$ is said to be **totally disconnected**. If $\mathbf{X}$ is both compact and totally disconnected it is called a **Boolean space**. A map $\varphi\colon \mathbf{X} \to \mathbf{Y}$ from a topological space $\mathbf{X}$ to a topological space $\mathbf{Y}$ is **continuous** if $\varphi^{-1}(U)$ is an open set in $\mathbf{X}$ whenever U is open in $\mathbf{Y}$. The following lemma gives some useful connections between these ideas.

B.1 Lemma *Let $\varphi\colon \mathbf{X} \to \mathbf{Y}$ be a continuous map from a compact space $\mathbf{X}$ to a Hausdorff space $\mathbf{Y}$.*

(i) *If Z is a closed subset of $\mathbf{X}$, then its image $\varphi(Z)$ is a closed subset of $\mathbf{Y}$.*

(ii) *If φ is one-to-one, then its inverse φ^{-1} is a continuous map from $\varphi(\mathbf{X})$ onto $\mathbf{X}$.*

Proof (i) Let $y \in Y \setminus \varphi(X)$. Since $\mathbf{Y}$ is Hausdorff we can find, for each point z of $\mathbf{Z}$, disjoint open sets separating $\varphi(z)$ and y. Since φ is continuous, the inverse images of these open sets containing each $\varphi(z)$ form an open cover of $\mathbf{Z}$. Since $\mathbf{X}$ is compact and $\mathbf{Z}$ is closed, $\mathbf{Z}$ is compact and therefore has a finite subcover. The corresponding finite collection of open sets containing y have an open intersection which contains y but does not intersect $\varphi(Z)$.

(ii) Let U be an open set in $\mathbf{X}$. The inverse image of U under φ^{-1} is the set

$$(\varphi^{-1})^{-1}(U) = \varphi(U) = \varphi(X) \setminus \varphi(X \setminus U).$$

By part (i), the set $\varphi(X \setminus U)$ is closed and consequently its complement in $\varphi(X)$ is open. ∎

We often arrive at a topology for a space $\mathbf{X}$ by identifying any collection whatsoever of subsets of X and then specifying that $\mathcal{T}$ is to be the smallest topology on X in which each member of that collection is open. This topology $\mathcal{T}$ is obtained from a collection $\mathcal{S}$ of subsets of X by taking (as open sets) all arbitrary unions of finite intersections of members of $\mathcal{S}$. Notice that $\mathcal{T}$ will include $\varnothing$, which is the union indexed over the empty set, as well as X itself, which is the intersection indexed over the empty set. When $\mathcal{T}$ is obtained from a collection $\mathcal{S}$ in this way we say that $\mathcal{S}$ is a **subbasis** for the topology $\mathcal{T}$. If $\mathcal{S}$ is a collection of open subsets of the space $\mathbf{X}$ and every open set in $\mathbf{X}$ is a union of members of $\mathcal{S}$, then we say that $\mathcal{S}$ is a **basis** for $\mathbf{X}$.

Consider, for example, a collection $\{\mathbf{X}_i \mid i \in I\}$ of topological spaces. We define the topology $\mathcal{T}$ on the product $\prod X_i$ as the smallest topology for which all of the projection maps are continuous. Said more carefully, $\mathcal{T}$ is the topology which has as subbasis all sets of the form $\pi_j^{-1}(U) = \prod\{X_i \mid j \neq i \in I\} \times U$ where $j \in I$ and U is an open set in $\mathbf{X}_j$. The space $\prod \mathbf{X}_i$ is called the **direct product** of the spaces $\{\mathbf{X}_i \mid i \in I\}$. (See Exercise B.3.)

In general a subbasis gives us a rather cumbersome description of the open sets in a topological space. For this reason the following result is rather surprising and unexpected. The proof depends upon the Prime Ideal Theorem 1.2.1 for distributive lattices given in Chapter 1.

B.2 Alexander's Subbasis Lemma *Let $\mathcal{S}$ be a subbasis for the topological space $\mathbf{X}$. Then $\mathbf{X}$ is compact provided only that every cover of $\mathbf{X}$ by members of $\mathcal{S}$ has a finite subcover.*

Proof Assume that every cover of $\mathbf{X}$ by members of $\mathcal{S}$ has a finite subcover, and suppose that $\mathcal{U}$ is an arbitrary open cover of $\mathbf{X}$ which has no finite subcover. Let x be the ideal in the lattice $\mathbf{P}(X)$ of all subsets of X consisting of all subsets of finite unions of members of $\mathcal{U}$. Since $\mathcal{U}$ has no finite subcover, X is not in x. In other words, x is disjoint from the one-element filter $\{X\}$. By the Prime Ideal Theorem there is a prime ideal z of $\mathbf{P}(X)$ containing x such that $X \notin z$. Thus z is a prime ideal of $\mathbf{P}(X)$ containing $\mathcal{U}$.

Now let t be an arbitrary point of the space $\mathbf{X}$. Since $\mathcal{U}$ covers $\mathbf{X}$, there is a member M of $\mathcal{U}$ containing t. Since M is a union of finite intersections of members of $\mathcal{S}$ and z is a prime ideal, t is in some member N_t of z that is in $\mathcal{S}$. Consequently the set $\{N_t \mid t \in X\}$ is a cover of $\mathbf{X}$ by elements of $\mathcal{S} \cap z$. By hypothesis this cover has a finite subcover whose union, which is X, must be in the prime ideal z, a contradiction. ■

We leave to the reader the proof of a simple and useful fact which asserts in another instance that verification for subbasic open sets is sufficient.

B.3 Lemma *Let φ be a map from a space $\mathbf{X}$ to a space $\mathbf{Y}$ and let S be a subbasis for $\mathbf{Y}$. Then φ is continuous provided only that $\varphi^{-1}(U)$ is open for each U in S.*

An important application of Alexander's Subbasis Lemma is Tychonoff's Theorem for direct products.

B.4 Tychonoff's Product Theorem *A direct product of compact spaces is compact.*

Proof Consider a direct product $\prod \mathbf{X}_i$ of compact spaces $\{\mathbf{X}_i \mid i \in I\}$ and a subbasic open cover $\mathcal{C}$ of $\prod \mathbf{X}_i$. Then each member of $\mathcal{C}$ has the form $\pi_j^{-1}(U)$ for some unique $j \in I$ and some open set $U \subseteq X_j$. For each $j \in I$ let $\mathcal{C}_j$ denote the collection of all $U \subseteq X_j$ such that $\pi_j^{-1}(U) \in \mathcal{C}$. Then, for some $j \in I$, the collection $\mathcal{C}_j$ must cover $\mathbf{X}_j$; otherwise $\mathcal{C}$ would not cover $\prod \mathbf{X}_i$. Taking $\mathcal{D}_j \subseteq \mathcal{C}_j$ to be a finite subcover of $\mathbf{X}_j$, we have that $\{\pi_j^{-1}(U) \mid U \in \mathcal{D}_j\} \subseteq \mathcal{C}$ is a finite subcover of $\prod \mathbf{X}_i$. ■

The topological spaces used in natural dualities are all Boolean. We ask the reader to verify that the clopen subsets of a Boolean space $\mathbf{X}$ form a basis for $\mathbf{X}$ (Exercise B.2). The following fact is an immediate consequence of Tychonoff's Theorem.

B.5 Lemma *The class of Boolean spaces is closed under the formation of isomorphic copies, closed subspaces and arbitrary direct products.*

Boolean spaces for natural dualities are generated using Lemma B.5 and beginning with very simple Boolean spaces. The **discrete topology** on a set X is the topology in which every set is open. Notice that a finite set M endowed with the discrete topology is an example of a Boolean space. Given a set $\{\mathbf{M}_i \mid i \in I\}$ of finite discrete spaces we can, by Lemma B.5, produce a new Boolean space $\mathbf{Z} = \prod_{i \in I} \mathbf{M}_i$ by taking their direct product. Given $X \subseteq Z$, we say that a map $\varphi\colon X \to M$ has **finite support** if there are a finite set $J \subseteq I$ and a map $\psi\colon \pi_J(X) \to M$ such that $\varphi = \psi \circ \pi_J{\restriction}_X$, where π_J is the restriction to J. Since these spaces are so fundamental to natural dualities, it is useful to give an explicit description of their properties. We leave the proofs to the reader.

B.6 Lemma *Let* $\mathbf{Z} = \prod_{i\in I} \mathbf{M}_i$ *be the product of a non-empty collection of finite discrete spaces* $\{\mathbf{M}_i \mid i \in I\}$*. For each* $i \in I$ *and each* $a \in M_i$ *define* $U_{i,a} = \{\, x \in Z \mid x(i) = a \,\}$*.*

(i) *The set* $\mathcal{S} = \{U_{i,a} \mid i \in I; a \in M_i\}$ *is a clopen subbasis for the topology on* $\mathbf{Z}$*.*

(ii) *A subset* $U \subseteq Z$ *is clopen if and only if there are a finite subset* $J \subseteq I$ *and a subset* $V \subseteq \prod_{j\in J} M_j$ *such that* U *is the set of all members of* Z *whose restriction to* J *is in* V*.*

(iii) *The closure of a subset* $X \subseteq Z$ *consists of all* $y \in Z$ *which agree with a member of* X *on each finite subset of* I*.*

(iv) *If* $\mathbf{M}$ *is a finite discrete space and* $\mathbf{X} \subseteq \mathbf{Z}$ *is closed, then a map* $\varphi\colon X \to M$ *is continuous if and only if it has finite support.*

Exercises

Exercises from the text. Prove Lemma B.5 and Lemma B.6.

B.1 A collection of sets has the **Finite Intersection Property** (FIP) if every finite subcollection has a non-empty intersection. Prove that a topological space is compact if and only if every collection of closed subsets with the (FIP) has itself a non-empty intersection.

B.2 Show that if $\mathbf{X}$ is a Boolean space, then the collection of clopen subsets of $\mathbf{X}$ forms a basis for $\mathbf{X}$.

B.3 Let $\mathbf{X}$ and $\mathbf{Y}$ be topological spaces. Show that the product $\mathbf{X} \times \mathbf{Y}$, as defined in this appendix, is also the product of $\mathbf{X}$ and $\mathbf{Y}$ in the category of topological spaces, as defined in Chapter 1.

B.4 Let $\mathbf{X}$ and $\mathbf{Y}$ be compact Hausdorff spaces and let $\psi\colon \mathbf{X} \to \mathbf{Y}$ be a continuous surjection. Show that there is a closed subset X_0 of $\mathbf{X}$ such that $\varphi(X_0) = Y$ and $\varphi(X')$ is a proper subset of Y for every proper, closed subset X' of X_0. [Hint: Use the order-theoretic dual of Zorn's Lemma.]

Bibliography

[AC90] M.E. Adams and D.M. Clark, Endomorphism monoids in minimal quasi primal varieties, *Acta Sci. Math.* (Szeged) **54** (1990), 37–52.

[AD96] M. Adams and W. Dziobiak (eds), *Studia Logica Special Issue on Priestley Duality*, *Studia Logica* **56**, Nos. 1–2, 1996.

[AKS83] M.E. Adams, V. Koubek and J. Sichler, Pseudocomplemented distributive lattices with small endomorphism monoids, *Bull. Austral. Math. Soc.* **28** (1983), 305–318.

[AKS84] M.E. Adams, V. Koubek and J. Sichler, Homomorphisms and endomorphisms in varieties of pseudocomplemented distributive lattices, *Trans. Amer. Math. Soc.* **285** (1984), 57–79.

[BP75] K.A. Baker and A.F. Pixley, Polynomial interpolation and the Chinese remainder theorem for algebraic systems, *Math. Z.* **143** (1975), 165–174.

[Bal67] R. Balbes, Projective and injective distributive lattices, *Pacific J. Math.* **21** (1967), 405–420.

[BD:DL] R. Balbes and Ph. Dwinger, *Distributive Lattices*, University of Missouri Press, Columbia, Missouri, 1974.

[BH70] R. Balbes and A. Horn, Injective and projective Heyting algebras, *Trans. Amer. Math. Soc.* **148** (1970), 549–559.

[Ban70] B. Banaschewski, Injectivity and essential extensions in equational classes, *Proc. Conf. on Universal Algebra* (Queens Univ., Kingston, Ont., 1969), Queens University, Kingston, Ontario, 1970, pp. 131–147.

[Ban71] B. Banaschewski, Projective covers in categories of topological spaces and topological algebras, *General Topology and its Relation to Modern Analysis and Algebra* (Proc. Kanpur Topology Conf., 1968), Academia, Prague, 1971, pp. 63–91.

[Ban81] H.J. Bandelt, Endomorphism semigroups of median algebras, *Algebra Universalis* **12** (1981), 262–264.

[Ber80] J. Berman, Algebraic properties of k-valued logics, *Proc. of the Tenth International Symposium on Multiple-valued Logic,* Evanston, Illinois, 1980, pp. 195–204.

[Bir33] G. Birkhoff, On the combination of subalgebras, *Proc. Camb. Phil. Soc.* **29** (1933), 441–464.

[Bir35] G. Birkhoff, On the structure of abstract algebras, *Proc. Camb. Phil. Soc.* **31** (1935), 433–354.

[Bir37] G. Birkhoff, Rings of sets, *Duke Math. J.* **3** (1937), 443–454.

[Bir44] G. Birkhoff, Subdirect unions in universal algebra, *Bull. Amer. Math. Soc.* **50** (1944), 764–768.

[BV:OA] T.S. Blyth and J.C. Varlet, *Ockham Algebras*, Oxford Science Publications, Oxford University Press, Oxford, 1994.

[Bur84] S. Burris, Model companions for finitely generated universal Horn classes, *J. Symb. Logic* **49** (1984), 68-74.

[BS:UA] S. Burris and H.P. Sankappanavar, *A Course in Universal Algebra*, Springer, New York, Heidelberg, Berlin, 1981.

[BW79] S. Burris and H. Werner, Sheaf constructions and their elementary properties, *Trans. Amer. Math. Soc.* **248** (1979), 269-309.

[BW87] S. Burris and R. Willard, Finitely many primitive positive clones, *Proc. Amer. Math. Soc.* **101** (1987), 427–430.

[CK:MT] C.C. Chang and H.J. Keisler, *Model Theory*, Studies in Logic and the Foundations of Mathematics, Vol. 73, North-Holland, Amsterdam, 1990.

[Cla89] D.M. Clark, The structure of algebraically and existentially closed Stone and double Stone algebras, *J. Symb. Logic* **54** (1989), 375–386.

[CD95] D.M. Clark and B.A. Davey, The quest for strong dualities, *J. Austral. Math. Soc.* (*Series A*) **58** (1995), 248–280.

[CD96] D.M. Clark and B.A. Davey, When is a natural duality 'good'?, *Algebra Universalis* **35** (1996), 265–295.

[CISSW] D.M. Clark, P.M. Idziak, L. Sabourin, Cs. Szabó and R. Willard, Natural dualities for commutative rings (to appear).

[CK76] D.M. Clark and P.H. Krauss, Para primal algebras, *Algebra Universalis* **6** (1976), 165–192.

[CK77a] D.M. Clark and P.H. Krauss, Varieties generated by para primal algebras, *Algebra Universalis* **7** (1977), 93–114.

[CK77b] D.M. Clark and P.H. Krauss, Relatively homogeneous structures, *Logic Colloquium '76*, (R. Gandy and M. Hyland, eds), North-Holland, Amsterdam, 1977, pp. 255–285.

[CK84] D.M. Clark and P.H. Krauss, Topological quasi-varieties, *Acta Sci. Math.* **47** (1984), 3–39.

[CS96] D.M. Clark and J. Schmid, The countable homogeneous universal model of $\mathcal{B}_2$, *Studia Logica* **56** (1996), 31–36.

[CP:ATS] A.H. Clifford and G.B. Preston, *The Algebraic Theory of Semigroups*, Vol. 1, Mathematical Surveys No. 7, American Mathematical Society, Providence, Rhode Island, 1961.

[Cor:AA] W.H. Cornish, *Antimorphic Action. Categories of Algebraic Structures with Involutions or Anti-endomorphisms*, Research and Exposition in Mathematics, **12**, Heldermann, Berlin, 1986.

[CF77] W.H. Cornish and P.R. Fowler, Coproducts of de Morgan algebras, *Bull. Austral. Math. Soc.* **16** (1977), 1–13.

[Dan97] C.S. Daniels, *The Fibred Ultraproduct of Structured Boolean Topological Spaces*, PhD thesis, La Trobe University, 1997.

[Dav76] B.A. Davey, Dualities for equational classes of Brouwerian algebras and Heyting algebras, *Trans. Amer. Math. Soc.* **221** (1976), 119–146.

[Dav77] B.A. Davey, Weak injectivity and congruence extension in congruence distributive equational classes, *Canad. J. Math.* **29** (1977), 449–454.

[Dav78] B.A. Davey, Topological duality for prevarieties of universal algebras, *Studies in Foundations and Combinatorics* (G.C. Rota, ed), Advances in Mathematics Supplementary Studies, Vol. 1, Academic Press, New York, 1978, pp. 61–99.

[Dav82] B.A. Davey, Dualities for Stone algebras, double Stone algebras and relative Stone algebras, *Coll. Math.* **46** (1982), 1–14.

[Dav93] B.A. Davey, Duality theory on ten dollars a day, *Algebras and Orders* (I.G. Rosenberg and G. Sabidussi, eds), NATO Advanced Study Institute Series, Series C, Vol. 389, Kluwer Academic Publishers, 1993, pp. 71–111.

[Dav96] B.A. Davey, Dualisability in general and endodualisability in particular, *Logic and Algebra* (A. Ursini, and P. Aglianò, eds), Lecture Notes in Pure and Applied Mathematics **180**, Marcel Dekker, New York, 1996, pp. 437–455.

[DHP95a] B.A. Davey, M. Haviar and H.A. Priestley, The syntax and semantics of entailment in duality theory, *J. Symb. Logic* **60** (1995), 1087–1114.

[DHP95b] B.A. Davey, M. Haviar and H.A. Priestley, Endoprimal distributive lattices are endodualisable, *Algebra Universalis* **34** (1995), 444–453.

[DHM95] B.A. Davey, L. Heindorf and R. McKenzie, Near unanimity: an obstacle to general duality theory, *Algebra Universalis* **33** (1995), 428–439.

[DILM97] B.A. Davey, P.M. Idziak, W.A. Lampe and G.F. McNulty, Dualisability and graph algebras, preprint, 1997.

[DP00] B.A. Davey and J.G. Pitkethly, Endoprimal algebras, *Algebra Universalis* (to appear).

[DP86] B.A. Davey and H.A. Priestley, Lattices of homomorphisms, *J. Austral. Math. Soc.* (*Series A*) **40** (1986), 364–406.

[DP87] B.A. Davey and H.A. Priestley, Generalized piggyback dualities and applications to Ockham algebras, *Houston Math. J.* **13** (1987), 151–198.

[DP:ILO] B.A. Davey and H.A. Priestley, *Introduction to Lattices and Order*, Cambridge University Press, Cambridge, 1990.

[DP93a] B.A. Davey and H.A. Priestley, Partition-induced natural dualities for varieties of pseudocomplemented distributive lattices, *Discrete Math.* **113** (1993), 41–58.

[DP93b] B.A. Davey and H.A. Priestley, Optimal natural dualities, *Trans. Amer. Math. Soc.* **338** (1993), 655–677.

[DP96a] B.A. Davey and H.A. Priestley, Optimal natural dualities II: general theory, *Trans. Amer. Math. Soc.* **348** (1996), 3673–3711.

[DP96b] B.A. Davey and H.A. Priestley, Optimal natural dualities for varieties of Heyting algebras, *Studia Logica* **56** (1996), 67–96.

[DQ95] B.A. Davey and R.W. Quackenbush, Bookkeeping duality for paraprimal algebras, *Contributions to General Algebra* **9** (H.K. Kaiser et al, eds), Hölder-Pichler Tempsky, Wien, 1995, pp. 19–26.

[DQ96] B.A. Davey and R.W. Quackenbush, Natural dualities for dihedral varieties, *J. Austral. Math. Soc.* **61** (1996), 216–228.

[DQS90] B.A. Davey, R.W. Quackenbush and D. Schweigert, Monotone clones and the varieties they determine, *Order* **7** (1990), 145–167.

[DR82] B.A.Davey and I. Rival, Exponents of lattice-ordered algebras, *Algebra Universalis* **14** (1982), 87–98.

[DR97] B.A. Davey and I. Rosenberg, Natural dualities and clones, *Advances in the Mathematical Sciences—CRM's 25 years* (L. Vinet, ed), Centre de recherches mathématiques Proceedings and Lecture Notes, Vol.**11**, American Mathematical Society, Providence, Rhode Island, 1997, pp. 385–402.

[DSW91] B.A. Davey, V. Schumann and H. Werner, From subalgebras of the square to the discriminator, *Algebra Universalis* **28** (1991), 500–519.

[DW79] B.A. Davey and H. Werner, Injectivity and Boolean powers, *Math. Z.* **166** (1979), 205–223.

[DW83] B.A. Davey and H. Werner, Dualities and equivalences for varieties of algebras, *Contributions to Lattice Theory* (Szeged, 1980) (A.P. Huhn and E.T. Schmidt, eds), Colloq. Math. Soc. János Bolyai **33**, North-Holland, Amsterdam, New York, 1983, pp. 101–275.

[DW85] B.A. Davey and H. Werner, Piggyback-dualitäten, *Bull. Austral. Math. Soc.* **32** (1985), 1–32.

[DW86] B.A. Davey and H. Werner, Piggyback dualities, *Lectures in Universal Algebra* (Szeged, 1983) (L. Szabó and A. Szendrei, eds), Coll. Math. Soc. János Bolyai **43**, North-Holland, Amsterdam, New York, 1986, pp. 61–83.

[Den86] K. Denecke, Varieties generated by two-element majority algebras and their equivalences, *Beiträge zur Algebra und Geometrie* **21** (1986), 35–56.

[DL89] K. Denecke and O. Lüders, Varieties generated by two-element near-unanimity algebras, their dualities, and their equivalences, *Potsdamer Forschungen, Reihe B* **62** (1989), 65–81.

[Fle55] I. Fleischer, A note on subdirect products, *Acta Math. Acad. Sci. Hungar.* **6** (1955), 463–465.

[Fos53] A.L. Foster, Generalized 'Boolean' theory of universal algebras. I: Subdirect sums and normal representation theorem, *Math. Z.* **58** (1953), 306–336; II: Identities and subdirect sums of functionally complete algebras, *Math. Z.* **59** (1953), 191–199.

[Fre66] P. Freyd, Algebra valued functors in general and tensor products in particular, *Coll. Math.* **14** (1966), 89–106.

[Gle58] A.M. Gleason, Projective topological spaces, *Ill. J. Math.* **2** (1958), 482–489.

[Gol81] M.S. Goldberg, Distributive Ockham algebras: free algebras and injectivity, *Bull. Austral. Math. Soc.* **24** (1981), 161–203.

[Grä:UA] G. Grätzer, *Universal Algebra*, Springer, New York, 1979.

[Hei93] L. Heindorf, Some examples of nondualizable algebras, *General Algebra and Applications* (Potsdam, 1992), Research and Exposition in Mathematics **20**, Heldermann, Berlin, 1993, pp. 143–151.

[HMS74] K.H. Hofmann, M. Mislove and A. Stralka, *The Pontryagin duality of compact 0-dimensional semi-lattices and its applications*, Lecture Notes in Mathematics **396**, Springer, Berlin, New York, 1974.

[Hu71] T.K. Hu, On the topological duality for primal algebra theory, *Algebra Universalis* **1** (1971), 152–154.

[Isb80] J. Isbell, Median algebras, *Trans. Amer. Math. Soc.* **260** (1980), 319–362.

[Jon:PS] G.T. Jones, *Pseudo Complemented Semi-Lattices*, PhD thesis, UCLA, 1972.

[Jón67] B. Jónsson, Algebras whose congruence lattices are distributive, *Math. Scand.* **21** (1967), 110-121.

[Lee70] K. B. Lee, Equational classes of distributive pseudocomplemented lattices, *Canad. J. Math.* **22** (1970), 881–891.

[Lüd92] O. Lüders, *Äquivalenzen und Dualitäten von Varietäten und Quasivarietäten die von endlichen Algebren erzeugt werden*, doctor rerum naturalium Dissertation, Potsdam, 1992.

[Mac:CW] S. Mac Lane, *Categories for the Working Mathematician*, Graduate Texts in Mathematics, Vol. 5, Springer, New York, Heidelberg, Berlin, 1971.

[Mag72] K.D. Magill, The semigroup of endomorphisms of a Boolean ring, *Semigroup Forum* **4** (1972), 411–416.

[Mal54] I.A. Mal'cev, On the general theory of algebraic systems (Russian), *Mat. Sb.* **35** (1954), 3–20.

[MP93] L. Márki and R. Pöschel, Endoprimal distributive lattices, *Algebra Universalis* **30** (1993), 272–274.

[MR65] W.D. Maurer and G.L. Rhodes, A property of finite simple nonabelian groups, *Proc. Amer. Math. Soc.* **16** (1965), 552–554.

[Max72] C.J. Maxson, On semigroups of Boolean ring endomorphisms, *Semigroup Forum* **4** (1972), 78–82.

[McK96] R.M. McKenzie, An algebraic version of categorical equivalence for varieties and more general algebraic categories, *Logic and Algebra* (A. Ursini and P. Aglianò, eds), Lecture Notes in Pure and Applied Mathematics **180**, Marcel Dekker, New York, 1996, pp. 211–243.

[MMT:UA] R.M. McKenzie, G.F. McNulty and W.F. Taylor, *Algebras, Lattices, Varieties.* Vol.1, Wadsworth & Brooks/Cole, Monterey, California, 1987.

[MT81] R.N. McKenzie and C. Tsinakis, On recovering a bounded distributive lattice from its endomorphism monoid, *Houston J. Math.* **7** (1981), 525–529.

[Mit78] A. Mitschke, Near unanimity identities and congruence distributivity in equational classes, *Algebra Universalis* **8** (1978), 29–32.

[Mon70] D. Monk, On equational classes of algebraic versions of logic, *Math. Scand.* **27** (1970), 53–71.

[Mur65] V.L. Murskiĭ, The existence in three-valued logic of a closed class without a finite complete system of identities (Russian), *Doklady Akad. Nauk SSSR* **163** (1965), 815–818.

[Mur75] V.L. Murskiĭ, The existence of a finite basis of identities, and other properties of 'almost all' finite algebras (Russian), *Problemy Kibernet* **30** (1975), 434–456.

[Pix63] A.F. Pixley, Distributivity and permutability of congruence relations in equational classes of algebras, *Proc. Amer. Math. Soc.* **14** (1963), 105–109.

[Pix70] A.F. Pixley, Functionally complete algebras generating distributive and permutable classes, *Math. Z.* **114** (1970), 361–372.

[Pon34a] L.S. Pontryagin, Sur les groupes abéliens continus, *C. R. Acad. Sci. Paris* **198** (1934), 238–240.

[Pon34b] L.S. Pontryagin, The theory of topological commutative groups, *Ann. Math.* **35** (1934), 361–388.

[Pon:TG] L.S. Pontryagin, *Topological Groups*, second edition, Gordon and Breach, New York, 1966.

[PK:FR] R. Pöschel and L.A. Kalužnin, *Funktionen- und Relationenalgebren*, VEB Deutscher Verlag der Wissenschaften, Berlin, 1979.

[Pos21] E.L. Post, Introduction to the general theory of elementary propositions, *Amer. J. Math.* **43** (1921), 163–185.

[Pos41] E.L. Post, *The two-valued iterative systems of mathematical logic*, Ann. Math. Studies **5**, Princeton University Press, Princeton, New Jersey, 1941.

[Pri70] H.A. Priestley, Representation of distributive lattices by means of ordered Stone spaces, *Bull. London Math. Soc. 2* (1972), 186–190.

[Pri72] H.A. Priestley, Ordered topological spaces and the representation of distributive lattices, *Proc. London Math. Soc. 3* **24** (1975), 507–530.

[Pri75] H.A. Priestley, The construction of spaces dual to pseudocomplemented distributive lattices, *Oxford Quart. J. Math. Ser.* (*2*) **26** (1975), 215–228.

[Pri84] H.A. Priestley, Ordered sets and duality for distributive lattices, *Orders, Descriptions and Roles*, Annals of Discrete Math. **23** (M. Pouzet and D. Richard, eds), North-Holland, Amsterdam, New York, 1984, pp. 39–60.

[Pri95a] H.A. Priestley, Natural dualities, *Lattice Theory and its Applications* (In Celebration of Garrett Birkhoff's 80th Birthday), Research and Exposition in Mathematics **23** (K.A. Baker and R. Wille, eds), Heldermann, Lemgo, Germany, 1995, pp. 185–209.

[Pri95b] H.A. Priestley, Natural dualities for varieties of n-valued Łukasiewicz algebras, *Studia Logica* **54** (1995), 333–370.

[Pri00] H.A. Priestley, Varieties of distributive lattices with unary operations I, *J. Austral. Math. Soc.* (to appear).

[PT:GSC] A. Pultr and V. Trnková, *Combinatorial, Algebraic and Topological Representations of Groups, Semigroups and Categories*, North-Holland, Amsterdam, 1980.

[Qua79] R.W. Quackenbush, Pseudovarieties of finite algebras isomorphic to bounded distributive lattices, *Discrete Math.* **28** (1979), 189–192.

[Ras:NL] H. Rasiowa, *An Algebraic Approach to Non-classical Logics*, Studies in Logic and the Foundations of Mathematics, Vol. 78, North-Holland, Amsterdam, 1974.

[Sch70] B.M. Schein, Ordered sets, semilattices, distributive lattices and Boolean algebras with homomorphic endomorphism semigroups, *Fund. Math.* **68** (1970), 31–50.

[Sch79] J. Schmid, Algebraically and existentially closed distributive lattices, *Z. für Math. Logik u. Grundl. d. Math.* **25** (1979), 525–530.

[Sch76] P. Schmitt, The model-completion of Stone algebras, *Ann. Sci. Univ. Clermont* **60** (1976), 135–155.

[Sto36] M.H. Stone, The theory of representations for Boolean algebras, *Trans. Amer. Math. Soc.* **4** (1936), 37–111.

[Sto37] M.H. Stone, Topological representations of distributive lattices and Brouwerian logics, *Časopis pěst. Mat. a Fys.* **67** (1937), 1–25.

[Str80] A. Stralka, A partially ordered space which is not a Priestley space, *Semigroup Forum* **20** (1980), 293–297.

[Tay72] W. Taylor, Residually small varieties, *Algebra Universalis* **2** (1972), 33–53.

[Urq79] A. Urquhart, Lattices with a dual homomorphic operation, *Studia Logica* **38** (1979), 201–209.

[Wer70] H. Werner, Eine Charakterisierung funktional vollständiger Algebren, *Arch. Math.* (Basel) **21** (1970), 381–385.

[Wer:DA] H. Werner, *Discriminator-Algebras. Algebraic representation and model theoretic properties.*, Studien zur Algebra und ihre Anwendungen, Band 6, Akademie, Berlin, 1978.

[Wer81] H. Werner, A duality for weakly associative lattices, *Finite algebra and multiple-valued logic* (Szeged, 1979), Colloq. Math. Soc. János Bolyai **28**, North-Holland, Amsterdam, New York, 1981, pp. 781–808.

[Wil77] R. Wille, A note on algebraic operations and algebraic functions on finite lattices, *Houston J. Math.* **3** (1977), 593–597.

[Zád95] L. Zádori, Natural duality via a finite set of relations, *Bull. Austral. Math. Soc.* **51** (1995), 469–478.

Notation

Index

Definitions in the main text are indicated by page numbers in boldface. References to the exercises are indicated by page numbers in italics.